W. Kaminsky H. Sinn (Eds.)

Transition Metals and Organometallics as Catalysts for Olefin Polymerization

With 245 Figures and 119 Tables

Springer-Verlag
Berlin Heidelberg New York London Paris Tokyo

Proceedings of an International Symposium, Hamburg (FRG),
September 21–24, 1987

Professor Dr. Walter Kaminsky
Professor Dr. Hansjörg Sinn

Universität Hamburg
Institut für Technische und Makromolekulare Chemie
Bundesstraße 45, 2000 Hamburg 13, FRG

ISBN-13:978-3-642-83278-9   e-ISBN-13:978-3-642-83276-5
DOI: 10.1007/978-3-642-83276-5

Preface

More than 30 years after the discovery of transition metals and organometallics as catalysts for olefin polymerization these catalysts did not have lost their fascination.

Since 1953 when Karl Ziegler has discovered the catalytic polymerization of ethylene leading to plastically formable polymers which are mechanically stable up to temperatures of about 100°C, synthetic polymers and rubbers have made their way right into private houses. This discovery has been a main impetus for the fast growing production of plastics. The stereoselective polymerization of propylene and other long-chain α-olefins first detected by Giulio Natta leads to an even broadened field of applications.

Another enforcing factor were the developments of Standard Oil of Indiana and Phillipps Petroleum Company who engaged in the polymerization of α-olefins supported molybdenum, cobalt and later on chromium catalysts which clearly indicates the wide variety of suitable systems. This kind of research acknowledged merit when in 1963 the Nobel prize of chemistry was awarded to Ziegler and Natta.

Although to a great extent there is a technical application for these catalysts, up to now the nature of the active centres and many reaction mechanisms are not completely known.

Now and again the customer is taken by surprise by new polymers or copolymers produced with the help of Ziegler-Natta catalysts. Several thousands of letters, patents and original publications provide with information about efforts and research in this field. In the first generation of Ziegler-Natta catalysts, systems came into use which displayed only minor activities of about 30 kg polyethylene per g titanium compound. Compared to those in the second generation of catalysts, i. e. the supported catalysts consisting of magnesiumdichloride or other magnesium compounds, titaniumtetrachloride and aluminiumtrialkyls, the activities could be raised by a factor of 100. Particularily via introduction of ethylbenzoate, the stereoselectivity could be increased dramatically. The complexity in the investigation of those systems is due to their heterogenity.

New impulses for olefin polymerizations have been brought about by the discovery of the very high cocatalytic activity of alumoxanes instead of aluminiumalkyls together with the soluble transition metal compounds of titanocene, zirconocene and hafnocene.

The way in which these soluble systems have found access to research and application can well be documented by looking at the extent to which they have been treated with on the last international symposia.

At the 'International Symposium on Transition Metal Catalyzed Polymerizations' held in Akron in 1981, only one contribution has dealt with these homogeneous catalysts. In 1985 at the 'Symposium on Catalytic Polymerizations of Olefins' in Tokyo already two reports were given on this topic. Finally, on the Hamburg Symposium in 1987 there were already nine lectures about this.

This 'International Symposium on Transition Metals and Organometallics as Catalysts for Olefin Polymerization' held in Hamburg from September 21–24, 1987 gives an overview of today's point of research for olefine polymerizations with metalorganic compounds. The goals of the symposium were to show the ways for further development of catalysts, to find proper methods in the determination of the number of active centres, to investigate the kinetics and the overall mechanisms of the catalysis with heterogeneous and homogeneous catalysts, to provide with synthetic pathways for new polyolefins and copolymers with different properties and finally to develop tools and techniques for a comprehensive characterization of the polymers.

In many aspects there remain a lot of yet unresolved problems for the future. Polyolefins are the only polymers that consist exclusively of carbon and hydrogen atoms. After combustion only carbondioxide and water are set free. Therefore also from an environmental point of view the polyolefins will become dominant in the future.

The symposium consists of 32 lectures and 15 posters. The editorial staff is pleased to announce that 42 manuscripts of lectures and posters could be included in this book.

The symposium itself and this proceedings book would not have been possible without the generous financial support of BASF, Bayer AG, Erdölchemie, Hoechst AG, Röhm GmbH, Shell Netherlands, Beiersdorf AG, Büchi Switzerland, and the University administration.

We like to thank all those contributors who made possible these proceedings with their manuscripts of recent advances in the polyolefin research field.

University of Hamburg   Walter Kaminsky
Institute for   Hansjörg Sinn
Technical and Macromolecular Chemistry

Table of Contents

---

* = Invited Lecturers

List of Contributors

Alarcon, C.
Centro de Quimica, Instituto Venezolano de Investigaciones Cientificas, Apartado 21827, Caracas 1020-A, Venezuela

Albornoz, L. A.
Centro de Quimica, Instituto Venezolano de Investigaciones Cientificas, Aptdo. 21827, Caracas 1020-A, Venezuela

Altomare, A.
Dipartimento di Chimica, Università di Pisa, Via Risorgimento 35, 56100 Pisa, Italy

Ammendola, P.
Dipartimento di Fisica, Università di Salerno, 84100 Salerno, Italy

Atwood, J. L.
Dept. of Chemistry, University of Alabama, Tuscaloosa, AL 35487, USA

Bark, A.
Institute of Technical and Macromolecular Chemistry, University of Hamburg, Bundesstr. 45, 2000 Hamburg 13, FRG

Bliemeister, J.
Institute of Technical and Macromolecular Chemistry, University of Hamburg, Bundesstr. 45, 2000 Hamburg 13, FRG

Bobichon, C.
CNRS-Laboratoire des Materiaux Organiques, BP 24, 69390 Lyon-Vernaison, France

Böhm, L. L.
Hoechst AG, PO Box 800320, 6230 Frankfurt/M. 80, FRG

Boleslawski, M. P.
Dept. of Chemistry, State University of New York at Binghamton, Binghamtom, NY 13901, USA

Brintzinger, H. H.
Fakultät für Chemie, Universität Konstanz, 7750 Konstanz, FRG

Bukatov, G. D.
Institute of Catalysis, Novosibirsk 630090, USSR

Busico, V.
Dipartimento di Chimica dell'Università, Via Mezzocannone 4, 80134 Napoli,
Italy

Cann, K. J.
Unipol Systems Dept., Union Carbide Corporation, PO Box 670,
Bound Brook, NJ 08805, USA

Carlini, C.
Dipartimento di Chimica Industriale e dei Materiali, Università di Bologna,
Viale del Risorgimento 4, 40136 Bologna, Italy

Cheng, H. N.
Research Center, Hercules Inc., Wilmington, DE 19894, USA

Chien, J. C. W.
Polymer Science & Engineering Dept., University of Massachusetts,
Amherst, MA 01003, USA

Ciardelli, F.
Dipartimento di Chimica, Università di Pisa, Via Risorgimento 35,
56100 Pisa, Italy

Cioni, P.
Swiss Fed. Institute of Technology, Institute for Polymers, Universitätsstr. 6,
8092 Zürich, Switzerland

Clausnitzer, D.
Institute for Technical and Macromolecular Chemistry,
University of Hamburg, Bundesstr. 45, 2000 Hamburg 13, FRG

Corradini, P.
Dipartimento di Chimica dell'Università, Via Mezzocannone 4, 80134 Napoli,
Italy

Cuffiani, I.
Dutral SpA, Ausimont Compo NV, Centro Ricerche G Natta, Ferrara, Italy

Dall'Occo, T.
Dutral SpA, Ausimont Compo NV, Centro Ricerche G Natta, Ferrara, Italy

Dias, A. R.
CQE-IST, 1096 Lisboa Codex, Portugal

Doi, Y.
Research Laboratory of Resources Utilization, Tokyo Institute of
Technology, Nagatsuta, Midori-ku, Yokohama 227, Japan

Drögemüller, H.
Institute for Technical and Macromolecular Chemistry,
University of Hamburg, Bundesstr. 45, 2000 Hamburg 13, FRG

Duranel, L.
Groupement de Recherches de Lacq, Atochem Groupe Elf-Aquitaine, BP 34
Lacq, 64170 Artix, France

Dyachkovskii, F. S.
Institute of Chemical Physics, Academy of Sciences of the USSR, Moscow, USSR

Eisch, J. J.
Dept. of Chemistry, State University of New York at Binghamton, Binghamton, NY 13901, USA

Elder, M. J.
Fina Oil, Box 1200, Deer Park, TX 77536, USA

Ewangelidis, S.
Technische Universität Berlin, Institut für Technische Chemie, Straße des 17. Juni 135, 1000 Berlin 12, FRG

Ewen, J. A.
Fina Oil, Box 1200, Deer Park, TX 77536, USA

Fink, G.
Max-Planck-Institut für Kohlenforschung, Kaiser-Wilhelm-Platz 1, 4330 Mülheim/R., FRG

Franke, R.
Hoechst AG, PO Box 800320, 6230 Frankfurt/M. 80, FRG

Fuentes, A.
IVIC, Apt. 21827, Caracas 1020-A, Venezuela

Galimberti, M.
Swiss Federal Institute of Technology, Universitätsstr. 6, 8092 Zürich, Switzerland

Garoff, T.
Neste Oy, Technology Centre, 06850 Kulloo, Finland

Giesemann, J.
Carl Schorlemmer Technical University, Dept. of Chemistry, 4200 Merseburg, GDR

Goodall, B. L.
Koninklijke/Shell Laboratorium, Badhuisweg 3, 1031 CM Amsterdam, The Netherlands

Grünig, H.
Institut für Chemische Technologie, Technische Hochschule Darmstadt, Petersenstr. 20, 6100 Darmstadt, FRG

Guerra, G.
Dipartimento di Chimica dell'Università, Via Mezzocannone 4, 80134 Napoli, Italy

Guyot, A.
CNRS-Laboratoire des Materiaux Organiques, BP 24, 69390 Lyon-Vernaison, France

Hanke, A.
Technische Universität Berlin, Institut für Technische Chemie,
Straße des 17. Juni 135, 1000 Berlin 12, FRG

Haspeslagh, L.
Fina Oil, Box 1200, Deer Park, TX 77536, USA

Heiland, K.
Institute for Technical and Macromolecular Chemistry,
University of Hamburg, Bundesstr. 45, 2000 Hamburg 13, FRG

Hoppin, C. R.
Amoco Chemical Company, PO Box 400, Naperville, IL 60566, USA

Hosaka, M.
Toho Titanium Co., 3-3-5 Chigasaki, Kanagawa 253, Japan

Hu, Y.
Institute of Chemistry, Academia Sinica, Beijing, Peoples Rep. of China

Iiskola, E.
Neste Oy, Technology Centre, 06850 Kulloo, Finland

Ishii, K.
Toho Titanium Co. Ltd., Chigasaki, Kanagawa 253, Japan

Johnson, B. V.
3M Center, St. Pauli, MN 55144, USA

Jones, P. J. V.
Research & Technology Dept., ICI Chemicals, The Heath, Runcorn,
Cheshire, UK

Kakugo, M.
Sumitomo Chemical Co., Ltd., Chiba Research Laboratory, 5-1 Anesaki
Kaigan, Ichihara, Chiba, 299-01, Japan

Kaminsky, W.
Institute for Technical and Macromolecular Chemistry,
University of Hamburg, Bundesstr. 45, 2000 Hamburg 13, FRG

Karayannis, N. M.
Amoco Chemical Company, PO Box 400, Naperville, IL 60566, USA

Karol, F. J.
Unipol Systems Dept., Union Carbide Corp., PO Box 670,
Bound Brook, NJ 08805, USA

Kashiwa, N.
Mitsui Petrochemical Industries Ltd., Waki-cho, Kuga-gun, Yamaguchi-ken,
740, Japan

Kataoka, T.
Toho Titanium Company, 3-3-5 Chigasaki, Kanagawa 253, Japan

Keii, T.
Numazu College of Technology, Ooka, Numazu, Shizuoka 410, Japan

Khelghatian, H. M.
Amoco Chemical Company, PO Box 400, Naperville, IL 60566, USA

Kimura, K.
Toho Titanium Co. Ltd., Chigasaki, Kanagawa 253, Japan

Kojima, K.
Sumitomo Chemical Co., Ltd., Chiba Research Laboratory, 5-1 Anesaki Kaigan, Ichihara, Chiba, 299-01, Japan

Kratochvila, J.
Chemopetrol, Research Institute of Macromolecular Chemistry, Tkalcovska 2, 65649 Brno, Czechoslovakia

Krauss, H. L.
Laboratorium für Anorganische Chemie der Universität Bayreuth, PO Box 101251, 8580 Bayreuth, FRG

Lacombe, J. L.
Groupement de Recherches de Lacq, Atochem Groupe Elf-Aquitaine, BP 34 Lacq, 64170 Artix, France

Leistner, A.
Carl Schorlemmer Technical University, Department of Chemistry, 4200 Merseburg, GDR

Lesná, M.
Chemopetrol, Research Institute of Macromolecular Chemistry, Tkalcovská 2, 65649 Brno, Czechoslovakia

Locatelli, P.
Istituto di Chimica delle Macromolecole del CNR, Via E. Bassini 15, 20133 Milano, Italy

Luft, G.
Institut für Chemische Technologie, Technische Hochschule Darmstadt, Petersenstr. 20, 6100 Darmstadt, FRG

Marques, M. M. V.
CNP, Ap. 287521, Sines Codex, Portugal

Mehner, R.
Institut für Chemische Technologie, Technische Hochschule Darmstadt, Petersenstr. 20, 6100 Darmstadt, FRG

Mejzlik, J.
Chemopetrol, Research Institute of Macromolecular Chemistry, Tkalcovská 2, 65649 Brno, Czechoslovakia

Menconi, F.
Dipartimento di Chimica, Università di Pisa, Via Risorgimento 35, 56100 Pisa, Italy

Miyake, H.
Research Laboratory of Resources Utilization, Tokyo Institute of
Technology, Nagatsuta, Midori-ku, Yokohama 227, Japan

Möller-Lindenhof, N.
Institute for Technical and Macromolecular Chemistry,
University of Hamburg, Bundesstr. 45, 2000 Hamburg 13, FRG

Muñoz-Escalona, A.
IVIC, Apt. 21827, Caracas 1020-A, Venezuela

Nesterov, G. A.
Institute of Catalysis, Novosibirsk 630090, USSR

Niedoba, S.
Institute for Technical and Macromolecular Chemistry,
University of Hamburg, Bundesstr. 45, 2000 Hamburg 13, FRG

Nunes, C. P.
CNP, Ap. 287521 Sines Codex, Portugal

Nunomura, M.
Research Laboratory of Resources Utilization, Tokyo Institute of
Technology, Nagatsuta, Midori-ku, Yokohama 227, Japan

Ojala, T. A.
Max-Planck-Institut für Kohlenforschung, Kaiser-Wilhelm-Platz 1,
4330 Mülheim/R., FRG

Oldman, R. J.
Research & Technology Dept., ICI Chemicals and Polymers Group, Wilton,
Cleveland, UK

Ostoja Starzewski, K. A.
Wissenschaftl. Hauptlaboratorium, BAYER AG, 5090 Leverkusen, FRG

Pakkanen, T. A.
University of Joensuu, Dept. of Chemistry, SF-80100 Joensuu, Finland

Pakkanen, T. T.
University of Joensuu, Dept. of Chemistry, SF-80100 Joensuu, Finland

Pasquet, V.
CNRS-Laboratoire des matériaux Organiques, BP 24, 69390 Lyon-Vernaison,
France

Piccolrovazzi, N.
Swiss Federal Institute of Technology, Institute for Polymers,
Universitätsstr. 6, 8092 Zürich, Switzerland

Pino, P.
Swiss Federal Institute of Technology, Institute for Polymers,
Universitätsstr. 6, 8092 Zürich, Switzerland

Piotrowski, A. M.
Dept. of Chemistry, State University of New York at Binghamton,
Binghamton, NY 13901, USA

Reichert, K. H.
Institut für Technische Chemie, Technische Universität Berlin,
Straße des 17. Juni 135, 1000 Berlin 12, FRG

Sacchi, M. C.
Istituto di Chimica delle Macromolecole del CNR, Via E. Bassini 15,
20133 Milano, Italy

Sadatoshi, H.
Sumitomo Chemical Co. Ltd., Chiba Research Laboratory,
5-1 Anesaki Kaigan, Ichihara, Chiba, 299-01, Japan

Sequera, J. A.
IVIC, Apt. 21827, Caracas 1020-A, Venezuela

Sinn, H.
Institute for Technical and Macromolecular Chemistry,
University of Hamburg, Bundesstr. 45, 2000 Hamburg 13, FRG

Soga, K.
Research Laboratory of Resources Utilization, Tokyo Institute of
Technology, Nagatsuta, Midori-ku, Yokohama 227, Japan

Sormunen, P.
Neste Oy, Technology Centre, 06850 Kulloo, Finland

Spiehl, R.
Institute for Technical and Macromolecular Chemistry,
University of Hamburg, Bundesstr. 45, 2000 Hamburg 13, FRG

Spitz, R.
CNRS-Laboratoire des Matériaux Organiques, BP 24,
69390 Lyon-Vernaison, France

Suzuki, S.
Research Laboratory of Resources Utilization, Tokyo Institute of
Technology, Nagatsuta, Midori-ku, Yokohama 227, Japan

Szczegot, K.
Institute of Chemistry, Pedagogical University, Opole, Poland

Tait, P. J. T.
Dept. of Chemistry, UMIST, Manchester M60 1QD, UK

Terano, M.
Toho Titanium Co., 3-3-5 Chigasaki, Kanagawa 253, Japan

Thiele, K. H.
Carl Schorlemmer Technical University, Dept. of Chemistry,
4200 Merseburg, GDR

Thum, G.
Hoechst AG, PO Box 800320, 6230 Frankfurt/M. 80, FRG

Tikwe, L.
Institute for Technical and Macromolecular Chemistry,
University of Hamburg, Bundesstr. 45, 2000 Hamburg 13, FRG

Tokuhiro, N.
Research Laboratory of Resources Utilization, Tokyo Institute of
Technology, Nagatsuta, Midori-ku, Yokohama 227, Japan

Tritto, I.
Istituto di Chimica delle Macromolecole del CNR, Via E. Bassini 15,
20133 Milano, Italy

Tsutsui, T.
Mitsui Petrochemical Industries Ltd., Waki-cho, Kuga-gun, Yamaguchi-ken,
740, Japan

Ulbricht, J.
Carl Schorlemmer Technical University, Dept. of Chemistry,
4200 Merseburg, GDR

Vähäsarja, E.
University of Joensuu, Dept. of Chemistry, SF-80100 Joensuu, Finland

Vasiliou, G.
Institut für Technische Chemie, Technische Universität Berlin,
Straße des 17. Juni 135, 1000 Berlin 12, FRG

Vasnetsov, S. A.
Institute of Catalysis, Novosibirsk 630090, USSR

Vozka, P.
Chemopetrol, Research Institute of Macromoleculare Chemistry,
Tkalcovská 2, 65649 Brno, Czechoslovakia

Wagner, B. E.
Unipol Systems Dept., Union Carbide Corp., PO Box 670,
Bound Brook, NJ 08805, USA

Weber, S.
BASF Aktiengesellschaft, Kunststoff-Laboratorium, Carl-Bosch-Str. 38,
6700 Ludwigshafen, FRG

Wei, J.
Swiss Federal Institute of Technology, Institut für Polymere,
Universitätsstr. 6, 8092 Zürich, Switzerland

Winter, H.
Institute for Technical and Macromolecular Chemistry,
University of Hamburg, Bundesstr. 45, 2000 Hamburg 13, FRG

Witte, J.
Wissenschaftliches Hauptlaboratorium, BAYER AG, 5090 Leverkusen, FRG

Yanagihara, H.
Oita Research Laboratory, Showa Denko Co. Ltd., 2 Nakanosu, Oaza, Oita
870 01, Japan

Yechevskava, L. G.
Institute of Catalysis, Novosibirsk 630090, USSR

Yokoyama, M.
Sumitomo Chemical Co. Ltd., Ehime Research Laboratory, 5-1 Sobiraki,
Niihama, 792, Japan

Yoshitake, J.
Mitsui Petrochemical Industries Ltd., Waki-cho, Kuga-gun, Yamaguchi-ken,
740, Japan

Zakharov, V. A.
Institute of Catalysis, Novosibirsk 630090, USSR

Zambelli, A.
Dipartimento di Fisica, Università di Salerno, 84100 Salerno, Italy

Zarncke, O.
Institute for Technical and Macromolecular Chemistry,
University of Hamburg, Bundesstr. 45, 2000 Hamburg 13, FRG

Zhang, H.
Dept. of Chemistry, University of Alabama, Tuscaloosa, AL 35487, USA

Zucchini, U.
Dutral SpA Ausimont Compo NV, Centro Ricerche G Natta, Ferrara, Italy

1. Studies of Active Sites, and Kinetics and Mechanisms in Heterogeneous Catalyst

Kinetic Behavior During Initial Stage (less than 1 s) of Propene Polymerization with Supported Catalyst.

Tominaga Keii
Numazu College of Technology, Ooka, Numazu, Shizuoka 410, Japan

Minoru Terano, Kouhei Kimura, Kazuhiro Ishii
Toho Titanium Co. Ltd., Chigasaki, Kanagawa 253, Japan

INTRODUCTION

The most reasonable method for determining values of the rate constants of propagation, transfer and the concentration of polymerization centers, k_p, k_{tr} and C^*, is that based upon the observation of both transitional changes of number average degree of polymerization $\overline{P}_n$ and polymer yield Y in the initial stages of polymerization. This method has been used for long time, however, the results reported before the development of GPC are not so precise because of the use of viscosity average molecular weight in place of number average molecular weight. In addition, the use of viscosity average molecular weight in place of number average molecular weight may be possible only in the case of constant polydispersity independent of polymerization time, as pointed out by the present author (Keii 1986). Furthermore some limitations in the application of the method should be noted. For example, the method can not be applied for polymerizations of rapid transfer reaction. The applicability of the method is determined by value of the mean lifetime of growing chain, τ, i.e. the reciprocal of transfer rate constant, $1/k_{tr}$, as below.

In the case of polymerization with a constant rate, the number average degree of polymerization is expressed by

$$\overline{P}_n = \frac{k_p[M]t}{1 + k_{tr}t} \qquad (1)$$

and the polymer yield by

$$Y = k_p[M]C^*t \qquad (2)$$

where [M] is monomer concentration. Then, from the change of $\overline{P}_n$ with time we can obtain values of k_{tr} and $k_p[M]$ by means of Eq.(1). Combining the value of $k_p[M]$ with the value of yield (in mol) at time t, we can obtain value of C^*. However, in the case of polymerization where $\tau << t$ or $k_{tr}t >> 1$, we can not obtain any transitional change of $\overline{P}_n$ but only its stationary value, $k_p[M]/k_{tr}$. In fact, we could confirm that the propene polymerization with a $MgCl_2$/Ethylbenzoate/$TiCl_4$/$Al(C_2H_5)_3$ at 40°C was the very case even for 5 s (Suzuki et al. 1979). On the contrary, propene polymerizations with

W. Kaminsky and H. Sinn (Eds.)
Transition Metals and Organometallics as
Catalysts for Olefin Polymerization
© Springer-Verlag Berlin Heidelberg 1988

4

the traditional TiCl$_3$ catalysts have been subjected for the method because of their rather small rate constants of transfer reaction (τ is in the order of minutes to hour).

Then, this method is useful for polymerizations under the condition that $\tau > t$ or $k_{tr}t < 1$. Such polymerizations may be called as "Quasi-Living Polymerizations". To establish such state of polymerization in the case of the propene polymerization with MgCl$_2$ supported catalyst, polymerization times must be taken as small as possible, e.g. far less than 5 s at 40°C.

Another limitation which has been recognized is the case of polymerization the rate of which is not constant. Furthermore, the case of polymerization where the rate constant k_p is a function of C*is also out of the application of the method. However, in this article the discussion is focused on the first case, i.e. the polymerization with MgCl$_2$ supported catalysts.

In order to apply successfully the method for the propene polymerization with a MgCl$_2$/Ethylbenzoate/TiCl$_4$/Al(C$_2$H$_5$)$_3$, a simple stopped flow method is developed.

A SIMPLE STOPPED FLOW POLYMERIZATION

The polymerization reactor used is illustrated in Fig. 1. The two solutions (A) and (B) kept under propene of atmospheric pressure are

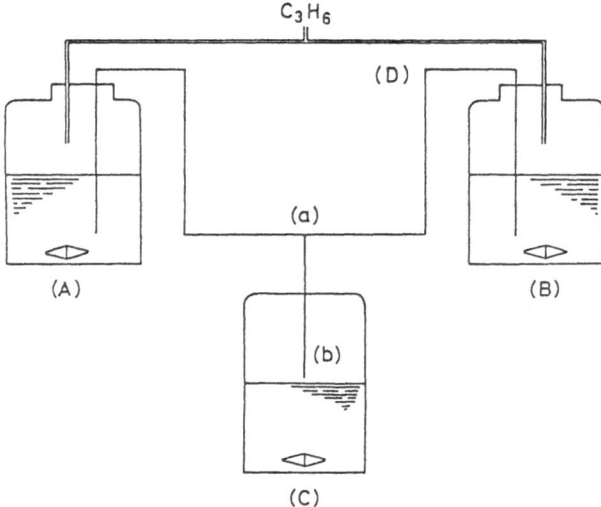

Fig. 1. A simple stopped flow polymerization apparatus. (A) and (B) are 250 ml flasks, containing 200 ml of catalyst suspension of propene saturated heptane and of Al(C$_2$H$_5$)$_3$ solution in heptane saturated by propene. (C) is a 1 l flask containing 400 ml of ethanol, and (D) is a Teflon tube of 2 mm inner diameter. (a) is a three necked connection.

forced flow out by imposing a little higher propene pressure and mixed at the three necked connection (a). The solution (A) is a suspension of the catalyst in heptane saturated by propene, while (B) heptane solution of $Al(C_2H_5)_3$ saturated by propene. (C) is reservoir of ethanol whose surface is adjusted close to the end of Teflon tube.

Polymerization time t is defined as

$$t = t_o \frac{v_r}{V_o} \qquad (3)$$

where V_o is the total volume of the solution mixture flowed out for t_o and v_r the inner volume of the Teflon tube from the connection (a) to its end (b). This definition is based on the assumption of complete mixing of the two solutions at the connection (a).

Using a Teflon tube of inner diameter 2 mm, c.a. 200 ml for V_o, and 30 s for t_o, we can have that 0.1 - 1.0 s for t with the length of Teflon from (a) to (b), 20 - 200 cm. For the sake of simple analysis of experimental data, the conversions of propene monomer were so adjusted to those of lower than 10% that both monomer concentration [M] and temperature of the polymerization mixture could be regarded as approximately constant in the Teflon tube from (a) to (b). The most important point of this stopped flow method is the mixing of the two solutions at (a). The above flow condition at 20°C corresponds to Reynolds number of c.a. 3200 (turbulent flow). Changing linear flow velocity from 200 cm s^{-1} to 42 cm s^{-1} which corresponds to Reynolds number of 700 (piston flow), we found no significant change in polymerization yield Y and number average molecular weight $\overline{M}_n$ at the same time t. Then, the assumption of complete mixing may be taken as the basis of the definition of polymerization time by means of Eq.(3). The complete mixing at this simple three necked connection might be due to the nature of the solution mixture which was slurry.

The catalyst used was prepared by a procedure that a coground mixture of $MgCl_2$ (30 g) and ethylbenzoate (6.8 g) was treated with $TiCl_4$ at 90°C and then washed with heptane.

The monomer concentration [M] was determined from the weight increase of heptane after propene saturation under 1 atm of total pressure. Then, the monomer concentration at T °C is that in equilibrium at the propene partial pressure, (1 atm - the vapor pressure of heptane at T °C).

The total polymer produced was obtained by drying the ethanol solution and was weighed and its molecular weights were determined by a Water's 150C GPC.

RESULTS AND DISCUSSION

The experimental results obtained are summarized in Table 1.

Table 1. Kinetic data of "quasi-living" polymerization of propene

T	[TEA][a]	[M][a]	t	Y[b]	$\overline{M}_n$[c]	$\overline{M}_w/\overline{M}_n$	k_p[d]	C^{*}[e]	k_t	$k_{p,a}$[d]	C_a^{*}[e]
°C			s						s^{-1}		
20	70	720	0.10	220	3.2	4.1				1.1	6.9
20	70	720	0.14	350	5.0	3.2				1.2	7.0
20	70	720	0.22	530	7.7	3.9				1.2	6.9
20	70	720	0.40	980	8.7	3.3	1.23	6.0	1.0	0.72	11.3
20	70	720	0.72	1900	13.8	4.4				0.63	13.7
20	70	720	1.36	4700	18.4	4.3				0.45	23.7
(20	70	720	10	30000	37.8	3.6				0.13	79.4)[f]
20	0.14	720	0.19	36	5.8	4.0				1.0	0.62
20	7	720	0.17	260	5.5	3.8				1.0	4.7
20	230	720	0.16	510	5.0	3.8				1.0	10.1
0	70	1480	0.21	250	4.9	4.5				0.38	5.1
10	70	1000	0.12	150	2.5	5.6				0.49	6.0
30	70	500	0.14	360	4.8	4.0				1.6	7.6
40	70	370	0.13	630	4.6	3.3				2.3	13.8
60	70	200	0.22	920	5.0	4.4				2.7	14.4

a mmol l^{-1}; b g mol-Ti^{-1}; c 10^3 g mol^{-1}; d 10^3 l mol^{-1} s^{-1};
e 10^2 mol mol-Ti^{-1};
f The data were obtained with a usual slurry polymerization in 500 ml flask of semi-batch.

As can be seen from Fig. 2, the polymer yield of polypropylene at 20°C is proportional to the polymerization time t, which assures the application of Eq.(1) with Eq.(2) for the data obtained at 20°C. Before making the application we note an approximate evaluation of k_p. The following approximation (living approximation) may be useful at the beginning stages of the polymerization.

$$k_{p,a} = \frac{\overline{M}_n}{42[M]t} \qquad (4)$$

$$C_a^{*} = \frac{Y}{\overline{M}_n} \qquad (5)$$

The approximation gives a lower limit of k_p and an upper limit of C^{*}, respectively. The approximate values are shown in Table 1. The results of the application of Eq.(1) to the data obtained at 20°C are illustrated in Fig. 3. From the linearity of the reciprocal number average degree of polymerization against the reciprocal time of polymerization we have 1230 l mol^{-1} s^{-1} as the value of k_p and 1 s^{-1} as that of k_{tr}. Using this value of k_p and [M], we have 0.06 mol mol-Ti^{-1} as the value of C^{*}.

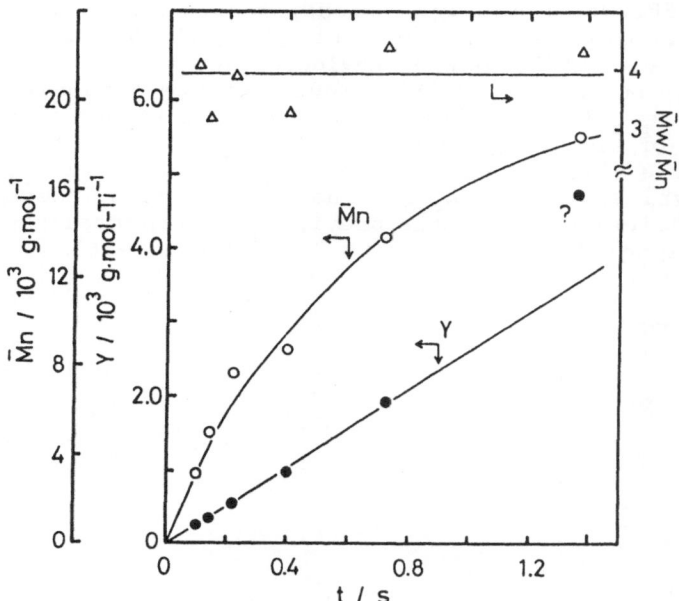

Fig. 2. Changes of polymer yield, molecular weight, and polydispersity with polymerization time at 20°C.

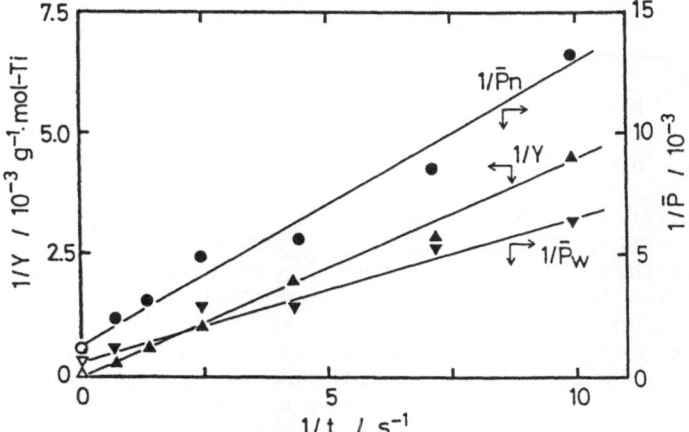

Fig. 3. The reciprocal values of Y and degree of polymerization versus reciprocal polymerization time.

As well-known, the activation energy of transfer rate constant is larger than that of propagation rate constant. Then, the values of k_{tr} at temperatures below 20°C may be neglected and the approximate values, $k_{p,a}$, can be regarded as k_p. Figure 4 shows Arrhenius plot of k_p (20°C) and $k_{p,a}$ (0 and 10°C) which gives 39.6 kJ mol^{-1} as the activation energy of k_p. Extrapolating the plot to 40°C, we have 3600 l mol^{-1} s^{-1} as the value of k_p and then 4.3 s^{-1} as that of k_{tr} at 40°C from the k_p value with the $k_{p,a}$ value at t = 0.13 s. These values appear to be coincident with the previous experimental results that $k_p \gg$ 970 l mol^{-1} s^{-1} and $k_{tr} \gg$ 0.2 s^{-1} (Keii 1986).

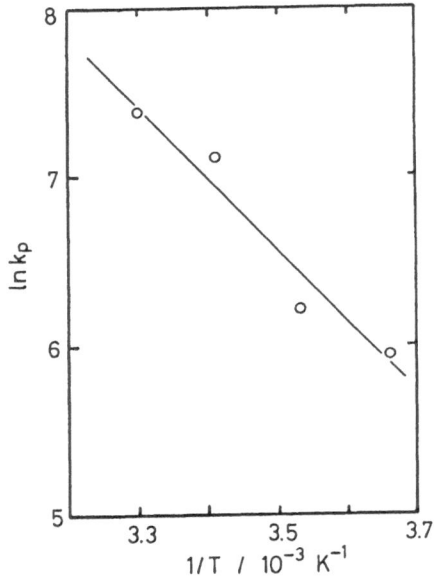

Fig. 4. The Arrhenius plot of k_p.

Figure 5 shows that Y/t (= $k_p[M]C^*$) depends on the concentration of triethyl aluminium, [TEA], whereas $\overline{M}_n/t$ (= $k_{p,a}[M]$) does not. Then, it may be considered that only C_a^* is a function of [TEA] which is a Frumkin-Temkin type (of Adsorption), C^* = AlogB[TEA], or roughly a Langmuir type on either [TEA] or [TEA]$^{1/2}$. This result does not agree with the previous result of polymerization for large polymerization time that Y/t is proportional to K[TEA]/(1 + K[TEA])^2 (Suzuki et al. 1979). The latter was explained on the basis that $k_p[M]$ was proportional to 1/(1 + K[TEA]) as well as C^* to K[TEA]/(1 + K[TEA]). The difference in k_p may be due to either the difference of the catalysts used or the slow coordination of alkylaluminium to the vacant site available for propene monomer. In fact the catalyst used here is extremely higher active than that in the previous experiment (Suzuki et al. 1979).

The most important result of this experiment appears the value of polydispersity, $\overline{M}_w/\overline{M}_n$, and its behavior in this initial stage of the

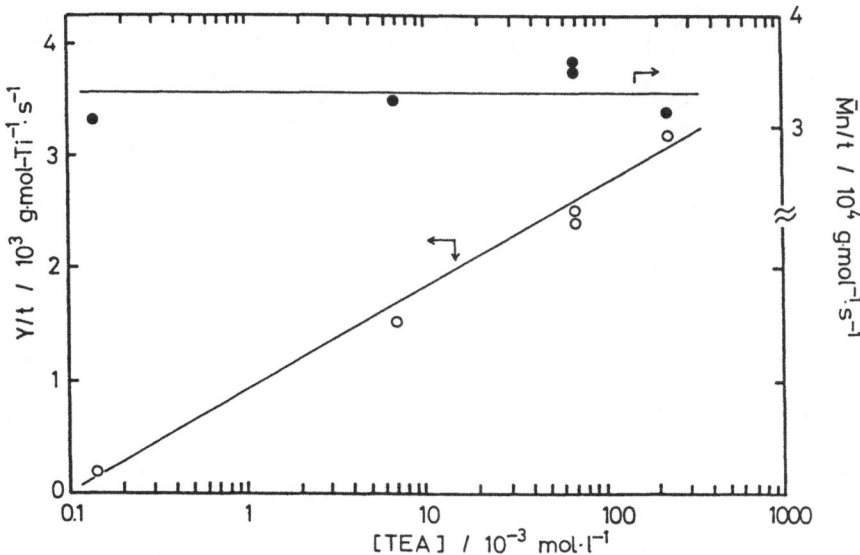

Fig. 5. The dependencies of Y/t and $\overline{M}_n$/t on [TEA] at 20°C.

polymerization. As can be seen from Table 1 and Fig. 2, the value of polydispersity is c.a. 4 and keeps constant from 0.1 s to 10 s. This value of 4 in this quasi-living polymerization, where transfer reaction can be neglected, suggests strongly that a non-uniformity of k_p value with polymerization centers causes the broadening of molecular weight distribution. The premise of this discussion is that there are only the three rival theories on the broadening of molecular weight distribution in heterogeneous polymerizations; the above mentioned non-uniformity of k_p (Keii et al. 1982, Keii et al. 1984), a non-uniformity of k_{tr} with chain length (Gordon 1961), and the diffusion control theory based upon polymer layer on catalyst surface (Singh and Merrill 1971). The present author has supported the non-uniformity of k_p (Clark and Bailey 1963) on the basis that the polydispersity remains constant with increasing amount of hydrogen, according to Roe (Roe 1961) and that the polymerization rate decreases non-linearly with amount of carbon monoxide, as a catalyst poison (Keii et al. 1982). The value of 4 for the polydispersity in the quasi-living polymerization is a new evidence of the non-uniformity of k_p.

However, it should be noted here that the constancy of the value of 4 in the course of the polymerization from 0.1 s to 10 s challenges us with a significant question that why not increase to 8, as expected at polymerization times larger than the mean lifetime of growing chain, 1 s, as shown in Appendix (A). If the transfer reaction in this polymerization is a second order with respect to growing chains, the polydispersity will change from 4 to 6 in the course of polymerization from the living to the stationary state as shown in Appendix (B). However, we can not find any trend of increase in the polydispersity in this experiment, though the experimental error appears to be large. It may be worth while to discuss this question.

APPENDIX

(A) Change of polydispersity in polymerization with first order transfer reaction.

Supposing the following elementary reactions on a polymerization center;

$$\text{cat-}P^*_{n-1} + M \xrightarrow{\ k_p\ } \text{cat-}P^*_n \tag{A1}$$

$$\text{cat-}P^*_n + T \xrightarrow{\ k_{tr}\ } \text{cat-} + P_n \tag{A2}$$

$$\text{cat-} + \text{initiator} + M \xrightarrow{\ \text{rapid}\ } \text{cat-}P^*_1 \tag{A3}$$

we have

$$d[P^*_1]/dt = k_{tr}\sum[P^*_n] - (k_p[M] + k_{tr})[P^*_1] \tag{A4}$$

$$d[P^*_n]/dt = k_p[M][P^*_{n-1}] - (k_p[M] + k_{tr})[P^*_n] \tag{A5}$$

$$d[P_n]/dt = k_{tr}[P^*_n] \tag{A6}$$

The total number of growing chains, $\sum[P^*_n]$, is constant during polymerization, which is compatible with this experiment of constant polymerization rate. The set of equations has been solved by Cabrerizo and Guzman (1979). Assuming a non-uniformity of k_p with polymerization centers, we have to use the average value of k_p, $\overline{k_p}$, and then we have the following modified solution.

$$\overline{P}_n = \frac{\overline{k}_p[M]}{k_{tr}} \frac{k_{tr}t}{1 + k_{tr}t} \tag{A7}$$

$$\overline{P}_w = \frac{2\overline{k_p^2}(k_{tr}t - 1 + e^{-k_{tr}t})}{\overline{k}_p k_{tr}^2 t} \tag{A8}$$

These equations have been simplified for the case with $\overline{k}_p/k_{tr} \gg 1$. From these equations it can be seen that the value of polydispersity increases to $2\overline{k_p^2}/(\overline{k_p})^2$ from $\overline{k_p^2}/(\overline{k_p})^2$ with increasing polymerization time from zero to $t \gg \tau$.

(B) Change of polydispersity in polymerization with second order transfer reaction.

In the case of second order transfer reaction we can apply the following set of equations

$$d[P_1^*]/dt = k_{tr}(\sum[P_n^*])^2 - (k_p[M] + k_{tr}\sum[P_n^*])[P_1^*] \qquad (A9)$$

$$d[P_n^*]/dt = k_p[M][P_{n-1}^*] - (k_p[M] + k_{tr}\sum[P_n^*])[P_n^*] \qquad (A10)$$

$$d[P_n]/dt = \frac{1}{2}k_{tr}\sum_r[P_{n-r}^*][P_r^*] \qquad (A11)$$

From these equations we obtain

$$d\sum[P_n^*]/dt = 0 \qquad (A12)$$

and then

$$\sum[P_n^*] = \text{constant} = I_o \qquad (A13)$$

which is also compatible with polymerization of constant rate.

From the solutions of the above set of equations, we have the number average and weight average degrees of polymerization as follows.

$$\overline{P}_n = \frac{(\overline{\rho}+1)(t/\tau) + 1}{1 + (t/\tau)/2} \qquad (A14)$$

$$\overline{P}_w = \frac{(3\overline{\rho^2}+5\overline{\rho}+2)(t/\tau)-(4\overline{\rho^2}-\overline{\rho})(1-e^{-t/\tau})+(\overline{\rho^2}/2)(1-e^{-2t/\tau})+1}{(\overline{\rho}+1)(t/\tau) + 1} \qquad (A15)$$

where $\overline{\rho} = \overline{k_p}[M]/k_{tr}$, $\overline{\rho^2} = \overline{k_p^2}[M]^2/k_{tr}$ and $\tau = 1/k_{tr}I_o$. Then, the value of polydispersity increases from $\overline{k_p^2}/(\overline{k_p})^2$ to $1.5\overline{k_p^2}/(\overline{k_p})^2$ with increasing polymerization time from smaller to larger than τ.

The second order transfer reaction taken here is that produce P_n from P_{n-r}^* and P_r^* by coupling reaction, which has been not yet prefer than that produce two dead polymers by disproportionation reaction. The latter transfer results in the same solution with Appendix (A).

REFERENCES

Cabrerizo JL, Guzman J (1979) Theoretical study of living polymerization reaction involving a transfer agent. Macromolecules 12: 526-530

Clark A, Bailey GC (1963) Formation of high polymers on solid surfaces I. theoretical study of mechanisms. J Catal 2: 230-240

Gordon M, Roe R-J (1961) Surface-chemical mechanism of heterogeneous polymerization and delivation of Tung's and Wesslau's molecular weight distribution. Polymer 2: 41-59

Keii T, Suzuki E, Tamura M, Murata M, Doi Y (1982) Propene polymerization with a magnesium chloride-supported Ziegler catalyst, 1. Principal kinetics. Makromol Chem 183: 2285-2304

Keii T, Doi Y, Suzuki E, Tamura M, Murata M, Soga K (1984) Propene polymerization with a magnesium chloride-supported Ziegler catalyst, 2. Molecular weight distribution. Makromol Chem 185: 1537-1557

Keii T (1986) Mechanistic studies on Ziegler-Natta catalysis -A methodological reconsideration-. In: Keii T, Soga K (eds) Catalytic polymerization of olefins. Elsevier, Amsterdam Oxford New York Tokyo, p1

Roe R-J (1961) A test between rival theories for molecular weight distributions in heterogeneous polymerization: The effect of added terminating agent. Polymer 2: 60-73

Singh D, Merrill RP (1971) Molecular weight distribution of polyethylene produced by Ziegler-Natta cayalysts. Macromolecules 4: 599-604

Suzuki E, Tamura M, Doi Y, Keii T (1979) Molecular weight during polymerization of propene with the supported catalyst system $TiCl_4/MgCl_2/C_6H_5COOC_2H_5/Al(C_2H_5)_3$. Makromol Chem 180: 2235-2239

ACTIVATION AND STEREOSPECIFIC CONTROL IN PROPYLENE POLYMERIZATION WITH MgCl$_2$ SUPPORTED ZIEGLER-NATTA CATALYSTS

A. GUYOT, C. BOBICHON, R. SPITZ
CNRS - Laboratoire des Matériaux Organiques
BP 24   69390 LYON-Vernaison France

L. DURANEL and J.L. LACOMBE
Groupement de Recherches de Lacq - ATOCHEM Groupe ELF-AQUITAINE
BP 34 Lacq   64170 ARTIX   France

ABSTRACT

In the isotactic polymerization of propene using MgCl$_2$ supported Ziegler catalysts the addition of small amounts of a Lewis base to the cocatalytic solution causes a severe reduction of the activity of the aspecific sites, but enhances the activity of the isospecific sites ; further addition of the same Lewis base cause moderate inhibition of all kinds of sites. These effects are observed with a variety of components of the cocatalytic solutions : both different alkylaluminiums and various aromatic esters or aromatic silanes. These results support the idea of alkylaluminium complexed with the Lewis base as major components of the isospecific sites.

INTRODUCTION

The purpose of this paper is to contribute to the knowledge of the activation processes for MgCl$_2$ supported Ziegler-Natta Catalysts. In addition to the alkyl-aluminium cocatalysts and related compounds, several organic compounds have been observed to enhance the rate of olefin polymerization. Among them, different families of Lewis bases were shown by Boor et al (1) to enhance the isospecificity of TiCl$_3$ catalysts in propylene polymerization and also the rate in some cases (2, 3). According to Boor (4), the best explanation for this activation effect was the ability of the electron donor to promote the disruption of TiCl$_3$ aggregates in smaller particles, and then to increase the number of active sites, while poisoning the more exposed aspecific titanium sites. Another explanation, based on the modification of the electrophilicity of the site, was suggested by Karayannis (3). Using simple MgCl$_2$ supported catalyst, Kashiwa (5) was the first to report an enhancement of the rate of isotactic polymerization upon addition of ethylbenzoate in the cocatalyst solution ; in the same, the rate of production of atactic polymer is strongly decreased, so that the aromatic ester acts simultaneously as a poison of the aspecific sites and an activator of the isospecific sites. We got a similar result (6) using a more complex catalyst containing cogrinded ethylbenzoate (EB) MgCl$_2$ solid support further impregnated with TiCl$_4$, used together with a solution of triethylaluminium partly complexed with ethylparatoluate (EPT) ; in that case, although the poisoning of the aspecific sites was very efficient, the whole polymerization rate was slightly enhanced by

W. Kaminsky and H. Sinn (Eds.)
Transition Metals and Organometallics as
Catalysts for Olefin Polymerization
© Springer-Verlag Berlin Heidelberg 1988

addition of moderate amounts of EPT ; i.e. the poisoning effect of the Lewis base on the aspecific sites was overcompensated by its activation effect on the isospecific sites.

The stereogulating effect of Lewis base has been thoroughly discussed by various authors, specially Pino et al (7-9), Barbe et al (10) and more recently Giannini et al (11). The increased production of isotactic polypropylene is attributed to various effects : poisoning of aspecific sites, transformation of aspecific sites with two coordination vacancies into stereospecific sites, and competition between Lewis bases and $TiCl_4$ for selective coordination with various faces of the $MgCl_2$ crystals. Increased molecular weights are obtained upon addition of EB to the $MgCl_2/TiCl_4/AlEt_3$ catalyst for both the propylene polymerization (5) and the ethylene polymerization (9). This fact, together with the transformation of aspecific sites to isospecific sites of increased activity may explain the enhancement of the overall activity observed.

We would like in this paper to report a more extensive study of these effects and to discuss then in terms of competitive association of the free and complexed alkylaluminium on the titanium sites.

EXPERIMENTAL

Preparation of the $MgCl_2/EB/TiCl_4$ catalyst

In a stainless-steel pot (80 cm^3 inside volume) containing 55 g of stainless-steel balls (3 to 10 mm diameter) 10.5 g of $MgCl_2$ are milled first 4 hours under argon, then comilled with ethylbenzoate (1.4 cm^3) during 12 hours. The resulting solid is treated by excess. $TiCl_4$ (10 cm^3) diluted heptane (14 cm^3) 2 h at 60° C, washed with heptane and dried in high vacuum (10^{-8} bars) at room temperature. The titanium contents of the catalyst is 5 % by weight. The other catalysts are prepared using similar procedures.

Polymerization : the polymerization are carried out in a one liter stainless steel reactor equipped with a high efficiency stirrer. The reagents are introduced at room temperature under argon stream in the order : heptane (500 ml), alkylaluminium (typical value : 5 mmole) Lewis base (typical value : 1.25 mmole) precatalyst (30mg). The reactor is then pressurized with propene (4 bars) and heated at the reaction temperature (60-62° C). The pressure is maintained constant during the reaction by monomer supply from a monomer reservoir, through a controlling valve ; the activity is deduced from the kinetic measurement of the pressure drop in the reservoir. The isotactic index is measured as the fraction insoluble in boiling heptane (kumagawa extraction for 2 hours).

RESULTS AND DISCUSSION

The increase of the productivity of isotactic polypropylene upon addition of a Lewis base seems to be a quite general phenomenon. It may be observed using different Lewis bases in the precatalyst. Corresponding results concerning $MgCl_2/EB/TiCl_4$ and $MgCl_2/EPT/TiCl_4$ catalyst used with cocatalytic solution of either $AlEt_3$ or Isoprenyl-Aluminium (IPRA) themselves complexed with a set of aromatic esters, are reported in Table 1. Table 2 is dealt with $MgCl_2/EPT/TiCl_4$ catalyst.

Table 1. Effect of addition of aromatic ester in the cocatalyst solution upon the stereospecificity of the catalytic system $MgCl_2/EB/TiCl_4$: Polymerization temperature 63° C
Polymerization time : 60 min - Monomer pressure : 4 bars

Run	Cocatalyst solution				Productivity g pol/g catalyst		
	Alkyl Al		Ester				
	nature	concent mmole/l		Ratio E/A	Overall	Isotactic	Atactic
1	TEA	3	none	-	1900	952	948
2	"	"	EPT	0.25	1850	1761	89
3	"	"	"	0.33	1460	1402	58
4	"	5	none	-	1500	766	734
5	"	"	EPT	0.25	1760	1646	114
6	"	"	"	0.33	1300	1263	37
7	"	10	none	-	1040	536	504
8	"	"	EA	0.25	1300	1220	80
9	"	"	EPT	0.25	1570	1450	120
10	"	"	EB	"	1270	1179	91
11	"	"	PTBMB	"	1630	1300	320
12	IPRA	10	none	-	1040	536	504
13	"	"	EA	0.16	710	615	95
14	"	"	"	0.25	550	507	43
15	"	"	"	0.33	300	280	20

TEA : triethylaluminium ; IPRA : isoprenylaluminium
EB : ethylbenzoate ; EA : ethylanisate ; EPT : ethylparatoluate ;
PTBMB : paratertiobutyle methylbenzoate

Table 2. Effect of the addition of aromatic esters in the cocatalyst solution on the stereospecificity of the catalytic $MgCl_2/EPT/TiCl_4$ (same conditions as in table 1 except for runs 6-10 : 90 min)

| Run | Cocatalyst solution | | | | Productivity g pol/g catalyst | | |
| | Alkyl Al | | Ester | | | | |
	Nature	mmole/L	Nature	Ratio E/A	Overal	Isotactic	Atactic
1	TEA	10	none	-	1040	614	426
2	"	"	EA	0.25	940	862	78
3	"	"	"	0.33	820	775	45
4	"	"	"	0.40	730	710	20
5	"	"	"	0.45	540	529	11
6	IPRA	10	none	-	1200	758	442
7	"	"	EA	0.25	280	251	29
8	"	"	EPT	"	530	464	66
9	"	"	EB	"	640	522	118
10	"	"	EAC	"	225	186	39

TEA : triethylaluminium ; IPRA : isoprenyl Aluminium ;
EA : ethylanisate ; EPT : ethylparatoluate ; EB : ethylbenzoate ;
EAC : ethylacetate

The isotactic productivity may be more than doubled, if moderate amounts of ester are added ; in that respect the actual amount of aromatic ester introduced is not the pertinent parameter ; it is better to state in terms of Ester/Al ratio ; small ratios are enough to strongly reduce the production of atactic polymers ; increasing the ratio causes a new decrease of that production, but also affects the isotactic productivity. When IPRA is used in the cocatalytic solution instead of TEA, the poisoning effect of the Lewis base is more important, and in some cases (Table 2 runs 6-10), the increase of isotactic productivity is not observed ; even if, in the absence of Lewis base in the cocatalytic solution, the isospecific character of the polymerization is a little higher when IPRA is used instead of TEA, the effect of addition of Lewis base in the cocatalyst solution is much smaller ; this is to be related to the complexation equilibrium between the ester and the organoaluminium compound ; we have shown previously (6) that the equilibrium is less displaced for IPRA than for TEA towards the formation of the complex ; thus in the case of IPRA, for the same ratio Ester/Al, more free ester remains uncomplexed and is most probably a strong poison of all the sites. When the amount of TEA increases, the productivity, in the absence of added ester, decreases with only a slight change in the tacticity ; upon adding EPT at a ratio EPT/Al of 0,25 the overall productivity decreases much less and the isotactic

productivity is strongly increased by a factor which may be more
than 2 ; by increasing the ratio up to 0,33, then the overall
productivity decreases again. In that case two effects are
observed : the first one is the effect of complexation of TEA by
EPT upon its reducing power versus titanium, which is made less
strong, so that the catalytic activity remain more stable during
the polymerization, because less sites are deactivated ; on the
other hand, the effect causes the initial polymerization rate to
be lower (Fig. 1), probably because less sites are activated but
the isotactic polymerization rate has the same behavior as the
productivity.

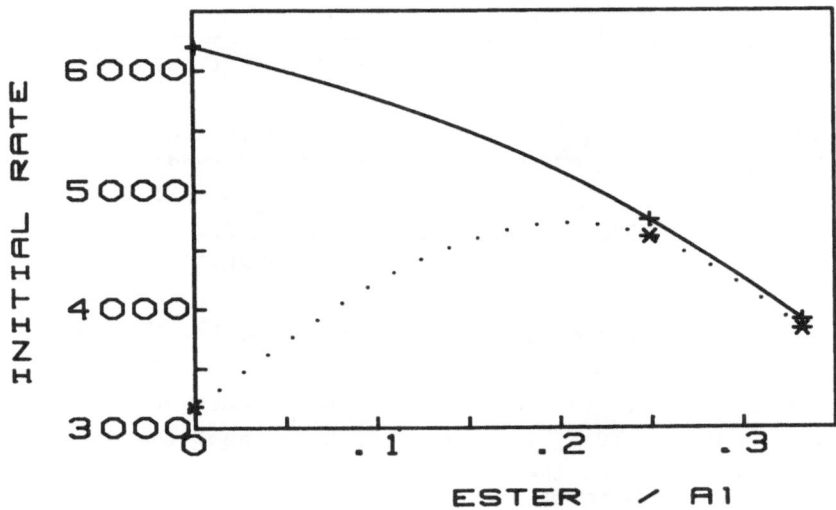

Fig. 1. Lowering of the initial rate (polymerization rate 2 min after
starting the process in g pol/g catalyst/h) by addition of EPT
Catalyst : $MgCl_2$/EB/$TiCl_4$, $Al(C_2H_5)_3$;
Monomer pressure 4 bars ; Polymerization temperature 63° C
—— whole polymer
.... isotactic polymer

According to the results reported by Kashiwa (5) these effects
should be rationalized upon stating that the initial Ti^{+4} becomes
active upon reduction to Ti^{+3} but are deactivated when they are
further reduced to Ti^{+2}, so that a lower reducing power of complexed
alkyl Al activate a smaller number of sites, but these sites are more
stable.

The second effect is due to the complexation-decomplexation
equilibrium ; even if it is strongly displaced towards the complexed
state, upon increasing the ratio EPT/Al, uncomplexed EPT is present as
a poison for both the isospecific and, of course preferably, the

aspecific sites. Depending on the stoiechiometry of the complex, the following equilibria are valid :

$$A + E \overset{K_1}{\rightleftharpoons} C_1 \quad \text{and} \quad 2\,A + E \overset{K_2}{\rightleftharpoons} C_2$$

where A and E are the concentrations of free alkylaluminium and aromatic ester respectively, C_1 and C_2 being the concentrations of the 1/1 and 2/1 complexes. Then if R is the initial ratio $\dfrac{E_o}{A_o}$, one has

$$K_1 = \frac{C_1}{A\,E} = \frac{E_o - E}{A\,E} \simeq \frac{E_o}{(A_o - C_1)E} \simeq \frac{E_o}{(A_o - E_o)E} = \frac{R}{(1-R)\,E}$$

because, when the equilibrium is displaced towards the right, E is small compared to E_o and C_1 and E_o are nearly equal.

$$\text{Similarly} \quad K_2 = \frac{C_2}{A^2 E} \simeq \frac{E_o}{(A_o - 2E_o)^2 E} = \frac{R}{A_o(1-2R)^2 E}$$

In the first case, the concentration of uncomplexed aromatic ester is not dependent upon the actual value of A_o, but depends only on the ratio E_o/A_o. In the second case, for a constant value of R, E decreases if A_o increases. In both case, the main parameter governing the poisoning power of the ester is the ratio R. The enhancement of the isospecific activity is very dependent on the nature of the aromatic ester (Table I - runs 8 to 11) but does not follow the order of basicity (EA > EPT > EB) of the ester. It seems that this enhancement is more dependent on the complexation equilibrium, i.e on the amount of residual uncomplexed ester ; when the amount of complexed ester is the higher (EPT) the reduction of the aspecific activity is the lower (run 9 Table I) and the enhancement of the isospecific activity is the higher. The case of EB is not exactly comparable, because with that ester the composition of the dominating complex is 2 Al/1 Ester instead of 1 Al/1 Ester (6). The case of p.tert.butyle methylbenzoate is also special, due to the large steric hindrance of the substituant ; in that case the aspecific activity is less reduced and the isospecific activity is not increased correspondingly. Also the situation in the presence of IPRA is more difficult to discuss due to the polydentate structure of the IPRA, which is not exactly defined ; however some similar trends are observed, for instance in the comparison between EA and EPT (run 7 and 8 of Table 2) ; in that case, as already stated, more uncomplexed aromatic ester is present and the productivities of both the aspecific sites and the isospecific sites are reduced. An aliphatic ester, ethyl acetate, has been used for comparison and is shown to be the strongest poison ; however its complexation equilibrium has not been studied in detail, but the complex is actually much more thermally unstable than in the case of aromatic esters.

A new set of catalysts, with higher performances in both catalytic activity and isospecificity have been more recently patented (12). They are based on the use of a difunctional ester in the precatalyst (solid part of the catalyst) and on the use of a silane derivative as complexing agent in the cocatalyst solution. Some results using a catalyst belonging to that class are reported in Table 3. Very small amounts of phenyltriethoxysilane (PTES) are enough to cause a very strong enhancement of the isospecific activity and also a strong reduction of the aspecific activity. Upon increasing the ratio Si/Al, both activities are reduced as in the previous cases. It is not the purpose of this paper to discuss thoroughly the peculiarities of these new catalytic systems, but it can be pointed out that the complex between PTES and TEA is thermally much more stable than the complexes between aromatic esters and alkylaluminium (13).

Table 3. Effet of the addition of aromatic silane in the catalytic solution on the stereospecificity of the catalytic system $MgCl_2/DBP/TiCl_4$
Polymerization temperature : 70° C ;
Polymerization time : 90 min ; Monomer pressure : 4 bars
TEA 6 mmole/L : Ratio Al/Ti : 250

Ratio Si/Al	Productivity g/g catalyst		
	Overall	Isotactic	Atactic
O	1700	1000	700
0.05	2770	2660	110
0.1	2000	1960	50
0.2	1600	1575	25

DBP : Dibutylphtalate   Silane : Phenyltriethoxysilane

In their recent review, Barbe et al (10) have discussed the various types of active centers, their possible structure and the mechanism of their formation. It is not possible to prove or to disprove the participation of the Lewis base to the structure of the active centers ; it seems clear however that its role cannot be reduced to a poisoning effect of the sites with an efficiency depending on the Lewis activity of these sites, the aspecific sites being more acidic and then more easily poisoned than the isospecific sites ; the enhancement of the isospecific productivity upon addition of the external Lewis base, involves a direct or an indirect modification of the sites. We have previously shown that these systems involve most probably bimetallic sites, due to the strong effects of the nature of the alkylaluminium (TEA or IPRA) and of the composition of the cocatalytic solution onto both the activity and the stereospecificity of the catalytic system (14). Exchanges between the cocatalytic solution and the solid precatalyst, which may cause changes in the precatalyst composition (Ref. 10 paragraph 5-2), and the rates of these exchanges are important factors in determining the behavior of the catalytic systems. The instantaneous composition of the cocatalytic solution is governed by the complexation equilibria between the aromatic esters and the alkylaluminium ; in terms of bimetallic sites, that means, as we have previously suggested (15), that the titanium sites may be associated with either uncomplexed or complexed alkylaluminium ; the later should be isospecific. We consider that all the data reported in this paper do support such a view, as shown by the direct relationship between the change of productivities and the Ester/Aluminium ratio. However some part may also be played by other kind of ternary sites, i.e those where the polydentate nature of the complexing agents -possibility of double complexation of EB leading, as an internal Lewis base to more active catalyst, diester character of the dibutylphtalate (DBP) in the last kind of catalyst, three functional silane complexing agent- may allow simultaneous complexation of these different kind of metal atoms (Mg, Ti, Al) involved in the system. More knowledge of the complexation equilibria should be then very important in the understanding of these systems.

Acknowledgments : The support by ATOCHEM (Groupe Elf-Aquitaine) was thoroughly appreciated.

REFERENCES

(1) Boor J.Jr. (1979)"Ziegler-Natta Catalysts and Polymerization" Academic Press, New York, Chapter 9 p. 213-243
(2) Boor J.Jr. (1963) J. Polymer Sci. Part C 1, 237, 257
(3) Karayamis N.M. and Lee S.S. (1982) Makromol chem. 183 1171
(4) Boor J.Jr. (1971) J. Polymer Sci Part A1, 617,
(5) Kashiwa N. and Yoshitake J. (1984) Makromol chem. 185 1133
(6) Guyot A., Spitz R., Duranel L. and Lacombe J.L. (1986) in "Catalytic Polymerization of Olefins" Ed. T. Keii and K. Soja, Kodansha-Elsevier, Tokyo-Amsterdam, p. 147
(7) Pino P. and Mulhaupt R., (1980) Angow. Chem.. Int. Ed. 19 857
(8) Pino P. and Rotzinger B., (1984) Makromol. Chem. Suppl. 7 41
(9) Pino P., Rotzinger B. and Von Achenbach E. (1986) in "Catalytic Polymerization of Olefins" Ed. Keii T. and Soga K., Kodancha-Elsevier, Tokyo-Amsterdam, p. 461
(10) Barbe P.C., Cecchin G. and Noristi L., (1987) Advances in Polymer Science 81 10
(11) Giannini U., Giunchi G., Albizatti E. and Barbe P.C. in "Advances on Mechanistic and Synthetic Aspects of Polymerization" Ed. Fontanille M. and Guyot A., Nato A.S.I. Serie Reidel (in press)
(12) Nocci R., Giannini U., Barbe P.C., Parodi S. and Scata U. (1981) Europ. Pat. 45977 to Montedison SpA
(13) Bobichon C., Duranel L., Spitz R. and Guyot A. (unpublished results)
(14) Spitz R., Duranel L. and Guyot A. (1985) Chicago ACS meeting Poly. Mat. Sci. Erg 53 p. 209 - to be published in ACS series
(15) Spitz R., Lacombe J.L. and Guyot A., (1984) J. Polymer Sci. Polymer Chem. Ed. 22, 2625 and 2641

Isotactic Polymerization of Propene Using Living Catalysts Originally Found by Hercules Inc.
——Activation, Stereospecific   Control and Model of isospecific Catalyst Centers ——

Kazuo Soga

Research Laboratory of Resources Utilization, Tokyo Institute of Technology, Nagatsuta, Midori-ku, Yokohama 227, Japan

Hisayoshi Yanagihara

Oita Research Laboratory, Showa Denko Co. Ltd., 2, Nakanosu, Oaza, Oita 870-01, Japan

INTRODUCTION

Recently it has been reported by Hercules Inc. that the catalyst system composed of $TiCl_3$ and $(RCp)_2TiMe_2$(R=Me, H) can catalyze isotactic living polymerization of propene (Hercules Inc. 1982; Hercules Inc. 1983). Although the polymerization activity of the catalyst is very low, such a simple catalyst system seems to contribute to our understanding of the detailed mechanism of stereoregulation of Ziegler -Natta catalysts, which has been hindered by the complexities arising from the multiplicity of the catalyst systems.

From such a viewpoint, we have prepared various kinds of similar catalysts by using various kinds of alkyl titanium compounds as well as solid catalysts, and propene polymerization was carried out with those catalyst systems. Some of the preliminary results of the polymerization have been already reported(Soga et al. 1986a; Soga and Yanagihara 1987). In this paper are summarized the polymerization results obtained with various catalyst systems and proposed the mechanism for the formation of isospecific polymerization centers.

EXPERIMENTAL PART

Materials

Propene (from Mitsubishi Petrochemical Co.) was purified by passing through NaOH and P_2O_5 columns. $MgCl_2$, $TiCl_3$(TA.-type) and $TiCl_4$ (from Toho Titanium Co.) were used without further purification. Research grade $AlMe_3$ and $AlEt_3$ were commercially obtained and used without further purification. Research grade styrene, heptane, toluene, ethylbenzoate (EB) and n-dibutyl phthalate (n-BP) were purified according to the usual procedures.

Preparation of alkyltitanium compounds

Cp_2TiMe_2, $(MeCp)_2TiMe_2$, $Cp_2TiMeCl$(Clauss and Bestian 1962), $Cp_2TiCH_2(CH_3)AlMe_2$, $(MeCp)_2TiCH_2(CH_3)AlMe_2$, $Cp_2TiCH_2(Cl)AlMe_2$ (Tebbe and Parshall 1978), Cp_2TiPh_2 (Rausch and Ciappenelli 1967) and

W. Kaminsky and H. Sinn (Eds.)
Transition Metals and Organometallics as
Catalysts for Olefin Polymerization
© Springer-Verlag Berlin Heidelberg 1988

$(C_5Me_5)_2TiMe_2$ (Bercaw et al. 1972) were prepared according to the literatures reported previously, diluted to approximately 0.3 mol/dm^3 in heptane or toluene, and restored as stock solutions.

Preparation of supported catalysts

Several kinds of supported catalysts were prepared according to the same procedures as described previously (Soga and Shiono 1986b).

Propene polymerization and analytical procedures

The polymerizaiton of propene was conducted at 25 to 60°C and 1 atm of propene with the same apparatus and procedures as described previously (Soga and Shiono 1986b). The isotactic index (I.I., wt% of polymer insoluble in boiling heptane) of the polymer produced was determined by fractionation with boiling heptane.

RESULTS AND DISCUSSION

Polymerization of propene was first conducted at 40°C by using several kinds of solid catalysts in the presence of Cp_2TiMe_2 as cocatalyst. As shown in Table 1, polymerization activity appeared when $TiCl_3$ or $MgCl_2$-supported Ti catalysts were used. Whereas, Cp_2TiMe_2 alone or the combinations of Cp_2TiMe_2 with $TiCl_2$, $TiCl_4$ and $MgCl_2$ did not show any activity for propene polymerization.

Table 1. Activity for propene polymerization with various solid catalysts and Cp_2TiMe_2

Alkyl titanium compound	Solid catalyst	Activity for propene polymerization
Cp_2TiMe_2	$TiCl_2$	No
	$TiCl_3$	Yes
	$TiCl_4$	No
	$TiCl_3/MgCl_2$	Yes
	$TiCl_4/MgCl_2$	Yes
	$MgCl_2$	No
	————	No

Propene polymerization was then conducted by using both $TiCl_3$ and supported catalysts in the presence of Cp_2TiMe_2 or $(MeCp)_2TiMe_2$ as cocatalyst (Table 2). Generally speaking, the polymerization activity was improved to a great extent by using supported catalysts in place of $TiCl_3$. It depended, however, strongly upon the internal donors containing in the supported catalysts; the activity increased in the following order, the reason of which will be discussed later:

$$TiCl_4/n\text{-}BP/MgCl_2 < TiCl_4/MgCl_2 < TiCl_4/EB/MgCl_2$$

It should also be noted here that the isotactic index of polypropylene produced with these catalyst systems is very high even in the absence of external donors.

Table 2.   The results of propene polymerization with various solid catalysts and $(RCp)_2TiMe_2$

Alkyl titanium compound	Solid catalyst	Activity[a] in g-PP/g-Ti·h	I.I.[b] in %
Cp_2TiMe_2	$TiCl_3$	20	95
$(MeCp)_2TiMe_2$	$TiCl_3$	27	98
$(MeCp)_2TiMe_2$	$TiCl_4/MgCl_2$	120	90
$(MeCp)_2TiMe_2$	$TiCl_4/EB/MgCl_2$	920	96
Cp_2TiMe_2	$TiCl_4/EB/MgCl_2$	770	95
Cp_2TiMe_2	$TiCl_4/n-BP/Mg(OEt)_2$	63	97
Cp_2TiMe_2	$TiCl_4/n-BP/MgCl_2$	90	96

Polymerization conditions ; $P(C_3^=)$ = 1 atm, heptane = 100 ml, [cocatalyst] = 5-7 mmol/l, [cocatalyst]/[$TiCl_3$] = 1, [cocatalyst]/[Ti in solid catalyst] = 10, 40°C, 2 h.

a) Average activity for the initial 2 h.
b) I.I. ; wt-% of boiling heptane insoluble polypropylene.

Then we have examined the dependence of both the polymerization activity as well as the isotactic index of polypropylene upon the cocatalysts. The results obtained (Table 3) clearly indicate that the polymerization activity is also strongly dependent upon the cocatalyst: in general, Tebbe type compounds increase the activity, while the compounds containing either phenyl or Cl ligands decrease it, respectively. On the other hand, the isotactic index of polypropylene did not depend so much upon the cocatalyst.

Therefore, polymerization of propene was conducted with various solid catalysts combined with Tebbe type cocatalysts. The results obtained are shown in Table 4. When $TiCl_3$ was used as solid catalyst, the polymerization activity was hardly improved even by using Tebbe type cocatalysts. The use of Tebbe type compounds, however, drastically increased the activity when supported catalysts were employed as solid catalyst. The activity in this case increased in the following order, which differs completely from the tendency found above with the use of Cp_2TiMe_2 as cocatalyst:

$$TiCl_4/MgCl_2 < TiCl_4/EB/MgCl_2 < TiCl_4/n-BP/MgCl_2$$

The polymerization result obtained by using a typical industrial catalyst system is also shown in Table 4 for a reference (Since propene polymerization was conducted without using external donors, the isotactic index was naturally low but the activity should be highest). As can be seen in Table 4, the polymerization activity of the $TiCl_4/n-BP/MgCl_2-Cp_2TiCH_2(CH_3)AlMe_2$ catalyst system is comparable with that of the industrial catalysts. In addition, the catalyst is highly isospecific even in the absence of external donors.

Table 3. Dependence of cocatalysts on the polymerization
activity and I.I. with $TiCl_4/EB/MgCl_2$ catalyst

Alkyl titanium compound	Polym.temp. in °C	Activity[a] in g-PP/g-Ti·h	I.I.[b] in %
Cp_2TiMe_2	40	770	95
$(MeCp)_2TiMe_2$	40	920	96
$(Me_5C_5)_2TiMe_2$	50	350	83
Cp_2TiPh_2	60	66	96
$Cp_2TiMeCl$	40	25	97
$(MeCp)_2TiCH_2(CH_3)AlMe_2$	40	1850	93
$Cp_2TiCH_2(Cl)AlMe_2$	40	560	94
$AlEt_3$	40	7500	60

Polymerization conditions ; $P(C_3^=)$ = 1 atm, heptane = 100 ml,
[cocatalyst]/[Ti in solid catalyst] = 10, [cocatalyst] =
5-7 mmol/l, 2 h.

a) Average activity for the initial 2 h.
b) I.I. ; wt-% of boiling heptane insoluble polypropylene.

Table 4. The results of propene polymerization with
various solid catalysts and $(RCp)_2TiCH_2(CH_3)AlMe_2$

Alkyl titanium compound	Solid catalyst	Activity[a] in g-pp/g-Ti·h	I.I.[b] in %
$(MeCp)_2TiMe_2$	$TiCl_3$	27	98
$(MeCp)_2TiCH_2(CH_3)AlMe_2$	$TiCl_3$	37	93
$(MeCp)_2TiCH_2(CH_3)AlMe_2$	$TiCl_4/MgCl_2$	940	87
$(MeCp)_2TiCH_2(CH_3)AlMe_2$	$TiCl_4/EB/MgCl_2$	1850	93
$Cp_2TiCH_2(CH_3)AlMe_2$	$TiCl_4/n-BP/Mg(OEt)_2$	2400	97
$Cp_2TiCH_2(CH_3)AlMe_2$	$TiCl_4/n-BP/MgCl_2$	3700	95
$AlEt_3$	$TiCl_4/n-BP/MgCl_2$	4000	76

Polymerization conditions ; $P(C_3^=)$ = 1 atm, heptane = 100 ml,
[cocatalyst] = 5-7 mmol/l, [cocatalyst]/[TiCl_3] = 1,
[cocatalyst]/[Ti in solid catalyst] = 10, 40°C, 2 h.

a) Average activity for the initial 2 h.
b) I.I. ; wt-% of boiling heptane insoluble polypropylene.

It may be considered that such a marked increase in the activity by using Tebbe type cocatalysts is attributable to the $AlMe_3$ formed by the following decomposition reaction Eq(1). To get a better insight into this point, polymerization of propene was carried out by adding trialkyl aluminum compounds to some of the present catalyst systems. The results obtained are shown in Table 5. Addition of trialkyl aluminum compounds caused a significant increase in the activity, but it drastically decreased the isotactic index, indicating that such a decomposition reaction hardly proceeds in the present catalyst systems.

$$
\begin{array}{c}
RCp \\
\searrow \\
Ti \\
\nearrow \\
RCp
\end{array}
\begin{array}{c}
CH_3 \\
\searrow \\
Al \\
\nearrow \\
CH_2
\end{array}
\begin{array}{c}
CH_3 \\
\nearrow \\
 \\
\searrow \\
CH_3
\end{array}
\quad\rightleftarrows\quad
\begin{array}{c}
RCp \\
\searrow \\
Ti = CH_2 \\
\nearrow \\
RCp
\end{array}
+ \ Al(CH_3)_3
\qquad Eq(1)
$$

Table 5. Effect of AlR_3 addition with the $TiCl_4/MgCl_2$ and $(RCp)_2TiMe_2$ catalyst system

Alkyl titanium compound	molar ratio vs. Ti in catalyst	Alkyl aluminum compound	molar ratio vs. Ti in catalyst	Activity[a] in g-PP/g-Ti·h	I.I.[b] in %
$(MeCp)_2TiMe_2$	10	———	———	120	90
$(MeCp)_2TiCH_2(CH_3)AlMe_2$	10	———	———	940	87
———	———	$AlMe_3$	10	2700	24
———	———	$AlEt_3$	10	3460	27
$(MeCp)_2TiMe_2$	10	$AlEt_3$	10	280	70
$(MeCp)_2TiMe_2$	10	$AlEt_3$	40	1100	60

Polymerization conditions ; [Ti in solid catalyst] = 5-7 mmol/l, $P(C_3^=)$ = 1 atm, heptane = 100 ml, 40°C, 2 h.
a) Average activity for the initial 2 h.
b) I.I. ; wt-% of boiling heptane insoluble polypropylene.

Then, we have briefly investigated the kinetic feature of the polymerization by using the $TiCl_3$-Cp_2TiMe_2 catalyst system. In Fig. 1 is shown the correlation between the polymerization activity and the concentration of cocatalyst. From the polymerization by changing the amount of $TiCl_3$ at a constant concentration of Cp_2TiMe_2, the polymerization rate was confirmed to be proportional to the amount of $TiCl_3$ used. Therefore, formation of the active species might be expressed by Eq(2) in Fig. 1. Fig. 2 illustrates the dependence of molecular weights of polypropylene upon the concentration of cocatalyst, which indicates that transfer reactions by Cp_2TiMe_2 hardly proceed in this catalyst system as previously reported by Hercules Inc (Hercules Inc. 1982; Hercules Inc. 1983).

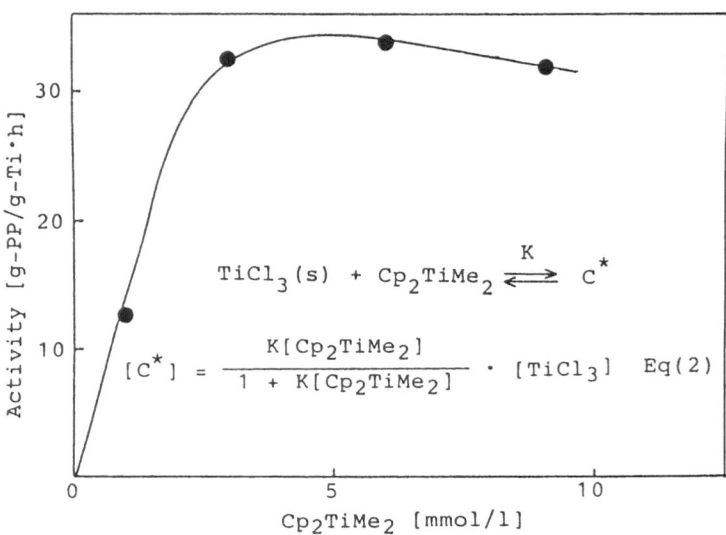

Fig. 1.　Dependence of polymerization activity on
the concentration of Cp_2TiMe_2

Polymerization conditions ; $P(C_3^!) = 1$ atm,
$TiCl_3 = 110$ mg, heptane = 100 ml, 25°C, 2 h.

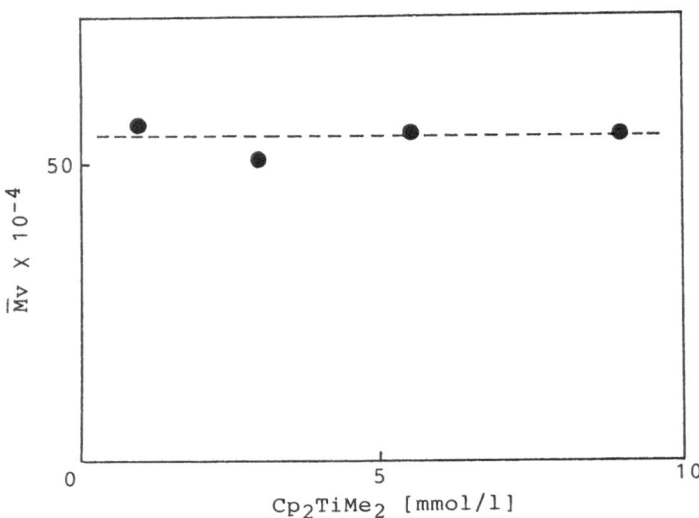

Fig. 2.　Plots of $\overline{M}v$ versus the concentration of Cp_2TiMe_2

Polymerization conditions were same as indicated
in Fig. 1.

Plausible models of the isospecific catalyst centers will be proposed on the basis of these results. In case of the original Hercules catalyst system ($TiCl_3$-Cp_2TiMe_2), Scheme 1, which is very similar to the mechanism proposed previously by Henrici-Olivé and Olivé (1969), may be most probable: An inactive intermediate complex is first formed between Cp_2TiMe_2 and the surface Ti(III) species having two Cl vacancies, followed by the ligand exchange between Cl and Me to yield an active species. Since neither the combinations of Cp_2TiCl_2 with $AlRnCl_{3-n}$ nor with $MgCl_2$ were inactive for propene polymerization, the active Ti is considered not to be $\boxed{Ti}$ but $\textcircled{Ti}$. The active Ti species have only one vacancy, and consequently they can selectively produce isotactic polypropylene.

On the other hand, when we use supported catalysts in place of $TiCl_3$, the majority of titanium is tetravalent, and so it may be very difficult for such Ti(IV) species to have two vacancies. Ligand exchange reactions should, therefore, proceed at first between the supported catalysts and cocatalysts. The methyl group attached to the Ti(IV) might be then eliminated as a methyl radical to give the Ti(III) having two vacancies. The following process forming the isospecific active Ti(III) species may be expressed as Schemes 2 and 3.

If EB or n-BP is used as an internal donor, one of the two vacancies of the intermediate Ti(III) species should be blocked by such a Lewis base. For the formation of isospecific active species, therefore, it is necessary to substitute such Lewis bases with cocatalysts. Thus, the polymerization activity of the present catalyst systems might markedly depend upon the internal donors as well as cocatalysts (Scheme 4). The results in Tables 2 and 4 can be fully understood by assuming that the strength of interactions between the intermediate Ti(III) species (having two vacancies) with EB, n-BP, $(RCp)_2TiMe_2$ and $(RCp)_2TiCH_2(CH_3)AlMe_2$ decreases in the following order:

$$(RCp)_2TiCH_2(CH_3)AlMe_2 > n\text{-BP} > (RCp)_2TiMe_2 > EB$$

For example, when $(RCp)_2TiMe_2$ is used as cocatalyst, EB can be replaced with $(RCp)_2TiMe_2$ but n-BP cannot, indicating that the activity of the $TiCl_4$/EB/$MgCl_2$ catalyst is much higher than that of the $TiCl_4$/n-BP/$MgCl_2$ catalyst. On the other hand, if we use $(RCp)_2TiCH_2(CH_3)AlMe_2$ as cocatalyst, not only EB but n-BP can be replaced with the cocatalyst, indicating that all the potentially active Ti species can be activated in this case.

Judging from the relative activity shown in Table 4, it may be said that the number of the potentially active Ti species increases in the following order for the present catalysts:

$$TiCl_4/MgCl_2 < TiCl_4/EB/MgCl_2 < TiCl_4/n\text{-BP}/Mg(OEt)_2 < TiCl_4/n\text{-BP}/MgCl_2$$

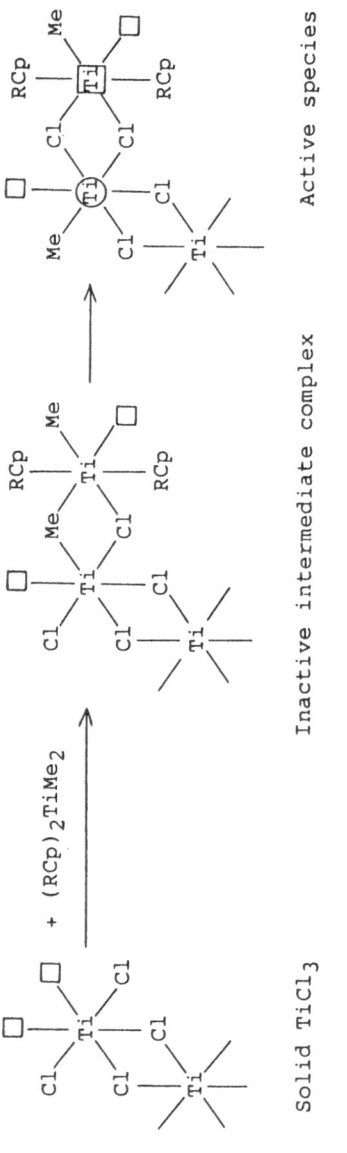

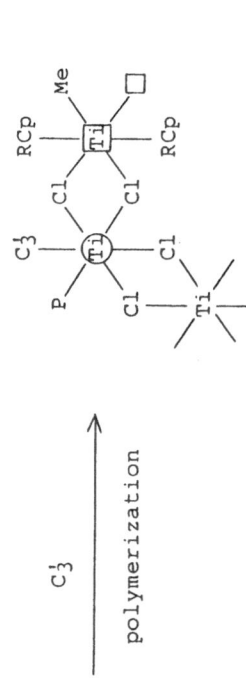

Scheme 1. Plausible mechanism for the formation of isospecific active species in the TiCl₃ and (RCp)₂TiMe₂ system

□ : inactive, ◯ : isospecific active species

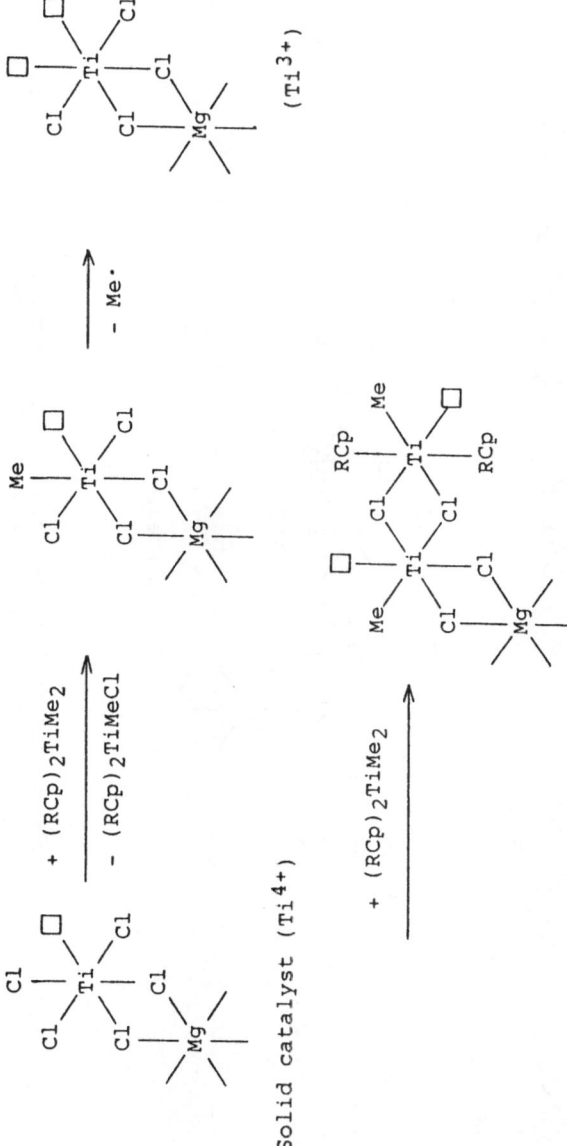

Scheme 2. Plausible mechanism for the formation of isospecific active species in the $MgCl_2$-supported Ti catalyst and $(RCp)_2TiMe_2$ system

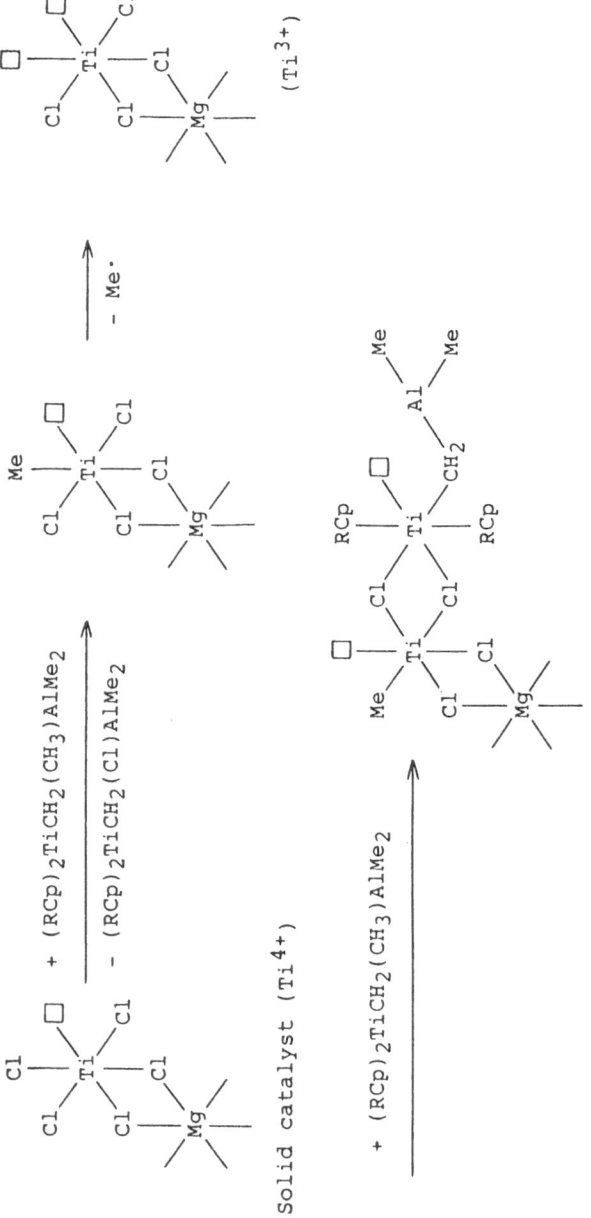

Scheme 3. Plausible mechanism for the formation of isospecific active species in the MgCl$_2$-supported Ti catalyst and (RCp)$_2$TiCH$_2$(CH$_3$)AlMe$_2$ system

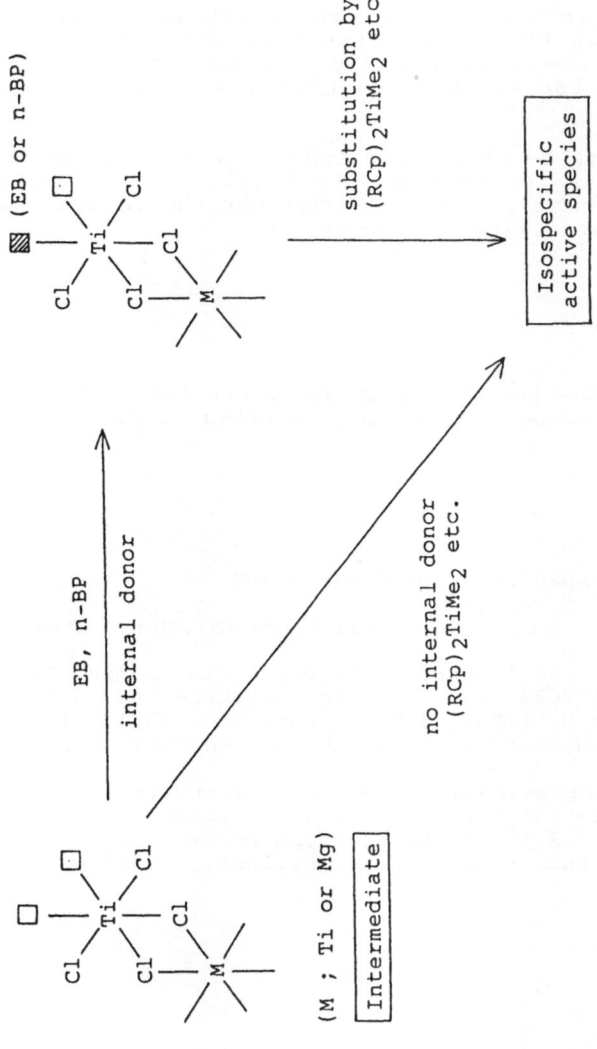

Scheme 4. Plausible mechanism for the formation of isospecific active species in the absence and presence of internal donors

Finally, we have carried out styrene polymerization at 40°C in heptane with the use of $TiCl_3-Cp_2TiMe_2$ and $TiCl_4/EB/MgCl_2-AlEt_3/EB$ catalysts. The polymers obtained were fractionated with boiling MEK for 5 h. The weight fractions of insoluble polymers, which were confirmed to be highly isotactic from the ^{13}C NMR spectra, were 99.2 % and 80.2 %, respectively. Thus, the present catalyst systems show extremely high isospecificity also for styrene polymerization (Soga and Yanagihara submitted).

In conclusion, we have developed new type of highly active and highly isospecific catalysts for olefin polymerization by modifying the Hercules catalyst, and a plausible model for the formation of iso-specific polymerization centers is proposed.

ACKNOWLEDGEMENT

The authors acknowledge the contribution of Dr. Minoru Terano of Toho Titanium Co. for preparing some of the supported solid catalysts.

REFERENCES

Hercules Inc.(1982) Japan Pat.(open):57-111307
Hercules Inc.(1983) U.S.Pat.:448019
Soga K, Shiono T, Yanagihara H (1986a) Makromol.Chem., Rapid Commun. 7:719-723
Soga K, Yanagihara H (1987) Makromol.Chem., Rapid Commun. 8:273-276
Clauss K, Bestian H (1962) Justus Liebigs Ann.Chem. 654:8-19
Tebbe FN, Ciappenelli DJ (1967) J.Organomet.Chem. 10:127-136
Bercaw JE, Marvich RH, Bell LG, Brintzinger HH (1972) J.Am.Chem.Soc. 94:1219-1238
Soga K, Shiono T "International Symposium on Transition Metal Catalyzed Polymerization (Akron, 1986b)" in press
Henrici-Olivé G, Olivé S (1969) Fortschr.Hochpolym-Forsch. 6:421-472
Soga K, Yanagihara H submitted to Makromol.Chem., Rapid Commun.

Olefin Polymerizations with a Highly Active $MgCl_2$ Supported $TiCl_4$
Catalyst System : Comparison on the Behaviors of Propylene, Butene-1
Ethylene and Styrene Polymerizations

N.Kashiwa, J.Yoshitake and T.Tsutsui

Mitsui Petrochemical Industries, Ltd.,
Waki-cho, Kuga-gun, Yamaguchi-ken, 740, Japan

ABSTRACT

Homopolymerizations of propylene, butene-1, ethylene and styrene were
performed with a highly active $MgCl_2$ supported $TiCl_4$ catalyst in conjunction with Et_3Al or Et_3Al and ethylbenzoate(EB). The stereospecificity of polypropylene(PP) and polybutene-1(PB-1) was remarkably
enhanced by EB and not affected by polymerization time, while that of
polystyrene(PSt) was affected a little with EB and significantly
enhanced by polymerization time, suggesting that the nature of active
centers for propylene and butene-1 monomers is more or less similar to
each other, but significantly different from that for styrene monomer.
In addition, a suitable amount of EB was found to increase the catalyst
activity in ethylene polymerization.

EXPERIMENTAL

Preparation of a $MgCl_2$ Supported $TiCl_4$ Catalyst

In a 800 ml stainless-steel pot containing 2.8 kg of stainless-steel
balls(15 mm diameter), 20 g of $MgCl_2$ were milled for 6 h under nitrogen.
The milled $MgCl_2$ was heated with 200 ml of $TiCl_4$ at 80°C for 2 h in a
400 ml frask. Subsequently, the solid product was separated by filtration and washed seven times with n-decane. Eight milligram of Ti atoms
were contained in 1 g of the $MgCl_2$ supported $TiCl_4$ catalyst.

Polymerization

Ethylene : In a 1 liter glass reactor equipped with a stirrer, 750 ml of
n-decane was added and the system was substituted by ethylene. Et_3Al, EB
and Ti catalyst were added at 50°C in this order([Al]=1.25 mmol/l,
[Ti]=0.05 mmol/l, [EB]=0 - 0.75 mmol/l). Polymerization was carried out
under atmospheric pressure at 50°C for 15 min. Ethylene was supplied
continuously to maintain atmospheric pressure. After 15 min, a small
amount of methanol was added to the system to terminate polymerization.
The resulting polymer was filtered and vacuum-dried at 80°C for 12 h.

Propylene and Butene-1 : Polymerizations of propylene(1) and butene-1(2)
were carried out as described in the previous papers.

Styrene : In a 500 ml glass reactor equipped with a stirrer, pure or
diluted styrene monomer by n-decane was added, and then the given amount
of Et_3Al, EB and Ti catalyst described in Figures were added at 50°C in

W. Kaminsky and H. Sinn (Eds.)
Transition Metals and Organometallics as
Catalysts for Olefin Polymerization
© Springer-Verlag Berlin Heidelberg 1988

this order. Polymerization was performed under atmospheric pressure at 50 °C for 5 sec - 30 min and terminated by the addition of a small amount of methanol into the system. The whole product was poured into a large amount of methanol. The resulting precipitate was collected by filtration and vacuum-dried at 80 °C for 12 h.

Fractionation of Polymer

PP and PB-1 : Fractionations of PP(1) by boiling n-heptane and PB-1(2) by n-decane were carried out as described in the previous papers.

PSt : One gram of polystyrene sample was extracted by boiling methyl ethyl ketone(MEK) with a Soxhlet's extractor for 5 h. MEK-insoluble polymer was vacuum-dried, while MEK-soluble polymer was recovered by evaporating MEK from the extract. I.I.(isotactic index, wt%) of polymer was defined as the weight fraction of MEK insoluble portion.

Molecular Weight

Molecular weight of the polymer sample was determined by GPC(Waters Associates, Model ALC/GPC150C) using polystyrene gel columns(10^7, 10^6, 10^5, 10^4 and 10^3 Å pore size) and o-dichlorobenzene as solvent at 150 °C for PP, PB-1 and PSt, and by viscometer(Fitz Simons type) using decalin as solvent at 135 °C for PE.

^{13}C-NMR Analysis

The polymer solution was prepared by dissolving 250 mg of the polymer sample at 100 °C in a mixture of 2 ml of hexachlorobutadiene and 0.3 ml of perdeuteriobenzene. ^{13}C-NMR spectrum was recorded on a JEOL-FX100 spectrometer operating at 25.05 MHz under proton decoupling in Fourier transform(FT) mode. Instrument conditions were as follows; pulse angle 45°, pulse repetition 4 sec, spectral width 4000 Hz, the number of pulse ca. 5000, temperature 100 °C.

RESULTS AND DISCUSSION

Comparison among Propylene, Butene-1 and Ethylene Polymerizations

Homopolymerizations of olefins, i.e. propylene, butene-1 and ethylene, were carried out with a highly active $MgCl_2$ supported $TiCl_4$ catalyst ($MgCl_2/TiCl_4$) prepared by the reaction of $TiCl_4$ and mechanically pulverized $MgCl_2$ in conjunction with Et_3Al or Et_3Al and ethylbenzoate (EB) at 50 °C and the obtained polypropylene(PP) and polybutene-1(PB-1) were fractionated into the soluble(nonstereospecific) and insoluble (stereospecific) portions by boiling n-heptane for PP(1) and by n-decane for PB-1(2) as described in our previous papers.

In Fig. 1, the yields of PP are plotted as a function of the EB/Ti molar ratio. As shown, the yield of stereospecific PP was increased to a maximum of two times upon addition of EB into the $MgCl_2/TiCl_4$-Et_3Al catalyst system and after a maximum, decreased upon further addition, while the yield of nonstereospecific PP, which was maximum at "without EB", was decreased rapidly, then moderately by EB addition. As a consequence, the yield of overall PP was decreased by EB.

The behavior of PB-1, as shown in Fig. 2, would be considered to be roughly similar to PP, suggesting the similarity of the nature of active

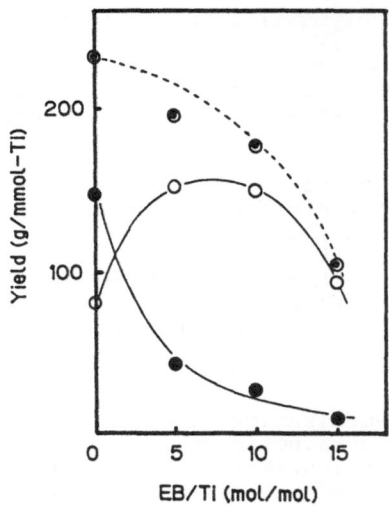

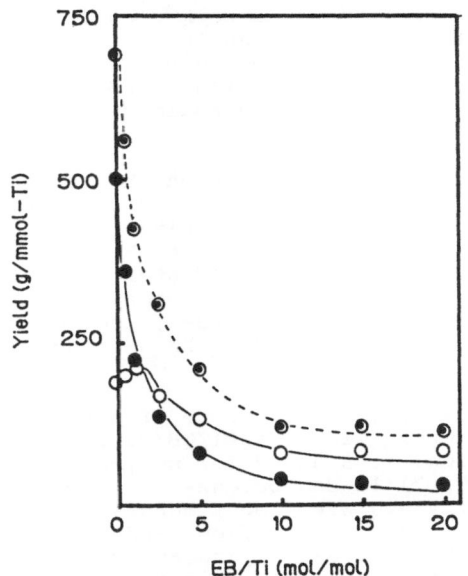

Fig. 1 Effect of EB/Ti molar
ratio on yield of stereospe-
cific(O), nonstereospecific
(●) and overall(◉) PP

Polymerization conditions ;
ref. 1)

Fig. 2 Effect of EB/Ti molar ratio
on yield of stereospecific(O), non-
stereospecific(●) and overall(◉) PB-1

Polymerization conditions ; ref. 2)

centers for two monomers, although a considerable difference in the
magnitude of the influence by EB can be seen.

The obserbed maximum peaks on the yields of stereospecific PP and PB-1
could be explained by assuming the following two kinds(positive and
negative) of effects of EB on the stereospecific active centers. As a
positive effect to increase the yield, EB would transform the nonstereospecific centers into the stereospecific ones by saturating one of
two vacant coordination sites in the nonstereospecific centers or would
increase the propagation rate constant at the stereospecific centers by
neighboring them. As a negative effect to reduce the yield, EB would
inactivate the stereospecific centers by saturating the single vacant
coordination site. The competition between these two opposing effects
would result in the maximum peaks in the profiles of stereospecific PP
and PB-1.

On the other hand, the behaviors of nonstereospecific PP and PB-1
suggest that there would be a rather broad distribution of the nonstereospecific centers having different acidity in this $MgCl_2/TiCl_4$-Et_3Al
catalyst system and those having stronger acidity would be selectively
and easily poisoned by a small amount of EB.

In Fig. 3, the result for ethylene polymerization is shown on the
relation between the yield and the EB/Ti molar ratio. As shown in Figs.
1 and 2, overall yields of PP and PB-1 were decreased by EB, due to the
very rapid decrease in nonstereospecific fractions. However, the result
in Fig. 3 shows that the yields of "overall PE" were apparently
increased by EB, as if ethylene would be polymerized only by the stereospecific centers for propylene or butene-1. Of course, the nonstereospe-

cific centers would work also for ethylene polymerization, therefore, the result would mean that in ethylene polymerization, the yield enhancement by EB would be large enough to overcome the yield decrease due to the inactivation by EB or that there would exist the additional active centers which would be active only for ethylene and be activated by EB.

Next, the effect of EB on the molecular weight of the obtained polymer was investigated. The relations between number average molecular weight (Mn) or intrinsic viscosity($[\eta]$) and the EB/Ti molar ratio are shown in Fig. 4 for PP, Fig. 5 for PB-1 and Fig. 6 for PE. With the increase of EB concentration, Mn of stereospecific PP and PB-1 were increased markedly, while those of nonstereospecific portions were decreased for PP and increased for PB-1, indicating the complexity of the nature of active centers. $[\eta]$ of PE was increased significantly upon addition of EB.

Moreover, the dependence of activity on polymerization time was examined with $MgCl_2/TiCl_4$-Et_3Al/EB catalyst system. The kinetic curves are shown in Figs. 7, 8 and 9 for polymerization of ethylene(EB/Ti=2.5), propylene (EB/Ti=7)(1) and butene-1(EB/Ti=5),respectively. The polymerization rate of ethylene was almost constant during polymerization, while in polymerizations of propylene and butene-1, the rapid activity decay was obserbed. Soga(3,4) and author(5) have reported that Ti^{4+} or Ti^{3+} species are active for polymerization of various olefins, while titanium species of lower valence state, probably Ti^{2+}, are active only for ethylene. During polymerization, $TiCl_4$ supported on $MgCl_2$ would be reduced to Ti^{2+} by Et_3Al(6,7,8). Therefore, the lower valence state of active Ti species might be one possible explanation for the activity decay in propylene and butene-1 polymerizations and no decay in ethylene polymerization.

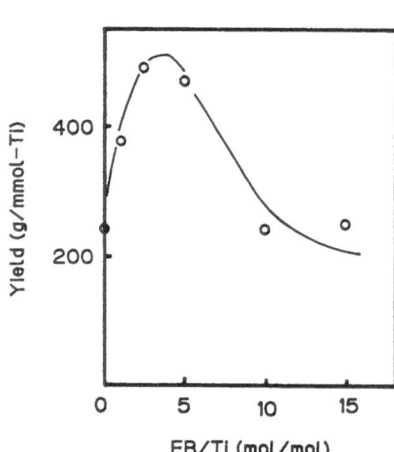

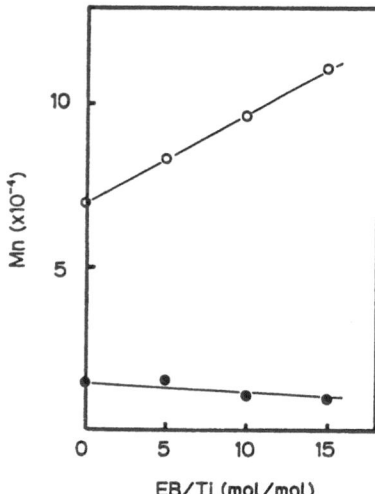

Fig. 3 Effect of EB/Ti molar ratio on yield of PE

Polymerization conditions ; experimental part

Fig. 4 Effect of EB/Ti molar ratio on Mn of stereospecific (O) and nonstereospecific(●) PP

Polymerization conditions ; the same as Fig. 1

Figure 10 shows the dependence of I.I.(isotactic index), expressed by
the weight fraction of stereospecific PP(1) or PB-1, on polymerization
time. I.I. remained almost unchanged during polymerization time for both
monomers in spite of the considerable activity decay, suggesting that
the stereospecific and nonstereospecific centers would be inactivated
with almost same rate.

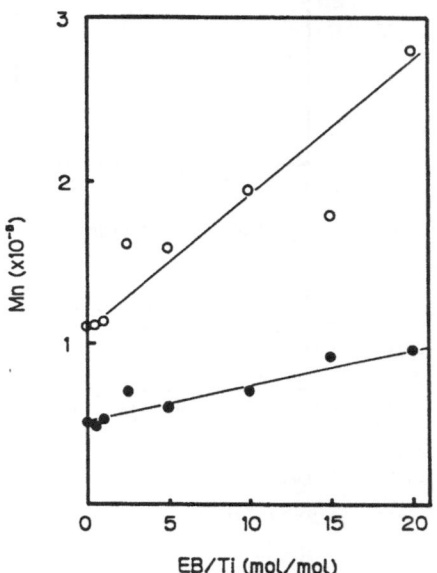

Fig. 5 Effect of EB/Ti molar
ratio on Mn of stereospecific
(O) and nonstereospecific(●)
PB-1

Polymerization conditions ;
the same as Fig. 2

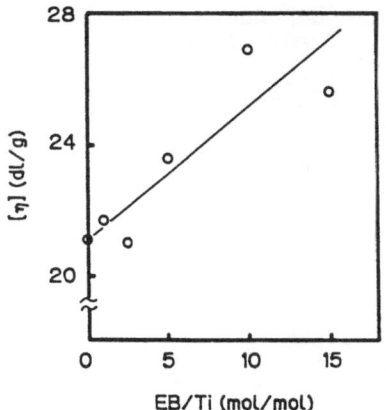

Fig. 6 Effect of EB/Ti molar
ratio on [η] of PE

Polymerization conditions ;
the same as Fig. 3

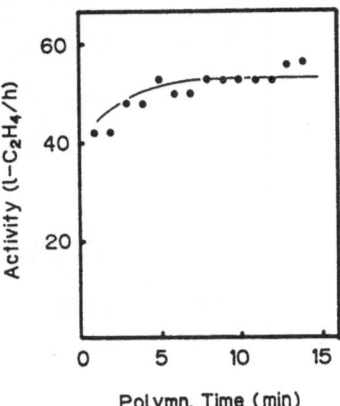

Fig. 7 Time dependence of
activity in ethylene polymerization

Polymerization conditions ;
EB/Ti=2.5 molar ratio in
experimental part

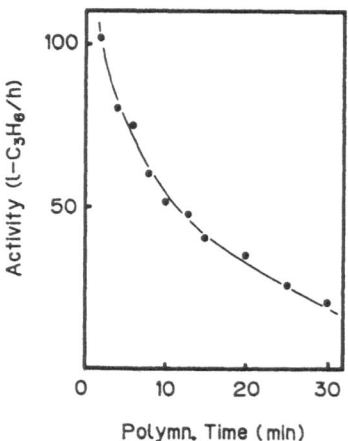

Fig. 8 Time dependence of
activity in propylene polymerization

Polymerization conditions ;
EB/Ti=7 molar ratio in ref.
1)

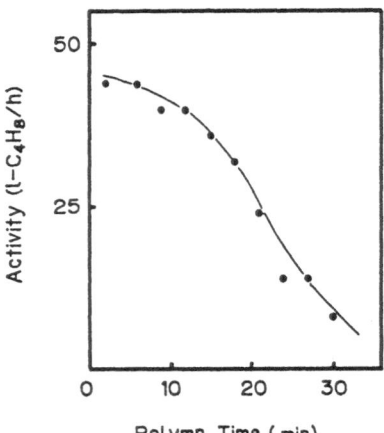

Fig. 9 Time dependence of
activity in butene-1 polymerization

Polymerization conditions ;
50°C under atmospheric pressure
in 500 ml of n-decane, 0.02
mmol of Ti, 0.5 mmol of Et_3Al
and 0.1 mmol of EB

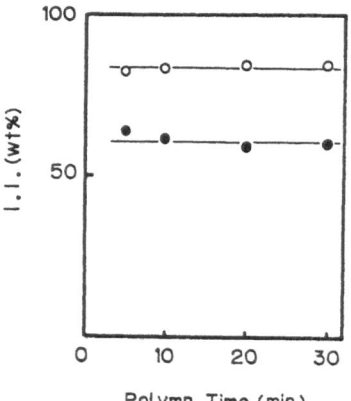

Fig. 10 Time dependence of I.I.
in polymerizations of propylene
and butene-1

Polymerization conditions ;
EB/Ti=7 molar ratio in ref. 1)
for PP(O), the same as Fig. 9
for PB-1(●)

Comparison Between Styrene and Propylene Polymerizations

Polymerization of styrene was performed with the said $MgCl_2/TiCl_4-Et_3Al$
or Et_3Al/EB catalyst system at 50°C for 30 min and the obtained poly-
styrene(PSt) was fractionated into soluble and insoluble portions by
boiling methyl ethyl ketone(MEK). The results were compared with those
of propylene polymerization described previously in this paper.

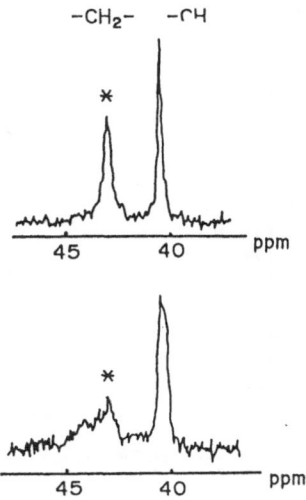

Fig. 11 ^{13}C-NMR spectra of MEK insoluble(above) and soluble(below) PSt with MgCl$_2$/TiCl$_4$-Et$_3$Al

* peak for mmm tetrad sequence

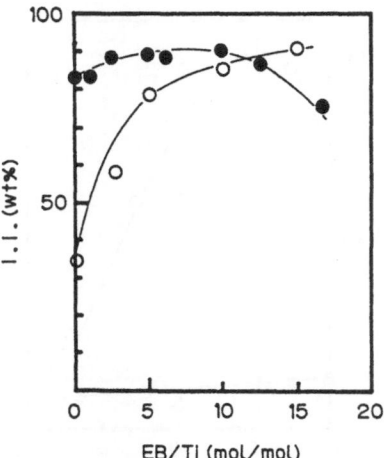

Fig. 12 Effect of EB/Ti molar ratio on I.I. of PSt(●) and PP(○)

Polymerization conditions ; 50°C for 30 min under atmospheric pressure, 0.17 mmol of Ti and 0.85 mmol of Et$_3$Al in mixture of styrene(50 ml) and n-decane(200 ml) for PSt, ref. 1) for PP

Figure 11 shows ^{13}C-NMR spectra of PSt with MgCl$_2$/TiCl$_4$-Et$_3$Al catalyst system for MEK soluble and insoluble portions. From the spectra, the mmm tetrad sequence content was determined to be more than 90 % for MEK insoluble portion, while less than 30 % for MEK soluble one, indicating that MEK extraction would be suitable as a method of the evaluation for the stereospcificity of PSt.

Figure 12 shows the change of I.I., defined as the weight fraction of MEK insoluble polymer, together with data on PP from Fig. 1. As shown, the effect of EB concentration on I.I. is apparently different between the two monomers. By EB addition, I.I. of PSt which was considerably high, more than 80 %, in the absence of EB, was increased a little, and then rather decreased upon further addition. Instead, the low I.I. of PP, ca.30 % in the absence of EB, was enhanced remarkably and monotonously to 90 % with the increase of EB concentration.

In order to understand the obserbed difference between polymerizations of styrene and propylene, the yields of MEK insoluble(sterepspecific) and soluble(nonstereospecific) portions are plotted as a function of the EB/Ti molar ratio in Fig. 13 and the relations were compared with those in propylene polymerization in Fig. 1. The yield of stereospecific PSt was increased to a maximum of 1.5 times upon addition of EB, and then decreased upon further addition, which was almost same trend as that in stereospecific PP.

On the other hand, the yield of nonstereospecific PSt was decreased only slightly by EB addition, rather different from the behavior of non-stereospecic PP which was sharply decreased by EB addition, suggesting the difference of the nature of the nonstereospecific active centers.

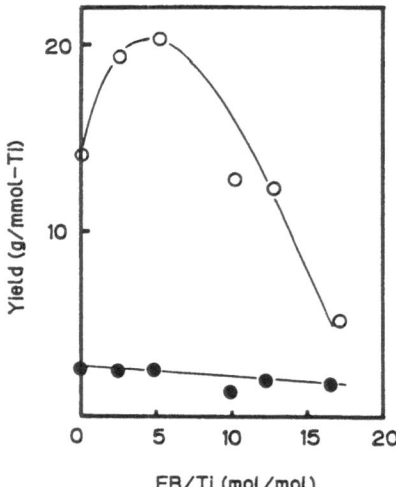

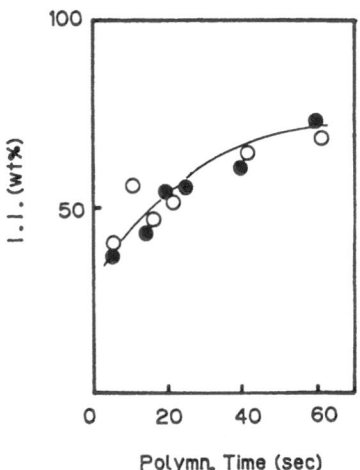

Fig. 13   Effect of EB/Ti molar
ratio on yield of stereospecific
(O) and nonstereospecific(●) PSt

Polymerization conditions ; the
same as Fig. 12

Fig. 14   Time dependence of
I.I. of PSt
● with EB   O without EB

Polymerization conditions ;
50°C under atmospheric pressure
in 100 ml of styrene, for "with
EB" 0.23 mmol of Ti, Al/Ti=25
molar ratio and EB/Ti=5 molar
ratio, for "without EB" 0.34
mmol of Ti and Al/Ti=25 molar
ratio

Could the increase of stereospecific PSt with EB addition be explained
by the same reason as PP or PB-1 ? What would the different behavior of
nonstereospecific polymers mean ?

In order to obtain the answer on these questions, further investigation
was carried out by a very short time polymerization in the range of 5-60
sec, using two kinds of catalyst systems, "without EB" and "with EB"
under the fixed concentration of EB/Ti=5.

In Fig. 14 is shown the relation between I.I. of PSt and polymerization
time. I.I. of PSt was increased significantly with polymerization time,
but was not changed at all by EB addition (I.I.=40 % at 5 sec, 70 % at
60 sec in both systems of "with" and "without" EB).

For comparison, data(9,10) on a very short time propylene polymerization
(7-60 sec) with the same catalyst systems are shown in Fig. 15. I.I. of
PP was not affected by polymerization time and was enhanced remarkably
by EB addition(EB/Ti=6.25) from ca.25 to 70 %, indicating the clear
difference from PSt.

In Figs. 16,17,18 and 19, the yields of the fractionated polymers are
plotted as a function of polymerization time for nonsterospecific PSt,
nonstereospecific PP, stereospecific PSt and stereospecific PP, respectively.

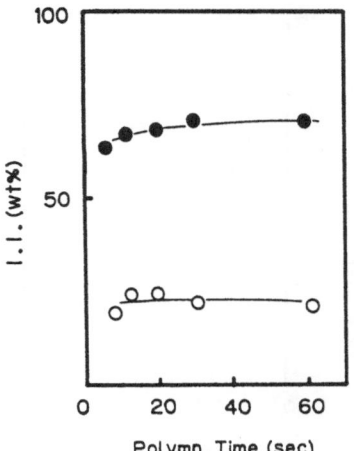

Fig. 15   Time dependence of
I.I. of PP
● with EB   ○ without EB

Polymerization conditions ;
ref. 9) and 10)

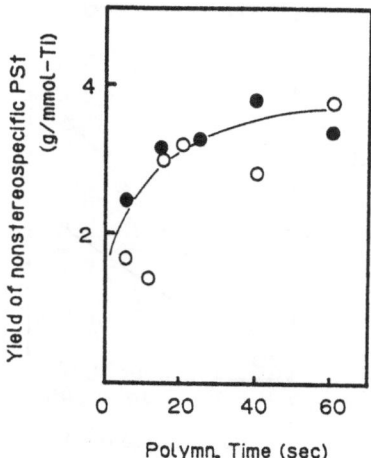

Fig. 16   Time dependence of
yield of nonstereospecific
PSt
● with EB   ○ without EB

Polymerization conditions ;
the same as Fig. 14

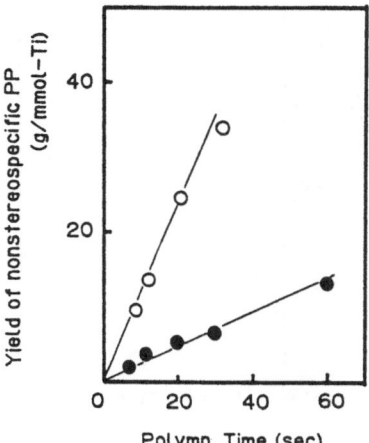

Fig. 17   Time dependence of
yield of nonstereospecific PP
● with EB   ○ without EB

Polymerization conditions ;
ref. 9) and 10)

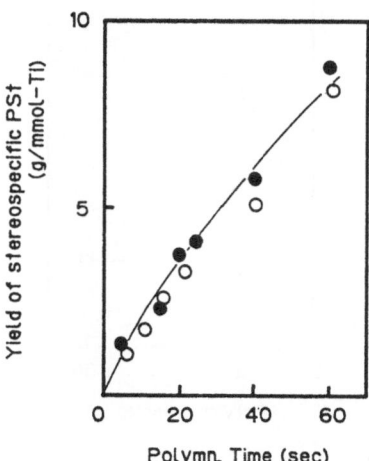

Fig. 18   Time dependence of
yield of stereospecific PSt
● with EB   ○ without EB

Polymerization conditions ;
the same as Fig. 14

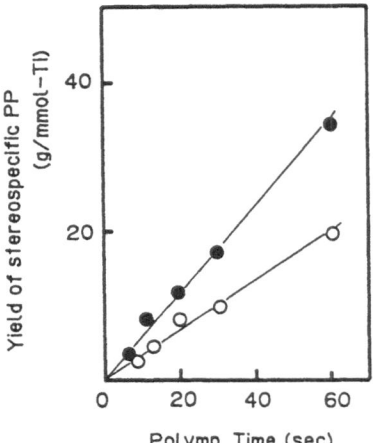

Fig. 19 Time dependence of
yield of stereospecific PP
● with EB   ○ without EB

Polymerization conditions ;
ref. 9) and 10)

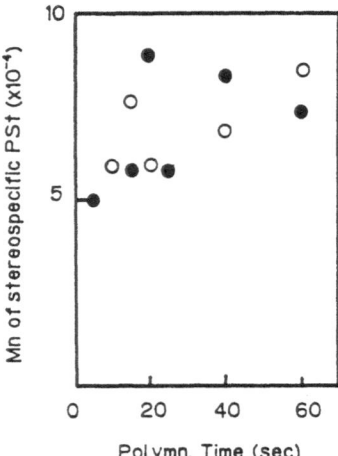

Fig. 20 Time dependence of
Mn of stereospecific PSt
● with EB   ○ without EB

Polymerization conditions ;
the same as Fig. 14

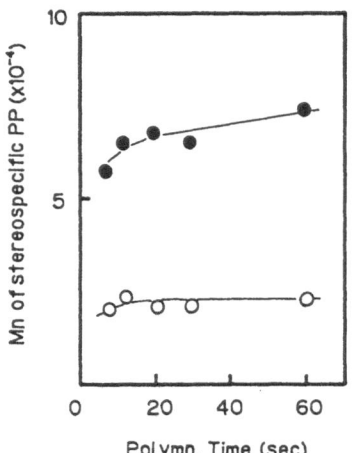

Fig. 21 Time dependence of
Mn of stereospecific PP
● with EB   ○ without EB

Polymerization conditions ;
ref. 9) and 10)

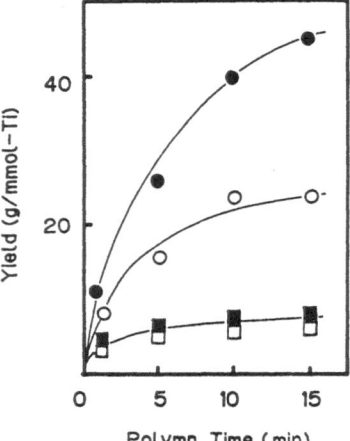

Fig. 22 Time dependence of
yield of stereospecific and
nonstereospecific PSt
● stereospecific with EB
○ stereospecific without EB
■ nonstereospecific with EB
□ nonstereospecific without EB

Polymerization conditions ;
the same as Fig. 14

From Figs. 16 and 17, one can see that polymerization rate for non-stereospecific PSt was decreased rapidly to almost zero within 20 sec and was affected hardly by EB addition, while that for nonstereospecific PP was unchanged by polymerization time, but lowered remarkably by EB addition. These findings strongly suggest that the nonstereospecific active centers for PSt would be quite different from those for PP, for example, cationic(11) or radical sites for the former and coordinative sites for the latter.

As for stereospecific polymers, Fig. 18 shows that the yield of stereo-specific PSt, unexpectedly from the results of 30 min polymerization (Fig. 13), was not be affected by EB addition within 60 sec polymeriza-tion, on the other hand, Fig. 19 shows that the yield of stereospecific PP was enhanced by EB addition as well as 30 min polymerization.

Moreover, as shown in Figs. 20 and 21, the molecular weight of stereo-specific PSt was unchanged by EB addition, while that of stereospecific PP was enhanced about threefold.

These facts would imply that the function of EB on stereospecific active centers would be also rather different between styrene and propylene polymerizations.

Finally, Fig. 22 which shows the time dependence of the yield of stereo-specific PSt in the range of 1-15 min reveals that EB would surpress the activity decay of stereospecific active centers to result in the signif-icant increase of stereospecific PSt in longer time polymerizations such as 30 min. This observation suggests that the increase in the yield of stereospecific PSt and PP upon addition of EB would be caused by differ-ent mechanisms.

REFERENCES

1) Kashiwa N, Kawasaki M, Yoshitake J (1986) Catalytic Polymerization of Olefins, Kodansha and Elsevier, Tokyo, p43
2) Kashiwa N, Yoshitake J, Mizuno A, Tsutsui T (1987) Polymer 28: 1227
3) Soga K, Sano T, Ohnishi R (1981) Polym. Bull. 4: 157
4) Soga K, Chen S.I, Ohnishi R (1983) ibid 10: 168
5) Kashiwa N, Yoshitake J (1984) Makromol. Chem. 185: 1133
6) Schindler A (1968) ibid 118: 1
7) Natta G, Pino P, Mazzanti G, Longi P (1957) Gazz. Chim. It. 87: 549
8) Natta G, Pino p, Mazzanti G, Longi P (1957) ibid 87: 570
9) Kashiwa N, Yoshitake J (1984) Polym. Bull. 12: 99
10) Kashiwa N, Yoshitake J (1983) Makromol. Chem. Rapid Commun. 4: 41

Determination of the Titanium Oxidation States in a $MgCl_2$-Supported
Ziegler-Natta Catalyst (CW-Catalyst) During Aging and Polymerization

S. Weber, J.C.W. Chien* and Y. Hu**

BASF Aktiengesellschaft, Kunststofflaboratorium, Carl-Bosch-Straße 38,
D-6700 Ludwigshafen/Rh., Federal Republic of Germany

SUMMARY

We have developed a precise method for the determination of Ti(+2),
Ti(+3) and Ti(+4) in a $MgCl_2$-supported Ziegler-Natta catalyst. Before
activation this CW-catalyst contains mostly Ti(+4) ions, with 6 % Ti(+3)
and 4 % Ti(+2) ions. Activation with $AlEt_3$ alone at room temperature reduces all of the titanium to lower valence states, consisting of 71 %
Ti(+3) and 29 % Ti(+2). Surprisingly, the reduction is incomplete when
methyl-p-toluate is present as external Lewis base. At $25^{\circ}C$ the distribution of Ti(+4):Ti(+3):Ti(+2) is 36 %:25 %:39 %; the distribution at
$50^{\circ}C$ is nearly the same. Aging of the activated catalyst causes little or
no change. Finally the distribution of Ti(+n) has been determined for the
CW-catalyst during the course of a decene-1 polymerization. It is found
to be Ti(+4):Ti(+3):Ti(+2) = 30 %:27 %:43 %, which does not change with
polymerization time.

INTRODUCTION

For a long time it has been known that one of the parameters affecting
the course of any transition metal catalyzed polymerization is the
oxidation state of the transition metal ions.

But surprisingly few determinations of the oxidation states have been
published [1]. In the case of titanium, which is the subject of the work
presented here, only a small number of publications exist; most of them
have appeared within the last twelve years [2-10]. These investigations
show the renewed interest in the oxidation states of titanium which is
a consequence of the advent of $MgCl_2$-supported high activity catalysts,
for in these supported catalysts a significant fraction of the titanium
is catalytically active.

Therefore the determination of titanium oxidation states, combined with
that of active site concentration, may identify the valencies of the
active sites.

Furthermore, these catalysts are characterized by a very rapid decay in
activity. Depending upon whether there is a parallel change in the dis-

---

 * Department of Polymer Science and Engineering, Department of Chemistry, University of Massachusetts, Amherst, MA 01003
** Institute of Chemistry, Academia Sinica, Beijing, People's Republic
   of China

W. Kaminsky and H. Sinn (Eds.)
Transition Metals and Organometallics as
Catalysts for Olefin Polymerization
© Springer-Verlag Berlin Heidelberg 1988

tribution of titanium oxidation states or not, it may be possible to gain an understanding of the mechanism of catalyst deactivation.

In the case of titanium there are four methods used for determining these oxidation states:

- Ti(+2) may be oxidized to Ti(+3) by water and the evolved hydrogen measured volumetrically.

- Ti(+3)-ions may be determined by quantitative EPR-spectroscopy.

- Titanium may be oxidized to Ti(+4) by adding an exactly known amount of Fe(+3). The iron ions which are not reduced are determined by complexometric titration.

- The last procedure is similar to the previous one, but instead of the excess of Fe(+3) the amount of Fe(+2) formed by reduction is determined by redox titration.

The first objective of our work was to develop a simple and accurate method for the determination of titanium oxidation states. The second goal was to measure the distribution of the titanium valencies for the CW-catalyst and also changes in the distribution both upon catalyst aging and during the course of a polymerization.

The results of these investigations [11] are presented here, and possible structures for the active sites are discussed as well.

EXPERIMENTAL PART

The first task was solved by modifying in two aspects a redox-titration set-up used by Prof. Chien and his group [9]. A cumbersome double-glove-bag technique was replaced by working with a Schlenk-apparatus, and new redox-indicators were introduced for both titrations. By working under optimal conditions, i.e. by adding an organic solvent and working at elevated temperatures to dissolve the indicator, sharp end-points within ±0.7 % of titre were now observed. This high accuracy is also reflected by the fact that precisely the same amount of titre was consumed in both titrations for freshly prepared $TiCl_3$-solution. The chemical background of this procedure is as follows:

Titration A

In the first experiment, called titration A, Ti(+2) and Ti(+3) in the reaction mixture are oxidized together to Ti(+4) by the addition of an excess of Fe(+3). The reduced iron ions are now determined by titration with chromate.

$$Ti^{2+} + 2\ Fe^{3+} \longrightarrow Ti^{4+} + 2\ Fe^{2+} \qquad (1)$$

$$Ti^{3+} + Fe^{3+} \longrightarrow Ti^{4+} + Fe^{2+} \qquad (2)$$

$$Fe^{2+} + Cr_2O_7^{2-} \longrightarrow Fe^{3+} + Cr^{3+} \qquad (3)$$

Titration B

In the second experiment, called titration B, Ti(+2) is converted to
Ti(+3) with aqueous sulfuric acid solution. Then all Ti(+3) are oxidized
to Ti(+4) by titration with chromate.

$$Ti^{2+}/Ti^{3+} + H^+/H_2O \longrightarrow Ti^{3+} + 1/2 \ H_2 \qquad (4)$$

$$Ti^{3+} + Cr_2O_7^{2-} \longrightarrow Ti^{4+} + Cr^{3+} \qquad (5)$$

The concentration of titanium in the different oxidation states is calculated from titres A and B. The total amount of titanium can be determined either by gravimetric or colorimetric analysis after all the
titanium ions have been oxidized to Ti(+4).

$$[Ti^{2+}] = A - B, \qquad [Ti^{3+}] = 2B - A, \qquad [Ti^{4+}] = \sum Ti - B$$

RESULTS AND DISCUSSION

Titanium Oxidation States in the CW-Catalyst

The first system we examined with this procedure was the CW-procatalyst.
This procatalyst is prepared by ball-milling of $MgCl_2$ with ethyl benzoate
as internal base (B_1) followed by treatment at 50°C with p-cresol,
at 25 °C with $AlEt_3$ and finally at 100°C with $TiCl_4$ [9].

The first column in Fig. 1 shows that about 90 % of the $TiCl_4$ incorporated in this procatalyst was still in the Ti(+4) state; only 10 % was
found as Ti(+3) and Ti(+2). The two lower valent species are formed in
comparable amounts.

Next we were interested in the effects of activation with alkylating
agents. When the procatalyst was activated with $AlEt_3$ alone, all of
the Ti(+4) ions were reduced in less than 5 minutes to Ti(+3) and Ti(+2)
in the ratio shown in column two. This ratio did not change with time
even though there was an excess of $AlEt_3$ present.

The next column shows the distribution of oxidation states obtained
by activation at room temperature with $AlEt_3$ and methyl-p-toluate (MPT)
as external base. The procatalyst was also rapidly reduced. However,
roughly 1/3 of the Ti(+4) was not reduced, and the concentration of
Ti(+2) was slightly increased.

It is known that the alkyl reduction of Ti(+4) is usually temperature
dependent. Therefore we were surprised by the results presented in
column four, which were obtained by activation at 50°C. Moreover there
is not the slightest effect of aging at this temperature on the distribution of the titanium valencies.

The intention of the studies described up to this point was to get a well
characterized catalyst system in hand which would allow us to study the
really interesting question, namely how polymerization influences the
distribution of titanium oxidation states. In order to obtain meaningful titration results on the conc. of Ti(+n) during a polymerization,
decene was selected as the monomer since poly(decene) is soluble in n-
heptane at 50°C. The distribution of titanium oxidation states after
2 min. of polymerization time differs slightly from the activated catalyst (column five) but remains almost constant during further polymerization; only a small decrease in Ti(+2) and increase in Ti(+3) is found.

Fig. 1: Distribution of the titanium oxidation states in the CW-catalyst

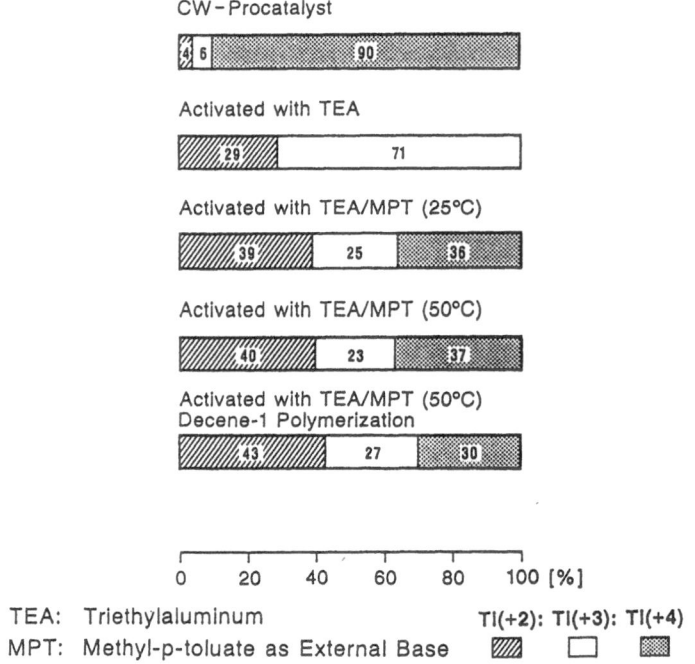

CW-Procatalyst

Activated with TEA

Activated with TEA/MPT (25°C)

Activated with TEA/MPT (50°C)

Activated with TEA/MPT (50°C)
Decene-1 Polymerization

0   20   40   60   80   100 [%]

TEA: Triethylaluminum
MPT: Methyl-p-toluate as External Base

TI(+2): TI(+3): TI(+4)

Structures of the Active Sites

A hypothetical model should explain the experimental observations during activation of the procatalyst.

Because the reduction of Ti(+4) occurs on the surface of the $MgCl_2$-support, two questions arise: what are the most probable crystallographic planes on the surface and how can $TiCl_4$ coordinate on these planes?

Several authors [12-14] have pointed out that the most probable surfaces of $MgCl_2$ crystallites are (100) and (110) planes having 5-coordinate and 4-coordinate Mg atoms, respectively. Treatment with $TiCl_4$ resulted in its epitaxial growth on these surfaces.

Three kinds of epitaxial placement of $TiCl_4$ in the (100) plane can in fact be envisioned.

Species Ia and IIa have $TiCl_4$ groups complexed in pairs, whereas an individual $TiCl_4$ is complexed to form species IIIa.

$TiCl_4$ in the (110) plane of $MgCl_2$ does not complex in the form of chlorine-bridged dimers. Nevertheless, there are $TiCl_4$ molecules adsorbed in clusters (i.e. IVa) and in isolation (Va) as shown in Fig. 2.

Fig. 2: Epitaxial placement of $TiCl_4$ in the (100) and (110) planes of $MgCl_2$.

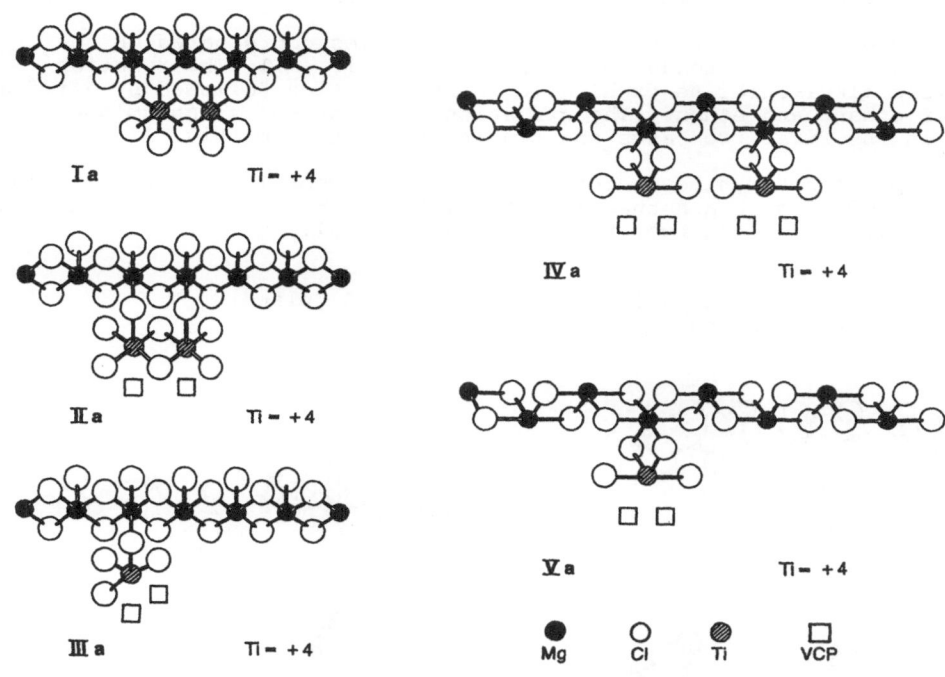

For activation of CW-procatalyst the following equilibria have to be considered.

$$Al_2Et_6 \rightleftharpoons 2\ AlEt_3 \tag{6}$$

$$AlEt_3 + Ti \cdot B_i \rightleftharpoons AlEt_3 \cdot B_i + Ti\text{-}\square \tag{7}$$

$$AlEt_3 \cdot B_e \rightleftharpoons AlEt_3 + B_e \tag{8}$$

B_i is the internal and B_e the external Lewis base and the square is a vacant coordination position (VCP). It may be assumed that the monomeric $AlEt_3$ is more effective than dimeric Al_2Et_6 in reactions with titanium-chloride-species.

It stands to reason that equation (8) is only relevant if an external base is present during the activation.

As already mentioned there are significant differences in the activation of CW-procatalyst in the presence of B_e and in the absence of B_e.

It is self-evident that those $AlEt_3$-molecules which are complexed with external base cannot undergo reaction (7) to produce titanium species with vacant coordination positions. Precisely those species should be alkylated and reduced by $AlEt_3$.

Three of the five types of coordinated titanium species (IIIa, IVa, Va) have two VCP and due to their high Lewis acidity it is most probable that they are complexed by B_i. Thus their reduction is affected by an external base because of the equilibria mentioned before. Consequently the reduction of $TiCl_4$ in the presence of external base is incomplete.

The treatment with $AlEt_3$ of the titanium-species coordinated on the (100) plane of $MgCl_2$ leads to the products I, II, III. The reaction comprises alkyl-chlorine exchange or $AlEt_3$-addition followed by elimination of ethane and ethene. The oxidation state of the titanium species is three in the first two cases and two in the last case (Fig. 3).

Similar processes can be depicted for titanium coordinated on the (110) plane of $MgCl_2$ to obtain the products IV and V. Here the oxidation states obtained are three and two, respectively (Fig. 3).

Fig. 3: Possible products in the (100) and (110) plane
        after reduction with $AlEt_3$.

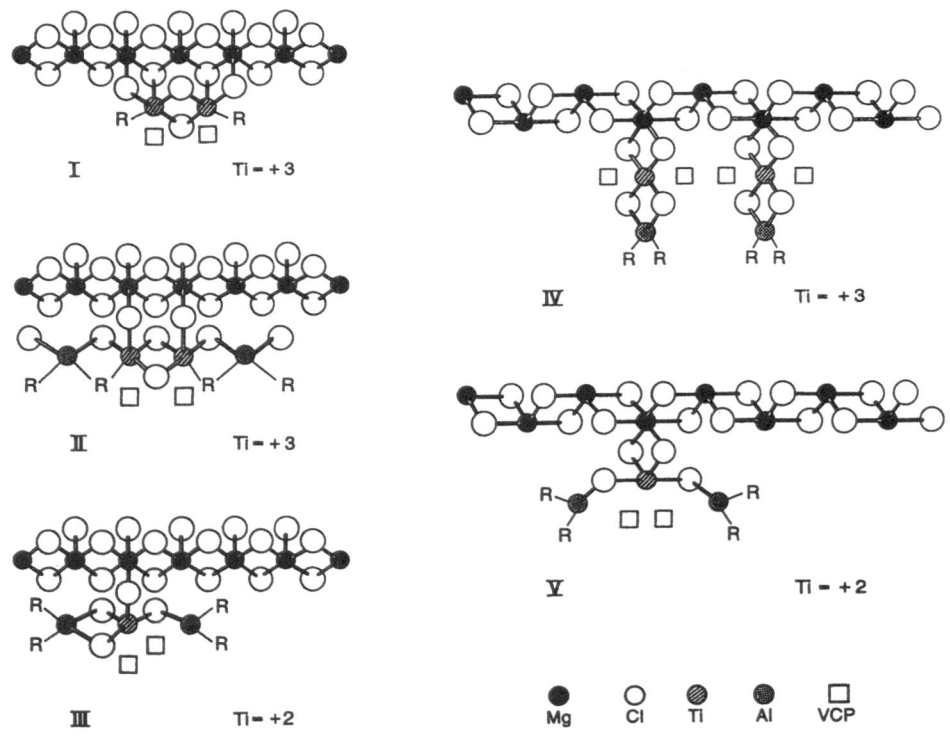

Comparing the results of this model, i.e. three species with Ti(+3) and two species with Ti(+2) after reduction, with the experimental data leads us to the following conclusion:

In the CW-catalyst 70 % of the reducible $TiCl_4$ is coordinated as species Ia, IIa and IVa, 30 % as species IIIa and Va.

Now we have an idea of how Ti(+4) can coordinate on MgCl$_2$-surfaces and what the reduced species may look like. But what we still do not know is in which oxidation state titanium is catalytically active for olefin polymerization.

It has already been shown by radioactive tagging [15-16] that in the absence of B$_e$ about 93 % of the Ti in the CW-catalyst is active for ethylene polymerization.

Comparing this with the result that the CW-catalyst reduced in the absence of B$_e$ contains 70 % Ti(+3) and 30 % Ti(+2) one can say with certainty that both Ti(+2) and Ti(+3) are catalytically active for ethylene polymerization.

In the presence of B$_e$ the active site concentration is 50 % of the total titanium. We have shown that in this case about 35 % Ti(+4) is present. Nevertheless it is not very probable that this Ti(+4) is catalytically active because it is a very strong Lewis acid and therefore should be complexed by the external base. Thus only Ti(+3) and Ti(+2) remain as active species.

The maximum concentration of active sites for propylene polymerization is 70 % when the CW-catalyst is activated in the absence of B$_e$, and this value is equivalent to the amount of Ti(+3) determined by titration. It was found previously that the maximum amount of total active sites in the presence of B$_e$ for propylene polymerization is 27 %. This result is in accordance with the finding that in the CW-catalyst activated in the presence of B$_e$ 23 % Ti(+3) and during decene polymerization about 27 % Ti(+3) are found. Therefore it is reasonable to conclude that it is the Ti(+3) species which are the active sites in this case as well.

It is well known that in the presence of an external base the propylene polymerization yields isotactic polypropylene. So the question arises which of the Ti(+3) species is responsible for isotactic polymerization. According to our model, three of the differently coordinated Ti-species can be reduced by TEA to Ti(+3). One of them has two VCP and should be complexed by an external base as mentioned above. Therefore only the species Ia and IIa should be reducible to Ti(+3). These species should be the centers for isotactic polymerization.

These findings are consistent with the proposal of Profs. Soga [17] and Kashiwa [8] that it is the Ti(+3) species which are active for propylene polymerizations.

Catalyst Deactivation

Four mechanisms have been proposed for the decay of polymerization rate:

1. monomer diffusion limitation due to encapsulation of the catalyst in the semicrystalline polymer,

2. a lowering of activity of active centers due to structural changes,

3. a decrease in the number of active centers by dissolution, and

4. deactivation of catalytic sites.

During the last few years experimental results have been obtained which almost rule out the deactivation processes 1, 2 and 3. This leaves the deactivation of the catalytic sites as the principal cause for decay in the polymerizataion rate. As mentioned above the most probable oxidation state for the active species in propylene polymerization is trivalent Ti, but there is no change in the conc. of Ti(+3) during either aging or decene polymerization. Therefore, the deactivation process cannot be a reductive metathesis.

Through quenching experiments with CH_3O^3H it has been found [18] that there is also no appreciable loss in the total number of metal polymer bonds during the period of rapid decay of the polymerization rate.

A mechanism which is consistent with these observations is represented in Fig. 4.

Fig. 4: Proposed mechanism for the deactivation of catalytic sites.

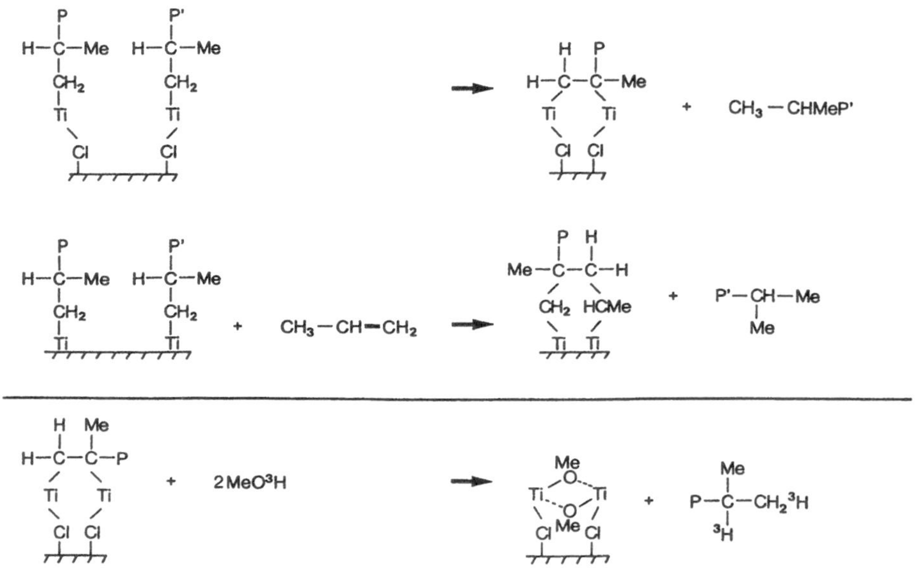

$$
\begin{array}{l}
\underset{\substack{| \\ H-C-Me \\ | \\ CH_2 \\ | \\ Ti \\ \backslash \\ Cl}}{\overset{P}{}} \quad
\underset{\substack{| \\ H-C-Me \\ | \\ CH_2 \\ | \\ Ti \\ / \\ Cl}}{\overset{P'}{}}
\longrightarrow
\underset{\substack{H \quad P \\ | \quad | \\ H-C-C-Me \\ / \quad \backslash \\ Ti \quad Ti \\ \backslash \quad / \\ Cl \quad Cl}}{}
\; + \; CH_3-CHMeP' \qquad (9)
\end{array}
$$

$$
\underset{\substack{P \\ | \\ H-C-Me \\ | \\ CH_2 \\ | \\ Ti}}{} \;
\underset{\substack{P' \\ | \\ H-C-Me \\ | \\ CH_2 \\ | \\ Ti}}{} \; + \; CH_3-CH=CH_2
\longrightarrow
\underset{\substack{P \quad H \\ | \quad | \\ Me-C-C-H \\ / \quad \backslash \\ CH_2 \quad HCMe \\ \backslash \quad / \\ Ti \quad Ti}}{} \; + \;
\underset{\substack{| \\ P'-CH-Me \\ | \\ Me}}{} \qquad (10)
$$

$$
\underset{\substack{H \quad Me \\ | \quad | \\ H-C-C-P \\ / \quad \backslash \\ Ti \quad Ti \\ \backslash \quad / \\ Cl \quad Cl}}{} \; + \; 2\,MeO^3H
\longrightarrow
\underset{\substack{Me \\ O \\ Ti \cdots Ti \\ O \\ Cl \; Me \; Cl}}{} \; + \;
\underset{\substack{Me \\ | \\ P-C-CH_2{}^3H \\ | \\ {}^3H}}{} \qquad (11)
$$

In reaction (9) the Ti valency is unchanged but the titanium carbon bond becomes inactive toward monomer insertion. The assistance by monomer in the catalyst deactivation may proceed via reaction (10). Neither process alters the specific activity of tritium in CH_3O^3H quenching because the inactive polymer chain will react with two CH_3O^3H molecules, as reaction (11) shows.

CONCLUSIONS

The main consequence of our findings is that the deactivation of the CW-catalyst during polymerization is not related to a change in the oxidation states of titanium.

The titration procedure with which the results presented here were obtained has been further developed. In its original form this procedure was restricted to clear and colorless aqueous solutions. Unfortunately a lot of supported catalysts give colored and/or opaque solutions. In a modified version of the titration procedure the end-points are now detected potentiometrically.

REFERENCES

[1]   Boor J (1979) Ziegler-Natta Catalysts and Polymerizations.
        Academic Press, New York, San Francisco, London, p 261-278
[2]   Ludlum DB, Anderson AW, Ashby CE (1958) J Am Chem Soc 80:
        1380-1384
[3]   Breslow DS, Newburg NR (1959) J Am Chem Soc 81: 81-86
[4]   Henrici-Olivè G, Olivè S (1967) J Polym Sci Part C 22: 965-970
[5]   Schindler A, Strong RB (1968) Makromol Chem 114: 77-91
[6]   Werber FX, Benning CJ, Wszolek WR, Ashby GE (1968)
        J Polym Sci Part A-1 6: 743-754
[7]   Baulin AA, Novikova YI, Mal'kova GY, Maksimov VL,
        Vyshinskaya LI, Ivanchev SS (1980) Polym Sci USSR 22: 205-214
[8]   Kashiwa N, Yoshitake J (1984) Makromol Chem 185: 1133-1138
[9]   Chien JCW, Wu JC, Kuo CI (1982) J Polym Sci Polym Chem Ed 20:
        2019-2032
[10]   Zakharow VA, Makhtarulin SI, Poluboyarov VA, Anufrienko VF
        (1984) Makromol Chem 185: 1781-1793
[11]   Chien JCW, Weber S, Hu Y (1987) J Polym Sci Poly Chem Ed,
        to be published
[12]   Giannini U (1981) Makromol Chem Suppl 5: 216-229
[13]   Corradini P, Barone V, Fusco R, Guerra G (1983)
        Gazz Chim Ital 113: 601-607
[14]   Galli P, Barbe PC, Noristi L (1984) Angew Chem Makromol 120:
        73-90
[15]   Chien JCW, Bres PL (1986) J Polym Sci Polym Chem Ed 24:
        2483-2505
[16]   Chien JCW, Bres PL (1986) J Polym Sci Polym Chem Ed 24:
        1967-1988
[17]   Soga K, Sano T, Ohnishi R (1981) Polym Bull 4: 157-164
[18]   Chien JCW, Kuo I (1985) J Polym Sci Polym Chem Ed 23: 731-760

Behavior of the Complexes of $TiCl_4$ with Various Carboxylic Acid Esters in $MgCl_2$-supported High Yield Catalysts

M. Terano, T. Kataoka, and M. Hosaka

Toho Titanium Company, 3-3-5 Chigasaki, Kanagawa 253, Japan

T. Keii

Numazu College of Technology, Ooka 3600, Numazu 410, Japan

INTRODUCTION

Electron donor compounds employed in Ziegler-Natta catalysts have various effects on the catalyst performance. For propylene polymerization, the most important role of an electron donor is to improve stereospecificity of the catalyst. Ever since a $MgCl_2$-supported high yield catalyst was developed by Shell International Research (Hewett, et al. 1964) and improved to a great extent by Montedison (Giannini, et al. 1977) and Mitsui Petrochemical (Kashiwa, et al. 1977), many types of the high yield catalysts as well as the electron donor compounds have been proposed and studied in both academic and industrial field (for example, Chien, et al. 1976).

Among the electron donor compounds, carboxylic acid esters such as ethylbenzoate(EB) are the most effective ones, the addition of which is essential to use the $MgCl_2$-supported catalyst in industrial plants. It may not be too much to say that the role of the ester is as important as that of $TiCl_4$ or $MgCl_2$ in the catalyst. However, the states of the ester and $TiCl_4$ in the catalyst have not been clarified yet. The study on them may provide some information which leads to the understanding of the structure of active species in the $MgCl_2$-supported high yield catalyst.

In this paper, thermal gravity-differential thermal analysis (TG-DTA) and IR spectroscopy, in combination with other methods, are applied to study the states of carboxylic acid ester, $TiCl_4$ and $TiCl_4$·ester complex in a primary type of the $MgCl_2$-supported high yield catalyst prepared by grinding $MgCl_2$ with $TiCl_4$·ester complex. The role of $MgCl_2$ is also discussed from the results obtained.

EXPERIMENTAL

Reagents: Extra pure heptane (from Toa Oil Co.), ethylpropanate(EP), ethylheptanate(EH), dibuthylphthalate(BP), and ethylbenzoate(EB) (from Kanto Chemical Co.) were used after passing through the molecular sieve 4-A column. γ-Al_2O_3 (from Nishio Industry Co.) was heated up to 800 °C before use. Anhydrous $MgCl_2$, $TiCl_4$(from Toho Titanium Co., respectively and triethyl aluminium(AlEt$_3$) (from Nippon Alkylalminium Co.) were used without further purification.

Preparation

TiCl$_4$·ester complex: In a 200 ml glass flask were placed 80 ml of Heptane and 0.10 mol of ester at 40 °C under nitrogen, followed by dropwise addition of 0.10 mol of $TiCl_4$. After the reaction at 40°C for 1 h, the

W. Kaminsky and H. Sinn (Eds.)
Transition Metals and Organometallics as
Catalysts for Olefin Polymerization
© Springer-Verlag Berlin Heidelberg 1988

yellowish solid product was separated by filtration washed with heptane and dried in a vacuum. The molar ratio of $TiCl_4$/ester in the resulting complex was found to be c.a. 1.0.
Grinding: 30 g of $MgCl_2$ with $11m^2/g$ or γ-Al_2O_3 and prescribed amount of each compound were placed in a 1 l stainless steel vibration mill pot with 50 balls (25 mmϕ) under nitrogen and vibrated at r.t..
cat-( )$_n$: 30 g of $MgCl_2$ and 15.4 g of $TiCl_4$·ester complex were mixed or coground as described above. The ester used is indicated in the parenthesis. The cogrinding time in hours is denoted by n.

Measurements: IR spectra were recorded under nitrogen on a Hitachi 270-30 spectrometer with a 2.5 mmϕ KBr pellet containing each sample. TG-DTA measurements were conducted on a Rigaku Thermoflex 8100 under nitrogen at a heating rate of 17 °C/min using γ-Al_2O_3 as a reference substance with DTA range of $\pm$ 50μV. The amount of each ester was measured by gas chromatography (Shimazu GC-7A) after the hydroulysis and extraction.

Polymerization: Polymerization of propylene was conducted in a 500 ml flask at 50 °C under a total pressure of 1 atm with 2.63 mmol of $AlEt_3$, prescribed amount of catalyst (Al/Ti molar ratio = 15) and 200 ml of heptane. The polymer obtained was washed with plenty of ethanol and dried in a vacuum. The isotacticity of the polymer was represented by the weight fraction of polymer insoluble in boiling heptane.

RESULTS AND DISCUSSION

Various electron donor compounds effective for $MgCl_2$-supported high yield catalysts have been reported in many patent literatures (for example, Giannini, et al. 1977, 1980). Among them, carboxylic acid esters seem to be the most effective to improve stereospecificity of the catalyst. In this study are employed three types of typical carboxylic acid esters including aromatic dicarboxylic acid diester such as BP that has been widely used as it makes possible to avoid a large decay of $MgCl_2$-supported catalyst system when used as a substitute for EB. All esters form an equimolar complex with $TiCl_4$ as described in the experimental section.

Figure 1 shows typical IR spectra of (1) EB, (2) $TiCl_4$·EB complex, (3) coground product of $MgCl_2$ and EB, (4) BP, (5) $TiCl_4$·BP complex, and (6) coground product of $MgCl_2$ and BP. The C=O band is shifted from 1720 and 1730 cm^{-1} in (1) and (4) to 1568 and 1596 cm^{-1} in (2) and 1584 and 1650 cm^{-1} in (5) by the complex formation. Both of the coground products have their peaks of the C=O band at 1688 cm^{-1}.

In Fig.2 are shown the TG-DTA curves of (1) $TiCl_4$·EB complex, (2) coground product of $MgCl_2$ and EB, (3) coground product of $MgCl_2$ and $TiCl_4$, (4) $TiCl_4$·BP complex, and (5) coground product of $MgCl_2$ and BP. The $TiCl_4$·EB complex (1) displays two peaks at 140 and 170 °C accompanied by weight decrease, while the $TiCl_4$·BP complex (4) shows two peaks at 110 and 200 °C. All these peaks may be caused by the decomposition of the complexes. The DTA curve of the coground product of $MgCl_2$ and EB in (2) has broad and sharp peaks at about 230 and 540 °C, respectively. In the case of BP in (5), there are two peaks at 240 and 530 °C. In each curve, two peaks at different temperature may give rise to the ester having weak and strong interaction with $MgCl_2$, respectively. The DTA curve of (3) shows a small peak at around 140 °C which seems to be caused by $TiCl_4$ having no or only weak interaction with $MgCl_2$, although the weight of the sample gradually decreases along with an increase in temperature. From these results, it may be possible to say that TG-DTA as well as IR spectroscopy is useful to study the states of the ester in $MgCl_2$-supported catalysts.

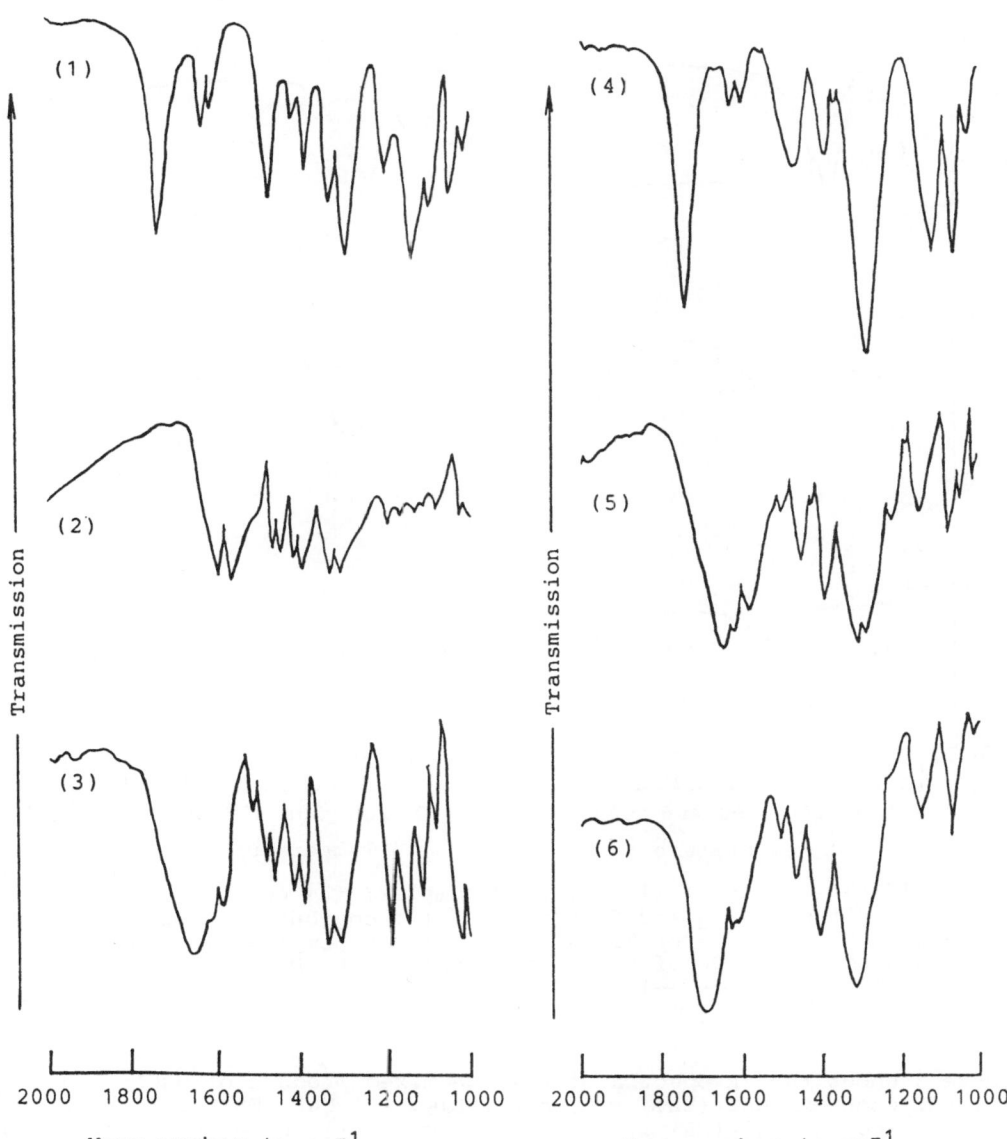

Fig. 1. IR spectra of (1) EB, (2) TiCl$_4$·EB complex, (3) coground product of 30g of MgCl$_2$ and 6.5ml of EB, (4) BP, (5) TiCl$_4$·BP complex, and (6) coground product of 30g of MgCl$_2$ and 6.5ml of BP.

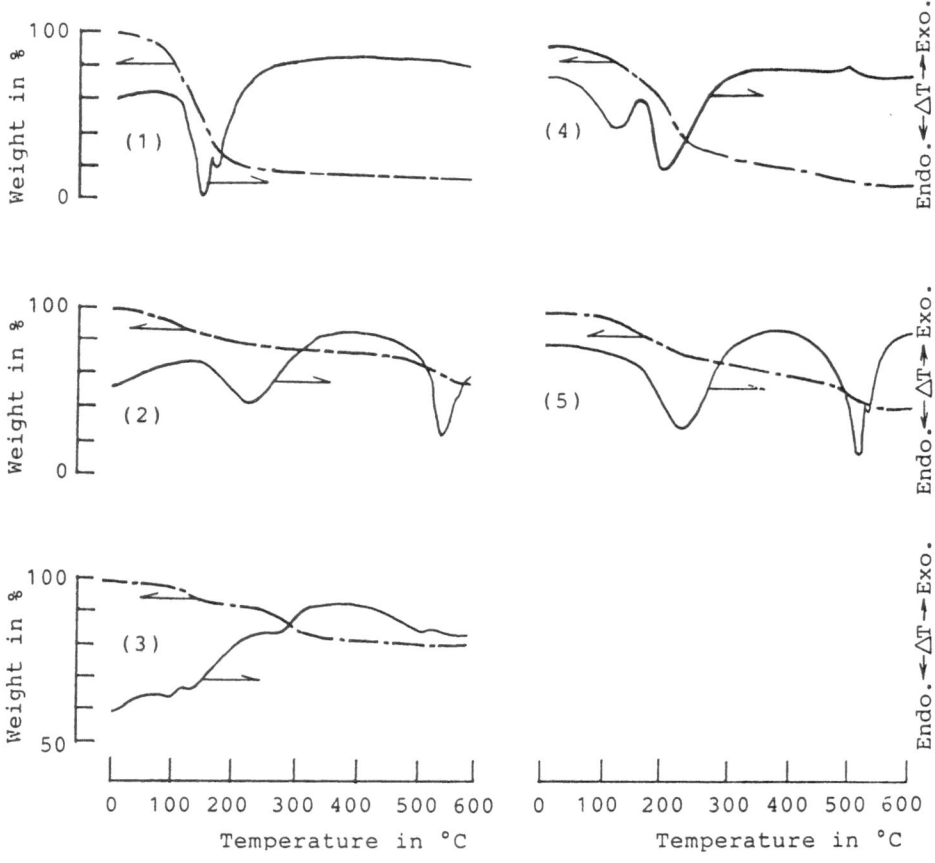

Fig. 2. TG-DTA curves of (1) $TiCl_4 \cdot EB$ complex, (2) coground
product of 30g of $MgCl_2$ and 6.5ml of EB, (3) coground product
of 30g of $MgCl_2$ and 5ml of $TiCl_4$, (4) $TiCl_4 \cdot BP$ complex, and (5)
coground product of 30g of $MgCl_2$ and 6.5ml of BP. DTA curve
(——————); TG curve (——·——)

Figure 3 shows the IR spectra of (1) the mixed product of $MgCl_2$ and
$TiCl_4 \cdot EB$ complex: cat-$(EB)_0$, (2) cat-$(EB)_5$, (3) cat-$(EB)_{15}$, (4) cat-
$(EB)_{30}$, and (5) cat-$(BP)_{30}$. The spectrum of (1) is the same as that of
$TiCl_4 \cdot EB$ complex as shown in Fig. 1-(2). The peak at 1680 cm^{-1}, which
indicates the existence of EB interacting with $MgCl_2$ appears in (2)
first, and then becomes drastically larger in (3) and (4). The peaks
at 1560 and 1590 cm^{-1}, owing to the original $TiCl_4 \cdot EB$ complex, becomes
smaller with an increase in the cogrinding time. The spectra of (4)
and (5) are quite similar to those of (3) and (6) shown in Fig. 1,
respectively, which correspond to the spectra of the coground product
of $MgCl_2$ and each $TiCl_4 \cdot$ ester complex. These results indicate that both
EB and BP in the $MgCl_2$-supported catalysts have the interaction with
$MgCl_2$, not with $TiCl_4$.

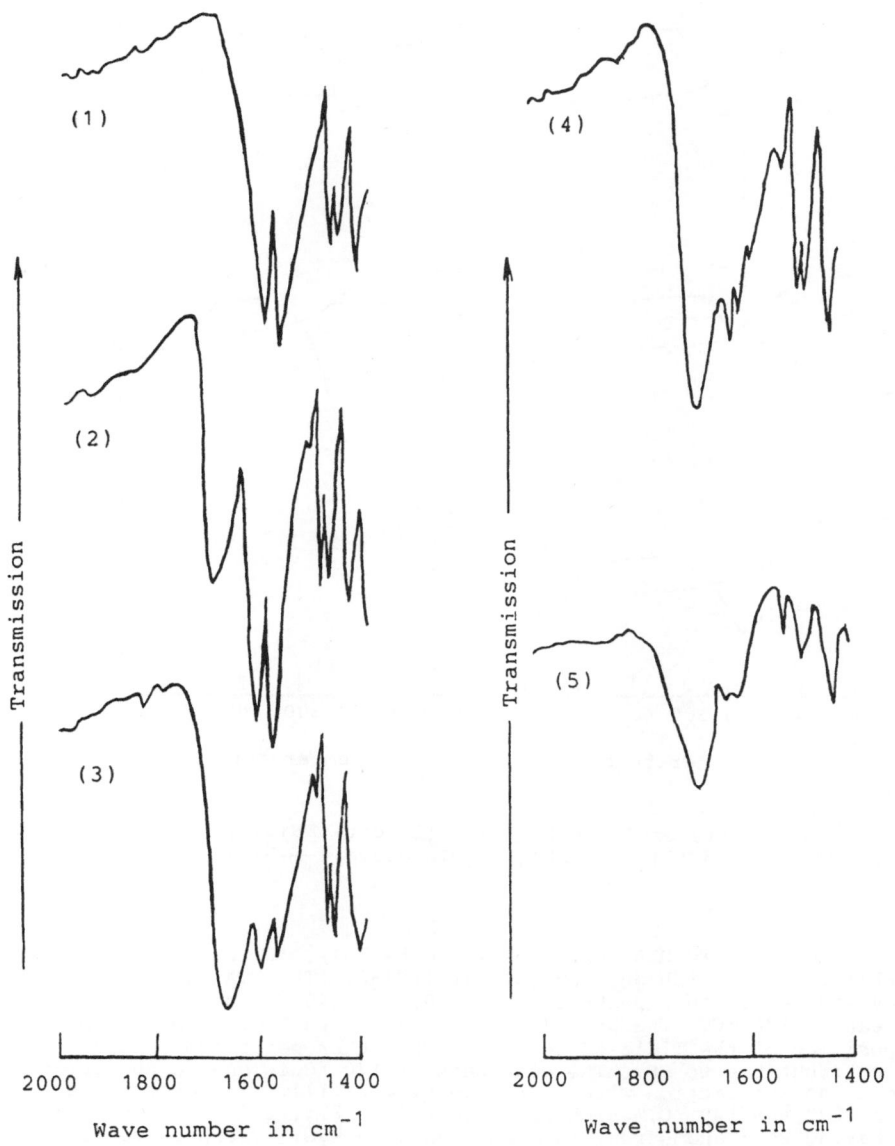

Fig. 3. IR spectra of (1) cat-(EB)$_0$, (2) cat-(EB)$_5$, (3) cat-(EB)$_{15}$, (4) cat-(EB)$_{30}$, and (5) cat-(BP)$_{30}$.

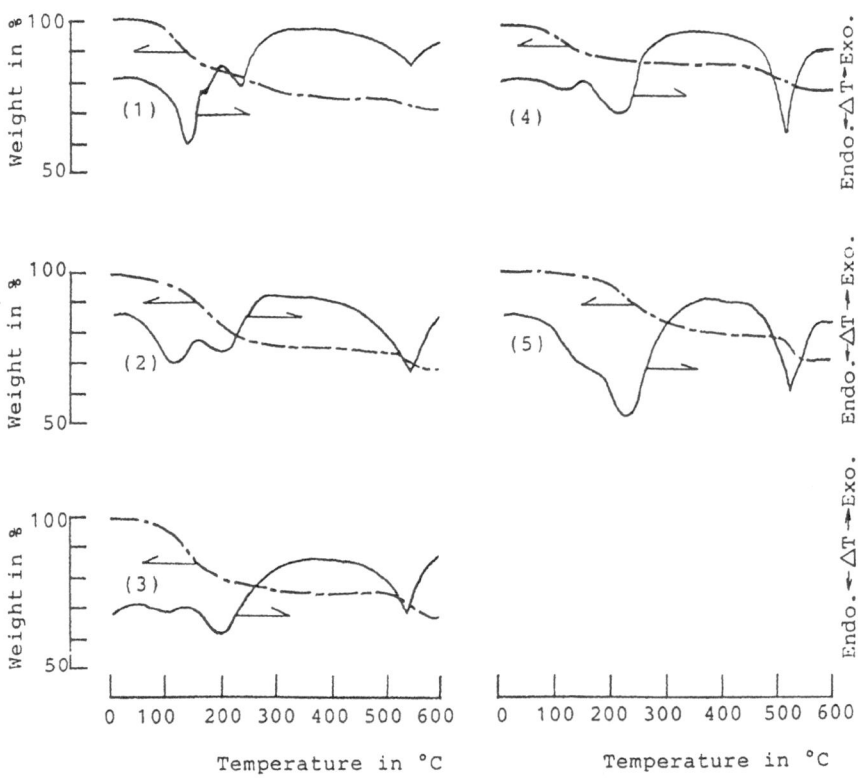

Fig. 4. TG-DTA curve of (1) cat-$(EB)_0$, (2) cat-$(EB)_5$, (3) cat-$(EB)_{15}$, (4) cat-$(EB)_{30}$, and (5) cat-$(BP)_{30}$. DTA curve (———); TG curve (— · —).

Figure 4 shows the TG-DTA curves of (1) cat-$(EB)_0$, (2) cat-$(EB)_5$, (3) cat-$(EB)_{15}$, (4) cat-$(EB)_{30}$, and (5) cat-$(BP)_{30}$. The DTA curve of (1) displays three distinct peaks at 140, 220, and 540 °C in addition to a weak peak at 170 °C. The peak at 140 °C is due to $TiCl_4$ produced by decomposition of the $TiCl_4 \cdot EB$ complex. The other peaks are considered to be attributable to EB. The weak peak at 170 °C is due to EB having little or no interaction with $MgCl_2$(see Fig. 2-(1)). Therefore, the peaks at 220 and 540 °C, which are observed in Fig. 2-(2), are caused by EB having weak and strong interactions with $MgCl_2$, respectively. They correspond to the DTA curve of the coground product of $MgCl_2$ and EB. This result strongly implies that EB produced by decomposition of the complex immediately interacts with $MgCl_2$.

In order to clarify the behavior of the complex, the IR spectrum was taken after heating cat-$(EB)_{30}$ up to 200 °C. The spectrum is shown in Fig. 5. A strong peak emerges at 1680 cm^{-1} which indicates the existence of EB interacting with $MgCl_2$. The peaks characteristic of the $TiCl_4 \cdot EB$ complex are not observed.

In Figs. 4-(2)~(4) are shown the changes in the TG-DTA curve when the duration of grinding was changed. The peak at 140 °C due to $TiCl_4$ decreased gradually and then disappeared almost completely as the

61

grinding proceeded. The weak peak at 170 °C in Fig. 4-(1) of cat-(EB)$_0$
cannot be observed in the ground catalyst. The peak at 220 °C decreased
gradually, while accompanied by an increase in the peak at 540 °C.
The DTA curve of cat-(EB)$_{30}$ is similar to that of the coground of MgCl$_2$
and EB (see Fig. 2-(2)). This may suggest that EB in cat-(EB)$_{30}$ exhibits
some interaction with MgCl$_2$, but no interaction with TiCl$_4$. Cat-(BP)$_{30}$
in Fig. 4-(5) shows a similar DTA curve to that of cat-(EB)$_{30}$ as well as
the coground product of MgCl$_2$ and BP in Fig. 2-(5). It follows from the
results of thermal analysis that BP seems to exist in the same states
as EB, in agreement with the results obtained from the IR spectra.
All these results lead to the conclusion that the TiCl$_4$·ester complexes
decompose by grinding with MgCl$_2$, and that TiCl$_4$ and each ester exist
separately in the MgCl$_2$-supported catalyst instead of remaining in the
form of a complex.

IR spectroscopy was extended to include the aliphatic carboxylic acid
esters. Figs. 6-(1)~(4) show IR spectra of EH, TiCl$_4$·EH complex,
coground product of MgCl$_2$ and EH, and cat-(EH)$_{30}$. Figs. 6-(5)~(8) are
IR spectra of EP, TiCl$_4$·EP complex, coground product of MgCl$_2$ and EP,
and cat-(EP)$_{30}$. Peak shifts of the C=O band in the IR spectra are
similar to what have been observed for EB and BP. TiCl$_4$·ester complexes
seem to decompose by cogrinding with MgCl$_2$ and the esters have the
interaction only with MgCl$_2$ in the catalysts.

In Table 1, the amounts of TiCl$_4$ supported on MgCl$_2$ and on EB/MgCl$_2$ are
summarized. The results indicate that the amount of supported TiCl$_4$ is
not affected by the presence of EB, even when concentrated TiCl$_4$ is used.
If we suppose that TiCl$_4$ were supported on MgCl$_2$ with the aid of EB as
a supporting agent, the Ti content should have been increased in the
cases of (3) and (4). The lack of difference between (1) and (3)/(4)
strongly suggests that EB is not necessary for TiCl$_4$ to be fixed on
MgCl$_2$.

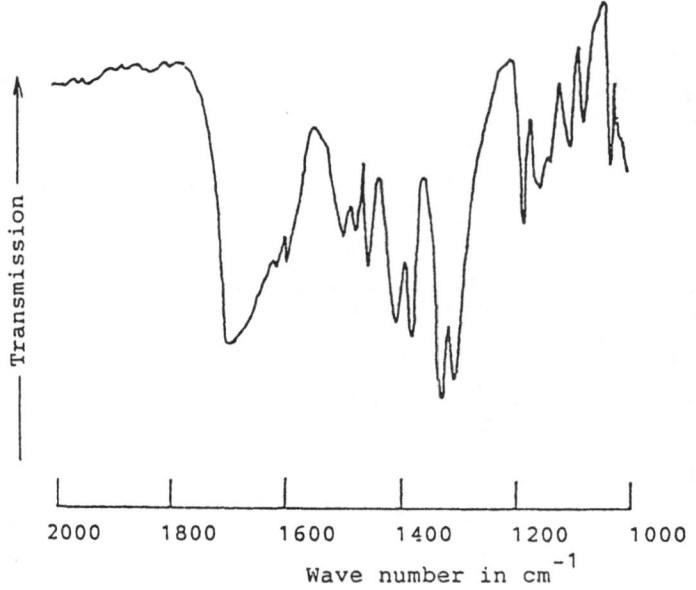

Fig. 5. IR spectrum measured after heating cat-(EB)$_0$ to 200°C

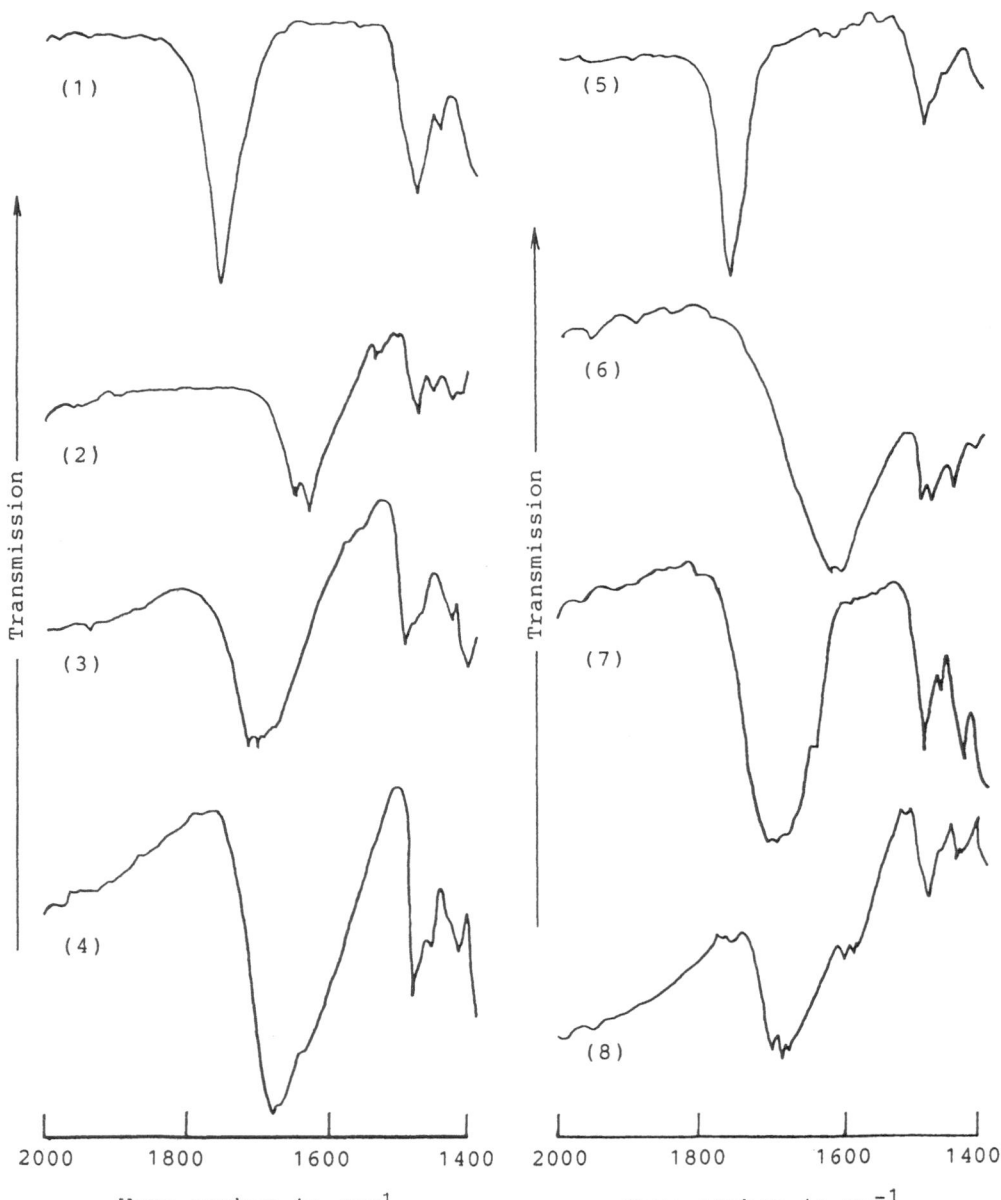

Fig. 6. IR spectra of (1) EH, (2) TiCl$_4$·EH complex, (3) coground product of 30g of MgCl$_2$ and 6.5ml of EH, (4) cat-(EH)$_{30}$, (5) EP, (6) TiCl$_4$·EP complex, (7) coground product of 30g of MgCl$_2$ and 6.5ml of EP, and (8) cat-(EP)$_{30}$.

Table 1. Difference of the amount of supported $TiCl_4$ on $MgCl_2$ or on $EB/MgCl_2$

	Ti $(mmol/g-MgCl_2)$	EB $(mmol/g-MgCl_2)$
(1) $TiCl_4/MgCl_2$ [a]	0.10	—
(2) $EB/MgCl_2$ [a]		0.15
(3) $TiCl_4/EB/MgCl_2$ [b]	0.09	0.11
(4) $TiCl_4/EB/MgCl_2$ [b]	0.08	0.07

a 5 g of $MgCl_2$(S.A.; 60 m^2/g) was treated with (1) 2.5 ml of $TiCl_4$ or
 (2) 1.1 ml of EB in (1) 27.5 ml or (2) 28.9 ml of heptane at 70 °C for
 2 h, followed by washing with 200 ml of heptane at 40 °C 10 times
 and then dried in vacuum.
b 5 g of the sample (2) was treated with (3) 2.5 ml of $TiCl_4$ in 27.5 ml
 of heptane or with (4) 30 ml of $TiCl_4$ at 70 °C for 2 h, followed by
 washing with 200 ml of heptane at 40 °C 10 times and then dried in
 vacuum.

Table 2. Adsorption of EB on $TiCl_4/MgCl_2$

	Ti $(mmol/g-MgCl_2)$	EB $(mmol/g-MgCl_2)$
(1) $TiCl_4/MgCl_2$ [a]	0.88	
(2) $EB/TiCl_4/MgCl_2$ [b]	0.88	0.69

a 30 g of $MgCl_2$ and 5 ml of $TiCl_4$ were coground for 30 h at r.t..
 5 g of the product was washed with 200 ml of heptane at 40 °C 10 times
 and then dried in vacuum.
b 5 g of the sample (1) was treated with 1.1 ml of EB in 28.9 ml of
 heptane at r.t. for 1 h and then dried in vacuum.

In Table 2, the amount of EB adsorbed on $TiCl_4/MgCl_2$ are shown. The
Ti content remains unchanged by the treatment with EB.

Figure 7 shows the TG-DTA curve and the IR spectrum of the sample (2)
in Table 2. The DTA curve shows the same two peaks as observed in the
coground product of $MgCl_2$ and EB shown in Fig. 2-(2). The IR spectrum
has a strong peak at 1680 cm^{-1} due to the interaction of EB with $MgCl_2$,
but it does not show the peak at 1560 cm^{-1} that has been assigned to
the absorption band of the $TiCl_4$ EB complex. It can be deduced that
EB interacts only with $MgCl_2$ and that it is not adsorbed on $TiCl_4$.
Therefore, it may be considered that $TiCl_4$ supported on $MgCl_2$ has no
vacant sites.

The results of propylene polymerization carried out for cat-$(EB)_n$ are
shown in Fig. 8. The activity of the catalyst increases in a liner
manner with the grinding time, while the degree of isotacticity
remarkably increases at first, followed by a slight increase. The
activity and the isotacticity of cat-$(EB)_0$, in which $TiCl_4$ exists with
EB in the form of a complex, are quite low. Therefore, the increase in
both the activity and the isotacticity must be achieved by the

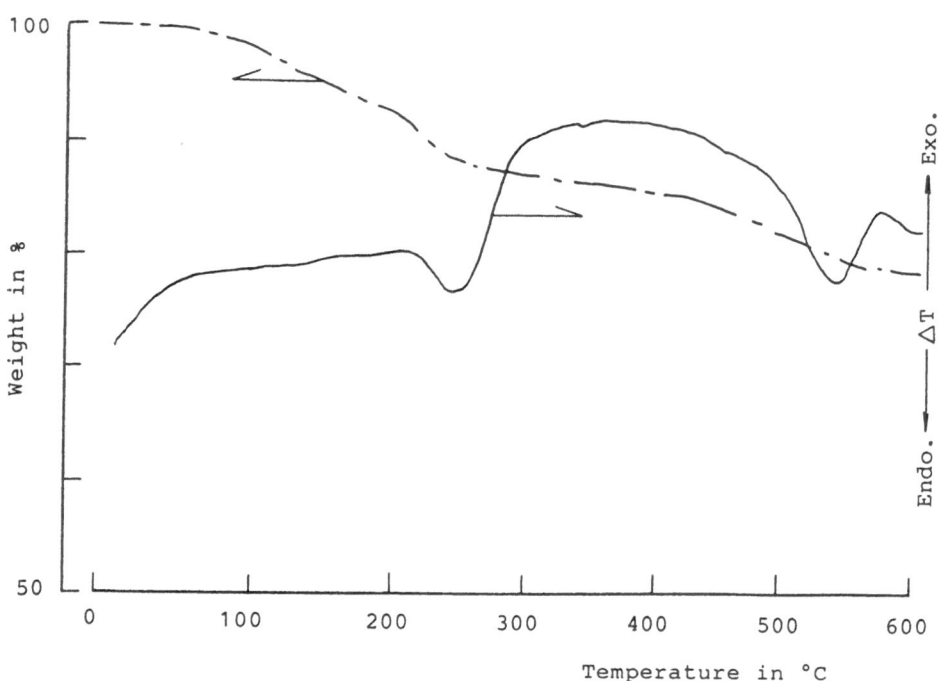

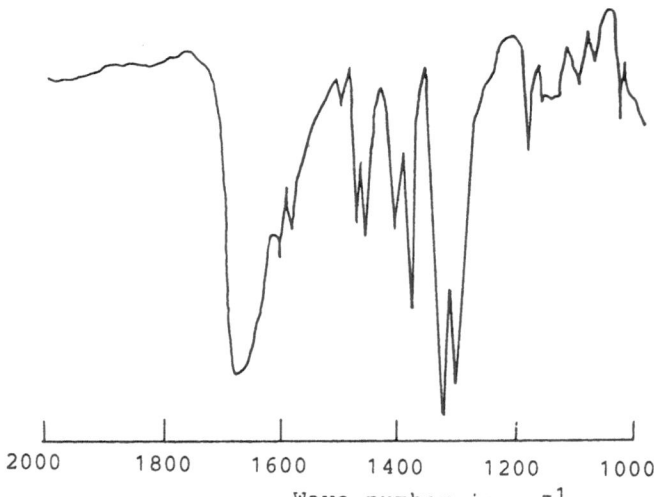

Fig. 7. TG-DTA curve and IR spectrum of EB/TiCl₄/MgCl₂ (Table 2-(2))

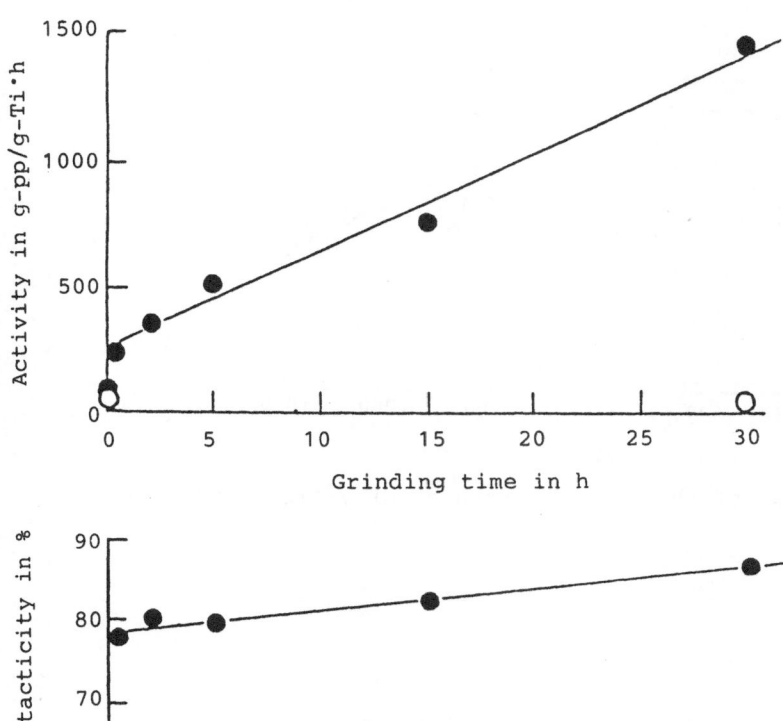

Fig. 8. Polymerization results by using various catalysts
  ○ : cat-(EB)$_n$
  ● : coground product of δ-Al$_2$O$_3$ and TiCl$_4$·EB complex

decomposition of the complex, thereby giving TiCl$_4$ in a desired state. This result, combined with analytical data of the catalytic systems, strongly implies that the active Ti moiety is likely to be free from EB.

To confirm the validity of the above argument, mixed and coground products of δ-Al$_2$O$_3$ and TiCl$_4$·ester complex were used for propylene polymerization. The results obtained are also included in Fig. 8. The activities are nearly null. While, the IR spectrum of the coground product of δ-Al$_2$O$_3$ and TiCl$_4$·EB complex is shown in Fig. 9. The existence of TiCl$_4$·EB complex on this material is evidenced by characteristic absorption peaks at around 1600 cm^{-1}. But, no clear peaks indicative of free EB can be seen at 1725 cm^{-1}. Correspondingly, the complex does not seem to decompose, even after coground for 30 h. Finally, it was clearly demonstrated that TiCl$_4$·EB complex cannot be active, even in a highly dispersed state.

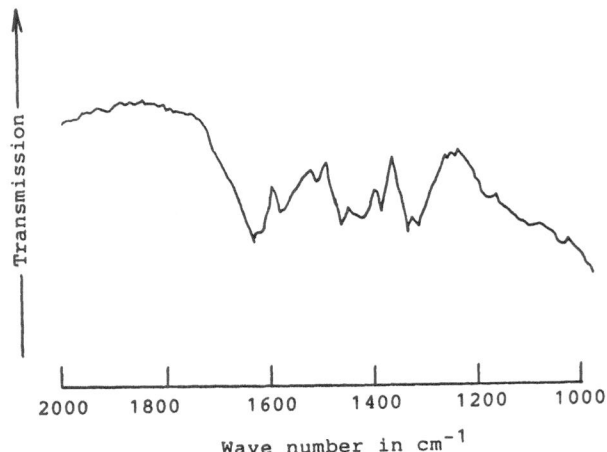

Fig. 9   IR spectrum of the coground product of γ-Al$_2$O$_3$ and TiCl$_4$·EB complex

CONCLUSIONS

1) A TiCl$_4$·ester complex decomposes by cogrinding with MgCl$_2$.
2) TiCl$_4$ and an ester exist independently having the interaction only with MgCl$_2$ in the catalyst.
3) TiCl$_4$ supported on MgCl$_2$ is coordinatively saturated.
4) One of the important roles of MgCl$_2$ is to decompose the complex, thereby stabilizing TiCl$_4$ and the ester in its matrix.
5) The complex itself is inactive, even after being ground to become highly dispersed on a support.

REFERENCES

Hewett WA, Shokal EC (1964) Japanese Patent 432533
Giannini U, Cassata A, Longi P, Mazzocchi R (1977) Japanese Patent 1076201; (1980) Italian Patent 24141A/80
Toyota A (1977) Japanese Patent 1014471
Chien JCW, Hsieh JTT (1976) J Poly Chem Ed 14: 1915
Munos-Escalone A, Villalba J (1977) J Polymer 18: 179
Soga K, Terano M, Ikeda S (1979) Poly Bull 1: 849
Suzuki E, Tamura M, Doi Y, Keii T (1979) Macromol Chem 180: 2235
Kashiwa N (1980) Polym J 12: 603
Galli P, Luciani L, Cecchini G (1981) Angrew Macromol Chem 94: 63
Sergeev SA, Bukatov GD, Moroz EM, Zakharov VA (1982) React Kinet Catal Lett 21: 403
Brockmeier NF, Rogan JB (1985) Ind Eng Chem Prod Res Dev 24: 278
Terano M, Kataoka T, Keii T (1986) Macromol Chem Rapid Commun 7: 725
Terano M, Kataoka T, Keii T (1987) Macromol Chem 188: 1477
Chien JCW, Wu JC, Kuo CI (1983) J Poly Sci Polym Chem Ed 725: 21
Kvisle S, Nirisen O, Rytter E (1980): in IUPAC Macro Florence preprints vol 2, p 32
Spitz R, Lacombe JL, Guyot A (1984) J Polym Sci Polym Chem Ed 22: 2641

Distinctive Features of Olefin Polymerization on the Surface of Supports

F.S.Dyachkovskii

Institute of Chemical Physics, Academy of Sciences of the USSR, Moscow, USSR

INTRODUCTION

Development and study of the catalytic systems for the polymerization on the surface of inorganic and organic supports have led to novel, highly efficient technological processes and a deeper understanding of the general problems of complex catalysis (1,2).

The use of highly active immobilized catalysts in technology has made it possible to eliminate both the energy-consuming washing of catalysts and, in some cases, the granulation of polymers (3).

From the point of view of polymer chemistry of special interest are the procedures developed for the synthesis of catalytic systems on polymeric supports and the processes based on their employment. Used as supports is a wide range of polymers, synthetic and natural compounds of different molecular weight, porosity, with various functional groups. The functional groups can be either located in a graft layer or uniformly distributed in the bulk of the support. Mosaic gels, in which fragments with grafted reactive functions are inserted into the bulk of an inert matrix, are also known.

The syntheses of gel-immobilized catalysts (4,5) and catalysts with metal complexes on the surface of various synthetic and natural polymers have demonstrated extensive possibilities of organic topochemistry, the chemistry of high-molecular compounds, and polymer-analogous transformations in creating various immobilized catalytic systems for the polymerization, hydrogenation, oligomerization, and dimerization of unsaturated compounds.

The present communication deals with polymerization using immobilized metal complexes, the peculiarities of catalytic systems and of the processes taking place on the surface, as well as some aspects of the synthesis and properties of polymeric compositions.

SYNTHESIS AND PROPERTIES OF CATALYTIC SYSTEMS IMMOBILIZED ON POLYMERS

Let us examine the synthesis and properties of catalytic systems immobilized on polymeric supports. Inorganic supports and the corresponding catalysts have been reviewed in (1,6,7). Such heterogeneous catalytic systems with immobilized transition-metal derivatives on inorganic surfaces usually retain for a long time a high catalytic activity in olefin polymerization even at elevated temperatures. Homogeneous catalysts, on the contrary, usually display a high activity during the first minutes of the process and are rapidly deactivated,

W. Kaminsky and H. Sinn (Eds.)
Transition Metals and Organometallics as
Catalysts for Olefin Polymerization
© Springer-Verlag Berlin Heidelberg 1988

especially at high temperatures. Immobilization of the components of homogeneous catalysts on a support results in a considerable stabilization of their activity. In the recent years immobilization of homogeneous catalytic systems is widely used in various types of catalytic processes. This enables one to regulate the reaction selectivity, the stereospecificity, and other properties of products.

Transition-metal complexes can now be bound to supports in many different ways (8-10). The results obtained in the recent years and summarized here concern the development and study of the catalytic systems for the polymerization of olefins immobilized on polymeric supports, the kinetics and mechanisms of polymerization in the presence of such catalysts, and the properties of the polymeric formed.

Carbo- and heterochain polymers containing hydroxy, carboxy, amino, hydrosulphide, and other groups have found extensive application for the fixation of transition metals. These polymers can be synthesized either from monomers with functional groups or by means of polymer-analogous transformations of ready polymers. Hydrolyzed styrene or ethylene copolymers with unsaturated esters, as well as the reduction products of methyl vinyl ketone or acrylonitrile copolymers proved to be convenient for this purpose. As a result, hydroxy and amino groups appear in the copolymers. Immobilization of transition-metal compounds on such supports is shown in the scheme (10):

$$-(\underset{Ph}{CH-CH_2})_p- (CH_2-\underset{X}{CH})_q \xrightarrow[H^+]{^-OH} -(CH_2 - \underset{Ph}{CH})_p- (CH_2 - \underset{YH}{CH})_q$$

where X = $OCOCH_3$, $-C\equiv N$, $-COCH_3$; Y = $-O-$, $-NH$

$$-(CH_2-\underset{Ph}{CH})_p - (CH_2-\underset{YH}{CH})_q - +qMX_nZ_m \longrightarrow (CH_2-\underset{Ph}{CH})_p - (CH_2-\underset{Y-MX_{n-1}Z_m}{CH})_q + HX$$

where MX_nZ_m = $TiCl_4$, VCl_4, $VOCl_3$, $(C_5H_5)_2TiCl_2$, $Ti(OR)_4$.

Application of vinyl chloride - vinyl acetate - vinyl alcohol terpolymers for similar reactions has been described. Reactions of transition-metal compounds with polymeric reagents have a number of peculiarities, as compared with the corresponding reactions of low-molecular analogues (11). These are mainly the differences in reactivity, functional groups, stereochemical effects associated with sets of the conformations of reacting macromolecules, mutual influence of the macromolecule on the transition-metal compound and vice versa.

Natural polymers with hydroxy groups, e.g. various cellulose derivatives, are sometimes used as polymeric supports for the catalysts of polymerization (12).

In essence, for the fixation of transition-metal compounds a polymer must possess an accessible reactive centre which can be a part of the main polymeric chain (heterochain polymers) or be located in a side chain (polymer of the $-CH_2-\underset{X}{CH}-$ type).

The reactive centres can be uniformly distributed over the polymer bulk or localized on its surface. Naturally, the surface fixation of a transition metal also results in the topochemical changes of the polymeric matrix itself.

Fixation of transition metals is actualized in the case of polymeric supports based on polyolefins or other carbochain polymers with a functionalized surface, resulting from the modification of the surface by grafting monomers with functional groups, donor centres, or chelate units (13). The general scheme of the synthesis of such supports can be presented as

$$
\text{polyolefin surface} \xrightarrow{\text{initiation}} m \; \overset{R''}{\underset{R'}{C}} = \overset{R'''}{\underset{Y}{C}} \longrightarrow -(\overset{R''}{\underset{R'}{C}} - \overset{R'''}{\underset{Y}{C}})_m -
$$

where $R' = H, CH_3, C_2H_5, C_3H_6$; $R'' = R''' = H, CH_3, C_6H_5$;

$\quad$ Y = $-OH, -NHR', -SR', -COCH_3, CH_2OH, -CH_2NHR', -CH_2SR', -COOH,$

$\quad\quad$ $-OCOCH_3, -COOCH_3, -C\equiv N, -NR'_2, -S-, R'SO_2, C_5H_5N-,$

$\quad$ m = 4-5000

To achieve this, a polymer-support (polyethylene, polypropylene, ethylene-propylene copolymer, polystyrene, etc.) is subjected to mechanical, chemical, or radiochemical action in the presence of gaseous grafted monomers.

Synthetic possibilities of graft polymerization in producing a functionalized surface, followed by the immobilization of transition-metal compounds, can be exemplified by types of fixation. The first one involves polymeric supports with a "protonolytic" surface capable of fixing the transition-metal compounds by covalent binding.

$$
-(CH_2 - CH)_t + t \; VO(OR)_3 \longrightarrow -(CH_2 - CH)_t
$$
$$
\quad\quad CH_2-OH \quad\quad\quad\quad\quad\quad\quad\quad\quad\quad CH_2 - OVO(OR)_2
$$

The functional groups here are hydroxylic, carboxylic, amino, amido, hydrosulphide, etc. Immobilized in this way are such transition-metal complexes that undergo "protonolysis" or ligand exchange, as in the case of PE-modified polyallyl alcohol and $VO(OC_2H_5)_3$. Acting as rather strong Lewis acids, the components of heterogeneous and pseudo-heterogeneous Ziegler systems can be fixed on supports with surface electron-donor groups or heteroaroms, such as carbonyl, ester, nitrile, sulphide, tertiary amine, phosphorus, and other functions.

A scheme of such immobilization of PE-gr-poly(4-vinylpyridine) is given below:

$$
-(CH_2 - CH)_t- + tVCl_4 \longrightarrow -(CH_2 - CH)_t -
$$
$$
\quad C_5H_4N \quad\quad\quad\quad\quad\quad\quad\quad\quad\quad\quad C_5H_4NVCl_4
$$

Transition-metal complexes are fixed on this type of supports in mild conditions, the degree of reaction completion depends both on the properties of grafted fragments and the nature of transition-metal compounds, while the bound-component content varies within a wide range $(1.5 \cdot 10^{-3} - 1.0 \cdot 10^{-5}$ mol/g of support).

Such polymeric supports are most effective for the heterogenization of typically homogeneous systems, e.g. $(C_5H_5)_2TiCl_2-Al(C_2H_5)_2Cl$, $VO(C_2H_5)_3-Al(C_2H_5)_2Cl$, or $Ti(OC_4H_9)_4-AlEt_2Cl$.

Of great interest for the immobilization of catalytic systems are mo saic gels (14), the accessibility of whose reactive centres depends on the degree of swelling. Thus, in the immobilization of transition--metal compounds on a polymeric support, covalent, donor-acceptor and ionic bonds can be formed, the reactive centre is chelated, etc.

Immobilization is simplified considerably when a single-stage proce-dure is used for the fixation of transition metals, based on the grafting of monomers containing a transition metal to the surface of carbochain polymers (15):

$$\text{initiation} \longrightarrow \text{.} + pMX_n(Y-CH_2-CH=CH_2)_{m-n} \longrightarrow$$

$$- (CH_2-CH)_p- \\ CH_2-Y-X_nM-(Y-CH_2-CH=CH_2)_{m-n-1}$$

$$M = Ti, V;$$
$$Y = -O-, -NH-, =N, \\ -COO^-.$$

Mono-, di-, tri- and tetraallylorthotitanates or vanadates, monoallyl-hydroxytitanocene, as well as metal-containing monomers with unsatu-rated carboxylic acid, amines, thiokol moieties can be used as tran-sition-metal compounds for such grafting. The main difficulty here consists in the synthesis of metal-containing monomers, but there is now a significant progress in this field.

A high activity of catalysts on inorganic supports has already been mentioned above. The main drawback of inorganic supports is, however, a limited assortment of functional groups, including mainly the hyd-roxyls. This restricts synthetic applications of inorganic carriers. One can overcome the difficulty by amination, sulphochlorination, phosphorylation, etc., of silica gel, magnesium oxide, or aluminium oxide surface (16), i.e. there exists a possibility of synthesizing reactive centres typical of organic carriers on the surfaces of inor-ganic compounds. The inverse problem, i.e. the formation of structu-ral elements of inorganic carriers on the surfaces of polymeric sup-ports, is also of great interest. To achieve this, a polyolefin with a functionalized surface is modified with an organosilicon, -alumi-nium, -magnesium, or other organometallic compound, whose subsequent transformations lead, e.g., to the formation of a hydroxy group at the modifying metal. Fixation of a transition metal on such mixed sup-ports takes place in the same way as on inorganic ones:

$$-(CH_2 - CH)_n- + nR_2SiCl_2 \longrightarrow \quad -(CH_2 - CH)_n- \longrightarrow \\ CH_2-OH \qquad\qquad\qquad\qquad CH_2-OSiR_2Cl$$

$$-(CH_2 - CH)_n- + MX_m \longrightarrow \quad -(CH_2 - CH)_n- \\ CH_2-OSiR_2OH \qquad\qquad\qquad CH_2-SiR_2OMX_{m-1}$$

Various ways of fixing transition- and nontransition-metal deriva-tives on the surface are available at present, as well as of creating practically any kind of ligand environment at them.

The surface or volume functionalization of polymers and immobiliza-tion of transition metals is sometimes accompanied by the destruction and cross-linking of a polymer matrix, and a change in support poro-sity. For instance, the grafting of allyl alcohol to polyethylene and the binding of triethoxyvanadate to hydroxyls is accompanied by the narrowing of pore radii distribution and a decrease in the total pore volume from 1.28 to 0.61 g/cm^3. These changes are indicative of predominant localization of transition metals inside the pores of a polymeric support.

The topochemistry of a functionalized polyethylene surface was studied by spin-label technique (17-19). Spin-labelled polymers were obtained by interacting polyethylene, whose surface had been treated with allylamine, with the iminooxyl radical derivative:

ESR study of such polymers has revealed that amino groups of the graft polymer interact with the radical over the whole graft layer (100-200 A) with the distance between the radicals equal to 20 A and dependent on the reaction conditions. Studies of the temperature dependence of the time of rotational correlation of radicals in the functionalized layer proved to be of interest. The Arrhenius plot of τ_c consists of two linear portions, the inflection point corresponding to the melting of the polymer-support. The activation energy above the melting point was equal to 13.5 kcal/mole, below it - to 5.0 kcal/mole. This means that, when the mobility of polymer segments is low (low temperatures), the rotation occurs without substantial energy barriers. An increase in the mobility of polymer segments, on the contrary, hinders the rotation. In the pre-melting state, the fixed radicals are already localized not only on the PE surface, but partially penetrate into the bulk of the support, leading to an increase in the rotation E_A.

The distribution of functional groups inside a polymer and on its surface determines the topochemistry of the immobilization of transition-metal compounds. Spin-labelled complexes were used to study the distribution of transition-metal compounds on the polymer surface (18). The complexes were obtained as shown below:

Transition-metal compounds with a spin-labelled ligand were bound to polyethylene surface with graft polydiallylamine (PE-gr-PDAA). The ultimately formed structures were of the following type:

Immobilized paramagnetic complexes of Ti(IV), V(VI), Al(III), Mo(V), etc., have been studied by ESR. It was first of all shown that the fixed components of catalytic systems were distributed over the surface layer of the support down to approximately 100-300 A. The fixation is a result of a gradual penetration of reagents into the surface layer. The process is, naturally, temperature dependent. Usually not all the functional groups react with MX_n. At low MX_n/functional groups ratios the content of MX_n on the surface is proportional to the number of graft fragments. The first to react are, obviously, the external functional groups. This is followed by MX_n penetration into the surface layer.

Despite the relatively homogeneous distribution of functional groups over the surface, the immobilized metal distribution is of dual nature. One observes both a relative uniformity of distribution for

transition-metal compounds over the depth of the graft layer (isolated Ti(IV) ions) and the presence of aggregates in the thin external
layer.

A study of the magnetic susceptibility of Mo(V), Ni(II), and Co(II)
immobilized on the surface (20) revealed spin-spin antiferromagnetic
interactions between the immobilized metal atoms. This also indicated
the formation of cluster compounds on the surface.

A study of VCl_4 distribution on polymeric supports by ESR corroborated
this conclusion (21). Isolated and clustered V(IV) ions were detected
in the surface layer. With increasing surface density of vanadium, the
fractions of isolated species sharply decreases, while that of the
clustered V(IV) ions increases (Fig.1).

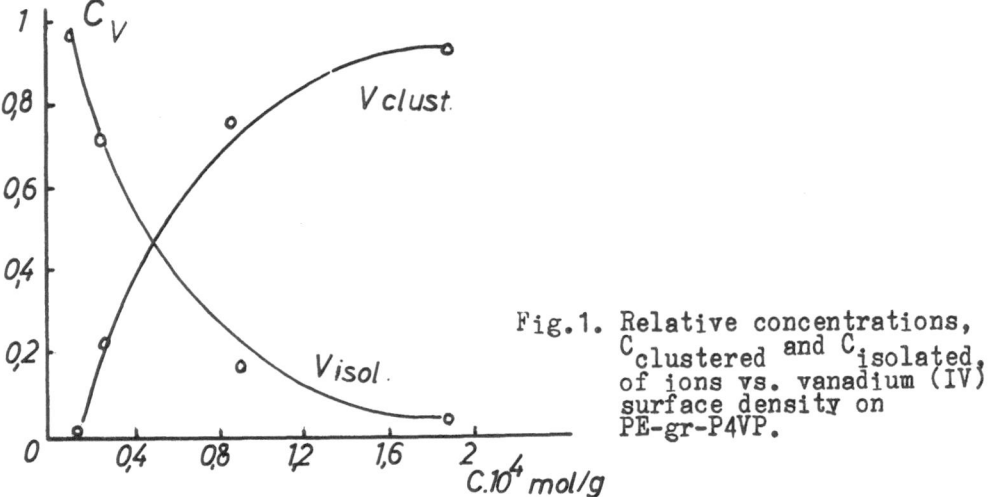

Fig.1. Relative concentrations,
$C_{clustered}$ and $C_{isolated}$,
of ions vs. vanadium (IV)
surface density on
PE-gr-P4VP.

Measured concentrations of the isolated and the clustered vanadium
ions allowed estimating the mean distances in these two groups of
ions: $r_{isol} = 22$ A   and $r_{clust} = 6.8$ A.

The clusters, probably, stabilize the isolated metal ions. This ex-
plains some kinetic peculiarities of catalytic processes involving
fixed systems.

A study of the structure and topochemistry of immobilized catalysts
has thus shown the presence of both isolated metal ions (homogeneous
distribution) and energetically stable cluster aggregates.

OLEFIN POLYMERIZATION IN THE PRESENCE OF IMMOBILIZED CATALYTIC SYSTEMS

Olefin polymerization in the presence of catalytic systems fixed on
the surface or in the bulk of polymers is accompanied by the stabilization of their catalytic activity, increasing the efficiency of the
catalysts, as compared with their homogeneous analogues. The temperature range of catalyst applicability expands, and the upper temperature limit rises (for some catalytic systems up to 100-150°C). These
properties of immobilized catalysts have led to their extensive application in polyolefin technology. Moreover, they possess some special

kinetic peculiarities that are important for the understanding of the general problems of catalysis mechanism.

A study of the influence of monomer concentration on the rate of ethylene polymerization by the $VO(OC_2H_5)_3$ system fixed on polyethylene with grafted polyallyl alcohol showed that within a wide range of 0.0.6-1.22 mol/l (ethylene pressure 1-20 atm) polymerization rate is of the first order with respect to the monomer. The absence of PE molecular-weight dependence on monomer concentration shows the chain transfer onto the monomer to be the main reaction limiting the growth of polymeric chains in immobilized systems. The polymerization rate dependence on the total concentration of the immobilized system is often also of the first order. However, the dependence of the catalytic process specific rate on the transition-metal surface density can be rather complicated. One often observes an extremal specific activity vs. the immobilized compound surface density dependence (Fig.2).

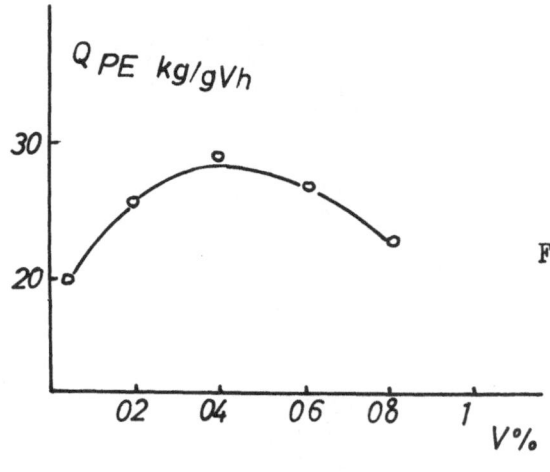

Fig.2. Specific yield vs. the surface density of $VO(C_2H_5O)_3$ immobilized on polyethylene grafted with polyallyl alcohol.

This is, possibly, due to the peculiarities of MX_n distribution in the polymeric matrix, and, specifically, the ratio between the isolated and the "clusterized" metal ions.

Another characteristic feature of immobilized catalytic systems is the stability of their catalytic action. What are the reasons for a high activity and stability of immobilized catalysts? The authors of (22) have concluded that n_p increase in immobilized systems is caused by the inhibition of the reactions of active centres deactivation. In the case of second-order decay of n_p the inhibition of their decay on the surface must be evident. The second-order decay of the centres is not, however, observed for all catalytic systems. Therefore, the rationale for the stabilization phenomenon is not so trivial. It seems that in immobilized organometallic systems the sharply retarded processes are those which necessitate a substantial rearrangement of the metal coordination sphere (intramolecular disproportionation, ß-elimination, etc.). Indeed, reduction of a transition metal which requires an intrasphere rearrangement proceeds much slower on the surface (23). It has been shown (24) that the cleavage of metal-carbon bond is sharply inhibited on the surface of macromolecular organic bases. Interestingly, the compensational effect is observed in these reactions (25), which is indicative of a pronounced role of entropic factors in the cleavage of metal-carbon bonds in general and on the surface of polymeric supports in particular.

A high stability of immobilized catalysts in these cases can be associated with cooperative interactions in the polymer - transition-metal complex system. Examples are known when one metal atom interacts with nine segments of a polymer (26,27). Therefore, the cleavage of a polymer-metal or a polymer-ligand bond requires a change in the state of adjacent segments of a polymeric chain. Restriction of the translational diffusion of active centres as a factor raising the thermostability of immobilized systems has been demonstrated by measuring ethylene polymerization rates at different temperatures (28) (Fig.3).

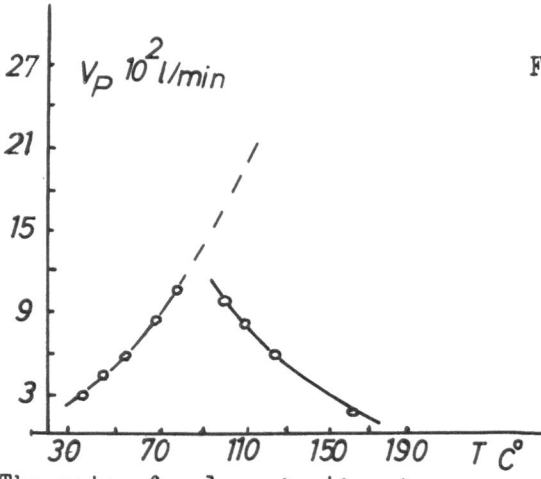

Fig.3. Temperature dependence of the rate of ethylene gas--phase polymerization in the presence of an immobilized catalytic system based on $TiCl_4 \cdot P4VP$ (dotted line indicated the polymerization rate calculated from Arrhenius law).

The rate of polymerization increases according to the Arrhenius law up to the temperature of pre-melting of the polymeric support (P4VP), E_A^{eff} being usually equal to 16 ± 2 kJ/mole for ionic-coordinative polymerization. In the temperature range within which chain segments are sufficiently mobile the temperature coefficient of the reaction becomes negative. The bimolecular law of n_p deactivation holds true at these temperatures.

Kinetic features of immobilized catalysts are thus mainly determined by the influence of polymeric matrix on the reactions of active centres.

A high activity of immobilized catalysts of olefin polymerization naturally makes it possible to exclude the polymer washing to remove catalyst traces and simplify the technology of synthesis. Application of immobilized catalytic systems on inorganic and organic supports opens up a possibility of performing the process in a fluidized bed in the gas phase. Another characteristic feature of fixed catalysts is the replication of the shape of the support by the polymer formed. The replication provides the synthesis of polymeric particles of a given size and shape. For instance, the use of microspherical mesoporous supports leads to the formation of large polyethylene and polypropylene particles of 1.5-4 mm diameter which copy the spherical shape of the support (29). The particle shape and szrface can vary. Depending on reaction conditions one can obtain almost spherical dense particles.

Their size increases proportionally to the reaction time. In other cases loose particles are formed. Both the density and the shape of the particles depend on numerous factors among which the most important are the polymerization rate, the rate of heat and mass transfer, polymer crystallization. If the rates of polymerization and of heat and mass transfer are well balanced, a close packing is observed. Large stresses inside the pores of the support, arising with polymer formation, may result in the decomposition of polymer granules and the formation of loose structures. Proper control over the main parameters of the polymerization process enables one to obtain a granule-like polymer directly in the course of synthesis. Energy-consuming granulation can thus be eliminated from the technological schemes. At the same time, arise the problems of introducing stabilizers, dyes, and other ingredients into the polymer. These problems are solved in two ways: by adding modifiers in the course of synthesis or by reprocessing synthetic granules into finished polymeric products.

The use of immobilized catalytic systems thus makes it possible to intensify and simplify radically the technology of polyolefin production. From the standpoint of polymer chemistry, the synthesis of polymeric catalytic systems has demonstrated extensive possibilities of the functionalization of polymer surfaces, grafting transition-metal compounds to macromolecules with the localization of these compounds either on the surface or in the bulk of a polymer with a given ligand environment.

POLYMERIZATION FILLING

Widespread application of immobilized catalytic systems in the synthesis of polyolefins has allowed up to propose a new method of producing filled polymers, that of polymerization filling. In this method the catalyst support simultaneously serves as a filler, being a component part of the composition formed (30). Surface polymerization on organic and inorganic fillers can be initiated by complex catalysts, organometallic compounds, free radicals, or ions fixed on the surface. Monomer polymerization on the surface of a filler forms a polymeric layer on its surface, i.e. the filler is actually incapsulated by the polymer (3). Polymerization filling opened up extensive possibilities for the synthesis of various polymers (homo- and copolymers, block copolymers, two different polymers, etc.) on the surface of fillers.

Polymerization filling produces compositions with a practically unlimited degree of filling (up to 95% of filler volume).

Filler activation requires the creation of active centres on its surface for subsequent polymerization which can occur via ionic-coordinative, radical, or ionic mechanisms. Naturally, the substances which inhibit polymerization (e.g. H_2O, SO_2, CO_2, CO, O_2, etc.) must be removed from the filler surface.

To perform the polymerization filling by ionic-coordinative polymerization, the already discussed metal-complex or organometallic catalysts must be fixed on the filler surface.

Different types of radiation can be also applied to initiate on the surface the active centres of the polymerization of vinyl chloride, vinyl acetate, acrylonitrile, and other monomers that polymerize via the radical mechanism.

The kinetics of the polymerization filling of polyolefins is governed by the specificity of the polymerization processes taking place on solid surfaces. Organometallic and metal-complex catalysts are known to be unstable in their highest valent state, which results in highly unsteady rates of olefin polymerization in their presence. As already mentioned, immobilization of catalytic systems on inorganic and organic supports stabilizes markedly their activity in time and allows increasing the temperature interval of the use of such catalysts and raising their efficiency.

The enhanced stability of fixed metal-complex catalysts, as compared with polymer-dispersed analogues, makes it possible to synthesize filled polyethylene at temperatures up to $100^{\circ}C$. The effective activation energy of polymerization on the surface of fillers activated by titanium or vanadium compounds is 11.0-11.8 kcal/mole. A gradual decrease in the polymerization rate with time can be caused both by the chemical deactivation of active centres or by their being blocked by the polymer layer (diffusive inhibition).

The nature of a filler exerts a noticeable effect on the catalytic properties of the system in ethylene and propylene polymerization. Naturally, a well developed surface increases the efficiency of catalysts. The pore size proved to be of significance. When the surface is large and the pore diameter small, the pores are rapidly clogged with the polymer, and the catalyst activity decreases, since most of it is no longer accessible to new portions of the monomer. That is why, carriers with a meso-porous structure proved to be the best.

An interesting phenomenon was discovered while polymerizing alpha-butylene from the gas phase in the presence of porous fillers. At P/P_s 0.5, where P_s is the saturated vapour pressure at a given temperature and P is the pressure of alpha-butylene under the reaction conditions, a sharp increase in the reaction rate was observed (the reaction order with respect to the monomer increased up to 3-3.5) (Fig.4).

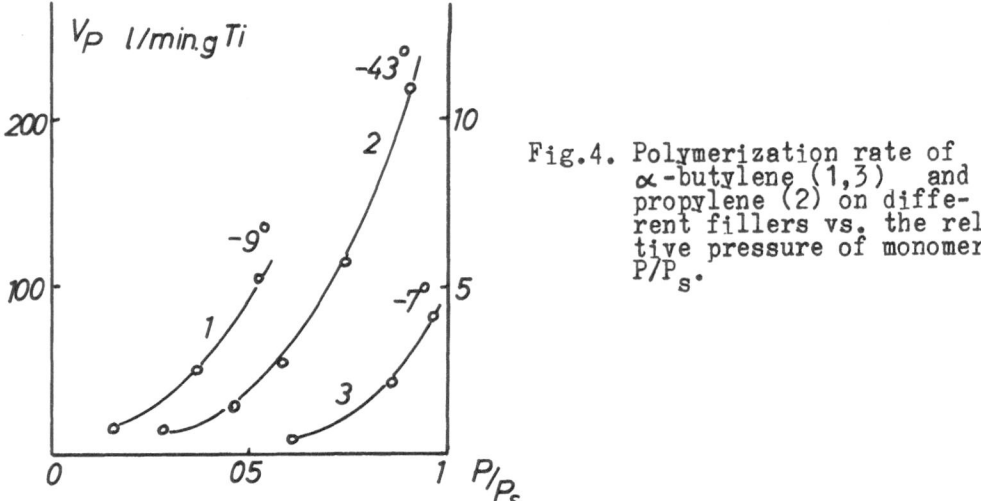

Fig.4. Polymerization rate of α-butylene (1,3) and propylene (2) on different fillers vs. the relative pressure of monomers P/P_s.

This fact is explained by monomer condensation in filler pores and the formation of a liquid phase of the monomer in a thin surface layer of the filler. Similar phenomena were also observed in heterogeneous acid-catalyzed reactions.

Also worthy of notice are the results obtained when graphite was used
as polypropylene filler. The $TiCl_3$ catalyst immobilized on graphite had the highest activity, as compared with $TiCl_3$ on other fillers, such as perlite, chalk, talc, etc. It is important that
polypropylene obtained on graphite in the presence of this catalyst
was highly isotactic (up to 98%). Different forms of Ti (Ti(II), Ti
(III), Ti(IV) were shown to be involved in stereospecific polymerization. It was also shown that immobilization of traditional homogeneous
catalysts (Ti(OBu)$_4$, Cp$_2$TiCl$_2$, etc.) on graphite results in stereospecific propylene polymerization.

As a result of polymerization on the surface, the filler particles are
covered by a polymeric layer of controlled thickness, the polymeric
layer, under optimal technological conditions being quite dense unin-
terrupted, and uniformly distributed over the surface of particles.
This provides a sufficiently homogeneous distribution of filler particles in the polymeric matrix and the composite material.

When polymerization proceeds on the filler surface, the polymer is also
formed inside the pores, microcracks, and defects, thus increasing the
contact surface of the filler with the polymeric matrix and, hence,
their interaction. This is supported by the data of scanning electron
microscopy used to study the failure surfaces of the samples of composites.

CONCLUSION

In conclusion, the following characteristic features of the processes
taking place on the surface of organic and inorganic supports should
be noted. From the point of view of polymer chemistry of great interest
is a wide variety of reactions that make it possible to introduce functional groups on the surface or in the bulk of polymers. By interacting transition-metal compounds with the functional groups of polymers
these compounds can be homogeneously fixed on the surface. Under certain conditions it is possible to achieve the required ratio of the
clustered and the isolated transition-metal ions fixed in a polymeric matrix.

Moreover, the fixation of a metal on the surface or in the bulk of a
polymer can induce the destruction of the polymeric chain and the
cross-linking of its fragments. In some cases it makes possible using
metal halides or alkoxides as cross-linking agents. Macromolecules of
inert polyolefins (polyethylene, polypropylene, etc.) can be involved
in the cross-linking. As a result, polymers gain new physico-mechanical properties.

From the standpoint of a high catalytic stability of fixed systems, of
great interest are the stable carbon-metal bonds in immobilized transition-metal complexes. The stabilization may be caused by the cooperative effects of macromolecule units on the rates of reactions in the
transition-metal coordination sphere, especially of those requiring its
considerable rearrangement. And, finally, if a support also serves as a
filler, surface polymerization leads to the formation of absolutely new
composite materials with a high degree of filling, superhigh-molecular
weight of the polymer, with new physico-mechanical and physical properties (conductivity, magnetic activity). At the same time, the characteristic features of polymerization on the surface area prominently
manifested: polymeric matrix anisotropy in the surface layer, a high
dispersion and encapsulation of the filler in the course of synthesizing the composite, etc. Therefore, polymerization filling not only
makes it possible to synthesize new highly filled polymers but also

opens up new aspects of the chemistry of high-molecular compounds.

REFERENCES

1. Ermakov YuI, Zakharov VA, Kuznetsov BI (1980) Immobilized Complexes on Oxide Carriers in Catalysis, Novosibirsk,
2. Dyachkovskii FS, Pomogailo AD (1980) J Polym Sci, Polym Sympos, No 68, p 97-108
3. Dyachkovskii FS, Novokshonova LA (1984) Usp Khim, 53, No.2, p 200-222
4. Kabanov VA, Smetanjuk BI, Popov VG (1975) Dokl.AN SSSR, 225, No 6, p 1377-1380
5. Ciardelli F. et al, (1982) J Mol Catal, 14, No 1, p 1-17
6. Bailar JC, (1974), Catal Rev, 10, No 1, p 17-36
7. Ballard DGH (1975) J Polym Sci. Polym Chem Ed, 13, No 10, p 2191-2212
8. Dyachkovskii FS, Pomogailo AD (1983) in Homo- and Copolymerization of Alpha-Olefins on Complex Catalysts. Moscow, p 72-83
9. Hodge P, Sherrington D (1983) Reactions on Polymeric Supports in Organic Synthesis, Moscow, (transl from Eng)
10. Dyachkovskii FS, Pomogailo AD, Lisitskaya AP, Gor'kova NS (1974) Dokl AN SSSR, 219, No 6, p 1375-1378
11. Davydova SL, Plate NA (1975) Coord Chem Rev, 16, p 195
12. Akelah A, Sherrington DC (1982) Europ Polym J, 18, No 4, p 301-305
13. Dyachkovskii FS, Pomogailo AD, Kritskaya DA, Lisitskaya AP, Ponomarev AN (1977) Dokl AN SSSR, 232, No 2, p 391-394
14. Kolotsei IN, Popov VG, Davydova SL, Kabanov VA, Vysokom Soed Kr Soobsch, 23, No 5, p 368-370
15. Pomogailo AD, Savostyanov VS (1983) in Complex Organometallic Catalysts in Olefin Polymerization, No 8, Ser 2, p 45-66
16. Grinenko SB, Belousov VM, Noskov AM, Lysova NN, Bucherenko EF, Chernyshov EA (1983) Ukr Khim Zhur, 49, No 2, p 136-140
17. Bravaya NM, Pomogailo AD, Dyachkovskii FS (1979) Vysokomol Soed, 21A, No 8, p 1781-1788
18. Dyachkovskii FS, Pomogailo AD, Bravaya NM (1980) J Pol Sci, Pol Chem Ed, 18, p 2615-2627
19. Dyachkovskii FS, Bravaya NM, Pomogailo AD (1983) Kinet Katal, 24, No 2, p 403-407
20. Pomogailo AD, Borod'ko YuG, Ivleva IN, Echmaev SB, et al (1980) Catalysts with Immobilized Complexes, Proceedings Sympos Tashkent Novosibirsk Publ, v 1, p 123-126
21. Pomogailo AD, Nikitaev AT, Dyachkovskii FS (1984) Kinet Katal, 25, No 1, p 166-170
22. Dyachkovskii FS, Golubeva ND, Pomogailo AD, Kuzaev AI, Ponomarev AN (1979) Dokl AN SSSR, 244, No 1, p 89-93
23. Baulin AA, et al (1980) Vysokomol Soed, A22, No 1, p 181-188
24. Khrusch NE, Chukanova OM, Serebryanaya IV, Dyachkovskii FS (1984) IV International Sympos on Homogeneous Catalysis, Abstracts of Papers, v 1, PI-123, p 195-196
25. Dyachkovskii FS, Chernaya LI, Khrusch NE, Matkovskii PE (1977) Zhur Obsch Khim, 47, No 8, p 1841-1847
26. Pomogailo AD, Baishiganov E (1983) in Complex Organometallic Catalysts in Olefin Polymerization, Chernogolovka, No 8, Ser 2, p 66-78
27. Lundbery RD, Bailey FE, Callard W (1966) J Polym Sci, Part A1, 4, p 1563
28. Dyachkovskii FS, Pomogailo AD, Irjak VI, Burikov VI, Enikolopyan NS (1982) Dokl AN SSSR, 266, No 5, p 1160-1163
29. Uvarov BA, Tsvetkova BI, Dyachkovski FS (1978) in Complex Organometallic Catalysts in Olefin Polymerization, Chernogolovka, No 7, p 7-23
30. Kostandov LA, Enikolopov NS, Dyachkovskii FS et al,(1976) Avt.Svid SSSR No 763379, Bull Izobr 1980, No 34, p 129

Recent Developments in the Determination of Active
Centers in Olefin Polymerization.

J. Mejzlík, P. Vozka, J. Kratochvíla, and M. Lesná

Chemopetrol, Research Institute of Macromolecular Chemistry,
Tkalcovská 2, 656 49 Brno, Czechoslovakia

INTRODUCTION

The active center determination has proved its importance in elucidating
the kinetics and mechanism of Ziegler-Natta polymerization. Yet considerable controversies exist concerning suitability and/or informing power of
the methods employed. This paper is intended to overview some developments
in the last two years. We have, however, tried to pick up most interesting
studies, thus sacrifying a bit of completeness.

A classification of methods will be reviewed and progress in each class
will be commented on. Some of the developments will be examplified by our
own results. The paper does not cover the structure of active centers,
which was reviewed in a very qualified way recently by Minsker et al. (1).

CLASSIFICATION OF METHODS FOR ACTIVE CENTER DETERMINATION

We shall adhere to the classification scheme used in a recent review (2).
The scheme divides typical methods into two main categories: (i) those
where the growing chain is labelled and "tags" in the polymer are monitored, (ii) those where a consumption of catalyst poison to block the
chain growth is measured. The first category is further subdivided according to the elementary reaction in which the "tag" is incorporated
(either during initiation or during chain growth). A finer division of
the latter sub-category can be made as shown in Table 1, where principles, advantages and shortcomings of each method are summarized.

DEVELOPMENTS WITHIN THE ABOVE CLASSES OF METHODS

As expected, some classes have run into dead end, while others are used
routinely without seeing really new developments. Still others have gained
considerable attention and their refinement and/or new approaches have
been witnessed during the recent years. The following chapters are intended to serve guides to these developments.

Labelling of Macromolecules

Labelling by Radioactive Organometals: These methods have never received
too much attention though they were the first in use (3). Their usefulness
became questionable when the incorporation of radioactive alkenes (formed
from the organometal) into the polymer chain was proved to occur (4-7).
No progress is seen in this category.

W. Kaminsky and H. Sinn (Eds.)
Transition Metals and Organometallics as
Catalysts for Olefin Polymerization
© Springer-Verlag Berlin Heidelberg 1988

Table 1. Classification scheme of methods

Classification scheme	Labelling of macromolecules				Consumption data of effective catalyst poison
	Labelling by radioactive organometal	Labelling of growing chain			
		Number of macromol.	Number of metal-polymer bonds	Selective tagging	
Designation	–	N-method	MPB-method	Select. stopper method	Cons. data method
Agents used	Labelled organometal	–	ROT, SO$_2$	^{14}CO, CS$_2$	Allene, CO
Species monitored	Labelled alkyl from organometal in polymer	$N = Q/\overline{P}_n$	MPB	Selective tags of growing chain in polymer	Unconsumed catalyst poison
Advantages	–	Universal method	Fairly universal method	Propagative species monitored; relatively simple technique; distribution of AC reactivities conditionally measurable	Propagative species monitored; distribution of AC reactivities measurable
Shortcomings	Isotopic laboratory required; non-propagative species also monitored; risk of increased values due to side reactions	Lowest ratio of C* to all species monitored; difficulties pertinent to M_n determination; non-propagative species also monitored	Kinetic isotope effect may be operative; isotopic laboratory required in most cases; non-propagative species also monitored risk of increased C* values due to side reactions with main chain	Isotopic laboratory required in most cases; risk of decreased C* values due to slow or non-quantitative insertion of the tag	Elaborate technique involved; corrections to consumption of the poison by side reaction necessary

Labelling of Growing Chains: A majority of methods in use belong to this class. Three types of methods, based on (i) the number of macromolecules, (ii) the number of metal-polymer bonds, and (iii) selective tagging of growing chains, are included in this class. Each of them will be dealt with separately:

(i) Methods based on the number of macromolecules

These, referred to as N-methods, rely on the calculation of the total number of chains (N) from the polymer yield and $\overline{M}_n$. Extrapolation to zero polymerization time or yield gives the value of $_n C^*$. Alternatively, $(M_n)^{-1}$ can be plotted against $(time)^{-1}$ and k_p value calculated from the slope of straight line (8). The N-methods have a universal applicability but suffer from low accuracy. A very recent application deviced by Terano et al. (9) has revealed their new horizonts. The authors were able to determine the number of macromolecules in sub-second polymerization times, using the stopped-flow procedure.

(ii) Methods based on the number of metal-polymer bonds

These methods have been in use since 1960 (10). Hydroxy-tritiated alcohols are usually employed as quenchers. The common shortcoming of these methods is that the quencher reacts not only with the growing chain but also with non-propagative metal-polymer bonds. Thus, the extrapolation of MPB to zero polymerization time or yield becomes necessary. Another shortcoming is seen in the kinetic isotope effect which may be operative when tritium labelled substances are employed. A common method to measure the effect is to compare the amounts of tritium incorporated into the polymer upon slow and fast additions of the quencher. This approach, however, was criticized (2) arguing that a slow addition of the quencher might modify the catalytic system. Another method for the kinetic isotope effect measurement based on varying the amount of ROT/ROH mixture was suggested by Chien and Kuo (11). The discussion about the applicability of MPB-methods is a subject of our recent review (2).

A rather thorough study aimed to compare the performance of several catalytical systems in the propylene polymerization has been carried out in our laboratory. The MPB-method with BuOT as a quencher was used. A simplified presentation of the results is shown in Table 2.

Table 2. Kinetic parameters of $TiCl_3$ - based catalytic systems in propylene polymerization

Catal. system	Temp. °C	$10^2 \times C^*$ max. mol mol^{-1} $TiCl_3$		k_p 1 mol^{-1} s^{-1}		Ref.
		non-iso	iso	non-iso	iso	
$TiCl_3$-H/AlEt$_2$Cl	70	0.07	0.04	0.96	11.3	(13)
$TiCl_3$-H/AlEt$_3$	50	0.05	0.03	6.5	43	(13)
$TiCl_3$-HA/AlEt$_2$Cl	-"-	0.38	0.17	1.3	15.0	(12)
- " -	70	0.60	0.30	2.1	15.3	(13)
$TiCl_3$-HA/AlEt$_3$	30	1.7	0.20	8.2	48	(13)
- " -	50	2.0	0.50	9.8	64	(12)
- " -	70	3.5	1.0	10.2	105	(13)
$TiCl_3$-AA/AlEt$_2$Cl	30	0.50	0.38	0.13	4.0	(13)
- " -	50	0.60	0.60	0.76	8.5	(12)
- " -	70	1.5	1.0	3.8	19.5	(13)

The systems varied widely in the total C^* and in a relative abundance of the centers of different specificities. The following general conclusions (some of them being already reported (12))can be made:

a) the k_p values within one catalytic system are by one order of magnitude higher for isospecific centers than for non-isospecific ones;

b) when the same type of centers in different catalytical systems are compared, the k_p values are much higher for the $AlEt_3$-than for $AlEt_2Cl$ - - cocatalyzed polymerizations;

c) there is no substantial difference in the k_p values for the $TiCl_3$-HA - and $TiCl_3$-AA-containing systems; this finding indicates that the active site structure is not significantly influenced by the presence of $AlCl_3$;

d) increases of both C^* and k_p are observed when polymerization is conducted at a higher temperature; the former phenomenon is probably due to an increased extent of the $TiCl_3$ alkylation by organometal at a higher temperature.

Mechanistic consequences arrived at from these conclusions are briefly discussed in the next Chapter.

Giesemann et al. (14) recently reported on the determination of C^* in ethylene polymerization catalyzed by the $TiBz_4/TiBr_4$ system. They used BuOT for quenching and observed the value of 1.5 for the apparent kinetic isotope effect. The authors compared the $TiBz_4/TiBr_4$ and $TiBz_4/AlBr_3$ systems and found very similar values for both C^* and k_p.

Ulbricht (15) studied ethylene polymerization catalyzed by the alumina supported $ZrBz_4$. The BuOT quenching revealed that up to 10 % of the transition metal was active in the polymerization. The addition of BuLi enhanced the catalyst activity four times due nearly exclusively to the increase of k_p.

(iii) Methods based on selective tagging of growing chains

The methods are based on the assumption (valid at least for some systems) that an efficient poison, such as CO, inserts selectively into the growing bond. The tag thus incorporated is then determined in the polymer. The methods deviced by a group of authors from the Institute of Catalysis, Novosibirsk (16,17), have gained a considerable support during recent years. Certain doubts, however, remain concerning this method (2).

A very interesting paper by Bukatov et al. (18) has revealed that CO used as a selective stopper is consumed progressively via side reactions. If e. g. a $MgCl_2$ supported catalyst is used in propylene polymerization, the extent of side reactions may be 10^2 times higher than that of AC tagging. Low-molecular-weight products thus formed are not removed easily from the isolated polymer and a reprecipitation is imperative to obtain correct C^* data. Similar conclusions concerning the extent of side reactions in CO retarded polymerizations have been arrived at by Caunt et al. (19).

Assuming that the above mentioned side reactions may be suppressed when using less efficient catalyst poisons, we have used CS_2 as a hopefully suitable tagging agent. Its insertion into transition metal-carbon bond is well documented (20,21). The advantage of CS_2 is that an isotopic laboratory is not required and a simple trace sulphur analysis of polymer can be employed. Propylene was used as a model monomer and polymerizations were conducted in heptane medium at $50^{\circ}C$ with TAC-144 $TiCl_3$ at $AlEt_2Cl/TiCl_3$ = 1.0. Details will be found elsewhere (25).

Preliminary experiments revealed that CS_2 reacted negligibly with $AlEt_2Cl$ and that the tagging of "dead", thoroughly deashed PP did not take place. A quantitative proof of the CS_2 insertion into the growing bond could not be obtained through IR and ^{13}C NMR analyses, though some indications of the appearance of S-H and C=S bonds were observed. A rather strong bonding of CS_2 to the polymer after a subsequent quenching the polymerization by alcohol follows from Fig. 1:

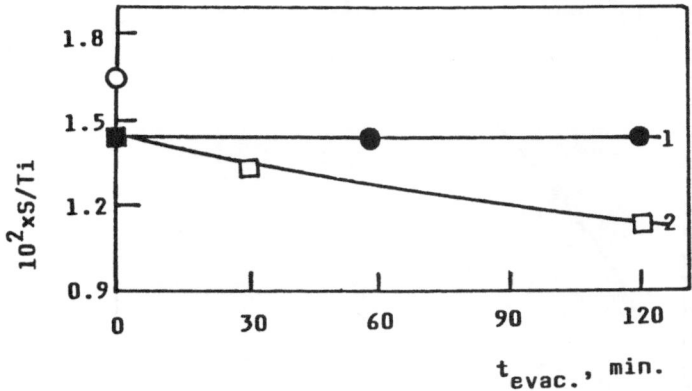

Fig. 1. Influence of evacuation time on sulphur content in polymer. Polymerization conditions: p_M = 83,2 kPa; t_{pol} = 7 min. Retardation conditions: $CS_2/TiCl_3$ = 80; t_{ret} = 20 min.; t_{evac} is counted from pumping-out all solvent. 1 - after alcohol quenching; 2 - without quenching; O - PP sample isolated using standard procedure.

Drying in vacuo after the quenching causes an abrupt decrease of the sulphur content by ca. 15 %. Then the sulphur level keeps constant while non-quenched PP loses the sulphur slowly but steadily. Reprecipitation of the tagged PP did not change the sulphur content confirming that low--molecular-weight sulphur-containing compounds were not present in PP.

Typical time dependences of R_p^r/R_p^o and S/Ti are shown in Fig. 2. A rather fast drop of the polymerization rate upon the addition of CS_2 is observed, followed by a slow rate decrease. The sulphur content exhibits an opposite pattern: a fast increase is followed by a slowing-down in the rate of accumulation of sulphur in the polymer; it cannot be expressed by a simple kinetic law, indicating a non-uniformity of the active center reactivities.

The best fit curve in Fig. 3 can be obtained by processing results of some 50 runs carried out under widely differing conditions. The curve can be interpreted in term of a rather wide distribution of the active site reactivities in polymerization. It is interesting to note that this curve is very similar to that obtained in the allene-retarded polymerization supposing that one molecule of allene consumed is equivalent to one molecule of CS_2 (i. e. two sulphur atoms) incorporated into the polymer.

One of the most important questions is whether the stopper does not influence the characteristics of the catalytic system. In fact, this question should be raised in case of any method where an agent is used to retard (or quench) the polymerization. The only evidence we can bring up to now is that the isospecificity of the centers does not change considerably even upon a strong retardation. This statement is based on the IR determination of the "isotactic index" of PP (22,23), though solubilities of PP in both cold and boiling heptane increased considerably after a strong retardation. The solubility increases are believed to be

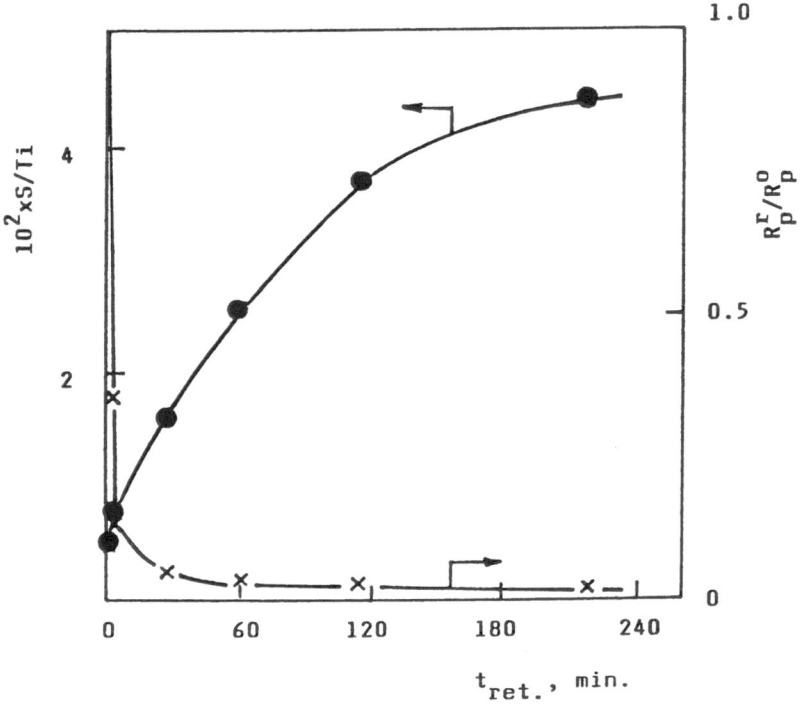

Fig. 2. Influence of CS$_2$ retardation time on sulphur content in PP and on relative polymerization rate. Polymerization conditions: p_M = 35,2 kPa; $t_{pol.}$ = 7 min. Polymerization retarded at CS$_2$/TiCl$_3$ = 135.

due to a much lower MW.

In the absence of monomer, the rate of CS$_2$ incorporation into the polymer is ca. three times lower. As shown in Fig. 4, it is connected with a lower retarding efficiency of CS$_2$ in the absence of propylene. Thus, a promoting ability of the monomer on the active site reaction with CS$_2$ must be postulated.

The following reactions which might account for an increase of the number of tags during the prolonged contact of CS$_2$ with the polymerization system were considered:

a) copolymerization of CS$_2$ with propylene;
b) exchange reaction between the CS$_2$-terminated chain and the organometal (similar as that postulated by Bukatov et al. (18,24) in CO terminated polymerizations):

$$L_x \ Ti-S-\overset{|}{\underset{\underset{S}{\|}}{C}}-\overset{|}{C}\sim\sim \quad + \quad >Al-R \quad --\blacktriangleright \quad L_x \ Ti-R \ + \ >Al-S-\overset{|}{\underset{\underset{S}{\|}}{C}}-\overset{|}{C}\sim\sim$$

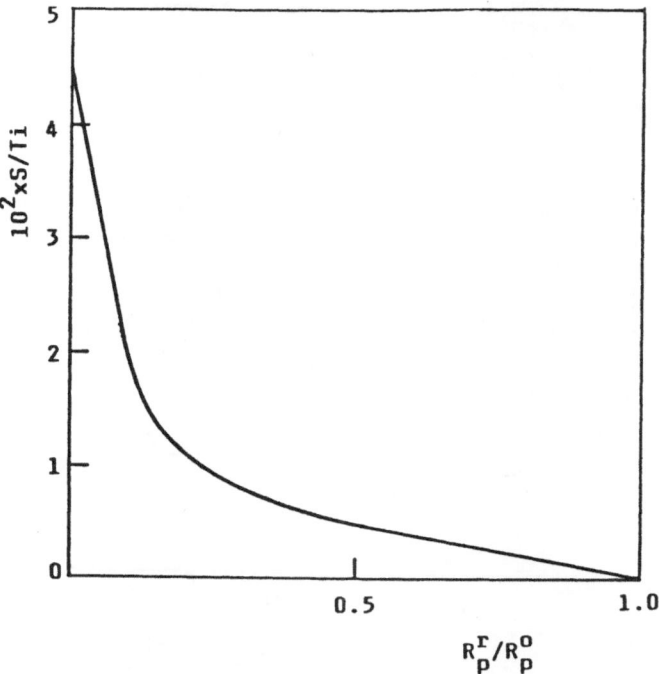

Fig. 3. Best fit curve of dependence of sulphur content in PP on relative polymerization rate under widely varying polymerization and CS_2 retardation conditions.

These reactions can be ruled out based on experimentation reported elsewhere (25). It will be shown in the same paper that the number of tags is not dependent on the elapsed polymerization time before the injection of CS_2. It evidences that CS_2 does not react with non-propagative Al-polymer bonds formed due to the transfer reaction with organometal. The absence of the above three reactions may classify CS_2 as a sound selective agent, at least for the catalytic system under study. If the common kinetic equation for the propylene polymerization rate (R_p = $= k_p C^* [M]$) is used and a total C^* is considered, the average value of $k_p \cong 2.3$ l $mol^{-1}s^{-1}$ can be obtained.

The action of CS_2 in blocking the polymer growth can be visualized either as a consecutive coordination to the active site followed by the insertion into the growing bond

$$L_x Ti - \overset{|}{\underset{|}{C}} + CS_2 \rightleftharpoons L_x \overset{CS_2}{Ti} - \overset{|}{\underset{|}{C}} \longrightarrow L_x Ti - S - \overset{|}{\underset{S}{C}} - \overset{|}{\underset{|}{C}}$$

or as a direct insertion. In the former case, the rate-controlling step may be either coordination or insertion. The following reasoning may help to answer the question of rate-controlling step:

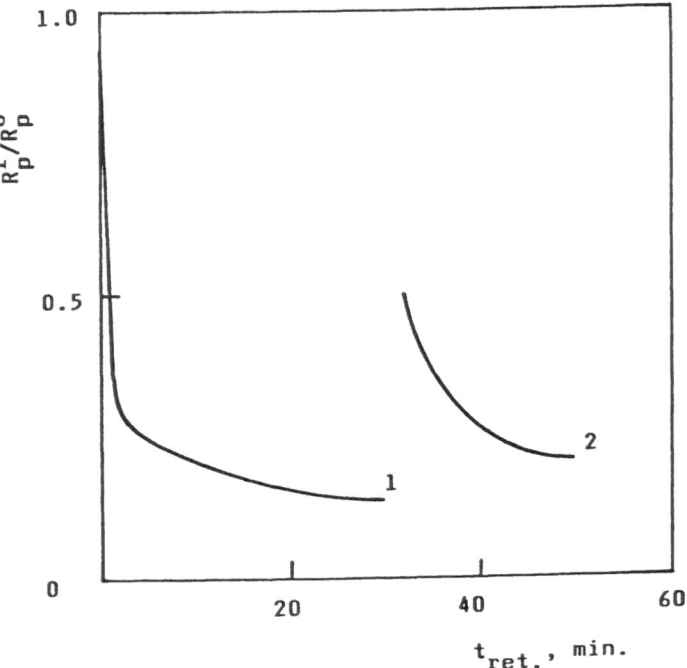

Fig. 4. Dependence of relative polymerization rate on CS_2 retardation time in the presence and absence of propylene. p_M = 83,2 kPa; $t_{pol.}$ = = 7 min. 1 - standard CS_2 retardation in the presence of propylene (CS$_2$/Ti = 100); 2 - propylene pumped-out, then CS_2 added at $t_{ret.}$ = = 0; propylene fed at $t_{ret.}$ = 30 min.

Let us consider as proven that the CS_2 insertion leads to its irreversible incorporation into polymer. As mentioned above, the patterns of CS_2 -
-retarded and allene-retarded polymerizations are very similar. Thus, we can postulate that only inserted CS_2 blocks the chain growth. Then we have to conclude that either the rate-controlling step is the CS_2 coordination or a direct insertion into the growing bond takes place. This being so, the CS_2-retardation method may be used to measure the AC reactivity distribution. In principle, the method can also serve to evaluate the kinetic characteristics of centers of differing stereospecificities, if one fractionates the polymer according its stereoregularity and determines the tags in the isolated fractions.

It may be argued that CO_2 should behave similarly as CS_2. A comparison with earlier data (26), however, reveals the following: Considering similar molar concentrations, CO_2 is much more efficient than CS_2 in retarding the propylene polymerization, while tagging (if any) is very restricted in the CO_2 retarded polymerization. If we rely on very limited data available, we can classify CO_2 as an agent which retards polymerization via coordination to active site (without appreciable insertion) while CS_2 inserts readily.

Consumption Data of Effective Catalyst Poisons

A stepwise retardation of polymerization using a strong catalyst poison
is a procedure characterizing this type of methods (27-29). In essence,
the decrease of the polymerization rate is correlated with the consumption
of the poison. The amount of poison consumed
is determined from a material balance and it is a measure of the number
of active centers. Allene (27-29) and carbon monoxide (28) have been
found to be suitable agents, though the former is more advantageous (19).
The merit of these methods is seen in that they allow a measurement of the
AC reactivity distribution. This characteristic is very important and
it can account for a large difference between data obtained employing
some other methods.

To explain certain kinetic and mechanistic aspects of monomer-poison-
catalyst interaction, a study of the propylene polymerization retardation
by allene was conducted (29). The catalytic system employed was
$TiCl_3 \cdot 1/3\ AlCl_3/AlEt_2Cl$. The main aim of this study was to describe
quantitatively the reactions involved and to use the data obtained to improve the reliability and accuracy of the active center determination.

The physically adsorbed amount of allene on the neat and alkylated
$TiCl_3 \cdot 1/3\ AlCl_3$ was found to be very small (<0.1 mmol $mol^{-1}_{TiCl_3}$ at $50^\circ C$ and
sorbate pressure 0.1 MPa) (30). Thus, it can be neglected in comparison
with the chemisorbed amount of allene (27,29).

The catalytic system $TiCl_3 \cdot 1/3\ AlCl_3/AlEt_2Cl$ induces a slow homopolymerization of allene, the rate of which is, however, by two orders of
magnitude lower than the propylene polymerization rate. If a substantial
amount of propylene is present, copolymer is formed with copolymerization
parameters $r_1 = 0.66 \pm 0.06$ and $r_2 = 40 \pm 30$ for propylene and allene,
respectively[1](29).

The allene retarded propylene polymerization can be visualized as a
competitive chemisorption of the monomer (M) and retarder (I). the polymerization rate (R_p) can be described as a sum of terms relevant to ACs
exhibiting different adsorption coefficients:

$$R_p = k_p' \Sigma C_i^* \frac{K_{1,i}\,[M]}{1 + K_{1,i}\,[M] + K_{2,i}\,[I]}$$

(where C_i^* is the number of i-type ACs; $K_{1,i}$ and $K_{2,i}$ are the adsorption
coefficients for propylene and allene, respectively, on i-type ACs).
If two types of active centers (α, β) are considered, the following
values of the allene adsorption coefficients and the numbers of active
centers can be calculated using the best fit method:

$$C_\alpha^* = 2\ \text{mmol mol}^{-1}_{TiCl_3} \qquad\qquad C_\beta^* = 20\ \text{mmol mol}^{-1}_{TiCl_3}$$

$$K_{2,\alpha} = 5.4 \times 10^4\ \text{l mol}^{-1} \qquad\qquad K_{2,\beta} = 400\ \text{l mol}^{-1}$$

When a more common kinetic equation for the polymerization rate
($R_p = k_p\ C^*\ [M]$) is used, the average value of k_p is 2.2 $\text{l mol}^{-1}\text{s}^{-1}$.

As to the methods based on the assumption that an effective poison reacts with ACs quantitatively, the following remarks should be made: From the classification point of view these procedures belong to the "consumption data methods". Considering a high extent of side reactions involved in the CO stopped polymerizations (18,19), the assurance that all CO has been used up (31) is not a proof that the agent reacts with ACs selectively. The C^* data thus obtained should be considered as their upper limit. If a quantitative consumption is even not documented, such as in an early work using allene (32), the measurement of meaningful C^* values is rendered impossible.

Other Methods

A short list of non-typical methods escaping from the classification scheme (2) can be enlarged by a method devised by Ammendola et al. (33). It is based on monitoring insertion of monomer units using the ^{13}C NMR analysis of E-P block copolymers. The k_p values of both ethylene and propylene polymerizations catalyzed by the $TiCl_3/AlMe_3$ system could be obtained. Rather high k_p values for the propylene polymerization were reported. A clear tendency of higher k_p values for more stereoregular chains was documented.

DISCUSSION

As shown above and in foregoing papers, each method has its merits and deficiencies. In addition to it, different methods may be determining non-equivalent species. The methods, however, have not been devised to be an end in themselves but rather as tools to answer basic kinetic and mechanistic questions. We would like to point to some of the problems which remain to be solved using the above methods.

A challenging task is to measure the AC reactivity distribution. The poison consumption data method allows such a measurement. Under certain conditions also the selective tagging method may serve this purpose.

A distribution of ACs according to their specificities is measurable using all the methods except for those based on the poison consumption data. The controversial views concerning the relevance between the k_p values and the center specificities have not been reconciled (cf. discussion in (2) and (34)). During the last years, however, a prevailing evidence has accumulated that

$$(k_p)_{iso} > (k_p)_{non-iso}$$

A never ending task is to relate a location and structure of an AC to its kinetic features. E. g. it has been almost generally accepted that the supported catalysts exhibit much higher k_p values that classical (such as $TiCl_3$-based) systems. This view, however, is not shared by Ammendola et al. (33) who found similar (and high) k_p values for both types of catalysts.

Organometal type may influence the kinetic performance of a particular type of AC. If such an influence is substantiated, a bimetallic mechanism of the chain growth is likely to be operative.

To give an example of the state of the art in the AC and k_p measurement, we collected and processed some recent data on propylene polymerization catalyzed by ($TiCl_4/MgCl_2$ + AlR_3) systems (Table 3). As seen, the k_p

Table 3. Selected k_p data for propylene polymerization catalyzed by $(TiCl_4/MgCl_2 + AlR_3^p)$ - based system. The k_p data refer to isotactic polymer and are related to $60°C$ and where necessary a tentative $E_p = 9.4$ kcal mol^{-1} (9) was used. When only k_p data for total polymer were reported, it was assumed that $(k_p)_{iso} > (k_p)_{total}$

Catalytic system	Exp. temp. °C	Method type	$10^{-2} \times (k_p)_{iso}$ at $60°C$ $1\ mol^{-1}s^{-1}$	Ref.
$TiCl_4/MgCl_2$(CW cat.) + $AlEt_3$ + MPT	50	Sel.tag.(^{14}CO)	1.3	(35)
- " -	- " -	MPB (MeOT)	2.5	(35)
$TiCl_4/MgCl_2 + AlEt_3$	70	Sel.tag.(^{14}CO)	4.9 - 5.4	(36)
- " -	- " -	- " -	>5.3	(37)
$TiCl_4.EB/MgCl_2 + AlEt_3$	- " -	- " -	5.7 - 8.3	(36)
$TiCl_4/MgCl_2/EB + AlEt_3$	41	Total cons.(CO)	>7.3	(31)
- " -	38	- " -	>9.1	(31)
$TiCl_4/MgCl_2/EB + AliBu_3$	60	Sel.tag.(^{14}CO)	12.6	(34)
$TiCl_4/MgCl_2 + AlEt_3$	- " -	N	5 - 15	(38)
$TiCl_4/MgCl_2/EB + AlEt_3 + EB$	- " -	N	>27	(39)
$TiCl_4/MgCl_2 + AlEt_3 + EB$	- " -	N	21 - 63	(40)
$TiCl_3/AlMe_3$ [a]	22	Kinetics based on ^{13}C NMR	49	(33)
$TiCl_4/MgCl_2/EB + AlEt_3 + EB$	50	N	>50	(41)
$TiCl_4/MgCl_2/EB + AlEt_3$	0-40	N	>85	(9)

[a] Listing of this case is based on the authors' statement that the $(TiCl_4/MgCl_2 + AlMe_3)$ system exhibits similar k_p values as the $TiCl_3/AlMe_3$ one; k_p value for fraction of E-P block copolymer insoluble in heptane was considered to be relevant to $(k_p)_{iso}$.

values ranging from 130 to >8600 $1\ mol^{-1}s^{-1}$ were reported for apparently similar systems. The methods employed should not be blamed for such a wide range of values: the upper edge of the range relies on usage of N-methods which tend to give lower k_p values that those based on selective tagging. Assuming that the reported k_p data do not suffer from methodical errors (though this condition may not necessarily be met), a large divergency of k_p values should be assigned to either of the followings:
a) The systems investigated differ so widely from one another as indicated by k_p values;
b) The systems exhibit so wide a distribution of AC reactivities that the k_p data obtained are much dependent on particular conditions of the AC determination.

Certainly, both cases may combine to make things even worse. This reasoning is not intended to disseminate pessimism but rather to call for a more precise experimentation, considering all possible pitfalls, and for a careful evaluation of the data obtained. It is our belief that this challenge will be overcome during the next decade.

REFERENCES

1. Minsker KS, Karpasas MM, Zaikov GE (1986) J Macr Sci- Rev Macromol Chem Phys C27: 1-90
2. Mejzlík J, Lesná M, Kratochvíla J (1986) Adv Polym Sci 81: 83-120
3. Natta G, Pasquon I (1959) Adv Catal 11: 1-66
4. Ketley AD, Mayer JD (1963) J Polym Sci Part A1: 2467-2476
5. Atarashi Y (1970) Polym Sci Part A-1, 8: 3359-3366
6. Ayrey G, Mazza RJ (1975) Makromol Chem 176: 3353-3370
7. Burfield DR (1978) J Polym Sci, Polym Chem Ed 16: 3301-3305
8. Natta G (1959) J Polym Sci 34: 21-48
9. Terano M, Kimura K, Ishii K, Keii T (to be published)
10. Feldman CF, Perry E (1960) J Polym Sci 46: 217-231
11. Chien JCW, Kuo CI (1985) J Polym Sci, Polym Chem Ed 23: 731-760
12. Mejzlík J, Lesná M (1987) in: Quirk RP (ed) Transition Metal Catalyzed Polymerization, Cambridge Univ Press, New York
13. Lesná M, Mejzlík J (to be published)
14. Giesemann J, Ernst E, Ernst A, Ulbricht J (1986) Makromol Chem 187: 1737-1744
15. Ulbricht J, paper presented at the 31th IUPAC Macromol Symp (1987) Merseburg
16. Zakharov VA, Bukatov GD, Ermakov YuI (1971) Kinet Katal 12: 263
17. Ermakov YuI, Zakharov VA, Bukatov GD (1973) in: Hightower JW (ed) Proc 5th Int Congr Catal, North Holland/American Elsevier, Amsterdam New York, p 399
18. Bukatov GD, Goncharov VS, Zakharov VA (1986) Makromol Chem 187: 1041-1051
19. Caunt AD, Davies S, Tait PJT (1987) in: Quirk RP (ed) Transition Metal Catalyzed Polymerization, Cambridge Univ Press, New York
20. Chandra G, Jenkins DA, Lappert FM (1970) J Chem Soc A: 2550-2558
21. Kukushkin NJ, Danilina IL (1980) Koord Khim 7: 163-200
22. Heinen W (1959) J Polym Sci 38: 545-547
23. Majer J (1961) Coll Czech Chem Commun 26: 1756-1762
24. Bukatov GD, Zakharov VA, Ermakov YuI (1978) Makromol Chem 179: 2097-2101
25. Vozka P, Mejzlík J (to be published)
26. Mejzlík J, Lesná M, Majer J (1983) Makromol Chem 184: 1975-1985
27. Caunt AD (1981) Br Polym J 13, 22-26
28. Abu-Eid M, Davies S, Tait PJT (1983) Polym Prepr (Am Chem Soc, Div Polym Chem) 24: 114-115
29. Kratochvíla J, Mejzlík J (1987) Makromol Chem 188: 1781-1794
30. Kratochvíla J, Mejzlík J (1987) Makromol Chem 188: 1773-1779
31. Doi Y, Murata M, Yano K (1982) Ind Eng Chem Prod Res Dev 21: 580-585
32. Ambrož J, Hamřík O (1963) Coll Czech Chem Commun 28: 2550-2555
33. Ammendola P, Zambelli A, Oliva L, Tancredi T (1986) Makromol Chem 187: 1175-1188
34. Tait PJT (1986) in: Keii T, Soga K (eds) Studies in Surface Science and Catalysis 25, Catalytic Polymerization of Olefins, Kodansha, Elsevier, Tokyo, Amsterdam - Oxford - - New York - Tokyo, p 305
35. Chien JCW, Kuo CI (1985) J Polym Sci, Polym Chem Ed 23: 761-786
36. Bukatov GD, Shepelev SH, Zakharov VA, Sergeev SA, Ermakov YuI (1982) Makromol Chem 183: 2657-2665
37. Ermakov YuI (1981) in: Ciardelli F, Giusti P (eds) Int Union Pure Applied Chem, Structural Order in Polymers, Pergamon Press, Oxford and New York, p 37-50
38. Kashiwa N, Yoshitake J (1983) Makromol Chem, Rapid Commun 4: 41-44
39. Kashiwa N, Yoshitake J (1982) Makromol Chem, Rapid Commun 3: 211-214
40. Kashiwa N, Yoshitake J (1984) Polym Bull 12: 99-104
41. Kashiwa N, Yoshitake J (1984) Polym Bull 11: 479-484

Active Centers of the Supported Organic and Hydride Transition Metal Catalysts for Ethylene Polymerization

V.A. Zakharov, G.A. Nesterov, S.A. Vasnetsov and K.H. Thiele*

Institute of Catalysis, Novosibirsk 630090, U.S.S.R.

INTRODUCTION

Catalytic polymerization of olefins is one of the fields of catalysis in which the concept of similarity of action mechanisms of homogeneous and heterogeneous catalysts has long been accepted. This similarity is based on the idea that in both homogeneous and heterogeneous catalysts the active center is an organometallic compound of
a transition element having a metal-alkyl bond via which the polymer
chain propagation occurs. On this basis a general approach to the preparation of heterogeneous polymerization catalysts is an intentional
synthesis of surface organometallic compounds which are active sites
of these systems. One of the methods of synthesis of such compounds
consists of the interaction of organic complexes with the L_xMR_y composition (M is a transition element, R is an organic ligand, and L
is an inorganic ligand) with surface hydroxyl groups of silica or
alumina. Using this approach, efficient supported catalysts for
ethylene polymerization have been obtained (Yermakov et al. 1972;
Karol et al. 1972; Ballard 1973; Yermakov et al. 1981). These catalysts contain surface organometallic compounds (I) of the type
$(E-O)_n-ML_xR_m$ (E = Si, Al; M = Ti, Zr, Hf, Cr; L is halide; R is allyl, benzyl, tetramethylsylil etc.; n = 1-3; x = 0-2; m = 1-3) as
active components. It might be expected that surface organometallic
compounds (I) serve as direct active sites of these systems. However,
such a simple approach is inconsistent with a series of experimental
data obtained:
1. The number of active sites in the systems of interest is significantly smaller than the total number of surface organometallic
complexes (I) (Yermakov et al. 1981, Ballard 1975). In many situations
the activity of these systems can appreciably be enhanced using additional activation by heating, hydrogen treatment or UV-irradiation
(Yermakov et al. 1981). In a number of cases the high activity is
revealed by the systems in which surface compounds of transition metals do not have initial organic ligands at all.

One may think that in these systems only part of surface organometallic compounds transforms to active sites of the polymerization process. Another serious disadvantage of these systems is low activity
and low stereospecificity in propylene polymerization (Ballard 1975).

*Technische Hochschule Carl Schorlemmer, Merseburg, D.D.R.

W. Kaminsky and H. Sinn (Eds.)
Transition Metals and Organometallics as
Catalysts for Olefin Polymerization
© Springer-Verlag Berlin Heidelberg 1988

Despite these restrictions, the supported organometallic catalysts have some advantages over traditional and supported Ziegler catalysts, since they provide the possibility of intentional synthesizing the active sites and applying physical methods for the determination of compositions of surface compounds and mechanistic studies of the catalytic polymerization process.

In this paper we discuss literature and our new data on the composition of surface hydride compounds formed during the additional activation of the catalysts, their reactivity as well as on the influence of Group IVB transition metal upon catalytic properties of these systems for ethylene polymerization.

RESULTS AND DISCUSSION

1. Synthesis and Reactivity of Surface Titanium, Zirconium
   and Hafnium Hydrides

The most important reaction that occurs upon activation of supported catalysts based on organic and tetrahydroborate compounds of Ti, Zr and Hf is formation of surface hydrides M(IV) and M(III) of these elements (Zakharov et al. 1977; Nesterov et al. 1986). The surface hydrides were identified using chemical and IR-spectroscopic methods. By way of example see in Fig. 1-4 IR spectra for the $Hf(CH_2Ph)_4/SiO_2(Al_2O_3)$, $Zr(CH_2Ph)_4/SiO_2$ and $Zr(C_7H_{11})_4/SiO_2$ catalysts which have never been presented before in the literature. In all systems the treatment of catalysts with hydrogen gives rise to the appearance of surface hydrides of hafnium and zirconium, as ascertained by IR-studies of deutero-hydrogen exchange and water adsorption. IR data on surface hydrides obtained earlier (Zakharov et al. 1977) and in this work are listed in Table 1. Note that optimal conditions for the formation of surface hydrides correspond to the activation conditions that provide a maximum catalytic activity. The position of absorption bands characteristic of surface Ti, Zr and Hf hydrides is independent of the nature of an organic ligand in the starting compound and is determined only by the nature of a metal (Table 1).

Table 1. IR spectroscopy data on surface Ti, Zr and Hf hydrides
          (catalysts were activated by hydrogen treatment at 150°C)

Catalyst	γ_{M-H}, cm^{-1}
TiR_4/SiO_2 [a]	1560
ZrR_4/SiO_2 [b]	1625
HfR_4/SiO_2 [c]	1670
$Hf(CH_2Ph)_4/Al_2O_3$	1600-1700

a R = benzyl, naphthyl.
b R = allyl, benzyl, naphthyl and norbornyl.
c R = benzyl, allyl.

For example, in all cases the treatment of ZrR_4/SiO_2 (R = allyl, benzyl, norbornyl, naphthyl) with hydrogen results in surface hydrides of zirconium characterized by an absorption band at 1625 cm^{-1} in

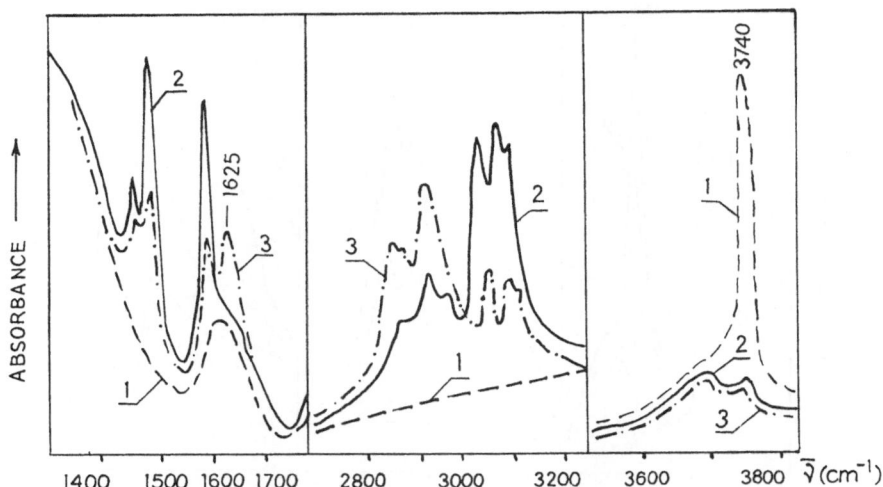

Fig. 1. IR spectra for the $Zr(CH_2Ph)$ /SiO_2 catalyst: 1 -- initial SiO dehydroxylated at 700°C; 2 -- $Zr(CH_2Ph)_4$ /SiO_2; sample (2) treated with H_2 at 120°C -- 3

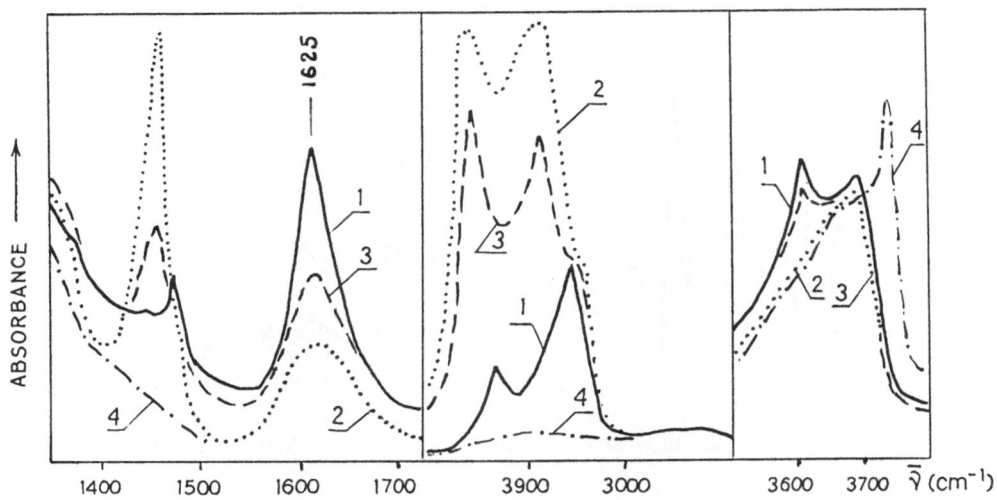

Fig. 2. IR spectra for the $Zr(C_7H_{11})$ /SiO_2 catalyst: $Zr(C_7H_{11})$ /SiO_2 heated in H_2 at 150°C -- 1; adsorption of C_2H_4 (10 Torr) on sample (1) for 10 min at 20°C -- 2; sample (2) heated in H_2 at 150°C -- 3; initial SiO_2 dehydroxylated at 450°C -- 4; C_7H_{11} -- norbornyl

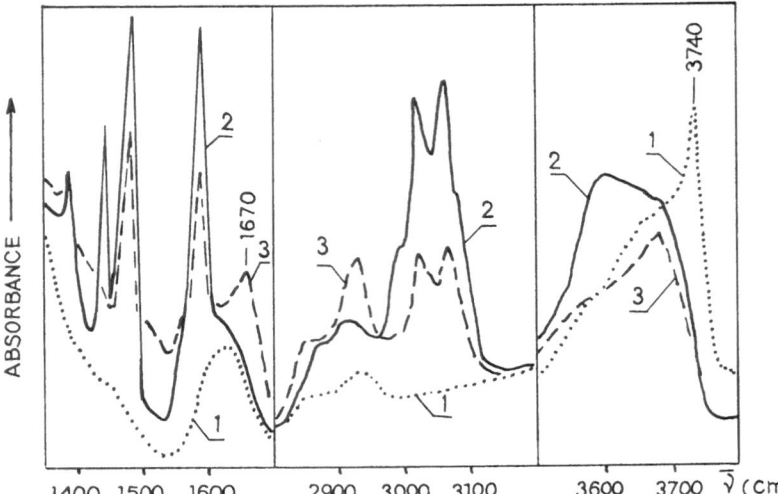

Fig. 3. IR spectra for the $Hf(CH_2C_6H_5)_4/SiO_2$ catalyst: SiO_2 dehydroxylated at 450°C -- 1; $Hf(CH_2C_6H_5)_4/SiO_2$ -- 2; sample (2) heated in H_2 at 150°C -- 3

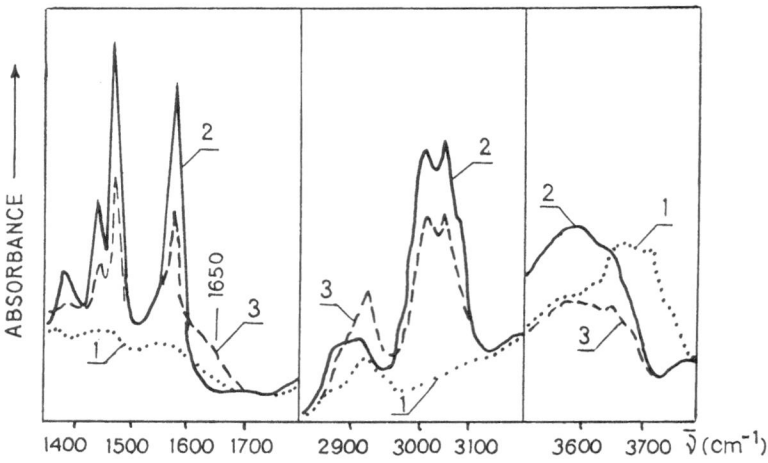

Fig. 4. IR spectra for the $Hf(CH_2C_6H_5)_4/Al_2O_3$ catalyst: initial Al_2O_3 dehydroxylated at 450°C -- 1; $Hf(CH_2C_6H_5)_4/Al_2O_3$ -- 2; sample (2) treated with H_2 at 150°C -- 3

the IR spectrum.

As has already been mentioned, the activation of supported organo-metallic catalysts by hydrogen treatment enhances their activity. As can be seen in Table 2, the increase in catalytic activity is associated with the increase in the number of active sites.

Table 2. The number of active sites (C_p) and propagation rate constants (K_p)[1] for catalysts in the initial state and after treatment with hydrogen. (80°C, ethylene pressure 5 atm)

Catalyst	Activation	Inhibitor	Activity, $\dfrac{g\ C_2H_4}{mmol\ Zr\ h}$	$C_p \cdot 10^2$, $\dfrac{mol}{mol\ Zr}$	$K_p \cdot 10^{-3}$, $\dfrac{mol}{mol\ s}$
$Zr(C_3H_5)_4/SiO_2$	–	CH_3O^3H	200	$2.7 \cdot \gamma$ [a]	$0.23/\gamma$
	H_2, 150°C	– " –	1200	$10 \cdot \gamma$	$0.32/\gamma$
$Zr(C_3H_5)_3Cl/SiO_2$	–	^{14}CO	650	1.1	1.9
	H_2, 150°C	^{14}CO	1300	2.1	2.0

a γ is kinetic isotope effect (γ = 1–3).

Surface hydrides are formed most easily when using silica as a support. Regarding alumina, surface hydrides of Ti, Zr and Hf were identified only for the systems based on tetrahydroborate (Nesterov et al. 1986) as well as for the $Hf(CH_2Ph)_4/Al_2O_3$ system (Table 1). Correspondingly, only for these latter catalysts an appreciable effect of additional catalyst activation on catalytic activity towards ethylene polymerization is observed (Tables 3, 4).

IR data evidence that surface hydrides of Ti, Zr and Hf interact with ethylene even at 130-170 K producing surface alkyl compounds which are active sites of ethylene polymerization. Further polymerization (ethylene insertion into the metal-alkyl bond) occurs at higher temperatures (200-230 K).

$$L_xM\text{-}H \xrightarrow[120\text{-}170\ K]{C_2H_4} L_xM\text{-}C_2H_5 \xrightarrow[200\text{-}230\ K]{n(C_2H_4)} L_xM\text{-}(C_2H_4)_n\text{-}C_2H_5 \quad (1)$$

Thus, surface Ti, Zr and Hf hydrides reveal a higher reactivity towards ethylene addition than the corresponding alkyl complexes. Note that in many cases for the polymerization of olefins with Ziegler catalysts the decrease in activity was observed when using hydrogen as chain transfer. Traditionally such a decrease is explained by the low reactivity of intermediate hydride complexes of titanium.

---

1 Data were obtained by G.D. Bukatov.

Table 3. Data on the activity of catalysts obtained using Ti, Zr and Hf tetrabenzyl complexes. (80°C, ethylene pressure 5 atm)

M	Support	M, mmol/g cat.	Activation conditions	Proportion of M(III)[a], molar %	Activity, $\frac{g\ PE}{mmol\ M \cdot h}$
Ti	SiO_2	0.44	–	0.5	70
– " –	– " –	– " –	H_2, 130°C	5	150
Zr	SiO_2	0.22	–	0.5	80
– " –	– " –	– " –	H_2, 130°C	2	200
Hf	SiO_2	0.48	–	0	20
– " –	– " –	– " –	H_2, 120°C	0.25	100
Ti	Al_2O_3	0.26	–	0.5	1850
– " –	– " –	– " –	H_2, 130°C	1.5	1600
Zr	Al_2O_3	0.22	–	0.5	7500
– " –	– " –	– " –	H_2, 130°C	1.0	10000
Hf	Al_2O_3	0.41	–	0	800
– " –	– " –	– " –	H_2, 120°C	0.5	2800

a as ascertained by ESR.

Table 4. Data on the activity of the catalysts obtained using tetrahydroborate complexes of Ti, Zr, Hf (M). (80°C, ethylene pressure 5 atm)

Catalyst	M $\frac{mmol}{g\ cat.}$	Activation temperature, °C	Proportion of M(III)[a], molar %	Activity $\frac{g\ PE}{mmol\ M \cdot h}$
$Ti(BH_4)_3/SiO_2$	0.34	150	100	230
$Zr(BH_4)_4/SiO_2$	0.38	220	5	1500
$Hf(BH_4)_4/SiO_2$	0.38	220	1	500
$Ti(BH_4)_3/Al_2O_3$	0.24	150	100	1100
$Zr(BH_4)_4/Al_2O_3$	0.28	220	5	4000
$Hf(BH_4)_4/Al_2O_3$	0.28	220	0.5	1400

a as ascertained by ESR.

The experimental data mentioned above indicate that such an explanation is not valid.

Surface hydrides of Ti, Zr and Hf formed in $M(CH_2Ph)_4/SiO_2/H_2$ system are also highly reactive towards interaction with carbon monoxide. As follows from IR data, at 100–170 K complexes of carbon monoxide with a transition metal ion are formed; these are characterized by the bands of the adsorbed carbon monoxide at 1970-2080 cm^{-1}. With increasing temperature of the interaction to 270 K carbon monoxide inserts into the M-H bond producing surface formyl complexes characterized by absorption bands at 1490-1540 cm^{-1}:

$$L_xM-H \xrightarrow{\text{CO}} L_x\overset{\text{CO}}{\underset{\downarrow}{M-H}} \longrightarrow L_xM-\overset{}{\underset{\underset{O}{\|}}{CH}} \quad (2)$$

In this case the carbon monoxide insertion into the M-H bond also occurs at lower temperatures (270 K) than the insertion into the metal-benzyl bond (300 K).

Surface hydrides of zirconium formed in the $Zr(CH_2Ph)_4/SiO_2/H_2$ system show a high activity towards isomerization and hydrogenation of olefins (Schwarts and Ward 1980). Recently it has also been established that surface titanium and zirconium hydrides formed in the $Ti(CH_2Ph)_4/SiO_2/H_2$ and $Zr(C_3H_5)_4/SiO_2/H_2$ systems are as highly active towards hydrogenation of benzene and cyclohexene as supported metals of Group VIII (Alekseev et al. 1987). According to the thermodesorption, ESR and EXAFS data obtained in this work the MR_4/SiO_2 systems treated with hydrogen at 423-523 K contain surface M(III)-H and M(IV)-H complexes which are chemically bound to the surface oxygen of the support.

Thus, surface Ti, Zr and Hf hydrides are characterized by a unique combination of a high thermal stability (200-300°C) and a high reactivity towards insertion of ethylene and carbon monoxide as well as hydrogenation of olefins, benzene and cyclohexene.

2. Influence of the Metal Nature on Catalytic Properties

To elucidate the influence of the nature of transition metal on catalyst composition and properties, Ti, Zr and Hf tetrabenzyl and tetrahydroborate complexes supported on silica or alumina were examined. The catalytic properties of these systems can be compared with thermal stabilities of surface compounds as well as with metal-carbon bond energies for Group IVB metals. Thermal stabilities of surface compounds, obtained by interacting benzyl and tetrahydroborate complexes of Ti, Zr and Hf with silica and alumina, can be estimated from the contents of M(III) compounds which form from the starting M(IV) compounds at the step of catalyst preparation and at the subsequent steps of additional activation. Data listed in Tables 3 and 4 indicate that stability of surface compounds increases in the order: Ti < Zr < Hf which corresponds to thermal stability of starting MR_4 compounds. As is known, the M-R binding energy increases in the same order (Table 5).

For the $M(CH_2Ph)_4/Al_2O_3$ and $M(BH_4)_4/Al_2O_3$ catalysts the activity towards ethylene polymerization changes in the order: Ti $\simeq$ Hf < Zr (Tables 3 and 4). The $M(CH_2Ph)_4/SiO_2$ catalysts in the initial form have close and low activities (Table 3). However, when activated, these catalysts show a markedly increased activity, the maximum activity being revealed by the zirconium-containing catalyst. The $M(BH_4)_n/SiO_2$ catalysts are initially inactive. Only after an additional activation by heating they show quite a high activity which increases in the order: Ti < Hf < Zr. Note that for a number of Ti-, Zr- and Hf-based systems data, close to ours, on the increase in activity when passing from titanium- to zirconium-organic catalysts were reported. In particular, during propylene polymerization on the $M(CH_2Ph)_4/Al_2O_3$ catalyst the activity was found to increase in the order: Ti < Zr < Hf (Ballard 1975). In the case of ethylene polymerization on homogeneous catalysts with the composition $(C_5H_5)_2MR_2$ + methylalumoxane the activity increases in the order: Ti < Hf < Zr (Kaminsky and Miri 1986).

Table 5. Data on M-C σ-bond energies for organometallic compounds MR_4 of Group IVB

Compound	Binding energy, kcal/mol	References
$Ti(CH_2Ph)_4$	63	Telnoj et al. 1977
$Zr(CH_2Ph)_4$	74	Lappert et al. 1975
$(C_5H_2)_2Ti(CH_3)_2$	60	Telnoj et al. 1977
$(C_5H_5)_2Zr(CH_3)_2$	65.6	Lappert et al. 1975
$Ti[CH_2C(CH_3)_3]_4$	44	Telnoj et al. 1977
$Zr[CH_2C(CH_3)_3]_4$	60	Lappert et al. 1975
$Hf[CH_2C(CH_3)_3]_4$	65.6	Lappert et al. 1975

According to the estimates of the number of active sites (C_p) and propagation rate constants (K_p) for ethylene polymerization on titanium- and zirconium-organic catalysts with the composition MR_4/Al_2O_3, K_p makes up 1×10^3 l/mol·s and 2.3×10^3 l/mol·s, respectively (Yermakov et al. 1981). Thus, the higher activity of zirconium-organic catalysts in comparison with titanium-organic catalysts is due to both the higher (2-4 times) number of active sites and the higher (ca. 2 times) value of K_p. The change in reactivity of active sites of these systems upon varying the nature of transition metal affects also the reactions of the polymer chain termination which determine the polyethylene molecular mass. Particularly, for ethylene polymerization at 170°C with the MR_4/Al_2O_3 catalysts the polyethylene molecular mass decreases in the order: Ti > Zr > Hf (Yermakov et al. 1981). For ethylene polymerization at 80°C in the presence of hydrogen the molecular mass changes in the order: Ti > Zr $\simeq$ Hf (Table 6). Data on the influence of the nature of transition metal on the molecular mass and melt index of polyethylene indicate that rate constants of reactions of the polymer chain termination tend to increase when passing from Ti to Zr.

Data on kinetics and mechanisms of the catalytic polymerization of olefins are customarily interpreted on the assumption that the reactivity of the active site is determined primarily by the properties of the metal-carbon active bond. It is assumed that the weakening of this bond should leads to the increase of the propagation and polymer chain transfer rate constants and of the catalytic activity.

Table 6. Melt indices of polyethylene for catalysts of different composition. (Polymerization at 80°C, H_2 content 40 vol.%)

Catalyst	Melt index, g/10 min
$Ti(CH_2Ph)_4/SiO_2$	0.1
$Zr(CH_2Ph)_4/SiO_2$	0.85
$Ti(CH_2Ph)_4/Al_2O_3$	0.01
$Zr(CH_2Ph)_4/Al_2O_3$	0.25
$Hf(CH_2Ph)_4/Al_2O_3$	0.20

However a comparison of the above data on the catalytic properties of supported organometallic catalysts with those on the metal-carbon bond energies (Table 5) indicates the absence of the direct correlation between the activity of catalysts, containing surface Ti, Zr and Hf organometallic complexes as active sites, and the metal-carbon bond energies which increase in the order: Ti < Zr < Hf. Hence, the catalytic activity is not determined directly by the properties of the transition metal-carbon active bond. In the general case, as follows from the theoretical calculations (Zakharov 1982), the absence of the correlation between the propagation rate constant and the metal-alkyl bond energy is due to the fact that upon insertion of the coordinated olefin into this bond the activation energy is determined primarily by the energy of the rupture of the π-bond of the coordinated olefin in its transient state. Besides, one should take into account that in most cases the catalytic activity depends upon the number of active sites. In this case the key step is the creation of conditions, providing an efficient coordination of ethylene prior to its insertion into the metal-alkyl active bond, rather than the activation of this bond (Zakharov 1983). Thus, the main method of increasing the catalytic activity is that via increasing the number of coordinatively unsaturated active sites that can effectively coordinate to the olefin.

REFERENCES

Alekseev OS, Volodin AM, Kochubey DI, Yudanov VF, Ryndin YuA, Yermakov YuI (1987) in: Preprints of the 2nd Soviet-Ital. Symposium on Catalysis in Solution of Energy Problems. Institute of Catalysis, Novosibirsk, pp 1-24
Ballard DGH (1973) Advan Catal 23:263-325
Ballard DGH (1975) J Polym Sci Polym Chem Ed 13: 2191-2212
Kaminsky W, Miri MH (1986) in: Homogeneous and heterogeneous Catalysis. VNU Science Press BV, Amsterdam, pp 327-342
Karol FI, Karapinka GL, Wu Ch, Dow AW, Johnson RN, Carrick WL (1972) J Polym Sci A-1 10:2621-2637
Lappert MF, Patil DS, Pedley JB (1975) J Chem Soc Chem Commun   830-

831
Nesterov GA, Zakharov VA, Volkov VV, Myakishev KG (1986) J Mol Catal
  36: 253-269
Schwarts JC, Ward HD (1980) J Mol Catal 8:465-473
Telnoj BI, Rabinovich IB (1977) Russ Chem Rev 46: 1337-1351
Yermakov YuI, Lazutkin AM, Demin EA, Zakharov VA (1972) Kinet Katal
  (Russ.) 13: 1422-1427
Yermakov YuI, Kuznetsov BN, Zakharov VA (1981) Catalysis by supported
  complexes. Elsevier, Amsterdam, pp 121-182
Zakharov VA, Dudchenko VK, Paukshtis EA, Yermakov YuI (1977) J Mol
  Catal 2: 421-435
Zakharov II, Zakharov VA (1982) J Mol Catal 14: 171-184
Zakharov II, Zakharov VA (1983) React Kinet Catal Lett 23: 61-66

COPOLYMERIZATION OF ETHYLENE WITH α-OLEFINS BY HIGHLY ACTIVE SUPPORTED CATALYSTS OF VARIOUS COMPOSITION

G.D. Bukatov, L.G. Yechevskaya and V.A. Zakharov

Institute of Catalysis, Novosibirsk 630090, USSR

INTRODUCTION

Copolymerization of ethylene with α-olefins is an important route for modifying of polyethylene properties and for obtaining of new materials based on ethylene copolymers (linear low-density polyethylene, ethylene-propylene elastomers et al.). The wide posibilities for that are opened by using of highly active catalysts of various composition (1-8).

This communication reports on the comonomer reactivity ratios for slurry and gas-phase copolymerization of ethylene with propylene, 1-butene and 1-hexene on highly active supported catalysts. Microstructure of ethylene-propylene copolymers, obtained with these catalysts, is studied by means of ^{13}C NMR spectroscopy.

RESULTS AND DISCUSSION

Slurry Copolymerization

With low concentrations of α-olefin in the copolymer the copolymerization equation can be simplified as follows (2):

$$(C_\alpha / C_2H_4)_{pol.} = \frac{1}{r_1} \left[C_\alpha \right] / \left[C_2H_4 \right] \qquad (1)$$

Here $(C_\alpha / C_2H_4)_{pol.}$ is the mole ratio of α-olefin to ethylene in the polymer; $\left[C_\alpha \right]$ and $\left[C_2H_4 \right]$ are the concentrations of α-olefin and ethylene in the reaction medium (hexane); $r_1 = K_{11}/K_{12}$ is the ratio of copolymerization constants (1-ethylene, 2-α-olefin).

Data for ethylene copolymerization with propylene, 1-butene, 1-hexene over various catalysts are presented in Fig. 1. Based on these data and equation (1) the reactivity ratio r_1 has been determined (Table 1).

Values of r_1 depend on the composition of both catalyst and comonomer. Thus, for copolymerization over titanium/magnesium (TMC) and vanadium/magnesium (VMC) catalysts the copolymerization ability of α-olefins, which is inversely proportional to r_1, decreases in the sequence: propylene > 1-butene > 1-hexene. In the case of copolymerization of ethylene with propylene copolymerization ability of the catalysts decreases in the sequence: VMC > TiCl$_3$-AA > TMC. Higher copolymerization ability of VMC compared to TMC is retained for all α-olefins (Table 1). Similar character of variations in the copoly-

W. Kaminsky and H. Sinn (Eds.)
Transition Metals and Organometallics as
Catalysts for Olefin Polymerization
© Springer-Verlag Berlin Heidelberg 1988

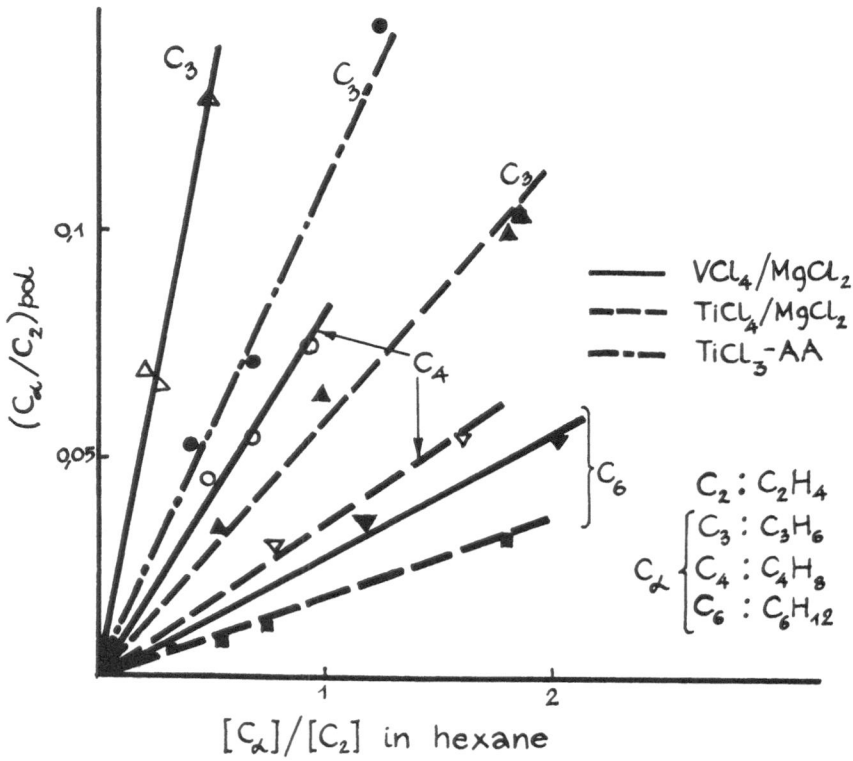

Fig. 1. Plot of the comonomer mole ratio in copolymer vs. comonomer mole ratio in hexane; 70°C, [Catalyst] = 0.03-0.1 g/l, [Al(i-Bu)$_3$] = 4 mmol/l, 3-18 vol.% of H$_2$

Table 1. Comonomer reactivity ratios (r_1) for ethylene-α-olefin slurry copolymerization

Catalyst	C_2+C_3	C_2+C_4	C_2+C_6
VCl$_4$/MgCl$_2$	3.7	12.7	38.5
TiCl$_3$-AA	9.1	–	–
TiCl$_4$/MgCl$_2$	16.9	28.6	55.5[a]

a) copolymerization temperature is 80°C.

merization ability is also observed for traditional V- and Ti-containing Ziegler catalysts; V-containing systems have a lower value of r_1 (9).

Two important peculiarities of ethylene copolymerization with α-olefins over the catalysts examined should be mentioned. First, in all cases the introduction of α-olefin increases the activity compared to ethylene homopolymerization. Second, α-olefins are sufficiently efficient chain transfer agents and their introduction noticeably increases the melt index of polymers.

Gas-Phase Copolymerization

Data on the **gas-phase** copolymerization of ethylene with propylene and 1-butene for TMC and Cr-containing catalysts are shown in Fig. 2. For TMC the data on slurry (in hexane) copolymerization are also shown. The reactivity ratios determined from these data and by equation (1) are given in Table 2. In the case, where reactivity ratios were calculated using the data on concentrations of ethylene and α-olefins in a gas phase (for gas-phase copolymerization, the r_1^* values in Table 2) or in hexane (for slurry copolymerization), the unusual results are obtained. (i) The r_1 values for slurry and gas-phase copolymerization, obtained on the same TMC and equal 3.2 and 13, respectively (Table 2), differ essentially. (ii) The close values of r_1^* for ethylene-propylene and ethylene-butene copolymerization on TMC are not in agreement with those known for Ziegler type catalysts. (iii) For Cr-containing catalyst the r_1^* value is close to unity, i.e. the reactivities of ethylene and α-olefin addition to the ethylene unit of polymer chain are close ($K_{11} \approx K_{12}$). This result, however, contradicts the known literature data on the higher reactivity of ethylene as compared with α-olefin. To obtain the correct values of reactivity ratios one should use the concentration ratio of comonomers dissolved in a polymer, which covers catalyst particles, rather than the comonomer concentrations in the gas phase of a reactor. According to Michaelis and Bixler (10) the solubility of olefins in polyethylene is determined by the content of the amorphous part in a polymer, and the concentration ratios of olefins dissolved in a polymer and in hydrocarbon solvent are close. Based on that and using for gas-phase copolymerization $[C_3]/[C_2] = 4$[1] and $[C_4]/[C_2] = 10$[1] (as in heptane) the values of r_1 were calculated (Table 2). These values differ essentially from the r_1^* values and correspond to the data for slurry copolymerization and to the ratio of copolymerization reactivities of various α-olefins.

It is **important** to note that the increasing ethylene solubility (ethylene concentration) in copolymer due to increasing of the content of amorphous part in it can explain the increasing of copolymerization rate compared with homopolymerization rate which is observed usually for various catalysts (1-8).

Data on r_1 values (Tables 1 and 2) show that the copolymerization ability of VMC and Cr-containing catalysts is higher in comparison with one of TMC.

---

[1] Coefficients for every experimental $[C_\alpha]/[C_2]$ ratio in the gas phase of a reactor(cf. two abscissas in Fig. 2).

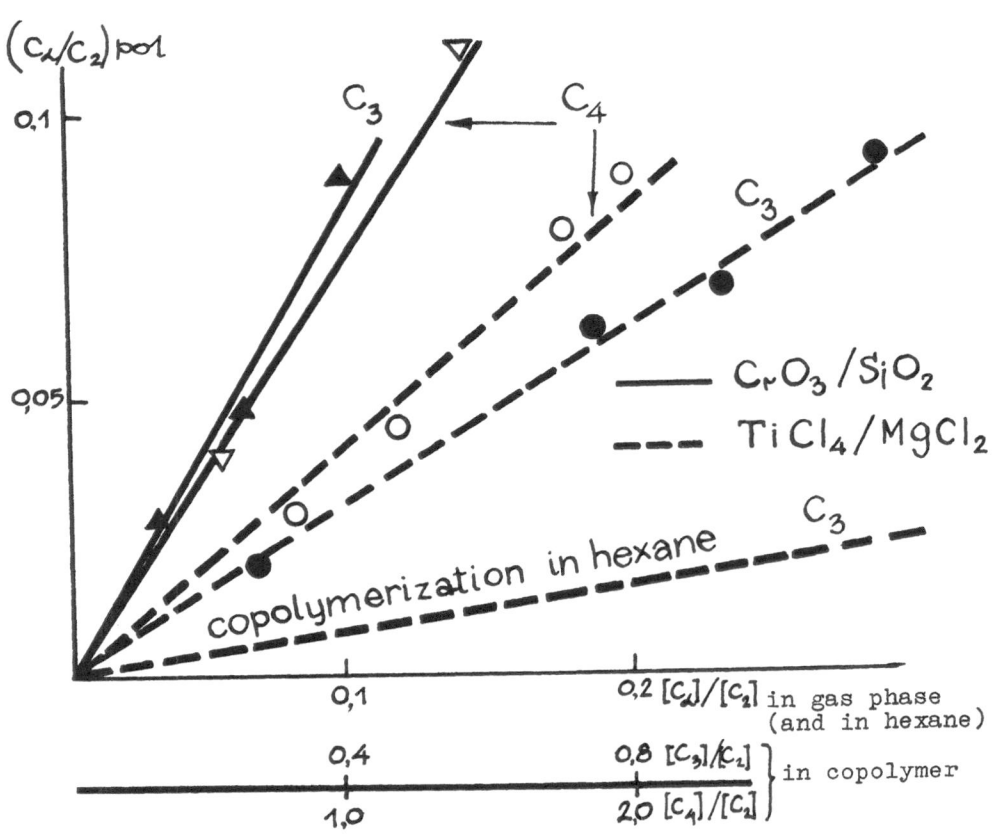

Fig. 2. Plot of the comonomer mole ratio in copolymer vs. comonomer mole ratio in reaction medium for gas-phase copolymerization; 80°C. Copolymerization was performed according to (14).

Table 2. The reactivity ratios[a] for ethylene-α-olefin gas-phase copolymerization

Catalytic system	Medium	r_1^*		r_1	
		C_2+C_3	C_2+C_4	C_2+C_3	C_2+C_4
CrO_3/SiO_2	gas	1.1*	1.3*	4.4	13
$TiCl_4/MgCl_2+AlR_3$ {	gas	3.2*	2.4*	12.8	24
	hexane	13	29	13	29

a) r_1^* and r_1 are calculated from the $[C_\alpha]/[C_2]$ ratio in gas phase and in polymer correspondingly.

Microstructure of Ethylene-Propylene Copolymers (by ^{13}C NMR)

The ^{13}C NMR data of ethylene-propylene copolymers (EPCs) allow to estimate the methylene sequence length distribution and to conclude on the comonomer distribution in copolymer (11). Table 3 shows the composition and the methylene sequence length distribution for EPCs obtained with various catalysts.

In the case of VMC and Cr-containing catalyst EPCs have the methylene sequences with n = 2 and n = 4. These sequences correspond to those with inverted propylene units (PP, PEP, PEEP). Hence, these catalysts have a lower regiospecificity compared with TMC and titanium trichloride which give EPCs without inverted propylene units (Table 3). Kashiwa et al. (12) associate a lower regiospecificity of the catalysts with their physical state, i.e. with the solubility of catalytic systems. The data in Table 3 show the formation of propylene inversions in the case of supported catalysts.

The contribution of methylene sequences with n = 3 that characterize the catalyst tendency towards alternating copolymerization is nearly the same for TMC and VMC and essentially higher for chromium-oxide catalyst (Table 3).

As can be seen in Table 3 the proportion of the block propylene sequences (methylene sequences with n = 1) is minimum for chromium-oxide catalyst and VMC and tends to increase for TMC and TiCl$_3$. Thus, the copolymers obtained on chromium-oxide catalyst and VMC are characterized by a more homogeneous distribution of propylene in comparison with EPCs obtained with TMC and TiCl$_3$. This phenomenon seems to be due to the fact that VMC and chromium-oxide catalysts have a lower regiospecificity than titanium-containing catalysts.

From ^{13}C NMR data on EPCs according to (5) we have determined the reactivity ratios, r_1 and r_2, for TMC and titanium trichloride (Table 3). For both catalysts the values of r_1 are close, but the values of r_2 are slightly different.

On comparing the values of r_1 and r_2 found by ^{13}C NMR and Fineman-Ross methods (Table 3) it can be seen that there is an essential difference between r_1 values in the case of TiCl$_3$ and between r_2 values in the case of TMC. For both catalysts the r_1r_2 value calculated from NMR data is higher than that found by using the Fineman-Ross method. The reason for such discrepancies seems to be connected with the copolymer inhomogeneity.

Data on some fractions of EPCs prepared with various catalysts are given in Table 4. For all catalysts the content of propylene in copolymer fractions decreases with decreasing solubility. In the case of VMC the difference in propylene content is observed to a smaller extent that indicates a more homogeneous propylene distribution in copolymer fractions in comparison with Ti-containing catalysts. The most probable reason for different content propylene in EPC fractions seems to be inhomogeneity of active sites.

For TMC and TiCl$_3$ (as opposed to VMC) the propylene insertion is highly regiospecific; in all fractions, including an ether soluble one, there are no inverted propylene units. Considering this fact, we have determined the monomer reactivity ratios, r_1 and r_2, for TMC

Table 3. Sequence length distribution of continuous methylene sequences in EPCs, prepared with various catalysts[a]

Catalyst	Content of C_3H_6, mol.%	$-(CH_2)_n-$ Proportion, mol.%						r_1	r_2	r_1r_2	By Fineman–Ross method		
		$n=1$	$n=2$	$n=3$	$n=4$	$n=5$	$n=6$				r_1	r_2	r_1r_2
$TiCl_4/MgCl_2$	36.2	46	≈0	21	≈0	11	22	16.8	0.082	1.6	15.8	0.031	0.5
$TiCl_3$(Solvay)	43.2	57	≈0	22	≈0	6	15	14.2	0.17	2.4	4.3	0.19	0.8
$VOCl_3/MgCl_2$	35.8	38	4.2	20	4	15	19	–	–	–	3.4	0.56	1.9
CrO_3/SiO_2	16.2	14.5	5.5	10.6	4.5	9	56	–	–	–			
$TiCl_4/MgCl_2$	14.5	35	≈0	5.5	≈0	16	44	–	–	–			

a) procedures of the preparation of catalysts and EPCs, recording of NMR spectra are given in ref. (8,13).

Table 4. Monomer composition and methylene sequence distribution for various fractions of EPC

Catalyst	Fraction of EPC[a]	Proportion of the fraction, % wt.	Content of C_3H_6 in the fraction, mol.%	$-(CH_2)_n-$Proportion, mol.%						r_1	r_2	r_1r_2
				n=1	n=2	n=3	n=4	n=5	n=6			
TiCl$_4$/MgCl$_2$	Ether	59.7	50.0	56	≈0	20	0	10	13	12	0.12	1.4
	Pentane	23.8	29.9	32	≈0	19	0	16	34	24.5	0.05	1.2
	Hexane	4.7	-	-	-	-	-	-	-	-	-	-
	Residue	11.7	14.5	35	≈0	5.5	0	16	44	65.9	0.08	5.3
	Total EPC	100	36.1	47	≈0	20	0	13	21	16.8	0.08	1.6
VOCl$_3$/MgCl$_2$	Ether	44	41.1	41	5.5	24	4	11	15	-	-	-
	Pentane	47	37.3	41	4.5	20	3.5	11	20	-	-	-
	Residue	8.1	20.2	28	3	20	3	14	31	-	-	-
	Total EPC	100	35.8	38	4	20	4	14	19	-	-	-
TiCl$_3$ (Solvay type)	Ether	31.8	53.5	58	≈0	22	0	11	10	7.6	0.19	1.4
	Pentane/+ether/	37.5	47.2	56	≈0	20	0	12	11	11	0.18	2
	Hexane	53.4	41.6	56	≈0	23	0	6	15	19.1	0.17	3.2
	Residue	9.1	18.4	44	≈0	15	0	15	27	52.8	0.04	2.2
	Total EPC	100	43.2	57	≈0	22	0	6	15	14.2	0.17	2.4

a) EPC fractions sequentially extracted by the solvents indicated at temperatures of their boiling.

and $TiCl_3$ from the NMR data for each copolymer fraction (Table 4). In accord with the decrease in propylene content in the fractions, the values of r_1 increase drastically with decreasing solubility.

Thus, the reactivity ratios determined for nonfractionated copolymer samples are average. For a more accurate estimate of monomer reactivity ratios and for an analysis of the copolymer microstructure one has to study fractions of the EPCs.

ACKNOWLEGEMENT

The authors are grateful to S.I. Makhtarulin, A.D. Khmelinskaya and T.M. Ivanova for catalyst preparation and to A.V. Nosov for recording ^{13}C NMR spectra.

REFERENCES

1   Finogenova LT, Zakharov VA, Buniat-Zade AA, Bukatov GD, Plaksunov
      TK (1980) Vysokomolek Soed A22: 404
2   Böhm LL (1980) Angew Makromol Chem 89: 1-32
3   Baulin AA, Ivanchev SS, Rodionov AG, Kreitzer TV, Goldenberg AL
      (1980) Vysokomolek Soed (1980) A22: 1486
4   Kissin YV, Beach DL (1983) J Polym Sci Polym Chem Ed 21: 1065
5   Doi Y, Ohnishi R, Soga K (1983) Makromol Chem Rapid Commun 4: 169
6   Abis L, Bacchilega G, Milani F (1986) Makromol Chem 187: 1877
7   Kaminsky W, Miri M (1985) J Polym Sci Polym Chem Ed 23: 2151-2164
8   Echevskaya LG, Zakharov VA, Bukatov GD (1987) React Kinet Catal
      Lett 34: 99
9   Cesca S (1975) J Polym Sci Macromol Rev 10
10 Michaels AS, Bixler HJ (1959) J Polym Sci 154: 393
11 Randall J (1978) Macromolecules 11: 33
12 Kashiwa N, Mizuno A, Minami S (1984) Polym Bull 12: 105
13 Yechevskaya LG, Bukatov GD, Zakharov VA (1987) Makromol Chem 188

14 Zakharov VA, Yechevskaya LG, Bukatov GD (to be published)

STEREOELECTIVE POLYMERIZATION OF RACEMIC α-OLEFINS BY HIGHLY-ACTIVE CHIRALLY MODIFIED ZIEGLER-NATTA CATALYSTS

Ciardelli F, Carlini C[*], Menconi F, Altomare A, Chien JCW[**]

Dipartimento di Chimica e Chimica Industriale, Università di Pisa, Centro di Studio del CNR per le Macromolecole Stereordinate ed Otticamente Attive, Via Risorgimento 35, 56100 Pisa, Italy

The possibility of inducing the preferential polymerization of a single enantiomer from a racemic α-olefin(stereoelective polymerization) by using optically active Lewis bases firstly shown for conventional Ziegler-Natta catalysts(1), was extended by Pino and coworkers(2,3) to highly-active systems such as $MgCl_2/TiCl_4 \cdot B^*/Al(i.Bu)_3 \cdot B^*$. In this last case (−)-menthyl anisate(B^*) was used both as "internal"(IB) and as "external" (EB) Lewis base. IB and EB can in general play different roles by interacting to different extent with the transition metal and the aluminum alkyl.

In order to contribute clarifying this point optically active (−)-menthyl benzoate[(−)MtB] was selectively used(4) as IB or EB in combination with p.methyl toluate(MT) or ethyl benzoate(EtB) in specific $MgCl_2$ supported highly-active systems $MgCl_2/IB/p.cresol/AlEt_3/TiCl_4/-/Al(i.Bu)_3/EB$, prepared by Chien and coworkers(5).

The preliminary data on the polymerization of racemic 3,7-dimethyl-1-octene(DMO) suggested that the chiral IB played a more important role in determining stereoelectivity, probably due to its more close interaction with the active sites(4).

In the present paper additional data are reported concerning polymerization experiments of DMO carried out in the presence of the overmentioned catalyst, but using different molar ratios of monomer versus internal chiral Lewis base(i.e. monomer/Ti) in order to attempt an improvement of stereoelectivity. The same goal was also pursued by replacing (−)MtB with a bidentate chiral Lewis base such as (−)-di-menthyl terephthalate [(−)MtP] which could provide more specific steric interactions with the active sites. Finally an achiral bidentate base, such as di-methyl terephthalate(MP), was used as EB in combination with (−)MtB as IB, in order to check possible exchange reactions between the two bases during the DMO polymerization process.

The different stereochemical role of IB and EB is clearly evidenced by comparing polymerization experiments where (−)MtB was used alternatively as IB or EB(Runs 1 and 4, Table 1). In fact when IB = (−)MtB and EB

[*]Dipartimento di Chimica Industriale e dei Materiali, Università di Bologna, Viale del Risorgimento 4, 40136 Bologna, Italy

[**]Polymer Science & Engineering Department, University of Massachusetts, Amherst, MA 01003, USA

W. Kaminsky and H. Sinn (Eds.)
Transition Metals and Organometallics as
Catalysts for Olefin Polymerization
© Springer-Verlag Berlin Heidelberg 1988

is MT the preferentially polymerized antipode of DMO has R absolute configuration, the opposite chiral discrimination occurring when IB = EtB
and EB = (-)MtB.

TABLE 1. Stereoelective polymerization of racemic 3,7-dimethyl-1-octene
(DMO) by chirally modified $MgCl_2$ supported highly-active Ziegler-Natta
catalysts[a]

Run	IB	EB	Chiral base/Ti (mol/mol) (B)	DMO/Ti (mol/mol)	Conv. (%) (C)	Resulting polymer		Stereo-electivity[b] efficiency (x 100)
						P_p[c]	Prev. chir.	
1	(-)MtB	MT	1	3,600	3.9	4.4	R	440
2	(-)MtB	MT	1	1,800	4.5	2.8	R	280
3	(-)MtB	MT	1	900	6.0	3.6	R	360
4	EtB	(-)MtB	56	3,600	19.7	0.25	S	0.4
5	EtB	(-)MtP	28	3,600	10.3	1.8	R	3
6	(-)MtB	MP	1	3,600	12.9	3.2	R	320
7	(-)MtB	(-)MtP	29	3,600	23.6	6.6	R	11

[a]Catalyst composition: $MgCl_2$/IB/p.cresol/$AlEt_3$/$TiCl_4$/-/Al(i.Bu)$_3$/EB;
Al(i.Bu)$_3$/ester group in EB = 3 mol/mol; Solvent: n.heptane; T = 50°C.
[b]Evaluated as P_p/B, where B is expressed in terms of chiral ester groups
[c]$P_p = P_m(1 - C)/C$, where P_m is the optical purity of unpolymerized DMO

In addition the stereoelectivity efficiency is three orders of magnitude higher in the presence of IB = (-)MtB than that occurring when (-)MtB
is used as EB, despite of the large excess of chiral base used in this
last case. These data can be explained by assuming either a more close
interaction between IB and the active sites or the formation of active
sites of a prevailing one single chirality.
As expected, no large variation of stereoelectivity efficiency is observed by decreasing the DMO/Ti(i.e. DMO/IB) molar ratio from 3,600 to
900(Runs 1-3, Table 1), thus suggesting that the stereoelectivity mainly depends on chiral IB/Ti molar ratio. From these data the stereochemical control seems to be better explained in terms of intrinsic chirality of active sites.
The replacement of (-)MtB by the chiral bidentate Lewis base (-)MtP,
when used as EB and still maintaining the achiral base EtB as IB, causes
both an improvement of the stereoelectivity efficiency(about one order
of magnitude) and an inversion of the chiral discrimination(Runs 4 and
5, Table 1). The possible explanation of the above data on the basis of
at least a partial exchange between (-)MtP and EtB, due to the bidentate
character of the former base, seems to be unlikely. Accordingly the stereoelectivity efficiency does not appreciably change when, in the pres-

ence of (-)MtB as IB, MT and MP are alternatively used as EB(Runs 1 and 6, Table 1).
When the chiral bases (-)MtB and (-)MtP are used in the same experiment as IB and EB, respectively, a remarkable improvement of the polymerized monomer optical purity is obtained(Run 7, Table 1).
The fractionation by boiling solvents of the crude polymers(Table 2) gives additional information on the nature of active sites and their stereospecificity.

TABLE 2. Fractionation data by boiling solvents of polymers obtained from stereoelective polymerization of racemic 3,7-dimethyl-1-octene(DMO) in the presence of chirally modified $MgCl_2$ supported highly-active Ziegler-Natta catalysts[a]

Run	IB	EB	Low stereoregular polymer (Diethyl ether sol. fract.)			High stereoregular polymer (Cyclohexane sol. fract.)		
			%	$[\alpha]_D^{25}$	Prev. chir.	%	$[\alpha]_D^{25}$	Prev. chir.
4	EtB	(-)MtB	18.7	+0.3	S	81.3	0.0	-
1	(-)MtB	MT	13.8	-2.3	R	86.2	-5.1	R
5	EtB	(-)MtP	20.1	+1.7	S	79.9	-2.8	R
6	(-)MtB	MP	11.1	-1.8	R	88.9	-3.9	R
7	(-)MtB	(-)MtP	13.4	-1.5	R	86.6	-7.6	R

[a]In all the above experiments DMO/Ti = 3,600 mol/mol was used.

When the chiral base is used as IB(Runs 1, 6 and 7, Table 2), the percentage of isotactic polymer is markedly improved and the stereospecific active sites display higher chiral discrimination.
When the chiral monodentate base is used as EB no chiral discrimination occurs by the stereospecific sites, the stereoelectivity being only due to the interaction between EB and the more exposed less stereospecific sites(Run 4, Table 2). On the contrary the use of the chiral bidentate base as EB induces a chiral discrimination on the stereospecific sites which is opposite to that occurring on the less stereospecific sites (Run 5, Table 2).

FINAL REMARKS

The data reported here allow to draw some concluding remarks about the use of chiral Lewis bases in highly-active $MgCl_2$ supported Ziegler-Natta catalysts:
- Chiral internal bases IB, as compared with chiral external bases EB, display a much higher stereoelectivity efficiency which involves the most stereospecific active sites

- Chiral internal bases IB cause an appreciable improvement of stereospe cificity of the catalyst system
- Internal(IB) and external(EB) bases are not exhanged
- Chiral bidentate bases are more effective than the chiral monodentate ones, at least when used as EB, for inducing stereoelectivity
- Stereoelectivity does not depend on monomer/Ti molar ratio, but it is expected to be affected by IB/Ti molar ratio. This result is better explained in terms of intrinsic chirality of active sites
- Chiral bidentate bases, when used as IB, are very promising for ob- taining a large improvement of stereoelectivity.

References

1.Carlini C, Ciardelli F (1981) Chim Ind(Milan) 63: 486-491
2.Pino P, Fochi G, Piccolo O, Giannini U (1982) J Am Chem Soc 104: 7381- 7383
3.Pino P, Guastalla G, Rotzinger B, Mülhaupt R (1983) in: Proc Internat Symp on "Transition Metal Catalyzed Polymerizations: Unsolved Problems" Midland(USA) pp 435-463
4.Ciardelli F, Carlini C, Altomare A, Menconi F (1987) J C S Chem Commun 94-95
5.Chien JCW, Wu JC (1982) J Polym Sci Polym Chem Ed 20: 2019-2032

Acknowledgement

Financial support from Ministery of Public Education(60%) and from NATO Research Grant N° 290/84 are gratefully acknowledged.

Solution NMR and FT-IR studies on the reactions and the complexes of silyl ethers with triethylaluminium

Eero Iiskola*, Pekka Sormunen and Thomas Garoff

Neste Oy, Technology Centre
SF-06850 Kulloo, Finland

Eila Vähäsarja, Tuula T. Pakkanen and Tapani A. Pakkanen

University of Joensuu, Department of Chemistry
SF-80100 Joensuu, Finland

SUMMARY

Complexation and reactions of silyl ethers $R_nSi(OMe)_{4-n}$ (n = 0-3, R = Ph, alkyl) with $AlEt_3$ were studied in benzene and followed by ^{13}C and ^{29}Si NMR. In the presence of excess $AlEt_3$, of five consecutive steps: alternating complexation and alkylation of silyl ethers, $Ph(Et)_nSi(OMe)_{3-n}$ (n = 1.2), and dialkyl-aluminium alkoxodes, $(Et_2AlOMe_3)_n$ (n = 2.3).

Because of the relaxation time effect in NMR spectroscopy the alkoxysilane-triethylaluminium complexes were studied by FT-IR spectroscopy to find out the degree of complexation in different Al/Si - molar ratios using about 10 volume-% solutions of complexes in heptane. Stereochemical effects clearly dominate the complexation number in silyl ethers containing more than one alkoxy group. IR spectra suggest that complexation is complete with only one alkoxy group because of sterical hindrance mostly caused by the bulky $AlEt_3$ molecule.

INTRODUCTION

The stereoselectivity of high activity supported catalysts for alpha-olefin polymerization is improved by the use of various types of organic donors as modifiers or coactivators. One of the functions of external donors is to complex or react with aluminium alkyl cocatalyst. The interaction mechanisms of organic esters and alcohols with Al-alkyls are fairly well known[1].

Silyl ethers, $R_nSi(OR')_{4-n}$, have recently been discovered to be highly active promoters in stereospecific alpha-olefin polymerization by Ziegler-Natta catalysts.

---

* To whom correspondence should be addressed

W. Kaminsky and H. Sinn (Eds.)
Transition Metals and Organometallics as
Catalysts for Olefin Polymerization
© Springer-Verlag Berlin Heidelberg 1988

EXPERIMENTAL

The manipulation of all components were carried out under dry, oxygen free, nitrogen (99.998 %).

Materials

The handling and purification of compounds is reported elsewhere[2].

NMR spectra

^{13}C and ^{29}Si NMR spectra were recorded on a Bruker AM-250 spectrometer operating at 62.9 MHz for ^{13}C-NMR and at 49.7 MHz for ^{29}Si-NMR. The quantitative proton-decoupled ^{13}C and ^{29}Si spectra were obtained with the gated decoupling technique. The ^{13}C and ^{29}Si spectra were referenced to the central line of C_6D_6 with shifts to lower field being positive.

FT - IR spectra

FT-IR spectra were obtained on a NICOLET 5 DXC spectrometer. A total of 128 scans of nominal resolution 4 cm^{-1} were signal averaged and Fourier transformed with Happ-Genzel apodization function to obtain spectrum in the range 5200 cm^{-1}. The detector used was DTGS (deuterated triglycine sulfate).

RESULTS AND DISCUSSION

A stable donor-acceptor complex is formed instantly at room temperature when Al/Si-molar ratio is 1:1. In the presence of excess AlEt$_3$ the complex decomposes and silyl ethers exchange their -OR with the alkyl groups of AlEt$_3$. In the case of PhSi(OMe)$_3$ the proposed mechanism (Scheme 1) was verified by independent synthesis of the alkylated silyl ethers and their complexes with AlEt$_3$ [2,3]. ^{29}Si NMR is a powerful method in characterizing solutions, which contain mixtures of different Si-compounds or complexes as compared to ^{13}C NMR (fig. 1 and fig. 2). The numbers in spectra refer to the complexes represented in Scheme 1.

Scheme 1

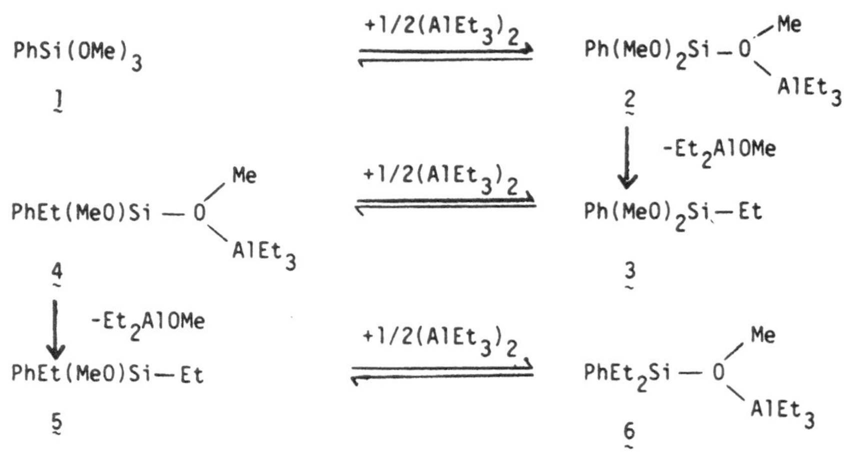

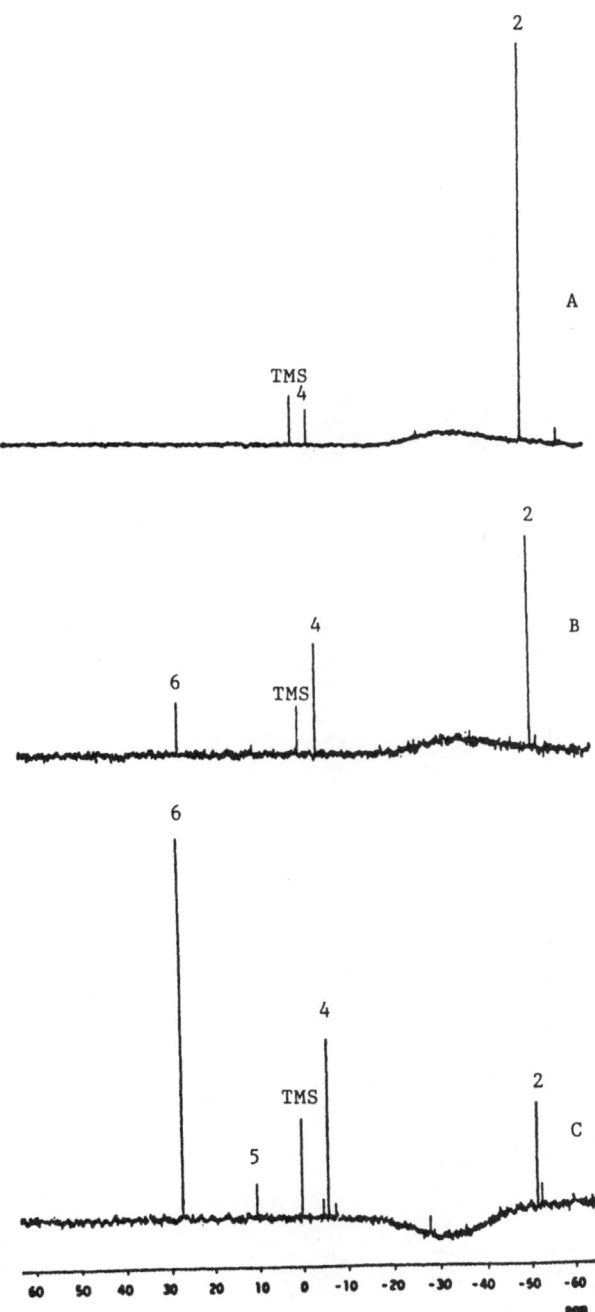

Fig. 1. 49.7 MHz ^{29}Si - {^1H} NMR spectra of 3:1 mixtures of AlEt$_3$ with PhSi(OMe)$_3$ at ambient temperature : (a) before heating, (b) after 2 hours at 75 °C, (c) after 20 hours at 75 °C.

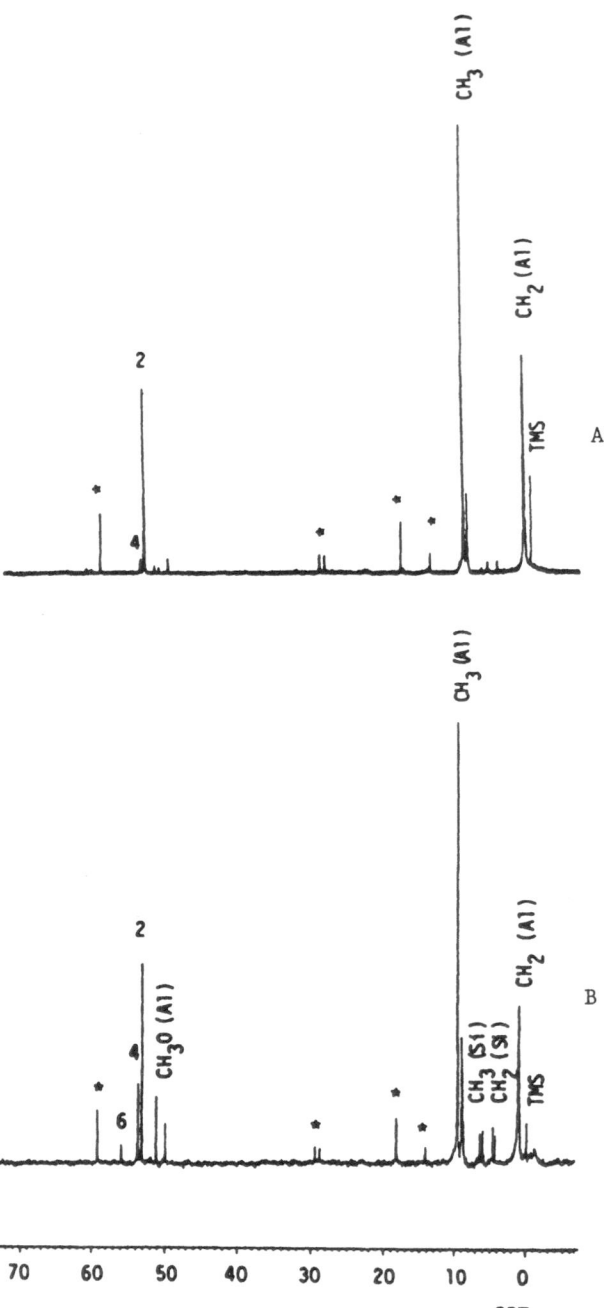

Fig. 2. 69.9 MHz $^{13}C - \{^1H\}$ NMR spectra of 3:1 mixtures of AlEt$_3$ with PhSi(OMe)$_3$ at ambient temperature : (a) before heating, (b) after 2 hours at 75 °C. (*impurities of AlEt$_3$).

The decomposition reaction is dependent on the structure of silyl ethers (fig. 3), high $AlEt_3$: silyl ether molar ratio (fig. 4) or heating (fig. 5).

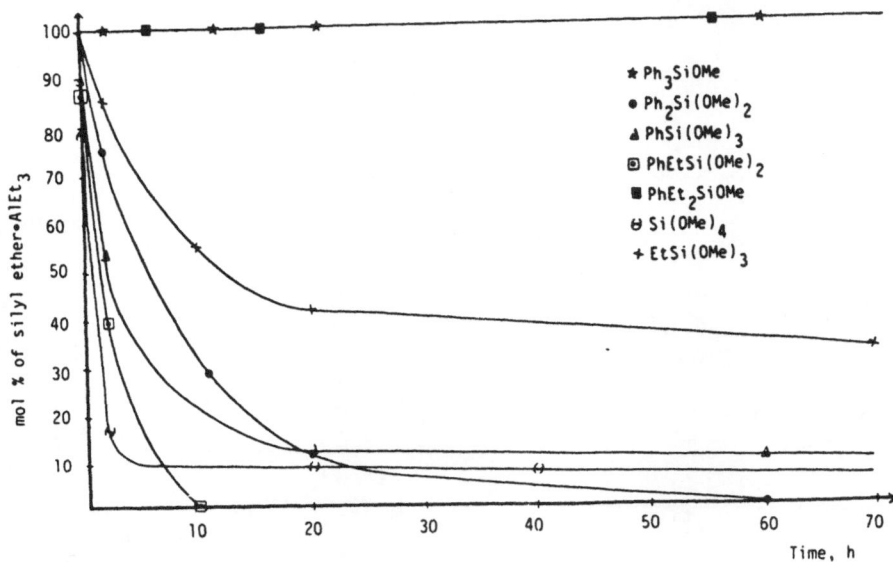

Fig. 3. Effect of the donor on decomposition of the silyl ether . $AlEt_3$ complex. ($AlEt_3$: silyl ether ration 3:1)

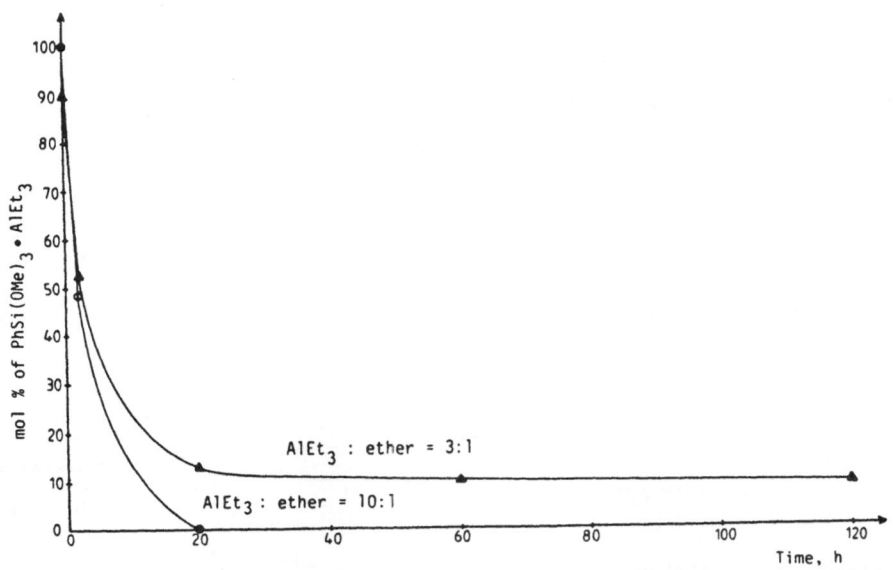

Fig. 4. Effect of molar ration on decomposition of the $PhSi(OMe)_3$. $AlEt_3$ complex. (Heating temperature 75 °C)

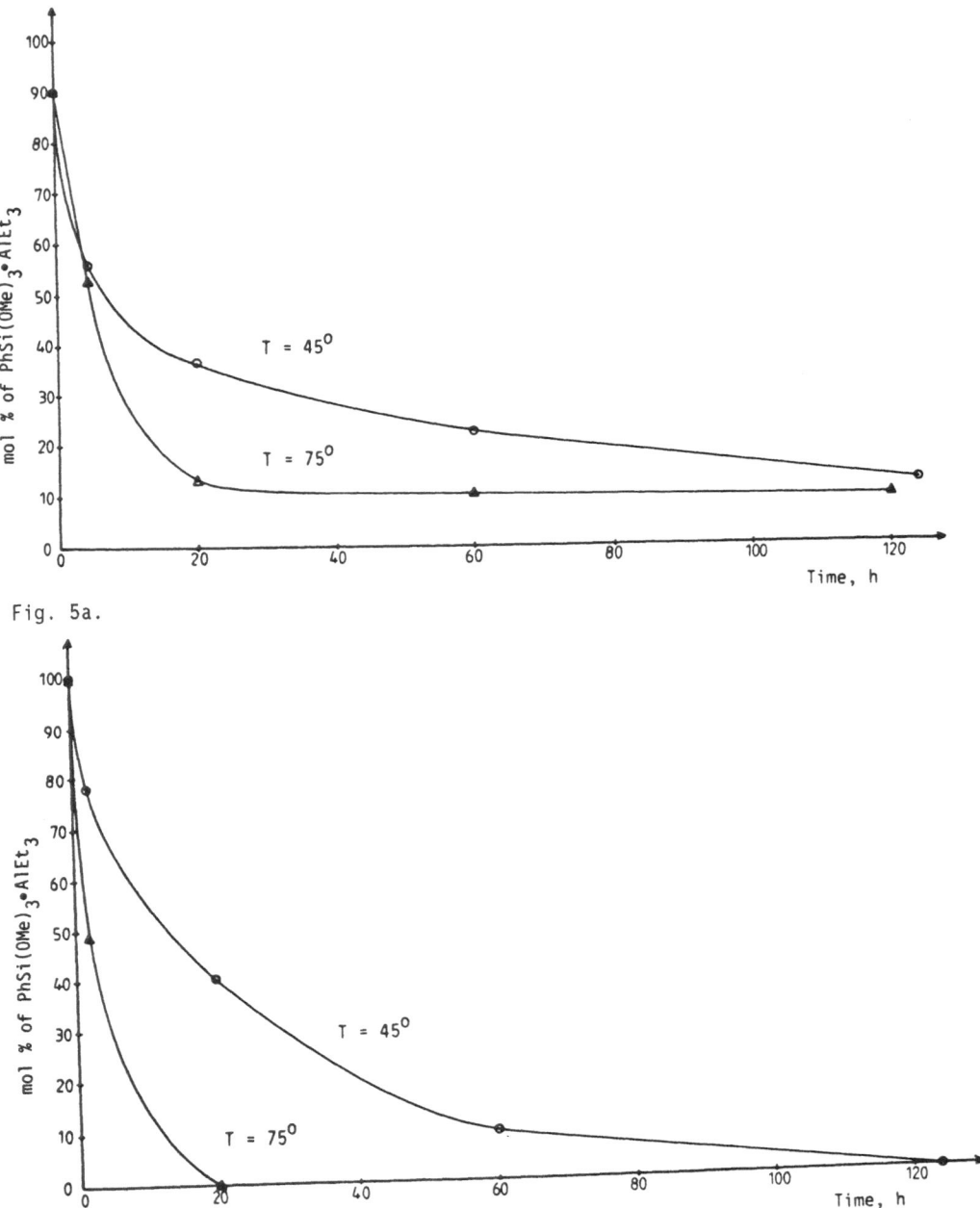

Fig. 5a.

Fig. 5b. Effect of temperature on decomposition of the $PhSi(OMe)_3 \cdot AlEt_3$ complex. $AlEt_3$: silyl ether ratios (a) 3:1, (b) 10:1.

Monoalkoxysilanes, $Ph_3Si(OMe)$ and $PhEt_2Si(OMe)$, show no decomposition after heating at 75 °C for 5 days.

In IR-spectra Si-alkoxy compounds have one or more strong bands in the 1200 - 1000 cm^{-1} range. In the case of the Si-OMe group, the complexation with $AlEt_3$, using different Al/Si - molar ratios, was followed by only one strong peak in the 1100-1080 cm^{-1} range at room temperature[4] (fig. 6 and fig. 7).

NESTE OY TECHNOLOGY CENTRE   FT-IR SPECTRUM

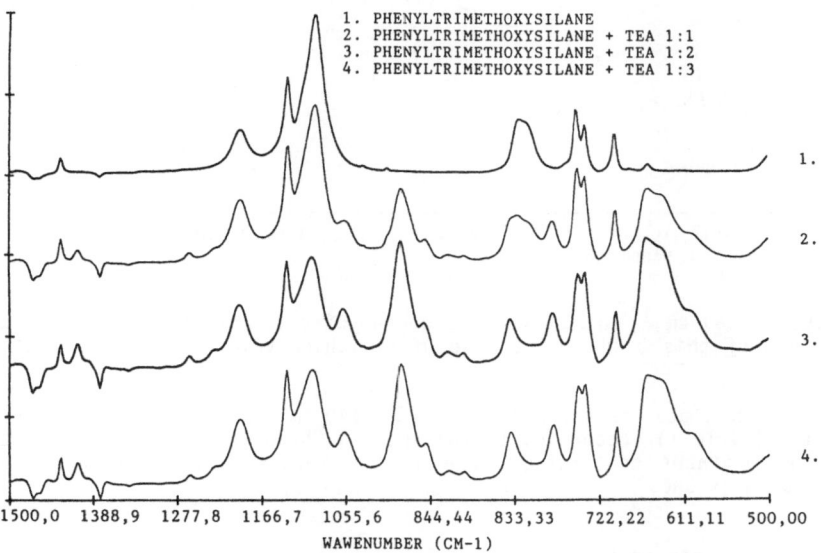

1. PHENYLTRIMETHOXYSILANE
2. PHENYLTRIMETHOXYSILANE + TEA 1:1
3. PHENYLTRIMETHOXYSILANE + TEA 1:2
4. PHENYLTRIMETHOXYSILANE + TEA 1:3

WAWENUMBER (CM-1)

Fig. 6.

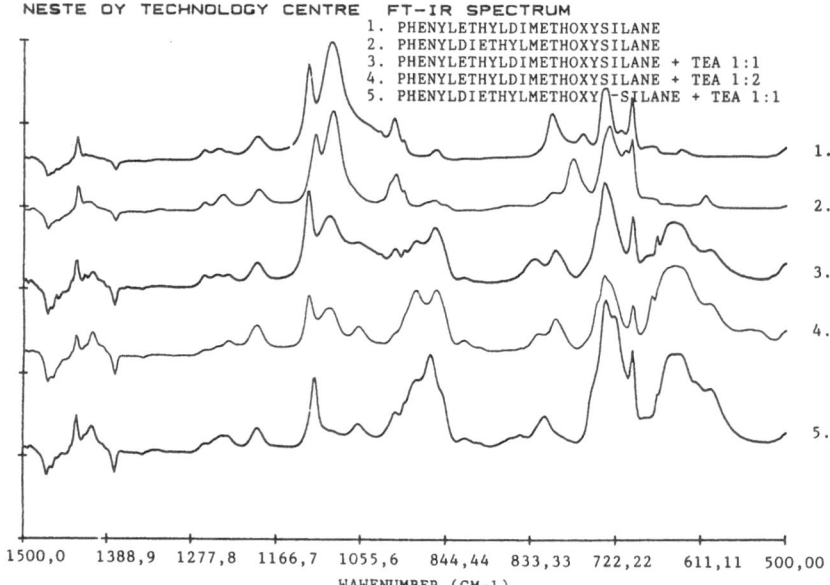

NESTE OY TECHNOLOGY CENTRE   FT-IR SPECTRUM
1. PHENYLETHYLDIMETHOXYSILANE
2. PHENYLDIETHYLMETHOXYSILANE
3. PHENYLETHYLDIMETHOXYSILANE + TEA 1:1
4. PHENYLETHYLDIMETHOXYSILANE + TEA 1:2
5. PHENYLDIETHYLMETHOXY-SILANE + TEA 1:1

| 1500,0 | 1388,9 | 1277,8 | 1166,7 | 1055,6 | 844,44 | 833,33 | 722,22 | 611,11 | 500,00 |

WAWENUMBER (CM-1)

Fig. 7.

The fact that 1100 - 1080 cm^{-1} band shows this degree of complexation, is evidenced by the disappearance of this band in the case of stoichiometric $PhEt_2Si(OMe)-AlEt_3$ complex.

In all cases the strongest peak is shifted about 100 - 140 cm^{-1} to the lower frequency due to complexation (table 1). According to the spectra there are still free Si-OMe groups left in silyl ethers containing three or two methoxy groups, though the Al/methoxy molar ratio is one.

Table 1. IR spectral changes in $PhSi(Et)_{3-x}(OMe)_x$ (x=1-3) and the $AlEt_2$-silyl ether complexes.

	Si-OMe (cm^{-1}) free	complexed Si/Al molar ratio
$PhSi(OMe)_3$	1093	-
	1094	1:1 ; 981
	1098	1:2 ; 982
	1098	1:3 ; 982
$PhEtSi(OMe)_2$	1092	-
	1095	1:1 ; 983 (sh), 958
	1096	1:2 ; 983 956 (doublet)
$PhEt_2Si(OMe)$	1091	-
	-	1:1 ; 983 (sh), 964

This suggests, that stereochemistry does not allow complete complexation, mainly due to bulky $AlEt_3$.

Sterical hindrance can be visualized by the computer produced molecular model[7] shown in figure 8.

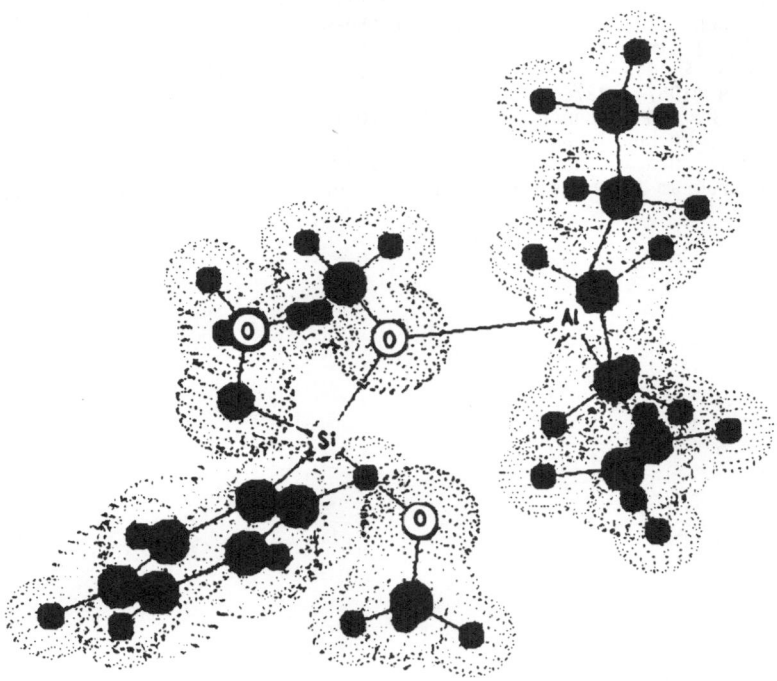

Fig. 8. A molecular model of $PhSi(OMe)_3$. $AlEt_3$ complex

Preliminary propylene polymerization tests using supported $MgCl_2$-$TiCl_4$ catalysts have shown, that the change in the type and number of R and OR' groups in silyl ethers has a considerable effect on activity and stereospecificity, when they are used as catalyst modifiers with triethylaluminium[5].

The real function of silyl ethers is perhaps still more complicated on a heterogeneous catalyst surface catalyst because of the following reaction[6]:

$$TiCl_4 + R_nSi(OMe)_{4-n} \longrightarrow TiCl_2(OMe)_2 + R_nSiCl_2(OMe)_{2-n}$$

R = Ph, Et
n = 0, 1

REFERENCES

1. a)Goodall, B.L., in Transition Metal catalyzed polymerizations, Quirk, R.P.,
   Harwood Academic Publishers, New York (1983), 355.
   b)Kissin, Y.V. and Sivak, A.J., J. Polym. Sci. 22 (1984), 3747.
   c)Chien, J.C.W. and Wu, J., J. Polym. Sci. 20 (1982), 2445.
2. Vähäsarja, E., Pakkanen, T.T., Pakkanen; T.A., Iiskola, E. and Sormunen, P., J.
   Polym. Sci., 25 (1987), in press.
3. Sormunen, P., Iiskola, E., Vähäsarja, E., Pakkanen, T.T. and Pakkanen, T.A., J.
   Organomet. Chem., 319 (1987), 327.
4. Iiskola, E., Sunila, P. and Sormunen, P., to be published.
5. Iiskola, E. et al., to be published.
6. Bradley, D.C. and Hill, D.A.W., J. Org. Chem. (1963), 2101.
7. Produced by CHEM-X program (Chemical Design Ltd. Oxford), University of Joensuu,
   Finland

A Study of the Active Site Structure in MgCl$_2$ Supported
Ziegler-Natta Catalysts by a Stereochemical Investigation of
the Initiation Step

M.C. Sacchi, I. Tritto, and P. Locatelli

Istituto di Chimica delle Macromolecole del C.N.R.
Via E. Bassini 15, 20133 Milano (Italy)

SUMMARY

The ^{13}C NMR analysis of the steric structure of the chain-end groups
resulting from the insertion of the first propene unit into the
Ti-^{13}CH$_2$CH$_3$ bond has been performed on polypropene samples prepared
in the presence of Ziegler-Natta catalytic systems supported on
differently activated MgCl$_2$. The data show that the kind of MgCl$_2$
activation does not influence the steric features of both isotactic
and atactic sites.
By means of the same ^{13}C NMR analysis of chain-end groups the
activation effect of ethyl benzoate is studied, by comparing
^{13}C NMR spectra of polypropene samples obtained in presence of the
catalytic systems: i) MgCl$_2$ / TiCl$_4$ / Al(^{13}CH$_2$CH$_3$)$_3$; ii) MgCl$_2$ / EB
/ TiCl$_4$ / Al(^{13}CH$_2$CH$_3$)$_3$ / EB.
The noticeable enhancement of the first step stereoregularity
observed in the isotactic fractions of the sample obtained in the
presence of ethyl benzoate gives evidence of the direct influence of
the base on the isotactic active sites.

INTRODUCTION

During the last few years we have studied by ^{13}C NMR analysis the
steric structure of the chain-end groups resulting from the insertion
of the first propene unit into the Ti-^{13}CH$_2$CH$_3$ bond in the presence
of traditional Ziegler-Natta catalysts.[1-4]

Table 1 shows the extent of the first step stereoregularity, in the
heptane insoluble fraction, evaluated as intensity ratio I$_e$/I$_t$ of the
^{13}C NMR methylene resonances related respectively to the isotactic
and syndiotactic placement of the first propene unit (Fig. 1). The
data show that the extent of the first step stereoregularity of the
most isotactic fractions, depending on the bulkiness and on the
mutual interactions of the alkyl and halide ligands as well as on the
crystal modification of TiCl$_3$, is a characteristic of each catalytic
system.

This kind of analysis has been applied to the study of the nature and
the distribution of the catalytically active sites of high yield
supported Ziegler-Natta catalysts. In the present paper two problems
are investigated: i) the effect of the way by which MgCl$_2$ is

W. Kaminsky and H. Sinn (Eds.)
Transition Metals and Organometallics as
Catalysts for Olefin Polymerization
© Springer-Verlag Berlin Heidelberg 1988

activated on the steric structure of atactic and isotactic sites;
ii) the mechanism by which Lewis bases influence the activity and
the stereospecificity of the isotactic sites.

Table 1. Chain-end stereoregularity of the heptane insoluble
fraction of polypropene fractions

Catalyst	[mm]	I_e/I_t[a]
$TiCl_3/Al(^{13}CH_2CH_3)_3$	0.98	3.3 [1,2]
$TiI_3/Al(^{13}CH_2CH_3)_3$	0.97	> 8 [3]
$TiCl_3 \cdot 0.3AlCl_3/Zn(^{13}CH_2CH_3)_2$	0.94	2.2 [4]
$TiCl_3 \cdot 0.3AlCl_3/Zn(^{13}CH_2CH_3)_2/Et_3N$	0.95	2.7 [4]
$TiCl_3 \cdot 0.3AlCl_3/Zn(^{13}CH_2CH_3)_2/TMPip$	0.97	5.5 [4]

[a] I_e/I_t = intensity ratio of resonances related to the
   isotactic (e) and syndiotactic (t) placement of the
   first propene unit.

Et_3N = triethylamine

TMPip = 2,2,6,6-tetramethylpiperidine

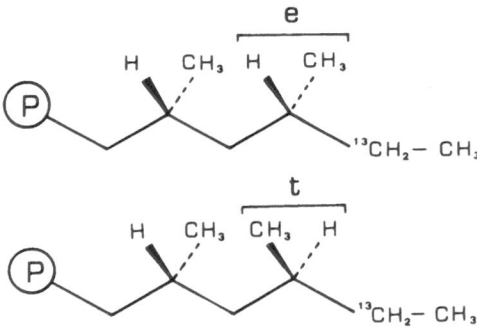

Fig. 1. Stereochemical placement of the isotactic (e) and
syndiotactic (t) first propene unit inserted into $Ti-^{13}CH_2CH_3$
bond

POLYMERIZATION OF PROPENE WITH DIFFERENTLY ACTIVATED $MgCl_2$

Polypropene samples were prepared in the presence of catalytic
systems supported on premilled $MgCl_2$ (samples A and B) and on $MgCl_2$
obtained by chlorination of a Grignard compound (sample C).

Table 2 shows that the kind of MgCl₂ activation influences the titanium content, the catalyst activity and the polymer fraction distribution. The nearly identical values of the I_e/I_t indicate that the steric features of the isotactic and atactic sites are the same in the three different systems, while their relative concentrations change. It is worthwhile to note that the isotactic active sites of all the observed supported catalysts, have a lower steric control on the first monomer insertion with respect to the traditional TiCl₃ based catalysts, probably connected to their more exposed location on the MgCl₂ surface.

Table 2.

Sa.[a]	Cat.	Ti%	Y[b]	polymer fractions no.[c]	wt%	[mm][d]	I_e/I_t[e]
A	1.	0.22	11	F.5	30	0.31	0.6
				F.7	35	0.76	1.2
				R.7	35	0.97	1.9
B	2.	0.34	21	F.5	26	0.31	0.7
				F.7	25	0.77	1.2
				R.7	49	0.96	2.2
C	3.	2.43	61	F.5	25	0.32	0.7
				F.7	33	0.75	1.2
				R.7	42	0.95	1.9

1. MgCl₂ / TiCl₄ / Al($^{13}CH_2CH_3$)₃; MgCl₂(type A), by 7 days milling.
2. MgCl₂ / TiCl₄ / Al($^{13}CH_2CH_3$)₃; MgCl₂(type B), by 10 days milling.
3. MgCl₂ / TiCl₄ / Al($^{13}CH_2CH_3$)₃; MgCl₂(type C), by Grignard.

[a] Sa. = polymer sample
[b] Yield in g of polym./g cat.·h
[c] F.5 diethylether soluble
   F.7 diethylether insoluble-heptane soluble
   R.7 heptane insoluble
[d] molar fraction of isotactic triads by NMR
[e] stereoregularity of first monomer insertion

On the other hand the "atactic" active sites show a prevailingly syndiotactic steric control ($I_e/I_t<1$) on the first monomer insertion and therefore are different from the atactic sites in the TiCl₃ based catalysts ($I_e/I_t=1$).[2] An interpretation of this fact may be that these sites are lacking in intrinsic chirality; as a consequence after the random first propene insertion, the second one is partially controlled by the configuration of the chiral carbon of the 2-methylbutyl end group.

```
              C                        C   C
              |                        |   |
 Mt-C-C-C-C-¹³C-C    >    Mt-C-C-C-C-¹³C-C
              |
              C
```

MODIFICATION EFFECT OF ETHYL BENZOATE

Propene was polymerized in the presence of the catalytic system MgCl$_2$/EB/TiCl$_4$ / Al(^{13}CH$_2$CH$_3$)$_3$/EB (sample D). Table 3 shows the data relative to sample D and sample C, obtained with the catalyst supported on the same MgCl$_2$, in the absence of base.

Table 3.

Sa.[a]	Cat.	Ti%	Y[b]	polymer fractions no.[c]	wt%	[mm][d]	I$_e$/I$_t$[e]
C	3.	2.43	61	F.5	25	0.32	0.7
				F.7	33	0.75	1.2
				R.7	42	0.95	1.9
D	4.+EB[f]	1.46	60	F.5	1	0.29	1.0
				F.7	4	0.72	2.6
				R.7	95	0.97	3.7

3. MgCl$_2$ / TiCl$_4$ / Al(^{13}CH$_2$CH$_3$)$_3$
   MgCl$_2$(type C)
4. MgCl$_2$ / EB / TiCl$_4$ / Al(^{13}CH$_2$CH$_3$)$_3$
   MgCl$_2$(type C)

[a] Sa. = polymer sample
[b] Yield in g of polym./g cat.·h
[c] F.5 diethylether soluble
   F.7 diethylether insoluble-heptane-soluble
   R.7 heptane insoluble
[d] molar fraction of isotactic triads by NMR
[e] stereoregularity of first monomer insertion
[f] EB = ethyl benzoate; EB/AlEt$_3$ = 0.1

In the presence of EB as internal and external base the overall catalyst productivity increases considerably. The fact that the content of the isotactic fraction becomes nearly double shows that a noticeable activation of the isotactic sites occurs. To account for this activation the following alternatives are conceivable:
i) Direct interaction between the base and the isotactic sites (producing for example an increase of the propagation rate or a decrease of chain transfer reactions);
ii) Increase of the number of isotactic active sites.
The fact that a high enhancement of the I$_e$/I$_t$ value (Fig. 2) accompanies the noticeable increase of isotactic productivity, is in keeping with the first hypothesis. Indeeed it is reasonable that the association of the ester with the catalytically active titanium produces such a change in the active site environment to explain the noticeable increase of the first step stereoregularity.

However the second hypothesis can not be ruled out. Indeed, in the presence of the base, along with the decrease of the atactic fraction, a change of the first step stereoregularity on the atactic sites is observed (Fig. 3), i.e. the observed I$_e$/I$_t$ value is =1. In other words the atactic sites with prevailingly syndiotactic first insertion seem to be absent. This absence could be due either to a poisoning or to a transformation. The observed noticeable increase of the isotactic fraction could be attributed to the presence of new isotactic sites resulting from the partial transformation of this kind of atactic sites to isotactic.

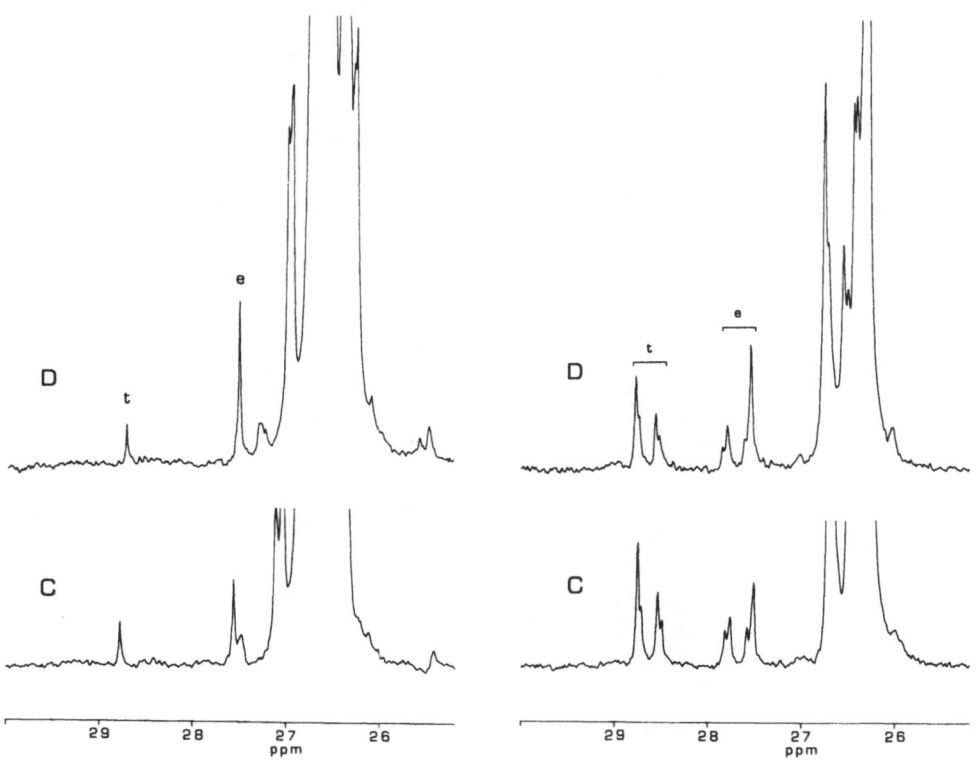

Fig. 2. ^{13}C NMR spectra of enriched methylene of $-^{13}CH_2-CH_3$ chain-end groups of isotactic polypropene fractions of samples

Fig. 3. ^{13}C NMR spectra of enriched methylene of $-^{13}CH_2-CH_3$ chain-end groups of atactic polypropene fractions of samples

REFERENCES

1   Zambelli A, Sacchi M.C, Locatelli P, Zannoni G (1982) Macromolecules 15: 212

2   Zambelli A, Locatelli P, Sacchi M.C, Tritto I (1982) Macromolecules 15: 831

3   Tritto I, Sacchi M.C, Locatelli P (1986) Makromol. Chem. 187: 2145

4   Tritto I, Sacchi M.C, Locatelli P (1988) Macromolecules in press

Ethylene Polymerization Process with a Highly Active ZieglerNatta
Catalyst - Kinetic Studies

*M.M.V. Marques, *C.P. Nunes, **P.J.T. Tait, ***A.R.Dias

*   C.N.P. - Ap. 287521 Sines Codex, Portugal
**   Dep. Chem. Umist - Manchester M60 1QD, U.K.
***   C.Q.E. - I.S.T. - 1096 Lisboa Codex, Portugal

ABSTRACT

Factores affecting the particular shape of kinetic rate-time profiles
in the polymerization of ethylene with a $MgCl_2$ - supported $TiCl_4$
catalyst activated by $Al(C_2H_5)_3$ have been investigated.

The number of active centres in the polymerization system was
estimated both by a kinetic method and a Radio-Tagging technique
using [14]CO. However the results obtained using the two methods
were too different to be reconciled.

RESULTS AND DISCUSSION

1 - Phenomenological Behaviour

The specific shapes of rate-time profiles, altough usually
characteristic of a particular catalyst system, are also affected
by operational conditions (1-3)

a) Pressure

Both the pressure of monomer and of hydrogen affected the observed
kinetic rate-time profiles as is shown in fig. 1 and fig. 2.

As can be seen from these figures, the present catalyst system
shows almost steady kinetic rate-time profiles for polymerizations
at atmospheric pressure (fig. 1) and decay type profiles at higher
monomer pressures (fig. 2). Similar results had already been found
by Keii (1) for the polymerization of propylene.

It was noticed also that when we increase significantly the partial
pressure of hydrogen, their shape becomes similar to the one found
for the polymerization at atmospheric pressure (Curve A - fig.
2).

b) Alkylaluminium

The kinetic rate-time profiles were affected both by type and
concentration of alkylaluminium compound which were used as is
shown in figs. 3, 4 and 5.

W. Kaminsky and H. Sinn (Eds.)
Transition Metals and Organometallics as
Catalysts for Olefin Polymerization
© Springer-Verlag Berlin Heidelberg 1988

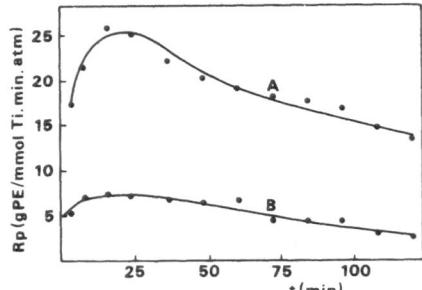

Fig.1. Rate versus time
$|Ti| = 0,0238$ mmol 1^{-1}; Al/Ti=84;
$P_t = 1$ bar; T=60°C; solvent EC-130
(A) $P_{H_2} = 0$; (B) $P_{H_2} = 0,14$ bar

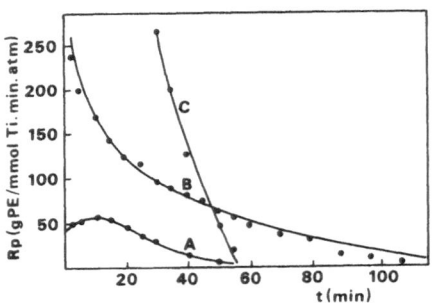

Fig.2. Rate versus time
$[Ti] = 0,006$ mmol 1^{-1}; Al/Ti=
=166; $P_t = 9$ bar T=70°C; n-hexane;
(A) $P_{H_2} = 5$ bar; (B) $P_{H_2} = 3$ bar;
(C) $P_{H_2} = 0$

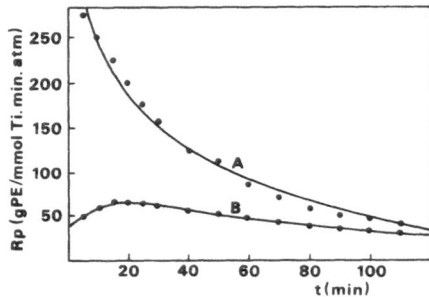

Fig.3. Rate versus time
$[Ti] = 0,0102$ mmol 1^{-1},
Al/Ti = 150 T = 80°C, n-hexane,
Pt = 6 bar, $P_{H_2} = 1,5$ bar;
(A) Et$_3$Al; (B) (i-Bu)$_3$Al

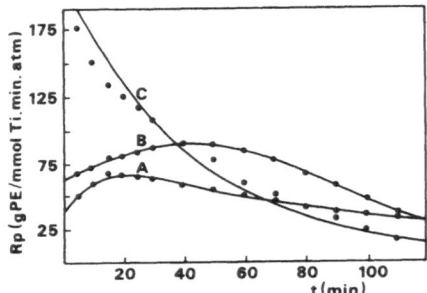

Fig.4. Rate versus time
$[Ti] = 0,0102$ mmol 1^{-1}; T =80°C;
n-hexane; cocatalyst (i-Bu)$_3$Al;
$P_t = 6$ bar; $P_{H_2} = 1,5$ bar
(A) Al/Ti = 150 (B) Al/Ti=300
(C) Al/Ti = 400

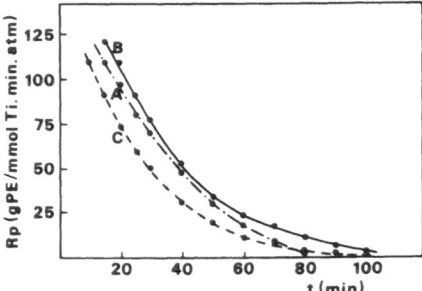

Fig.5. Rate versus time
$[Ti] = 0,01$ mmol 1^{-1}; $P_t = 6$ bar; $P_{H_2} = 1,5$ bar, T=80°C; n-hexane; Et$_3$Al;
(A) Al/Ti = 50 (B) Al/Ti = 100 (C) Al/Ti = 300

Variation of the trialkylaluminium concentration brings about profound changes in the kinetic rate-time profiles (fig. 4) and the effect is more pronounced when $(i-Bu)_3Al$ (fig. 4) rather than Et_3Al is used (fig. 5).

c) Temperature

Although our results were limited to the temperature range of 50-80°C the overall effect of increasing temperature led to an increase in the rate of polymerization.

2 - Formulation of the Rate Law

2.1 - Elimination of the Effect of Monomer Diffusion

2.1.1 - Internal difusion limitations

The Thiele modulus, which is the ratio of the characteristic diffusion time to the characteristic reaction time is given by

$$\phi = \left(\frac{k_p c_o}{D_m}\right)^{1/2} S_o \tag{1}$$

where k_p is the propagation rate constant, C_o the initial concentration of active centres, D_m the diffusivity of monomer and S_o the initial radius of the catalyst particle. ϕ can be used to estimate the degree of diffusion control; a larger value of ϕ ($\phi=10$) will indicate a diffusion control and a small one a kinetic control. Although the application of the Thiele modulus concept is complicated by the fact that Ziegler-Natta catalyst tends to fragment into smaller crystallities thus changing the values of S_o during polymerization and also by the fact that it is not easy to know exactly the value of D_m, we have applied this concept to estimate the extent of diffusion control.

By using $D_m = 35 \times 10^{-6}$ cm^2 s^{-1} (4) $S_o = 15 \times 10^{-4}$ cm we obtained a value of $\phi = 1.3$ for $R_p = 300$ gPE/mmol. Ti.min.atm., which was one of the highest polimerization rates found; this value of ϕ is, in fact, very small. So, with this catalytic system no internal diffusion control is observed.

2.1.2 - External diffusion limitations

We estimated the effect of monomer diffusion on the overall polymerization rate by two methods:

i) The transfer rate of ethylene through the "film" at the surface of the catalyst was evaluated using the equation:

$$\overline{R}_p = \beta_n ([M]_o - [M]_s) = k' [Ti][M]_s \tag{2}$$

where $\overline{R}_p$ is the average polymerization rate at the stirring speed n, β_n the mass transfer coefficient, $[M]_o$ the monomer concentration in solution, $[M]_s$ the monomer concentration at the surface of catalyst and k' the polymerization rate constant per unit weight of titanium.

ii) Method of stirring speed switch (5)

Figure 6 shows the relation of $1/R_{p\ 700}$ and $1/$ Ti . We can see that the deviation from linearity occurs only for Ti > 0,0156 mmol l⁻¹ which corresponds to polymer concentration inside the reactor > 400 gl⁻¹.

It appears that the amount of polymer accumulated in solution is responsible for such deviation and that no other kind of diffusion problem exists, namely diffusion control through the polymer "film" at the surface of the catalyst, since the straight line obtained for Ti < 0,0156 mmol l⁻¹ passes through the origin.

The switch of one stirring speed to another resulted in a rapid change in the absorption rate of monomer followed by gradual relaxation. The deviation of the rate after the transition to a different stirring speed, is due to the monomer diffusion effect.

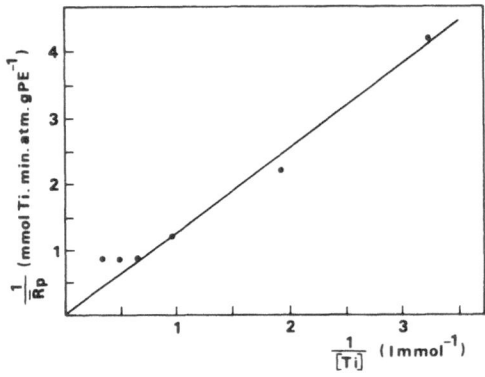

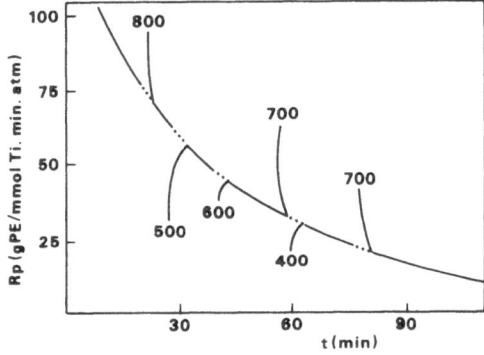

Fig. 6. $\dfrac{1}{\overline{R}_p}$ versus $\dfrac{1}{[Ti]}$

Fig. 7. Effects of stirring speeds on the overall polymerization

We were able to determine the polymerization condition free of the effect of monomer diffusion through the gas-liquid interface.

Figure 7 shows the influence of changing the stirring speeds from 700 rpm to 800 or 400 rpm, on the overall polymerization rate at the polymerization conditions indicated in the fig. 7.

It can be seen that under these polymerization conditions (including catalyst concentration < 0,015 mmol l⁻¹) the effect of monomer diffusion through the gas liquid interface is non-existent.

Subsquently, a kinetic analysis of ethylene polymerization under conditions where the stirring speed and polymer concentration inside the reactor do not affect the overall polymerization rate, could be carried out.

We analised the following affects:

2.2. effect of Titanium Concentration

2.3. Effect of Alkylaluminium Concentration

2.4. Effect of Monomer Concentration

2.5. Effect of the Partial Pressure of Hydrogen

2.6. Effect of Temperature

We found that the resulting polymerization rate could be represented by Eq. (3) or Eq. (4).

$$R_p = k_p \frac{K_A K_M [A] [M] C_p}{(1+K_A [A] + K_M [M])^2} \qquad (3)$$

under conditions where $K_A [A] \gg K_M [M]$ and $[M]$ constant

$$R_p \alpha \frac{C_p [A]}{(1+K_A [A])^2} \qquad (4)$$

3 - Active Center Concentration

In order to elucidate the kinetic feature of the $MgCl_2$ supported catalyst, the number of active centers and the values of the propagation rate constant, k_p were determined.

We used kinetic and molecular weight methods to estimated C_p and k_p in the runs carried out at pressure of 6 bar and T = 80°C and a Radio-Tagging method in those carried out at atmospheric pressure and T = 60°C.

Values for C_p and k_p obtained using both procedures are summarized in table 1.

The values of rate constants and active centre concentration obtained by using a "kinetic" method differ greatly from the values found by using a "Radio-tagging" method. the discrepancy between the data obtained by using these two methods is too large to be reconciled.

Table 1 - k_p and C_p values as a function of time obtained by different methods.

t (min)	C_p (mol/mol Ti)			$k_p \times 10^{-3}$ (1 mol^{-1} s^{-1})			
	(a)	(b)	(c)	(d)	(e)	(f)	(g)
10	0,0155				107		
30	0,0132	0,052			91	3,81	10,9
60	0,0104	0,100	0,0166	115	72	1,64	4,7
90	0,0083				58		
120	0,0064	0,166			45	0, 5	1,4

a C_p values obtained from k_p value calculated using a kinetic methods

b C_p values obtained by a Radio-Tagging method using ^{14}CO.
   (Experimental conditions: $[Ti] = 0,0238$ mmol l^{-1}, T = 60°C, Al/Ti = 84, $P_t = 1,2$ bar, $P_{H_2} = 0$, EC-180).

c C_p average value over the time obtained by using a molecular weight method

d k_p value obtained by using a kinetic method

e k_p values obtained from C_p value calculated by using a molecular weight method

f k_p values calculated from C_p values obtained by using the Radio-tagging method indicated in b

g k_p values obtained as indicated in f but extrapolated to T = 80°C

CONCLUSIONS

The kinetic study of a high activity magnesium chloride supported Ziegler-Natta catalyst when used for the polymerization of ethylene established the following characteristics.

(i) The shape of the kinetic rate-time profiles is strongly dependent on the operational conditions of pressure, temperature and concentration and type of alkylaluminium cocatalyst.

(ii) The variation of the rate of polymerization with alkylaluminium concentration is consistent with a Langmuir-Hinshelwood type of relationship.

(iii) Only a fraction of the available titanium atoms are active as polymerization centres (5-16% using ^{14}CO radio-tagging and much lower if a kinetic and molecular weight method is used).

(iv) The number of active centres is time dependent, incraesing with the time of polymerization (0-2h).

(v) The rate constant for propagation decreases with time and is thus not a time rate coefficient.

(vi) The kinetic data confirm a two-step mechanism for the propagation step, i.e., coordination of the olefin to a metal ion with subsquent insertion of the coordinated monomer molecule into the metal-carbon bond.

REFERENCES

1 - Keii T. (1972) Kinetic of Ziegler-Natta Polymerization, Kodansha, Tokyo
2 - Burfield D.R., McKenzie I.D. and Tait P.J.T. (1976) Polymer 17:130
3 - Tait P.J.T. Conference presented in Tokyo in June/85
4 - Chien J.C.W. (1979) J Polym Sci Polym Chem ED 17:2555 (1985) J Polym Sci Polym Chem Ed 23:723
5 - Keii T., Doy Y. , Kobayashi H. (1973) J Polym Sci Polym Chem Ed 11:1881

2. New Aspects in the Heterogeneous Catalysts Polymerization of Olefins

ALUMINIUM-FILLED POLYETHYLENE.
ZIEGLER-NATTA POLYMERIZATION ON FILLER SURFACE

S. Ewangelidis, A. Hanke, K.H. Reichert

Technische Universität Berlin
Institut für Technische Chemie
Straße des 17. Juni 135, D-1000 Berlin 12

ABSTRACT

A new route to aluminium-filled polyethylene is reported in which a Ziegler-catalyst is fixed on the surface of aluminium powder. In the presence of this supported catalyst, ethylene is polymerized in suspension in a stirred tank reactor. The polymerization takes place on the surface of the filler forming a uniform shell of polyethylene around each aluminium particle.

The surface of the aluminium powder used was characterized by BET-surface measurement, X-Ray Photoelectron Spectroscopy and Auger-Electron-Spectroscopy.

For preparation of the catalyst $TiCl_4$ was fixed on the surface of the aluminium powder by addition of $TiCl_4$ to a slurry of aluminium powder in hydrocarbons at 80 °C. The adsorption isotherme of $TiCl_4$ on aluminium powder indicates that $TiCl_4$ is chemically bonded to the surface and the maximum coverage can be determined at the conditions given.

The supported $TiCl_4$ was activated with $Al(iBu)_3$ and the polymerizations were conducted in the temperature range of 40 - 80 °C at ethylene pressures less than 5 bar.

The metal content of the filled polyethylene was varied in the range from 14 to 99 % by weight.

Some thermal, electrical and mechanical properties of the metalloplastics were measured as a function of aluminium content. The electrical properties of the metalloplastics which were formed by physical and chemical blending show great differences. The electrical resistivity of the products formed by in situ polymerization is some decades larger than those formed by mechanical blending.

It could be shown that the polymerization of ethylene in the presence of aluminium powder leads to metalloplastics with good thermal conductivity and simultaneously very low electrical conductivity.

A simple theoretical model for the prediction of electrical conductivity of metalloplastics will be presented under consideration of agglomeration of metal particles.

INTRODUCTION

Metal-filled polymers are materials with interesting properties. Some properties of interest are:
-Thermal and electrical conductivity
-Magnetizability
-Heat capacity
-Radiation absorption
-Thermal expansion

The properties mentioned above lead to the following applications:
-Repair and maintenance products
-Electrically conductive adhesives
-Magnetic tapes and disks

W. Kaminsky and H. Sinn (Eds.)
Transition Metals and Organometallics as
Catalysts for Olefin Polymerization
© Springer-Verlag Berlin Heidelberg 1988

-Corrosion resistant coatings
-Specialized uses for electronic industries
-Solid rocket propellants
-Body soldes, etc.

There are different ways to prepare metal-filled polymers:
-Blending of metal powder and polymer powder or polymer melt
-Formation of metal particles in the presence of polymer melt or solution
-Polymerization in the presence of metal particles

The first and second way leads to products with relatively low electrical resistivity because the metal particles in the polymer matrix can contact each other and can form an electrical conductive path through the material. The fixation of a Ziegler-catalyst on the surface of metal particles and the following polymerization leads to products with enhanced thermal conductivity and simultaneously very high electrical resitivity because every metal particle is covered with a stable shell of polymer which prevents electrically conductive contacts of the particles.
In this paper the Ziegler-Natta polymerization of ethylene with the catalyst system $TiCl_4 / Al(iBu)_3$ in the presence of aluminium powder and some properties of the material are reported.

EXPERIMENTAL

Characterization of the Aluminium Powder

The aluminium powder (Eckart-Werke, Ecka AS 081) which was used as filler was characterized by different methods. The results are listed below:

    Particle size distribution: 5µm - 45µm
    BET - surface: 0.41 m²/g

The powder is nonporous and contains oxydic and hydroxydic oxygen on the surface as detected by X - Ray Photoelectron Spectroscopy. The powder contains little amount of the elements Fe, Mg, Si and Zn. The total concentration of the four metals is less than 2 % by weight.

Adsorption of TiCl₄ on Filler Surface

In order to avoid free $TiCl_4$ during polymerization the adsorption isotherme of $TiCl_4$ on the surface of the aluminium powder was determined.
The investigations were carried out in a 250 ml glas reactor equipped with magnetic stirrer under nitrogene atmosphere. The diesel oil (ESSO AG, Exsol D 140/170) used as suspension fluid was dried by molecular sieves (0.4 nm). The aluminium powder was pre-treated for 5 hours under vacuum at 250 °C. The powder was suspended and $TiCl_4$ was added in little portions. The concentration of $TiCl_4$ in the suspension fluid was measured by atomic absorption spectroscopy (AAS). The saturation concentration of $TiCl_4$ adsorbed on the powder at given conditions is $6.3 \cdot 10^{-6}$ mol $TiCl_4$ per m² of surface. It could be shown by the method of Wave Dispersive X - Ray Analysis with microprobe that the distribution of $TiCl_4$ on the surface of molded pieces of aluminium particles is rather uniform. Figure 2 shows the adsorption isotherme of $TiCl_4$ on aluminium powder.

Polymerization

The aluminium powder was pre-treated as described above, before dispersing it into dry diesel oil in a stirred tank reactor under nitrogene atmosphere. After the suspension was heated to the reaction temperature the $TiCl_4$ was added. Free $TiCl_4$ should not be present in the suspension fluid, therefore the added amount of $TiCl_4$ was kept below the saturation concentration on the surface of the suspended powder. The reactor was saturated with ethylene and the polymerization started by the addition of $Al(iBu)_3$ to the suspension. The powdery product of the polymerization was filtered and washed with acetone. Figure 1 shows the flow sheet of the polymerization process.

Time Dependency of Ethylene Absorption Rate

The ethylene absorption rate during polymerization was measured by a thermal mass flowmeter (Brooks Instruments B.V.). Figure 3 shows a characteristic curve of the absorption rate versus time.

Determination of Activation Energy

The activation energy of the ethylene polymerization with the surface fixed catalyst system $TiCl_4/Al(iBu)_3$ was determined by polymerizations at different temperatures. The value of the activation energy is 29.7 kJ/mol. The Arrhenius - plot for the determination of the activation energy is shown in Fig. 4.

Product Morphology

In Fig. 5 two Scanning Electron Microscope photos of aluminium particles before and after polymerization are shown. The surface of the aluminium particles is smooth and nonporous (right photo). After polymerization the particles are covered by a shell of polyethylene produced on the surface of the filler particles (left photo).

Electrical Properties of the Filled Polymer

Figure 6 shows the comparision of the electrical resistivity of aluminium filled polyethylene prepared in two different ways as a function of aluminium content. The material obtained by mechanical blending of aluminium powder and polyethylene powder shows a rapid decrease of the electrical resistivity at filler contents in the region of 10 - 20 % by volume of aluminium.

The product of the in situ polymerization can be filled up to 70 - 80 % by volume of aluminium without significant loss of electrical resistivity. This can be explained by the existence of a closed shell of polymeric material around each aluminium particle which prevents the filler particles from forming electrical conductive contacts. At filler contents higher than 80 % by volume the polymeric shell is very thin therefore it cannot avoid disruptive breakdown of the material.

Thermal Conductivity of the Filled Polymer

The plot of the measured thermal conductivity of the in situ polymerization products versus filler content is shown in Fig. 7. The dotted lines are calculated curves for the thermal conductivity which were obtained by two different theoretical models. Many models for the prediction of the thermal conductivity and other properties of filled polymers are described in literature. Those of Bruggemann [4] and Lewis-Nielsen [5] show the best correspondence to the experimental results of this work.

Mechanical Properties of the Filled Polymer

Figure 8 indicates the relative tensile strength of aluminium filled polyethylene as a function of filler content. There is only a little difference between the curves of the compounds prepared by in situ polymerization and the mechanical blending of the components. As expected for a non-reinforcing filler material the relative tensile strength of the material decreases with increasing filler content. The plateau for filler contents in the range of 20 - 70 % by volume is unexpected.

MODEL FOR THE PREDICTION OF THE ELECTRICAL CONDUCTIVITY OF METAL FILLED POLYETHYLENE

In a given metalloplastic matrix a cubic volume element is regarded. The volume element is devided into smaller sections by an orthogonal lattice. These are partly filled with metal spheres by a random procedure layer by layer.

The following assumptions were made:
- all the metal particles have the same diameter

- two metal particles which contact each other are electrical conducting; contact resistants are neglected
- a constant number of metal particles for each layer is given, i.e. the occupation density is the same for all layers
- the electrical conductivity is directly proportional to the number of connected particles, which have simultaneously contact with the first and the last layer of the cube.
- the coordination number of the spherical particles is six.

In order to simulate agglomeration of the metal particles within the polymer matrix the random generation of lattice sites was repeated until a contact of this lattice site with an occupied lattice site in its neighborhood is occuring. The number of repetitions is a direct measure of agglomeration probability. For details see PhD Thesis of W. Schöppel [1] and S. Ewangelidis [7].

The following results were achieved from the simple model:
1. With increasing agglomeration probability of the metal particles electrical conductivity starts at lower metal content (see Fig. 9)
2. In the case of a random distribution of metal particles in the lattice (P(AG) = 1) the results of this model are in agreement with those of the percolation theory (see Fig. 9)
3. As a consequence of equal occupation densities of all layers of the lattice with metal particles an anisotropy of the electrical conductivity is observed in the case of agglomerized particles.
4. One advantage of the model is that loops of conducting paths are also taken into account. By this consideration the lattice can be analyzed completely (see Fig. 10).

References:

1. Schöppel W (1985) PhD Thesis, Technische Universität Berlin
2. Reichert K-H, Schöppel W, Ewangelidis S, Bednarz J, Leute U (Siemens AG)
    EP 167 000 (1985), JA 60 - 1223 80 (1985), US 74 1351 (1985), KA 483 105 (1985)
    „Gut wärmeleitende, elektrisch isolierende Thermoplaste"
3. Reichert K-H, Ewangelidis S, Hanke A, Leute U, Bednarz J   Symposium
    proceedings of the International Symposium on Transition Metal Catalyzed
    Polymerizations in Akron(USA) (1986), Cambridge University Press, New York (in press)
4. Bruggeman DAG (1935) Ann Phys 24: 636-664
5. Nielsen LE (1973) J Appl Polym Sci 17: 3819-3820
6. Shante VKS, Kirkpatrick S (1971) Adv Phys 20: 325
7. Ewangelidis S (in preparation) PhD Thesis, Technische Universität Berlin

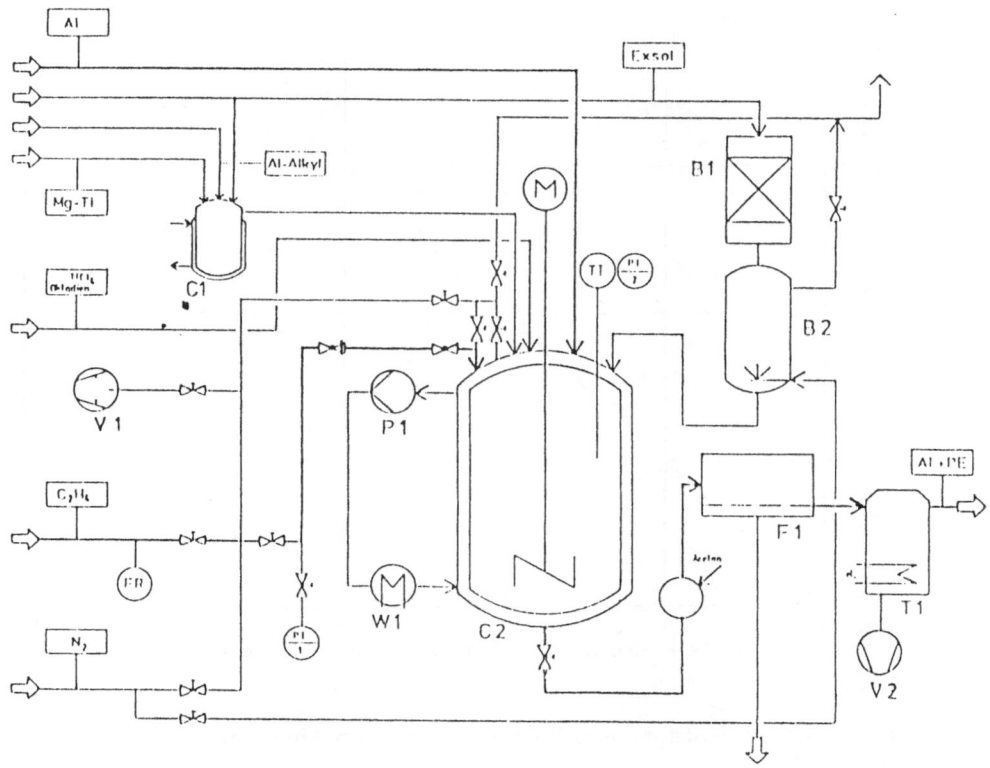

Fig. 1. Flow sheet of the ethylene polymerization process

B1:   molecular sieve column
B2:   nitrogene saturation vessel
C1:   chemical reactor for catalyst formation
C2:   polymerization reactor
F1:   filter
M: stirrer motor
P1:   pump
PI: manometer
T1:   dryer
Ti: thermocouple
V1,V2:   vacuum pump
W1:   heat exchanger

142

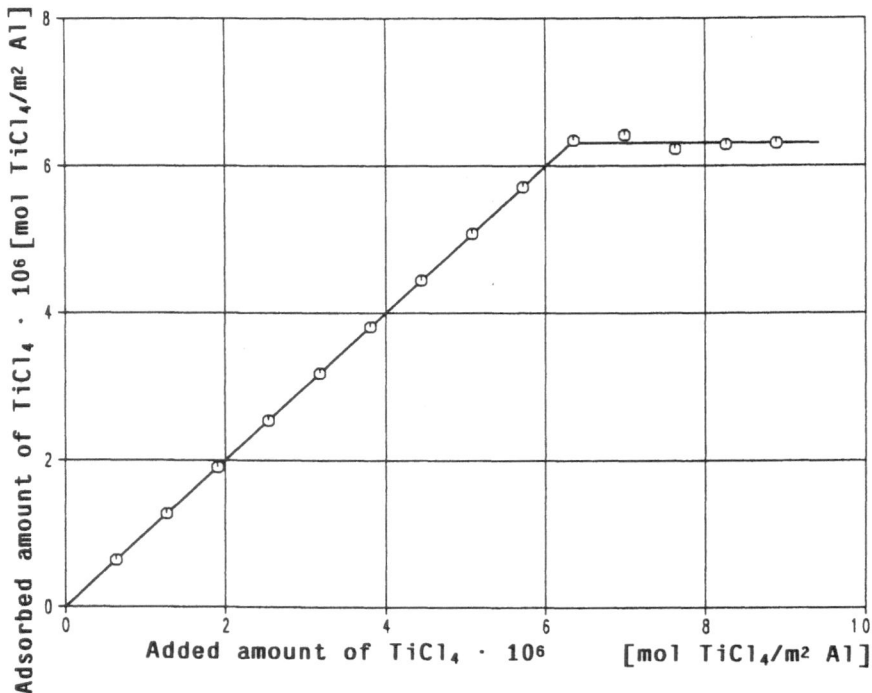

Fig. 2. Adsorption isotherme of TiCl₄ on aluminium surface.
Reaction conditions:
Temperature: 80 °C
Suspension fluid: Diesel oil (Exsol D 140/170)
Aluminium content of suspension :   6.3 % by volume

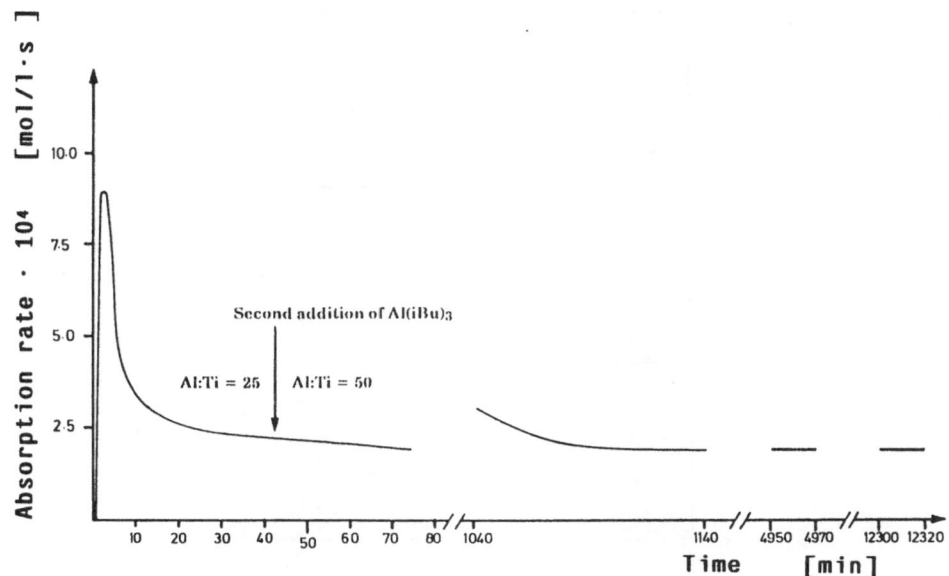

Fig. 3. Time dependency of ethylene absorption rate.
Polymerization conditions:

Temperature:	62 °C
Ethylene pressure:	2.45 bar
Stirring speed(Intermig):	1000 rpm
Suspension fluid:	Cyclohexane
Titanium concentration:	$7.15 \cdot 10^{-4}$ mol/l
Aluminium alkyle:	Al(iBu)$_3$
Molar ratio Al:Ti in catalyst:	25
Al-powder content of suspension:	9.8 % by volume

The polymerization was interrupted three times by venting and cooling of the reactor. The polymerization could be started again by saturation of the suspension at reaction temperature with ethylene. The interruptions had no influence upon the ethylene absorption rate. Addition of Al(iBu)$_3$ during the polymerization is without influence upon the ethylene absorption rate.

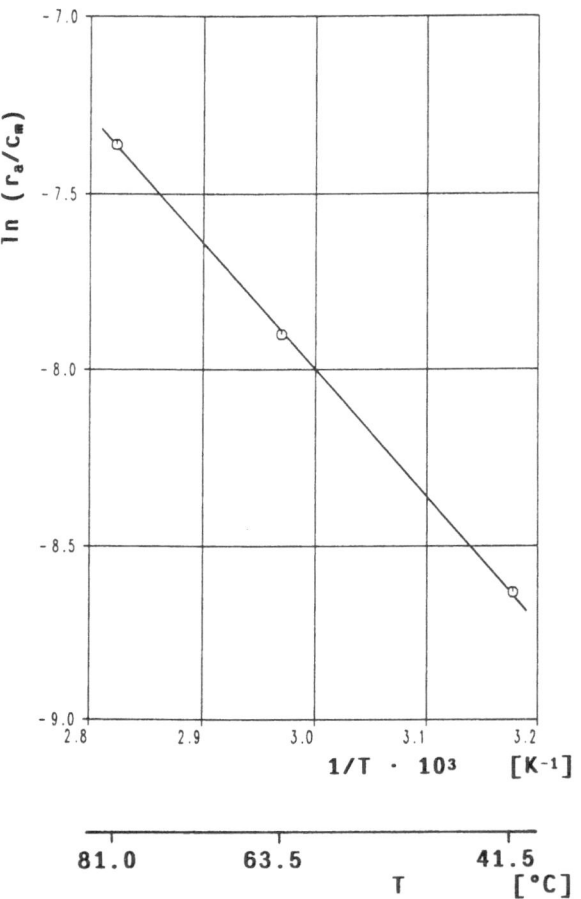

Fig. 4. Temperature dependence of ethylene polymerization rate at 1 hour reaction
time

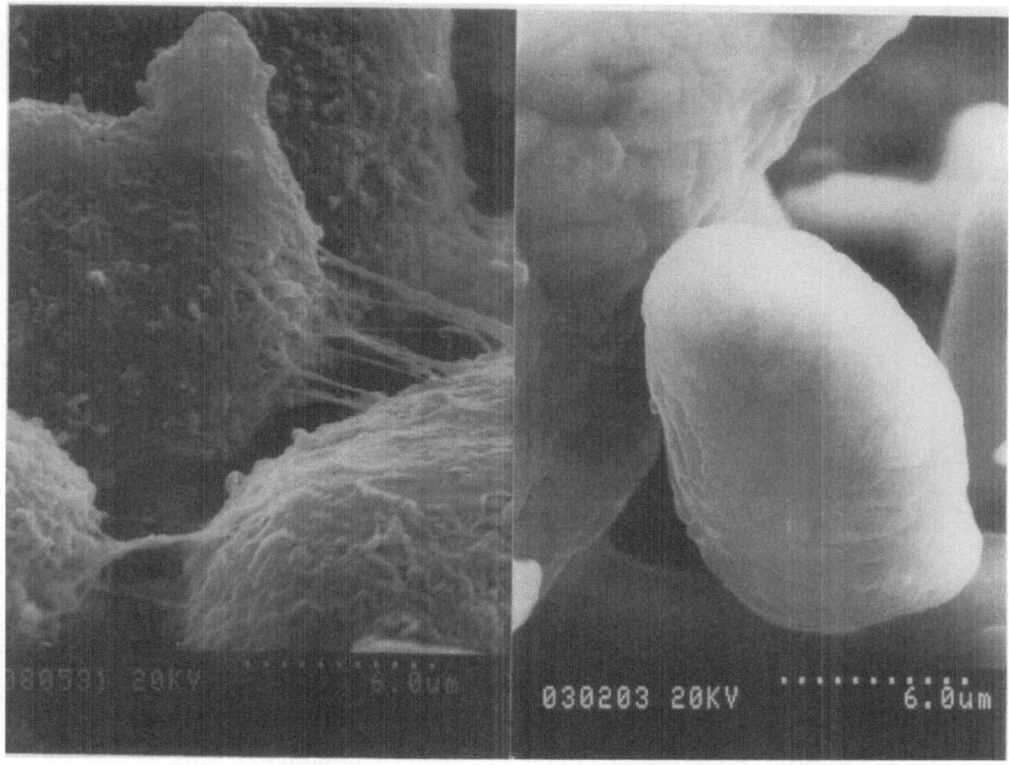

Fig. 5. Product morphology.
Right hand: SEM-photo of the aluminium particles before polymerization
Left hand: SEM-photo of the aluminium particles after polymerization

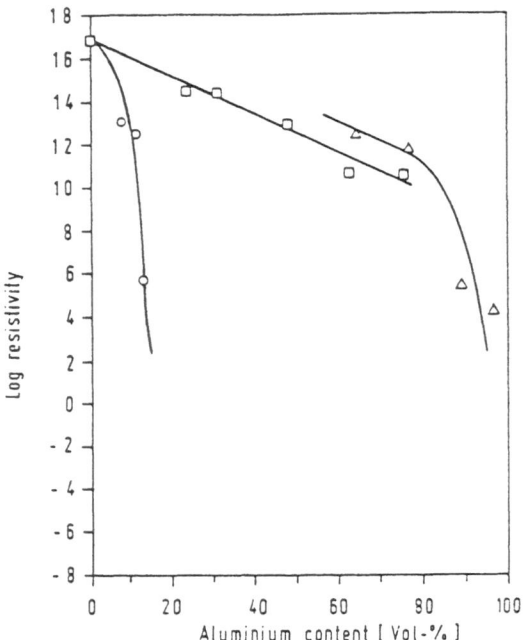

Fig. 6. Specific electrical resistivity [Ω · cm] at room temperature of polyethylene composites as a function of aluminium content for composites prepared by
— O —   mechanical blending
— □ —   in situ polymerization (100 V measuring d.c. voltage)
— Δ —   in situ polymerization (10 V measuring d.c. voltage)

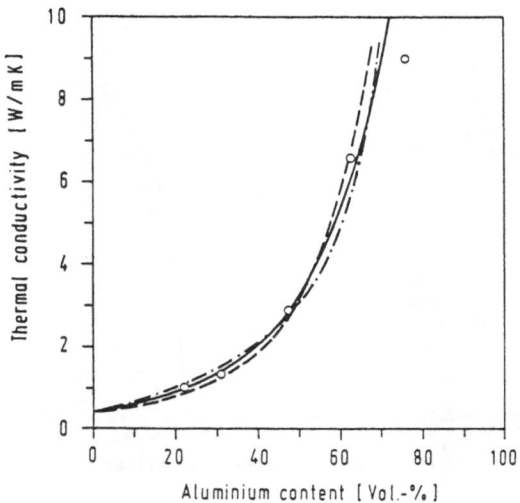

Fig. 7. Thermal conductivity of aluminium / polyethylene composites with different aluminium content at room temperature
—— O ——   experimental data
- - - - - - - - - - - -   Bruggeman - Model
· - ·· - · - · -··   Lewis - Nielsen - Model

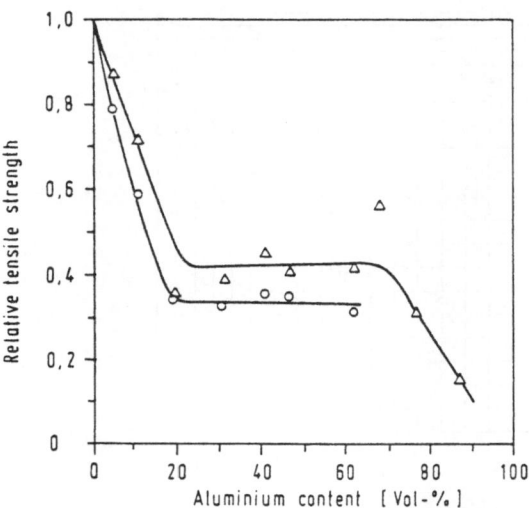

Fig. 8. Relative tensile strength (tensile stregth of composite / tensile strength of polyethylene) at different aluminium content at room temperature
—— O ——   mechanical blending
—— Δ ——   in situ polymerization

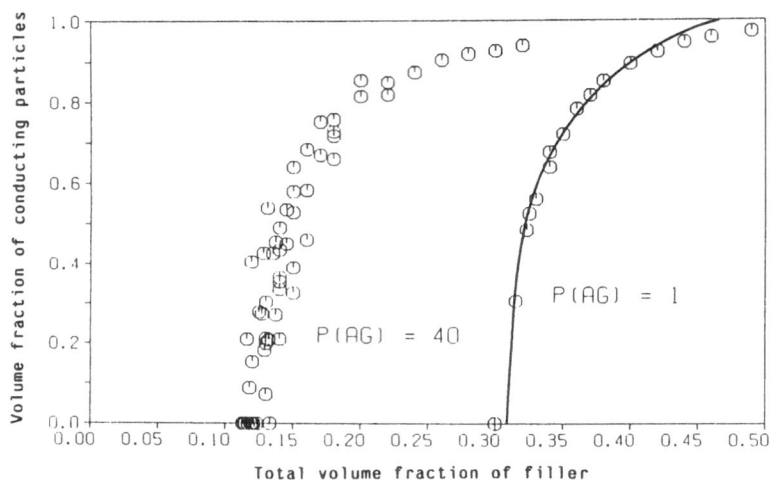

Fig. 9. Volume fraction of filler which participates in electrical conduction as function of total volume fraction of filler modeled for two agglomeration probabilities (1 and 40)
Ⓞ   determined by modelling (according to this work)
────   calculated by percolation theory (simple cubic, site problem) [6]

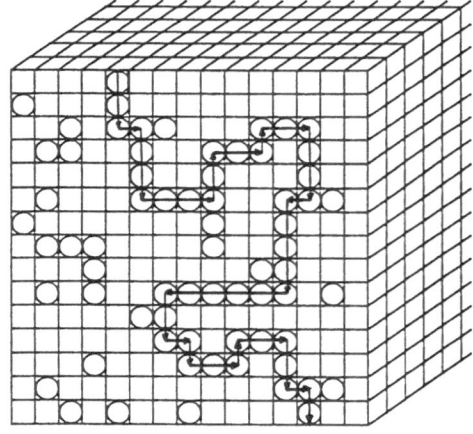

Fig. 10. Schematical description of loops of conducting paths

Developments with High-Activity Titanium, Vanadium, and Chromium Catalysts in Ethylene Polymerization

Frederick J. Karol, Kevin J. Cann, and Burkhard E. Wagner

UNIPOL Systems Department, Union Carbide Corporation, P.O. Box 670, Bound Brook, New Jersey 08805 USA

SUMMARY

Recent developments with high-activity catalysts based on titanium, vanadium, and chromium compositions continue to illustrate the unique behavior of each metal center in ethylene polymerization catalysis. Bimetallic complexes containing magnesium-titanium- or magnesium-aluminum-electron donor complexes have been identified and characterized. Interestingly, studies with $TiCl_3$ reveal that complex formation with $MgCl_2$ is not required for obtaining high-activity catalysts.

Supported vanadium catalysts, in the presence of halocarbons as promoters, display high activity. Highly saturated polyethylenes of intermediate or broad molecular weight distribution are usually produced. Studies with bimetallic vanadium-zinc complexes, and valence measurements in the presence of promoters, strongly indicate that divalent vanadium centers comprise the active sites.

Chromium catalysis has focused on support modifications, changes of ligand about chromium, and redox cycles as ways to expand the scope of these catalysts. The development and fine tuning of each catalyst type as a means to manipulate polymer properties remain active areas of research in laboratories throughout the world.

INTRODUCTION

Catalyst research in ethylene polymerization since the 1950s has focused for the most part on titanium, vanadium, and chromium compositions. By the early 1960s much of catalyst research and development became compartmentalized into three distinct camps. One camp directed attention toward the commercialization of titanium catalysts for high density polyethylenes. Another camp, initially led by the developments at Phillips Petroleum, pursued the commercialization of polyethylenes based on chromium oxide catalysts. For those interested in pursuing production of ethylene-propylene-diene rubbers, vanadium catalysts became the preferred catalytic route. Much of the energies of researchers in the 1960s focused on improving existing processes with each of the particular catalyst systems. An outgrowth of such directed research has been the identification of many major research laboratories with a focus on a particular metal (Ti, V, Cr) for ethylene polymerization catalysis.

W. Kaminsky and H. Sinn (Eds.)
Transition Metals and Organometallics as
Catalysts for Olefin Polymerization
© Springer-Verlag Berlin Heidelberg 1988

Recently research has become more directed toward the unique process and product opportunities that different metal centers offer in ethylene polymerization. This paper discusses some recent investigations in ethylene polymerization catalysis with titanium, vanadium, and chromium centers. The product and process opportunities offered through use of more than one metal center provide a challenge for the commercial development of catalyst technologies.

DISCUSSION

Bimetallic Complexes of Magnesium and Titanium

Bimetallic complexes containing magnesium, titanium, and electron donor molecules, when combined with aluminum alkyls, show high catalytic activity in ethylene polymerization (1-3). Bimetallic halide complexes of magnesium and titanium can be prepared by reacting magnesium halides with titanium(IV) halides at a temperature of from 25-150°C in electron donor solvents (1). Unlike surface complexes formed by chemical anchoring of a complex to a substrate, bimetallic complexes are well-defined compounds with characteristic properties (2,4-6). The product of interaction between the components $TiCl_4$, $MgCl_2$, and $CH_3COOC_2H_5$, namely $MgTiCl_6(CH_3COOC_2H_5)_4$, has been investigated by means of single-crystal, X-ray diffraction methods (4,5).

Likewise, reaction between $[MgCl_2(THF)_2]$ and $[TiCl_4(THF)_2]$ in tetrahydrofuran (THF) yields a yellow crystalline salt $[Mg_2Cl_3(THF)_6]^+$ $[TiCl_5(THF)]^-$. The crystal structure of this salt has been defined in our laboratories and by another group (6). The coordination geometry about the titanium atoms is octahedral with five chlorine atoms and one THF molecule bound to the titanium atoms. The coordination geometry about each magnesium atom in the cationic magnesium dimer is octahedral. Each Mg(II) ion is coordinated to three chloride ions and the oxygen atoms in three THF molecules. In addition, the three chlorine atoms are shared and bridge the two magnesium atoms.

Other bimetallic complexes can be prepared by reacting $TiCl_4$, $VOCl_3$, $MoOCl_4$, $WOCl_4$, or $AlCl_3$ with Be, Mg, Ca, or Sr chlorides in the presence of electron donors such as $POCl_3(L)$ or $C_6H_5POCl_2(L')$ (2). These complexes show well-defined stoichiometry, ionic character, and crystalline structure. Complexes $(TiCl_6)MgL_6$, $(TiCl_5L')_2MgL_6'$, and $(Ti_2Cl_{10})MgL_6$ when treated with $(i-C_4H_9)_3Al$ were found very active in ethylene polymerization.

Although many bimetallic-electron donor complexes containing titanium(IV) are known, magnesium-titanium(III)-THF complexes do not form. One study (7) examined a series of reactions between $[TiCl_3(THF)_3]$ and $[MgCl_2(THF)_2]$, under various conditions. These compounds did not mutually react. This lack of reactivity was attributed to a similarity in the acid-base properties of $MgCl_2$ and $TiCl_3$. Other workers have investigated the reaction of $TiCl_3$ with $MgCl_2$ in the presence of $POCl_3$ (2). The resultant complex $[MgL_6]^{2+}[TiCl_6]^{2-}$ strongly indicates a chlorination reaction exerted by $POCl_3$, and accompanied by oxidation of Ti(III) to Ti(IV). Hence, no Ti(III) complex with magnesium forms in the presence of $POCl_3$.

Our results using conductometry, wet solution methods, infrared, and X-ray diffraction provide strong evidence for the absence of complex formation between $MgCl_2$[1] and $TiCl_3$ in THF (eq. 1) or in the solid state (8).

$$MgCl_2(THF)_{1.5} + TiCl_3(THF)_3 \xrightarrow{\text{THF}} \text{no complex formation} \qquad (1)$$

The conductive capacity of THF solutions containing $MgCl_2$ and $TiCl_3$ (hydrogen-reduced) was measured (Table 1). Very low levels of conductance were measured for separate THF solutions of $MgCl_2$ and $TiCl_3$ (hydrogen-reduced). THF solutions containing both $MgCl_2$ and $TiCl_3$ also exhibited a low level of conductance. These measurements suggest that solvated ions or ion-pairs are not formed in THF containing $MgCl_2$ and $TiCl_3$.

However, solutions containing $MgCl_2$ and $TiCl_4$ in THF show large increases in conductive capacity when compared to the conductance of the individual components. The high conductance of $MgCl_2$ and $TiCl_4$ in THF can be attributed to salt formation as illustrated in equation 2.

$$2MgCl_2(THF)_2 + TiCl_4(THF)_2 \xrightarrow{\text{THF}} [Mg_2Cl_3(THF)_6]^+[TiCl_5(THF)]^-$$

$$(2)$$

The high levels of conductance of THF solutions containing aluminum compounds and $MgCl_2$ are attributed to formation of highly ionic magnesium-aluminum-THF salts, such as illustrated in equation 3.

$$[MgCl_2(THF)_2] + AlCl_3 \xrightarrow{\text{THF}} MgAlCl_5 (THF)_6 \qquad (3)$$

Another investigation (7) found that $AlCl_3$ reacts easily with magnesium-titanium compounds to provide the salt shown in equation 3.

When $TiCl_3$ (aluminum activated) was mixed with $MgCl_2$ in THF, a high level of conductance was measured. This high conductance level can be attributed to formation of magnesium-aluminum salts. Large differences in conductance between hydrogen-reduced and aluminum-reduced $TiCl_3$ provide support for this conclusion.

Infrared spectroscopy proved useful for characterizing new structural complexes and for providing evidence that new compositions are formed. This technique was used to study the interaction of $MgCl_2$ with $TiCl_4$ or $TiCl_3$ in the presence of THF. The extent of metal-ligand interaction was measured by monitoring key absorptions in the infrared. By comparing the absorption of the uncomplexed ligand and the degree this absorption was shifted to lower wave numbers, it was possible to determine the extent of complexation.

---

1 It is possible to prepare a wide range of products with the overall composition $MgCl_2 \cdot xTHF$ with values of x in the range of 0-4. All these products are highly soluble in THF.

Table 1. Conductivity of Magnesium, Titanium, and Aluminum Halides
in THF

$MgCl_2$, Molar Conc.	Substrate/Molar Conc.	Conductance,[a] $ohm^{-1} \times 10^6$
0.125	-	2.6
--	$TiCl_3/0.051$	1.8
--	$TiCl_4/0.060$	2.2
--	$(C_2H_5)_2AlCl/0.089$	3.0
0.125	$TiCl_3/0.051$	7.0
0.094	$TiCl_4/0.052$	53
0.075	$(C_2H_5)_2AlCl/0.12$	290
0.104	$TiCl_3 \cdot AA/0.13(Ti)$ $/0.044(Al)$	320
0.104	$AlCl_3/0.046$	490

[a] Measured at room temperature using a 20 ml conductivity cell with a
cell constant of one.

The infrared absorptions for the magnesium-titanium(IV)-THF complex
$[Mg_2Cl_3(THF)_6]^+[TiCl_5(THF)]^-$ shown in Table 2 provide evidence
for complex formation with new absorptions appearing at 1030
cm^{-1}, 890 cm^{-1}, and 875 cm^{-1}. When a similar study was carried
out using $TiCl_3$ in place of $TiCl_4$, no new absorptions appeared in
the infrared, indicating no new complex was formed. The $TiCl_3$
samples used in this study were prepared by physically blending
$MgCl_2(THF)_{1.5}$ and $TiCl_3(THF)_3$ and by total residue isolation of
$MgCl_2$ and $TiCl_3$ from THF solution.

The mechanism by which magnesium enhances catalyst activity is
frequently attributed to the similar layer lattice structures of
$MgCl_2$ and $TiCl_3$. The similarity in size of magnesium and titanium
ions probably plays a role. Whatever the importance of these factors
it seems clear that formation of a discrete chemical complex between
magnesium and titanium is not necessary for achieving a boost in
polymerization activity in the presence of $MgCl_2$.

Although no bimetallic complex formation occurs between $MgCl_2$ and
$TiCl_3$ in the presence of THF, polymerization performance with a
$TiCl_3(THF)_3$ catalyst was highly dependent on the Mg/Ti ratio
(0/1-4/1). Under the conditions studied the $TiCl_3$ catalyst, in the
absence of magnesium, had essentially no activity. A 0.5/1 Mg/Ti
catalyst had poor polymerization performance with low activity, poor
hydrogen response and comonomer incorporation. Catalysts with higher
Mg/Ti ratios up to 4/1 had improved behavior with much higher activity
and better hydrogen and comonomer response. These results show that
the presence of sufficient $MgCl_2$ in the vicinity of the active
titanium sites is essential for reaching high rates of polymerization
activity and improved catalyst performance.

One speculation (7) on the role of $MgCl_2$ associates its behavior to
the ability of $MgCl_2$ to form complexes with aluminum halides during
the polymerization process (eq. 3). Formation of these complexes
prevents the aluminum halides such as $C_2H_5AlCl_2$ from causing
catalyst deactivation by reaction with the active titanium centers.
The extent to which such complex formation is responsible for the
boost in polymerization activity in the presence of magnesium halides
has not been determined.

Table 2.   Infrared Data on Magnesium-Titanium-Electron Donor
            Compositions

Compositions

$[Mg_2Cl_3(THF)_6]^+[TiCl_5(THF)]^-$ Complex A
$MgCl_2(THF)_{1.5}/TiCl_3(THF)_3$ Mixture B

Diagnostic Infrared Absorbances (cm^{-1})

THF	1070, 913
$MgCl_2(THF)_{1.5}$	1037, 893
$TiCl_4(THF)_2$	994, 825
$[TiCl_4 \cdot THF]_2$	988, 821
Complex A	1030, 890, 875
$TiCl_3(THF)_3$	1010, 855
Mixture B	1037, 1010, 855

Bimetallic Complexes of Magnesium and Aluminum

Magnesium chloride reacts with Lewis acids such as $AlCl_3$, $C_2H_5AlCl_2$, $(C_2H_5)_2AlCl$, $(C_2H_5)_3Al$, and BCl_3 in THF or $CH_3COOC_2H_5$ and the reaction studied using ^{13}C-NMR, conductometry, ultraviolet, and physical isolation techniques (9). Several complexes can be isolated by crystallization from solution (Table 3). These complexes activate $TiCl_4$ to provide high-activity catalysts for ethylene polymerization.

Table 3.   Complex Formation with $MgCl_2$, Electron Donor, and Lewis
            Acid

$MgCl_2 \cdot 2AlCl_3 \cdot nTHF$
$MgCl_2 \cdot 2C_2H_5AlCl_2 \cdot nTHF$
$2MgCl_2 \cdot (C_2H_5)_3Al \cdot nTHF$
$MgCl_2 \cdot 2BCl_3 \cdot 6CH_3CO_2C_2H_5$
$MgCl_2 \cdot 2C_2H_5AlCl_2 \cdot nCH_3CO_2C_2H_5$

It is uncertain as to the actual structures of the complexes listed in Table 3. The bonding interactions of these species probably mimic those reported for the analogous $MgCl_2$-$TiCl_4$-electron donor adducts.

The increase in the conductive capacity of THF solutions of $MgCl_2$ with Lewis acids in THF and $CH_3CO_2C_2H_5$ suggests some type of ionic interaction has occurred (Table 4). Conductivity values for a number of Lewis acids with $MgCl_2$ were measured. These values fit well with the known Lewis acidity of the different organoaluminum species examined,

$$C_2H_5AlCl_2 > (C_2H_5)_2AlCl > (C_2H_5)_3Al$$

Most likely the formation of ionic species in solution relates to the ability of the aluminum compound to abstract a halide anion from $MgCl_2$.

Table 4. Conductance of Alkylaluminum Halides with $MgCl_2$ in
Electron Donor Solvents

$MgCl_2$, Molar Conc.	Substrate/Molar Conc.	Solvent	Conductance,[a] ohm^{-1} x10^6
0.10	-	THF	3
--	$(C_2H_5)_3Al/0.34$	"	3
0.094	$(C_2H_5)_3Al/0.057$	"	49
0.075	$(C_2H_5)_2AlCl/0.12$	"	290
0.075	$C_2H_5AlCl_2/0.12$	"	440
0.16	-	$CH_3CO_2C_2H_5$	67
--	$(C_2H_5)_3Al$	"	<1
0.13	$(C_2H_5)_3Al/0.13$	"	620
0.13	$(C_2H_5)_2AlCl/0.13$	"	650

[a] Measured at 22°C using a 20 ml conductivity cell with a cell
constant of one.

The magnesium-aluminum-electron donor complexes activate titanium
compounds when the complex is impregnated in a silica support. To
achieve this end, the magnesium halide and aluminum compound were
dissolved in the electron donor solvent and slurried with the silica
support. The excess solvent was then removed under reduced pressure.
The resulting impregnated complexes were slurried with, for example, a
tetravalent titanium compound in hexane, followed by washing and
drying. These impregnated catalysts, in the presence of
$(C_2H_5)_3Al$, were found to be highly active in ethylene polymerization.

The identification of the existence of magnesium-aluminum-electron
donor complexes means that species of this type could potentially be
important intermediates in the catalytic cycle for olefin
polymerization.

Developments with Supported Vanadium Catalysts

While extensive research has been carried out with high-activity
catalysts based on titanium, chromium, and zirconium, high-yield
supported vanadium catalysts have not received much attention. Since
vanadium catalysts have been known for some time, this neglect is
surprising. Soluble vanadium catalysts based on compounds such as
$VOCl_3$ and $VO(OC_2H_5)_3$ have been studied rather extensively in the
polymerization of ethylene with propylene (10-12). Unlike soluble
vanadium catalysts that usually produce polyethylenes of very narrow
molecular weight distribution ($\bar{M}w/\bar{M}n$ = 2-3) (10-14), supported
vanadium catalysts provide polyethylenes of intermediate or broad
molecular weight distribution (15-17). Polymerization activities and
productivities with supported vanadium catalysts are comparable to
high-activity catalysts based on titanium or chromium. These high
activities are reached when the vanadium catalysts are used in
conjunction with both an alkylaluminum cocatalyst and halogenated
hydrocarbon promoters. Typical promoters include chloroform,
fluorotrichloromethane, 1,2-difluoro-tetrachloroethane, benzyl chloride,
and ethyltrichloroacetate.

Monsanto, and later Cities Service, have described several vanadium
catalysts for producing polyethylenes (15). U.S. Patent 3,956,255
(15c) discloses an ethylene polymerization catalyst composition

consisting essentially of a vanadium compound combined with an alkylaluminum alkoxide and a trialkylaluminum. The catalyst composition is deposited on silica which has been reacted with an alkylaluminum alkoxide. A halogenated alkane may be used as a promoter. The focus of other patents to Monsanto and Cities Service (15) and others (16,17) is summarized in Table 5. Considerable attention in these patents relates to methods for control of molecular weight distribution, routes to high catalytic activity, methods to improve polymer morphology, and ways to provide more stable operations.

U.S. Patent 4,508,842 (16) describes an ethylene polymerization catalyst comprising a supported precursor of vanadium trihalide-electron donor complex-alkylaluminum or boron halide. The precursor when combined with an alkylaluminum cocatalyst and an alkyl halide promoter provided enhanced polymerization and productivity (Table 6). These high productivity, supported catalysts showed a high response to hydrogen and provided polyethylenes with minimal unsaturation. Polyethylenes of intermediate or broad molecular weight distribution were produced. Incorporation of α-olefins such as butene-1 and hexene-1 was very efficient, allowing a wide range of copolymers to be produced. Addition of $MgCl_2$ to the supported precursor did not significantly raise catalytic activity, and was therefore a nonessential component.

Table 5. Content of Patents on Supported Vanadium Catalysts

Operation	Results	References
Silica Modifications	High Activity	(15b, c)
Divalent V-Zn Salts	High Activity	(17, 18)
Promoter Selection	Lower Halide Content; Improved Reactor Operations	(15d, k)
Alcohol Addition	Narrow MWD	(15f, g, h, k)
Mixed V Compound	Control MWD	(15i)
Aluminum Alkyls/Halides in Catalyst Preparation	Improved Polymer Morphology	(16)
Catalyst Preparation Procedures	More Stable Operations	(15e, j)

Table 6. Features of Supported $VCl_3(THF)_3/SiO_2$ Catalysts

Cocatalysts		R_3Al
Promoters		Chlorocarbons ($CHCL_3$, CCl_4, CH_3CCl_3, CF_2ClCCl_3, etc.
H_2 Response	Outstanding	H_2/C_2 ~0.03 (FI = 460)
Copolymerization	Outstanding	$\bar{C}_4/\bar{C}_2$ ~0.14 (ρ = 915)
MWD	Broad	$\bar{M}w/\bar{M}n$ >8
Productivity	High	2-6 ppm V
Unsaturation	Very Low	<0.01 C = C/1000C

U.S. Patent 4,508,842 (1985)

Bimetallic Complexes in Vanadium Catalysis

Some studies were focused on the preparation (17-20) of bimetallic vanadium(II)-electron donor complexes containing zinc and aluminum. Divalent vanadium-THF-zinc or aluminum complex salts have been characterized. A comparison of the zinc complex $[V_2(\mu-Cl)_3(THF)_6]_2Zn_2Cl_6$ and the aluminum complex $[V_2(\mu-Cl)_3(THF)_6]AlCl_2(C_2H_5)_2$ shows the divalent vanadium cation in both complexes. Ethylene polymerizations with the zinc complex have been described (17,18). These complexes, in the presence of an aluminum alkyl and halogenated hydrocarbon such as $CFCl_2CFCl_2$ show high polymerization activities. However, within a series of zinc complexes, no clear correlations between structure and polymerization activity could be drawn.

Oxidation State of Vanadium

The oxidation state of active vanadium in ethylene polymerization has received attention by various investigators (11,12,21-23). Some investigators (11,21,22) have suggested that V(III) forms the active site and deactivation occurs by reduction to V(II) by the aluminum alkyl used as cocatalyst or by interaction of the active sites. However, a more recent study found no correlation between the concentration of V(III) and polymerization activity (23). Conclusions from one study (11) indicated that halogenated hydrocarbons, acting as promoters for vanadium catalysts, oxidize inactive vanadium species to an active trivalent form. Investigations with divalent vanadium-zinc salts strongly suggest that halocarbons do not oxidize vanadium salts, even under forcing conditions (18). Although earlier publications favored oxidation by activators, such reactions do not seem reasonable, and additional studies are needed to clarify the situation.

Our studies lend strong support to the proposal that V(II) species represents the important oxidation state in ethylene polymerization. In a comparison of the extent of reduction of $VCl_3(THF)_3$ impregnated in silica, we found that dialkylaluminum halides were stronger reducing agents than the corresponding trialkylaluminum compounds (Table 7). The ethyl derivative was a stronger reducing agent than the isobutyl compound. Addition of $VCl_3(THF)_3/SiO_2$ to a mixture of excess $(i-Bu)_3Al$ and $CHCl_3$ resulted in a V(II) concentration of 50%. This reaction without promoter resulted in a reduction of 10%. Such results show that the promoter actually contributes to the extent of reduction of V(III) to V(II) and is not behaving as an oxidant. Even with a large excess of $CHCl_3$ relative to $(i-Bu)_3Al$, only 50% of the total vanadium appeared as V(II). Similar observations on the role of the promoter were made using $(C_2H_5)_2AlCl$ as reducing agent.

Features relating to promoter-cocatalyst chemistry are summarized in Table 8. One proposal relates to the formation of a carbene or carbenoid species in the reaction of diiodomethane with $(C_2H_5)_3Al$. Trapping experiments in the presence of cyclohexene led to formation of bicyclo[4.1.0] heptane. A highly convenient and versatile method for the synthesis of cyclopropanes involves treatment of olefins with a variety of trialkylaluminum and alkylidene iodide (25). The proposed intermediate, dialkyl(iodomethyl) aluminum is a powerful methylene transfer agent as demonstrated by the facile cyclopropanation of olefins at low temperature.

Table 7. Extent of Reduction of $VCl_3(THF)_3/SiO_2$ with Organoaluminum Compounds (20X)[a]

	% V(II) Formed
$(C_2H_5)_3Al$	50
$(C_2H_5)_2AlCl$	85
$(i-Bu)_3Al$	<10
$(i-Bu)_2AlCl$	70

[a] Results were obtained on samples refluxed in hexane containing 20 equivalents of alkylaluminum per vanadium. The treated solids were isolated, washed with hexane, and dried. Divalent vanadium concentrations were determined by ESCA.

Table 8. Features of Promoter-Cocatalyst Chemistry

$$V(II) \xrightarrow{\text{Trichloroacetate}} V(III) \quad \text{Christman (11)}$$

$$R_3Al + -CX_3 \longrightarrow \text{Halogen-Metal Exchange, Carbenoid Reaction} \quad \text{Miller (24)}$$

$$\text{\Large >}C = C\text{\Large <} + RCHI_2 \xrightarrow{R_3'Al} \overset{CHR}{\underset{\text{\Large >}C \longrightarrow C\text{\Large <}}{\triangle}} \quad \text{Yamamoto, et al. (25)}$$

$$[V_2Cl_3(THF)_6]_2Zn_2Cl_6 \xrightarrow{CH_2Cl_2} V(II) \quad \text{Smith, et al. (18)}$$

$$VCl_3(THF)_3/SiO_2 + (C_2H_5)_2AlCl \xrightarrow{CHCl_3} V(II) \quad \text{This Paper}$$

Developments with Chromium Catalysts

Much of the focus of chromium catalysts such as supported chromium oxide has been directed toward support changes or modification, changes of ligand environment about chromium, or control of the redox cycles involving the oxidation states of chromium (26-28). Some studies have been carried out on copolymerization and isomerization kinetics with different chromium catalysts (29). One study (30) showed that chemical reaction of the surface silanol groups of silica offered an alternative to the high temperature (up to 800°C) dehydroxylation of silica supports frequently used with chromocene-based catalysts, $(C_5H_5)_2Cr/SiO_2$. Two chemical modifications that were effective in activating silica supports for chromocene-based catalysts were hydrosilane treatment and fluoride treatment.

The unique behavior of chromocene-based catalysts was recently highlighted when a comparison of this catalyst with its open ring analog, dimethylpentadienyl was made (31). Unlike the chromocene catalysts, the open ring chromium catalysts produced polyethylenes high in unsaturation and broad in molecular weight distribution.

Changes in the support for chromium oxide catalysts can affect the polymerization features. Polyethylenes of broader molecular weight distribution were obtained using $CrO_3/AlPO_4$ in place of CrO_3/SiO_2 as catalysts (27). A wider diversity of active chromium sites is probably present in the aluminophosphate carrier.

Comparisons of High-Activity Titanium, Vanadium, and Chromium Catalysts

Table 9 provides a summary of the focus of much research with titanium, vanadium, and chromium catalysts. Extensive worldwide research has shown that the specific metal composition of the catalyst exerts an important effect on polymer molecular weight and molecular weight distribution, comonomer incorporation, polymerization kinetics, and extent of catalyst decay (32,33). One goal of industrial research in olefin polymerization catalysis centers on the chemistry and technology necessary to obtain simultaneously favorable catalyst responses in all of the areas described above.

Table 10 displays a comparison of the features of certain titanium, vanadium, and chromium catalysts in ethylene polymerization. Currently, all three metal centers have the capability of providing acceptable catalyst productivities, not requiring any post-treatment of the polyethylenes. Typically, titanium catalysts provide polyethylenes of fairly narrow molecular weight distribution, while polyethylenes from supported vanadium and chromium catalysts are typically intermediate to broad in molecular weight distribution. Chromium oxide and vanadium catalysts display outstanding comonomer incorporation. Generally, titanium catalysts show a moderate level of catalyst decay during the polymerization process, while chromium oxide catalysts have a very low level of decay with time. Polyethylenes produced with titanium or vanadium contain low levels of unsaturation, while chromium oxide catalysts provide polyethylenes containing approximately one double bond per molecule.

Table 9. Focus of Metal Catalysis in Ethylene Polymerization

Metal	Active Center[a]	Cocatalysts, Promoters	Focus
Ti	$\begin{bmatrix} R-Ti(II,III) \\ \| \\ Cl \end{bmatrix}$	R_3Al	$MgCl_2$, Electron Donors, Complexes
V	$\begin{bmatrix} R-V(II) \\ \| \\ Cl \end{bmatrix}$	$R_3Al + RCCl_3$	Promoters, Modifiers, Complexes
Cr	$\begin{bmatrix} R-Cr(II) \\ \| \\ O \end{bmatrix}$	--	Support Modification and Treatment, Ligands, Redox Cycles

[a] The bracketed active sites are intended to provide only a partial description of the active sites which may be coordinated with other metal centers or reside on different substrates.

Table 10.   Comparisons of Silica-Supported Titanium-Vanadium-Chromium
Catalysts

Parameter	Mg/Ti/ED[a]	VCl$_3$(THF)$_3$[b]	CrO$_3$[c]
Productivity	High	High	High
H$_2$ Response	Moderate	High	Low
MWD	Narrow	Intermediate-Broad ------------>	
α-Olefin Incorporation	Moderate	High	High
Decay Rate	Moderate	Low-Moderate	Very Low
Polymer Unsaturation	Low	Very Low	One C = C Per Molecule

[a] Activated with R$_3$Al; ED = Electron Donor
[b] Activated with R$_3$Al + Halocarbon Promoter
[c] Thermal Activation

High catalyst productivities with the different catalysts are reached
through distinctly different chemical approaches. Chromium oxide
catalysts (CrO$_3$/SiO$_2$) are rendered highly active by a special high
temperature activation process during which the chromium oxide becomes
chemically anchored to the silica support. Magnesium compounds,
particularly MgCl$_2$, properly incorporated within a titanium halide
matrix provide an important route for reaching high activity with
titanium catalysts. Supported vanadium catalysts usually require a
halogenated promoter as a productivity booster to maintain high
polymerization productivity.

Complete control of molecular weight distribution by way of a single
catalyst family continues to be a formidable challenge. Often
multiple reactors, particularly with titanium catalysts, are required
to provide polyethylenes of intermediate molecular weight
distribution. Titanium, vanadium, and chromium catalysts each provide
polyethylenes that display their own characteristic molecular weight
distribution.

Different rates of α-olefin incorporation occur with titanium,
vanadium, and chromium catalysts. Generally, higher rates of
comonomer incorporation are desirable, since higher rates require a
lower comonomer/ethylene ratio to achieve the same comonomer
incorporation level. Most likely different catalysts will have a
significant effect on the degree of compositional heterogeneity of
copolymers produced with each catalyst type.

Understanding the differences in behavior of titanium, vanadium, and
chromium catalysts offers a unique challenge and opportunity in
polymerization catalysis. Frequently we build reaction models based
on limited studies with one specific transition metal center.
Comparative studies should provide the basis of a broader outlook and
a more comprehensive view of olefin polymerization catalysis.

REFERENCES

1. (a) Giannini U, Albizzati E, Parodi S, Pirinoli, F U.S. Patents 4,124,532 (1978) and 4,174,429 (1979); (b) Yamaguchi K, Kanoh N, Tanaka T, Enokido N, Murakami A, and Yoshida S U.S. Patent 3,989,881 (1976).
2. Greco A, Bertolini G, Cesca S J. Appl. Polym. Sci. (1980) 25:2045-2061.
3. Karol FJ, Goeke GL, Wagner BE, Fraser WA, Jorgensen RJ, Friis N U.S. Patent 4,302,566 (1981).
4. (a) Albizzati E, Bart JCJ, Giannini U, Parodi S Preprints of IUPAC International Symposium on Macromolecules (Sept. 7-12, 1980) 2:40-43; (b) Albizzati E, Giannetti E, Giannini U Makromol. Chem., Rapid Commun. (1984) 5:673-677.
5. Bart JCJ, Bazzi IW, Calcaterra M, Albizzati E, Giannini U, Parodi S Z. Anorg. Allg. Chem (1981) 482:121; (1983) 496:205.
6. Sobota P, Utko J, Lis T J. Chem. Soc., Dalton Trans. (1984) 2077-2079.
7. Sobota P Paper Presented at International Symposium on Transition Metal Catalyzed Polymerizations (June 16-20, 1986) Akron, Ohio.
8. Handlir K, Holecek J, Klikorka J, Bocek V International Polymer Science and Technology (1986) 13(8):1-5.
9. Cann KJ, Miles DL, Karol FJ U.S. Patent 4,670,526 (1987).
10. Cesca S J. Polym. Sci., Macromolecular Reviews (1975) 10:1-230.
11. Christman DL J. Polym. Sci., A-1 (1972) 10:471-487.
12. Karol FJ, Carrick WL J. Amer. Chem. Soc. (1961) 83:2654-2658.
13. Keim GI, Christman DL, Kangas LR, Keahey SK Macromolecules (1972) 5:217.
14. Phillips GW, Carrick WL J. Amer. Chem. Soc. (1962) 84:920.
15. Ort MR U.S. Patents (a) 3,784,539 (1974), (b) 3,925,338 (1975), (c) 3,956,255 (1976), (d) 4,232,140 (1980); Rogers TK (e) 4,426,317 (1984); Roling RV, Veazey RL, Aylward DE (f) 4,434,242 (1984); Pennington BT, Roling PV, Hsieh JT (g) 4,435,518 (1984); Veazey RL, Pennington BT (h) 4,435,519 (1984); Aylward DE (i) 4,435,520 (1984); Ahluwalia MS, Junker ML (j) EPA 0,099,660 (1986); Roling PV, Veazey RL, Aylward DE (k) EPA 0,196,830 (1986).
16. Beran DL, Cann KJ, Jorgensen RJ, Karol FJ, Maraschin NJ, Marcinkowsky AE U.S. Patent 4,508,842 (1985).
17. Smith PD, Martin JL U.S. Patent 4,559,318 (1985).
18. Smith PD, Martin JL, Huffman JC, Bansemer RL, Caulton KG Inorg. Chem. (1985) 24(19):2997-3002.
19. Cotton FA, Duraj SA, Roth WJ Inorg. Chem. (1985) 24:913-917.
20. Cotton FA, Duraj SA, Manzer LE, Roth WJ Inorg. Chem. (1985) 107:3850-3855.
21. Lehr MH Macromolecules (1968) 1:178.
22. Lehr MH, Carman CJ Macromolecules (1969) 2:217.
23. Kashiwa N, Tsutsui T Macromol. Chem., Rapid Commun. (1983) 4:491-495.
24. Miller DB Tetrahedron Lett. (1964) 17:989-993.
25. Maruoka K, Fukutani Y, Yamamoto H J. Org. Chem. (1985) 50:4412-4414.
26. McDaniel MP Adv. Catal. (1985) 33:47-98.
27. McDaniel MP, Johnson MM Macromolecules (1987) 20:773-778.
28. Karol FJ, Karapinka GL, Wu C, Dow AW, Johnson RN, Carrick WL J. Polym. Sci., A-1 (1972) 2621-2637.

29. Karol FJ Paper Presented at International Symposium on Transition Metal Catalyzed Polymerizations (June 16-20, 1986) Akron, Ohio.
30. Noshay A, Karol FJ Paper Presented at International Symposium on Transition Metal Catalyzed Polymerizations (June 16-20, 1986) Akron, Ohio.
31. Freeman JW, Wilson DR, Ernst RD, Smith PD, Klendworth DD, McDaniel MP J. Polym. Sci., A-1 (1987) 25:2063-2075.
32. Karol FJ Catal. Rev.-Sci. Eng. (1984) 26(3&4):557-595.
33. Yermakov YI, Kuznetsov VN, Zakharov VA (1981) Catalysis by Supported Complexes, Vol. 8 of Studies in Surface Science and Catalysis, Elsevier, New York.

SURFACE COMPOUNDS OF TRANSITION METALS, XXXIII (1)

PHILLIPS-LIKE SYSTEMS AND THEIR REACTION WITH OLEFINS

H.L.Krauss
Laboratorium für Anorganische Chemie der Universität Bayreuth, F.R.G.

Abstract. Supported Cr, V, Mo, Fe in form of coordinatively unsaturated ions of low oxidation state show an inhomogeneity due to the surrounding surface structure. A population profile can be determined: chromium e.g. occurs in two, molybdenum in three different types. The ions show high spin magnetism. - Olefins interact with support surface groups and with the metal sites reversibly prior to polymerization, which is of first order in sites and in olefin without discrimination of 1-olefins (copolymerization!). The yields exceed 95 %. At $T > 120°$ we observe $<CH_2>$ transfer together with a deactivation of the sites by redox reactions.

More than 30 years ago, Paul Hogan published the basic paper on the polymerization of ethylene at medium pressure and moderate temperatures, catalyzed by Cr(VI) fixed to a silica-alumina support and activated by thermal treatment - the so-called Phillips catalyst (2). Meanwhile a lot of research has been done, most in industry, some in the academic area. In spite of a long controversy there seems to be agreement that the active centres of the catalyst are coordinatively unsaturated chromium ions of low valency state in the support's surface (3,4). They are produced usually from the activated Cr(VI) precursor by reduction (4) to Cr(II) by different means (olefin, solvent, CO) - or the contact is made from Cr(II) compounds (5), e.g. by removal of ligands.

Of course this concept of "coordinatively unsaturated surface compounds" is not restricted to Cr(II), or even to Cr at all, or to silica-alumina:

- any suitable support (thermally stable, high surface aera, reactive surface groups)
- any suitable transition metal (e.g. Ti, V, Cr, Mn, Fe, Co, Ni, Mo, Re (6))

may form such sites, and many of these combinations are indeed able to polymerize 1-olefins. We may define these catalysts here as "Phillips-like systems", primarily differing from Ziegler Natta catalysts by their ability to act without additional alkyl source. In the following we will focus the discussion on the Cr/silica catalyst if not otherwise stated.

The coordinatively unsaturated chromium compound is showing an enormeous reactivity, especially vs. Lewis bases (6); the low O.N. may involve subsequent redox reactions, e.g. following the addition of O_2 or N_2O to Cr(II) (7,8).

Many problems in this field are still unresolved, mainly due to the unevitable disadvantage of all heterogeneous systems: the difficulty

W. Kaminsky and H. Sinn (Eds.)
Transition Metals and Organometallics as
Catalysts for Olefin Polymerization
© Springer-Verlag Berlin Heidelberg 1988

or even impossibility to produce (and to work with) uniform surface compounds (9).

With respect to the reaction between Phillips-like catalysts and olefins, we may ask: what is the character of a catalytic site, and: is there only one type of catalytically active centres?

Even if we restrict ourselves essentially to the reduced system Cr(II)/silica it shows to be inhomogeneous with respect to the metal centres: The mean oxidation number after reduction is actually always somewhat higher than 2,0 (best value 2.06 (10)), since there is some Cr(III) formed by thermal decomposition of Cr(VI) off from the stabilizing surface; the mean coordination number after reduction is not as low as 2, rather we find - inevitably formed - a mixture (11,12) of

Cr(II)A with a coordination number of 2 (vs. oxide ions)
Cr(II)B with a coordination number of 3 (vs. SiOH, SiOSi in addition)
Cr(II)C with a coordination number of 4 (vs. SiOSi in addition).
The Cr(III) mentioned and the Cr(II)C are not discussed in the following, since they show no reactivity vs. olefins.

The A/B ratio is mainly a function of activation temperature and - less predominantly - of the chromium concentration (13); it can be determinded by

- chemical reactions (different for A and B, e.g. with CO (11) or with N_2O (8))
- optical methods (A is green, B is blue with absorption bands at 6,8 ; 11,9; 13,7 and 8,6; 13,5; 17,9 kK respectively)
Both A and B are high spin d^4 systems with μ eff values of 4,7 B.M.

If we measure the optical absorption intensities of the real catalysts as function of T_{act}, $600°C$ is the critical temperature: the ratio A/B has reached its final value with a predominance of Cr(II)A (ca. 70 %). The same behaviour is seen with many chemical reactions, including the polymerization of olefins. But there are exceptions: the bright chemiluminescence from the reaction with O_2 (relaxation of excited surface Cr(VI) - see (7)) shows increasing intensity up to activation temperatures of ca. 850 °C. Obviously the luminescent species is a subspecies of Cr(II)A. So we have to conclude that our A + B model is oversimplified. Instead of discrete species we better should assume ensembles of similar centres, which form population profiles. Indeed these profiles were verified by thermal desorption of benzene or by the thermal decomposition of N_2O; both experiments show almost identical results with rather broad bands for the two types A and B (8).

It should be mentioned here, that the system Mo/silica comprises at least three species, found by photoluminescence studies with the activated Mo(VI) surface compound (14). The mean O.N. of the reduced product shows a remarkable deviation from the value 2 (best value 2.6) since one of these species obviously stops at 4 - the value found in Kazanskij´s samples (15,16).

All these results show that Phillips catalysts are not uniform and contain metal ions with different oxidation state and coordination number. The latter property especially will be of importance for the reaction with olefins, discussed in the following.

First we tried here the usual attempt: reaction of different catalysts with ethylene at higher pressure, varying the reaction temperature and using some additives. But this system shows to be complicated: After a more or less rapid increase the reaction rate (uptake of monomer per time unit) passes a maximum during the first 20 to 50 min and then decreases with almost first order (17). In our experiments, polyethylene itself could have been responsible for the latter effect, blocking the surface. So we tried high reaction temperatures to get low molecular weight products, hopefully well soluble. The results (18) were rather strange: besides polymer and oligomers we got at 200 - 300 $^\circ$C
- isomerized monomers (shift of double bound and branching)
- redox-products like H_2 and the alkanes from CH_4 to $n-C_4H_{10}$ (with built-in deuterium from deuterized support catalysts)
- transfer of <CH_2> (in case of cycloolefins: ring narrowing, formation of exo methyl groups).

Not surprisingly, we found a different reactivity of Cr(II)A and Cr(II)B: while Cr(II)B shows to be mainly responsible for the isomerization (and works indefinitely), the Cr(II)A centres do the <CH_2> transfer and undergo the redox process mentioned; they are finally converted to a bis-(π-allyl) Cr(IV) complex (19). Polymerization and oligomerization are terminated with the disappearence of the Cr(II)A centres. Since the Cr(IV) complex can be split off and extracted from the support, we have got here another way to determin the A/B ratio (20).

But after all: is there a common type of intermediate? Indeed, a way can be sketched which starts the reaction of Cr(II) with a the olefin by the formation of a carbene complex,[1] followed by the addition of a second olefin under formation of a metalla cyclobutane ring (like in metathesis (22) which indeed can be observed with the molybdenum system (15)). All the reaction products found may then be explained by different cleavage of this intermediate or by its redox behaviour, e.g. dehydrogenation by forming the π allyl ligand (23). This model (18) would also allow the formation of uneven C numbers in the oligomeric products - but there is no trace of <CH_2> transfer at T < 120°C. (The usual polymerization temperature is ca. 110 $^\circ$C.) The experiments show indeed an almost linear increase of <CH_2> transfer for increasing reaction temperatures starting from 120 $^\circ$C (24). Furthermore, carbene complexes of surface Cr(II) made via diazo compounds form cyclopropane derivatives and give Wittig analogous reactions at low temperatures (25), both not found with the Cr(II)/olefin system. So we may suppose the carbene model providing a good explanation for a deactivation of the catalyst but probably not for the steps of the usual polymerization.

An other approach to get movable, i.e. removable products was found with the use of "higher" l-olefins as monomers (26,27): the resulting comb polymers are easily soluble in hydrocarbons at room temperature.

[1] Similar intermediates are proposed for Fischer-Tropsch reaction (21).

With the Cr^{II}/silica systems we found

- Cr^{II} polymerizes higher 1-olefins (C_3-C_{14}) with > 95 % yield at room temperature to comb polymers (MW $\approx 10^4$)
- the reaction is first order both in monomer and in catalyst
- there is no decrease of the activity during the reaction (for weeks in contineous flow experiments)
- there is an induction period - repeatedly with repeated runs
- there is no discrimination of 1-olefins (copolymerization easily possible)
- no polymerization with inner or sterically hindered olefins (vanadium works here! (28))
- the polymers are mono-olefins
- the products are atactic and partially branched
- there is a broad MW distribution; MW is decreasing (and narrowing) with increasing polymerization temperature, but there is
- no regulation of MW by hydrogen: H_2 hydrogenates the monomer (29)
- and of course: lower specific rates if there is a lower A/B ratio.

Most important seems the fact, that there is no natural, intrinsic decrease of the catalyst´s activity during the reaction; all 1-olefins seem to undergo the same process at the catalytic center, obviously going through the same mechanistic step. On the other hand, the induction period is an intrinsic property of the system and needs further examination. The easy access to kinetic and thermodynamic data provided by the polymerization procedure with higher olefins will be helpful to get further information on mechanistic questions.

A principally different approach to these problems was based on IR measurements (FTIR Digilab), using the systems
- (H,D) silica support
- Cr(II), Fe(II) (since the polymerization with Fe proceeds rather slowly)
- ethylen, 1-penten, cyclopenten.
For comparison, the behaviour of n-alkanes was as well examined. The investigation was concentrated to the SiOH(SiOD) and the CH_2/CH_3 vibrations.

The following results were obtained:

- alkanes "react" with support and with catalyst by interaction with isolated SiOH (vibration at 3747/cm): a new absorption at 3700/cm appears - reversibly if the alkane is volatile, irreversibly in case of polymer CH_2 chain
- olefinic double bounds interact reversibly with the same SiOH under formation of a broad band at 3600/cm, while the 3747/cm vibration decreases
- for the CH part of the spectra, the examination of the Fe system is most suitable due to its slow reaction (30). First we find the reversible formation of a complex which already shows a CH_3 component besides the two CH_2 absorptions; after a while a faint irreversible band structure remains after evacuation of the olefin - again with CH_3 vibration; finally the grown CH_2 chain dominates the spectra with a decreasing contribution of CH_3.

These results could suggest a model for the starting step in analogy to the Ziegler pathway with formation of an alkyl chromium compound - here by "borrowing" a hydrogen from the support (31).

Even if we argue that the IR bands may represent the behaviour of a majority of rather inactive centres instead of the action of the few really active sites - even then the participation of the support should be carefully examined.

After all, in place of a concluding remark, we better should recall the fact that there is a basic "ignoramus" in the field of Phillips-like systems: unknown is the real structure of the active sites. Other questions rise from this central point, e.g.

- how many active sites?
- why broad MW distribution of the polymer?
- why atactic polymer?
- why no H_2 regulation of MW?
- why induction period?

Obviously for Ziegler-Natta- and for metathesis catalysts the knowledge is more advanced by far. It is a tempting hope that the research on Phillips systems - and maybe even on Fischer-Tropsch-contacts - will profit from this and that related patterns will be detected in these fields.

This work was supported by the Sonderforschungsbereich 213 of Deutsche Forschungsgemeinschaft and the Fonds der Chemischen Industrie.

References
(1)   Previous paper: H.L.Krauss, J.Mol.Cat., in press
(2)   A.Clark, J.P.Hogan, R.Banks and W.Lanning, Ind.Eng.Chem.48, 1152 (1956)
(3)   L.L.van Reijen and P.Cossee, Discuss.Faraday Soc.41, 277 (1966)
      D.D.Eley, C.H.Rochester and M.S.Scurrel, J.Cat.29, 20 (1973)
      L.K.Przhevalskaya, V.A.Shvets and V.B.Kazanski, J.Cat.39, 363 (1975)
      K.G.Miesserov, J.Cat.22, 340 (1971)
      D.D.Beck and G.H.Lunsford, J.Cat.68, 121 (1981)
      J.Karra and J.Turkevitch, Discuss. Faraday Soc.41, 310 (1966)
      C.Groeneveld, P.P.M.M.Wittgen, H.P.M.Swinnen, A.Wernsen and G.C.A.Schuit, J.Cat.83, 346 (1983)
      R.Spitz, A.Revillon and A.Guyot, J.Cat.35, 335 (1974)
(4)   H.L.Krauss and H.Stach, Inorg.Nucl.Chem.Letters 4, 393 (1968); Z.anorg.allg.Chem. 366, 280 (1969)
      D.Naumann, Dissertation Freie Universität Berlin, 1979
      G.Ghiotti, E.Garrone, S.Coluccia, C.Morterra and A.Zecchina, J.Chem.Soc.Chem.Comm.22, 1032 (1979)
      R.Merryfield, M.P.McDaniel and G.Parks, J.Cat.77, 348 (1982)
(5)   B.Horvath, J.Strutz, E.G.Horvath, Z.anorg.allg.Chem.457, 38 (1979)
(6)   H.L.Krauss and H.Stach, Z.anorg.allg.Chem.366, 34 (1969)

(7)   P.Morys, R.Gerritzen and H.L.Krauss, Z.Naturforsch.31b, 774
      (1976); P.Morys, U.Görges and H.L.Krauss, Z.Naturforsch.39b,
      458 (1984)
(8)   R.Höpfl, Dissertation Universität Bayreuth, 1981
(9)   A.Schimmel, Dissertation Universität Bayreuth, 1985
(10) H.L.Krauss and R.Dumler, unpublished results
(11) B.Rebenstorf, Dissertation Freie Universität Berlin, 1975
      H.L.Krauss, B.Rebenstorf and U.Westphal, Z.anorg.allg.Chem.414,
      97 (1975)
      P.Blümel, Dissertation Freie Universität Berlin, 1979;
      M.P.McDaniel and M.B.Welch, J.Cat.82, 98 (1983)
(12) B.Fubini,   G.Ghiotti,   L.Stradella,   E.Garrone,   C.Morterra,
      J.Cat.66, 200 (1980)
      E.Garrone, G.Ghiotti, C.Morterra and A.Zecchina, Z.Naturforsch.
      42b, 728 (1987)
(13) H.L.Krauss and U.Westphal, Z.anorg.allg.Chem.430, 218 (1977);
      Z.Naturforsch.33b, 1278 (1978)
(14) P.Morys and S.Schmerbeck, Z.Naturforsch.42b, 756 (1987)
(15) S.Schmerbeck, Dissertation Universität Bayreuth, 1987
(16) V.B.Kazanskij, A.N.Pershin, B.N.Shelimov, Proc. 7th Int.Cong.
      Cat., 1210, 1980; B.N.Shelimov, J.V.Elev, V.B.Kazanskij, J.Cat.
      98, 70 (1986)
(17) B.Hanke, Dissertation Universität Bayreuth, 1983
(18) H.L.Krauss and E.Hums, Z.Naturforsch.34b, 1628 (1979); 35b, 848
      (1980); 38b, 1412 (1983)
(19) H.L.Krauss, K.Hagen and E.Hums, J.Mol.Cat.28, 233 (1985)
(20) K.Hagen, Dissertation Universität Bayreuth, 1982
(21) G.Henrici-Olive and S.Olive, Angew.Chem.88, 144 (1976); J.Mol.
      Cat.4, 379 (1978); ibid. 24, 7 (1984)
      E.Hums and H.L.Krauss, Z.anorg.allg.Chem.537, 154 (1986)
(22) J.L.Herisson, Y.Chauvin, Makromol.Chem.141, 161 (1970)
(23) M.Ephritikhine, M.L.H.Green, J.Chem.Soc.Chem.Comm.926 (1976)
(24) H.L.Krauss and G.Zeitler-Zürner, unpublished results
(25) K.Weiss and K.Hoffmann, J.Mol.Cat.28, 99 (1985)
(26) K.Weiss and H.L.Krauss, J.Cat.88, 424 (1984)
(27) G.Langstein, Dissertation Universität Bayreuth, 1986
(28) G.Guldner, Dissertation Universität Bayreuth, 1986
(29) H.L.Krauss and B.Janocha, unpublished results
(30) H.L.Krauss and R.Merkel, unpublished results
(31) K.Weiss, J.Mol.Cat., in press

Copolymerization of Ethylene and Higher α-Olefins with MgH$_2$ Supported Ziegler Catalysts

G. Fink and T.A. Ojala

Max-Planck-Institut für Kohlenforschung,
Kaiser-Wilhelm-Platz 1,
D-4330 Mülheim a.d. Ruhr, FRG

INTRODUCTION

Investigations into ethene/propene[1] and ethene/1-butene[2,3] copolymerization using Ziegler catalysts have been reported. In spite of the technical importance of these copolymers, still not enough is known about the copolymerization of ethene with higher α-olefins.[4,5] Consequently, we have carried out investigations into ethene copolymerization with C$_4$ to C$_{16}$ α-olefins using the highly active MgH$_2$/TiCl$_4$/AlEt$_3$ Ziegler catalyst system. In a previous publication[6], in which the homopolymerization of ethene had been reported, we demonstrated that the above catalyst system, which is highly dispersed and starts with a high surface area of ca. 140 m^2/g, may have a model character for the investigation of the elementary steps in heterogeneous Ziegler catalyst systems.

COPOLYMERIZATION PARAMETERS IN DEPENDENCE ON THE CHAIN-LENGTH OF THE α-OLEFIN

The simplest kinetic scheme of the copolymerization of two monomers, M$_1$ and M$_2$, with heterogeneous Ziegler catalysts is

$$R\text{-}M_1\text{-}Kat + M_1 \xrightarrow{k_{11}} R\text{-}M_1\text{-}Kat$$

$$R\text{-}M_1\text{-}Kat + M_2 \xrightarrow{k_{12}} R\text{-}M_2\text{-}Kat$$

$$R\text{-}M_2\text{-}Kat + M_1 \xrightarrow{k_{21}} R\text{-}M_1\text{-}Kat$$

$$R\text{-}M_2\text{-}Kat + M_2 \xrightarrow{k_{22}} R\text{-}M_2\text{-}Kat$$

The reaction of a monomer in this scheme is a composed quantity of several elementary steps (diffussion and adsorption of the olefin and finally insertion into the Ti-C-bound); hence, the propagation

W. Kaminsky and H. Sinn (Eds.)
Transition Metals and Organometallics as
Catalysts for Olefin Polymerization
© Springer-Verlag Berlin Heidelberg 1988

constants k_{ij} in the above scheme, are likewise composed quantities. In the above scheme the two different catalyst centers, M_1–Kat and M_2–Kat, are formulated. For the estimation of the copolymerization parameter, according to the Mayo-Lewis equation[7], it is assumed that these two centers have the same reactivity.

Böhm[3] demonstrated in the copolymerization of ethene with 1-butene using a heterogeneous Ziegler catalyst that $r_1 \gg r_2$. In other words, the copolymerization parameters differ by orders of magnitude and therefore, for small comonomer contents in the copolymer, the Mayo-Lewis equation can be simplified to

$$\frac{d[M_2]}{d[M_1]} = \frac{1}{r_1} \cdot \frac{[M_2]}{[M_1]}. \qquad (1)$$

For the different olefin pairs, the data are plotted according to this equation of a proportionality ranging between

$$\frac{d[M_2]}{d[M_1]} \quad \text{and} \quad \frac{[M_2]}{[M_1]} \quad \text{(Fig. 1)}.$$

The entire evaluated data are collected in Table 1:

Table 1.   r-Parameters of copolymerization of ethene with α-olefins for 20°C

α-Olefin	r_1	r_2	$r_1 r_2$
C_4	55 ± 5	0.02 ± 0.01	1.1
C_5	50 ± 5	0.02 ± 0.01	1.0
C_6	47 ± 5	0.02 ± 0.01	0.94
C_7	75 ± 10	0.013 ± 0.007	0.98
C_8	90 ± 10	0.01 ± 0.005	0.90
C_{10}	550 ± 50[*]	0.004[**]	
C_{12}	1500 ± 150	0.0004[**]	
C_{14}	1400 ± 140[*]	0.0004[**]	
C_{16}	1050 ± 100[*]	0.0008[**]	

[*]   values determined by IR analysis

[**]   formal evaluation

Indeed, the copolymerization parameters for ethene r_1 and for all of the α-olefins r_2, are different by orders of magnitude and additionally, there is a dependence on the chain-length. The reactivity of the α-olefin decreases with increasing carbon chain-length. According to this fact, a random distribution of the comonomer units in the polymer chain, was expected. If the comonomer

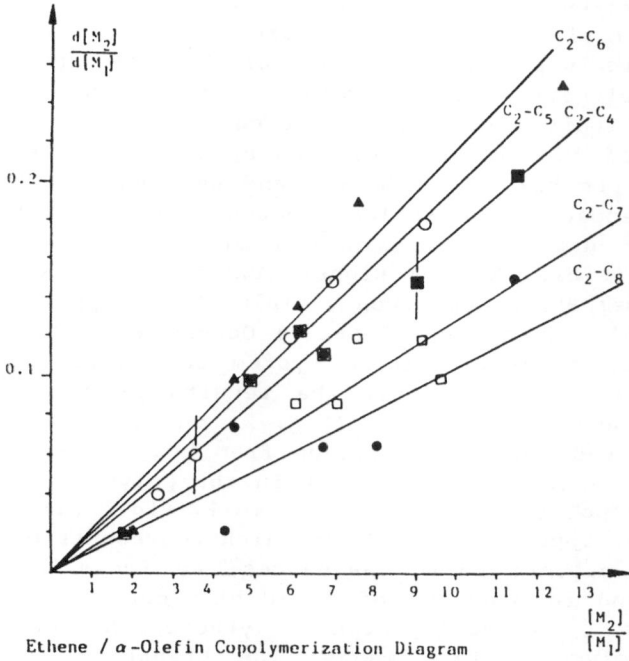

Ethene / α-Olefin Copolymerization Diagram

■ Ethene / 1-Butene Copolymer
○ Ethene / 1-Pentene Copolymer
▲ Ethene / 1-Hexene Copolymer
□ Ethene / 1-Heptene Copolymer
● Ethene / 1-Octene Copolymer

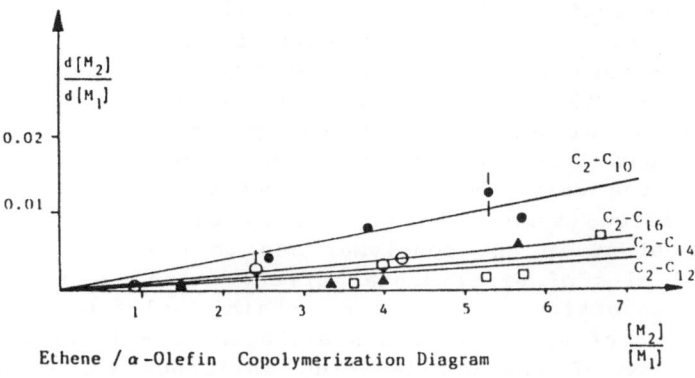

Ethene / α-Olefin Copolymerization Diagram

● Ethene / 1-Decene Copolymer
□ Ethene / 1-Dodecene Copolymer
▲ Ethene / 1-Tetradecene Copolymer
○ Ethene / 1-Hexadecene Copolymer

Figure 1

content is relatively low, then the branches should be largely isolated from one another along the polymer chain.
Fig. 2 shows, in the upper part, peaks of ^{13}C-NMR spectra of
ethene/1-butene copolymers with different butene contents. A
surprising result of this study is that even at very low butene
contents, sequences of two and more olefins appear. The resonances
at 11.0 ppm and 10.8 ppm belong to the EBE and the EBB sequences in
the polymer chain, respectively. At low 1-butene content in the copolymer, the shoulder next to the main peak attributed to the EBB
sequences, is well visible. With a higher content of 1-butene the
triads of the comonomer BBB become even visible at 10.6 ppm as shown
in Figs. 2(5) and 2(6), in the upper part. 1-Octene was found to be
less reactive in this copolymerization system than 1-butene by a
factor of 2 (see Tab. 1), therefore, the results of the ^{13}C-NMR
study of ethylene-1-octene copolymers (Fig. 2, in the lower part)
are even more unexpected. Diads and triads from hexyl branches can
already be found with a low octene content in the polymer. When the
octene content in the polymer is as low as 3 mol%, Fig. 2(2), in the
lower part, a shoulder appears at 26.9 ppm which represents diads of
octene (EOO) in the polymer chain. Sequences, as those discussed
above, were also found in products of other ethylene and α-olefin
copolymers such as ethene/1-pentene, ethene/1-hexene, and
ethene/1-heptene copolymers. This means, the branching in these
ethene/α-olefin copolymers was not statistical. Consequently, it can
be concluded that the α-olefin, once reacted, supports the insertion
of the next α-olefin molecule.

We have used the above hypothesis in terpolymerization studies.
Higher α-olefins $\geq C_{10}$ react very slowly and their insertion into
the polymer chain is very difficult. However, in terpolymerization
of ethene with short chain α-olefins the long chain α-olefins polymerized as well (Fig. 3). The upper part in Fig. 3 shows the
influence of temperature on the ethene/1-butene/1-hexadecene terpolymerization. From 40°C to higher temperatures 1-hexadecene
(symbols ●) inserts in higher amounts in the polymer chain, thus it
can be concluded that 1-butene supports the insertion of 1-hexa-
decene compared to the very low hexadecene content in ethene-hexa-
decene copolymerization (symbols ▲, Fig. 3, in the upper part). At
20°C the butene content in the copolymer and in the terpolymer is
equal within the limits of experimental error, which means that on
the contrary the insertion of the long chain α-olefin does not
support the insertion of the short chain α-olefin. The lower part of
Fig. 3 shows a study of the accelerating influence of different
short chain α-olefins on the insertion of long chain α-olefins. All
the short chain α-olefins, starting from 1-butene until 1-octene,
show a supporting influence on the insertion of long chain α-olefins
(1-dodecene). Seppälä[8] was the first one who recently published
this short chain accelerating effect on the insertion of a long
chain α-olefin in terpolymerization and called it a "synergistic
effect". He suspects that the short chain α-olefin, for instance

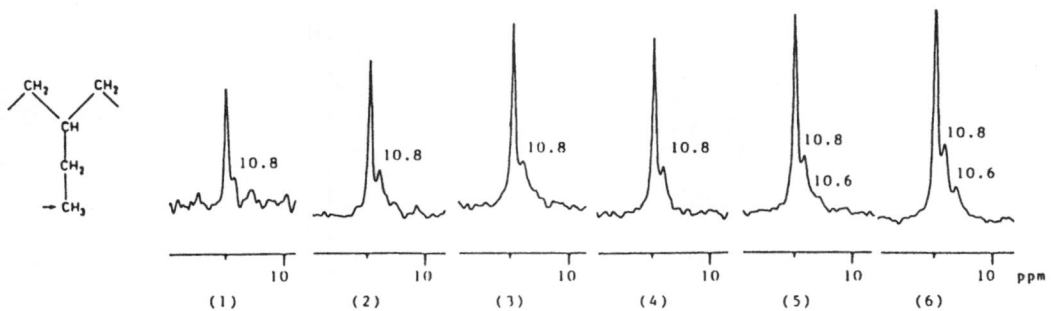

Details of ^{13}C NMR spectra of ethene / 1-butene copolymers with different contents of butene in polymer: (1) 2 mol%, (2) 9 mol%, (3) 11 mol%, (4) 10 mol%, (5) 13 mol%. (6) 17 mol%. CH$_3$(EBE) at 11.0 ppm, CH$_3$(EBB) at 10.8 ppm and CH$_3$(BBB) at 10.6 ppm.

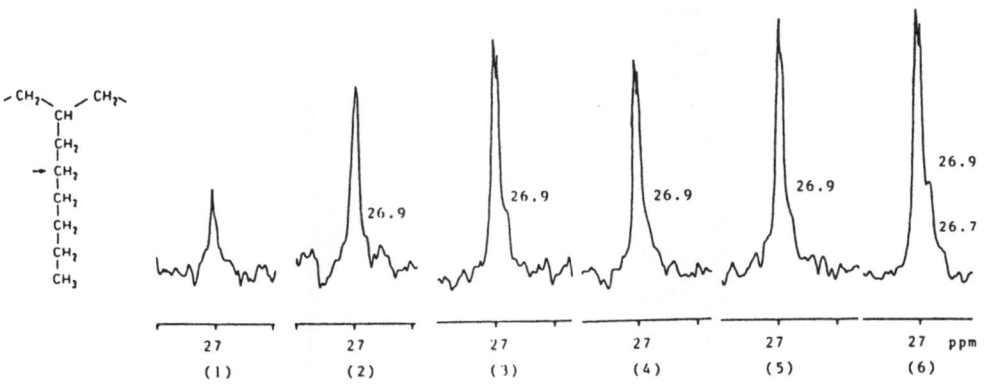

Details of ^{13}C NMR spectra of ethene / 1-octene copolymers with different contents of butene in polymer: (1) 2 mol%, (2) 3 mol%, (3) 6 mol%, (4) 6 mol%, (5) 7 mol%, (6) 13 mol%. CH$_2$(EOE) at 27.0 ppm, CH$_2$(EOO) at 26.9 ppm and CH$_2$(OOO) at 26.7 ppm.

Figure 2

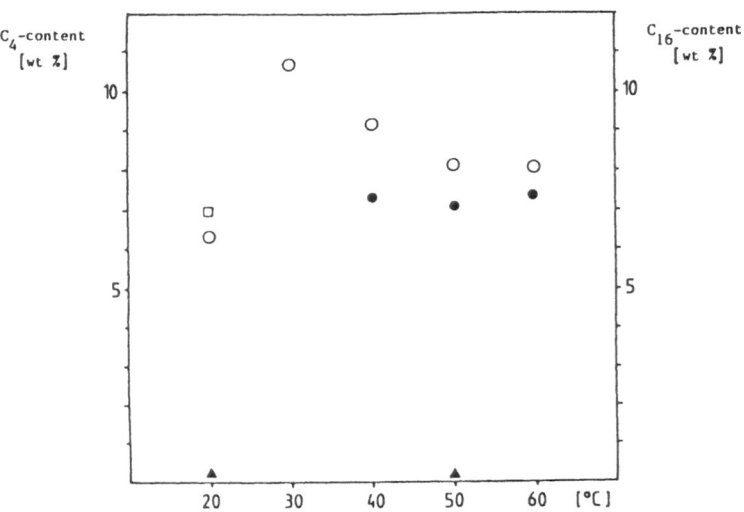

Terpolymerization of ethene, 1-butene and 1-hexadecene

[Ti] = 2,9 mmol/l; [Al] = 14,5 mmol/l

$[C_2]$: $[C_4]$: $[C_6]$ = 1 : 2 : 2

▲ hexadecene content in ethene / 1-hexadecene copolymer

● hexadecene content in ethene / 1-butene / 1-hexadecene terpolymer

○ butene content in ethene / 1-butene / 1-hexadecene terpolymer

□ butene content in ethene / 1-butene copolymer

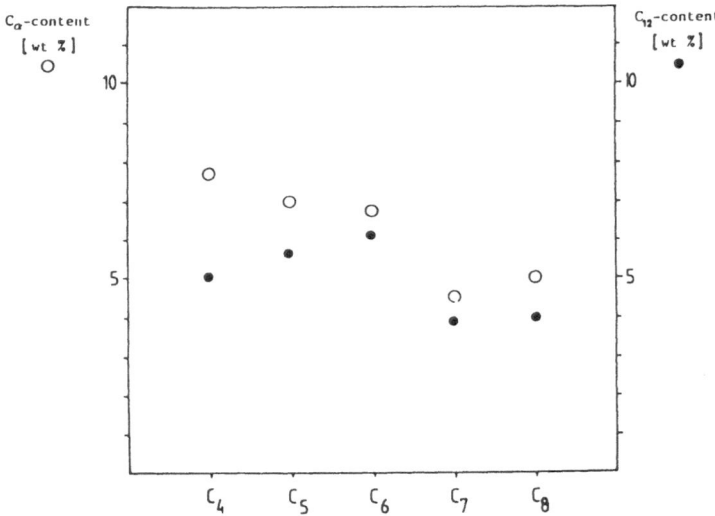

Terpolymerization of ethene, α-olefin and 1-dodecene.

[Ti] = 2,9 mmol/l; [Al] = 14,5 mmol/l

$[C_2]$: $[C_\alpha]$: $[C_{12}]$ = 1 : 2 : 2

○ α-olefin content in ethene / α-olefin / 1-dodecene terpolymer

● dodecene content in ethene / α-olefin / 1-dodecene terpolymer

Figure 3

1-butene, decreases the steric hindrance in the active centers.

ANALYSIS OF THE ACTIVATION PARAMETERS

A detailed kinetic analysis was carried out, as an example, for the ethen/1-octene copolymerization. The upper part of Fig. 4 shows the ethene/1-octene copolymerization diagram in dependence on temperature. The evaluated copolymerization parameters are compiled in Table 2.

Table 2. Dependence of copolymerization parameters on temperature; Copolymerization of ethene with 1-octene, Al : Ti = 5 :1

temperature [°C]	r_1	r_2
20	115±12	0.01±0.005
30	100±10	0.01±0.005
40	89±9	0.01±0.005
50	73±7	0.02±0.01
60	64±6	0.03±0.01

Ziegler Catalyst Sytem MgH_2 / $TiCl_4$ / $AlEt_3$.

The temperature dependence can be described within experimental error by the following equations: $\log r_1 = -0.03 + 615/T$ and $\log r_2 = -4.38 + 700/T$. These lines are shown in the lower part of Fig. 4. For the determination of the rate constants, we followed the procedure and derivation as had first been shown by Böhm[2,3]:

$$\frac{F}{v_p} = \frac{1}{k_{12}} + \frac{1}{k_{21}} \cdot \frac{[M_2]}{[M_1]} \qquad (2)$$

with

$$F = [r_1 + 2 \cdot \frac{[M_2]}{[M_1]} + r_2 \frac{[M_2]^2}{[M_1]^2}] \cdot [M_1] \cdot [C^*] \qquad (3)$$

Differently from Böhm (who works with the catalyst efficiency n^*/n_k) we used $[C^*]$, the concentration of active centers in mol/l, and the polymerization rate v_p in mol/l·s (for derivation and detail see[9]). To calculate the function F, the concentration of active centers must be known. This value has only been estimated for the homopolymerization process with the $MgH_2/TiCl_2/AlEt_3$ catalyst system by Kinkelin[10]; hence, we assumed $[C^*]$ in equation 3 to 1% of the Ti-concentration used, according to a ratio of AL : Ti = 5 : 1.
The entire copolymerization data, according to equations (2) and (3), are plotted in Fig. 5, the upper part. Furthermore, the Arrhenius plot of the four rate constants of the binary copolymerization of ethene (1) and 1-octene (2) is shown in the lower part of

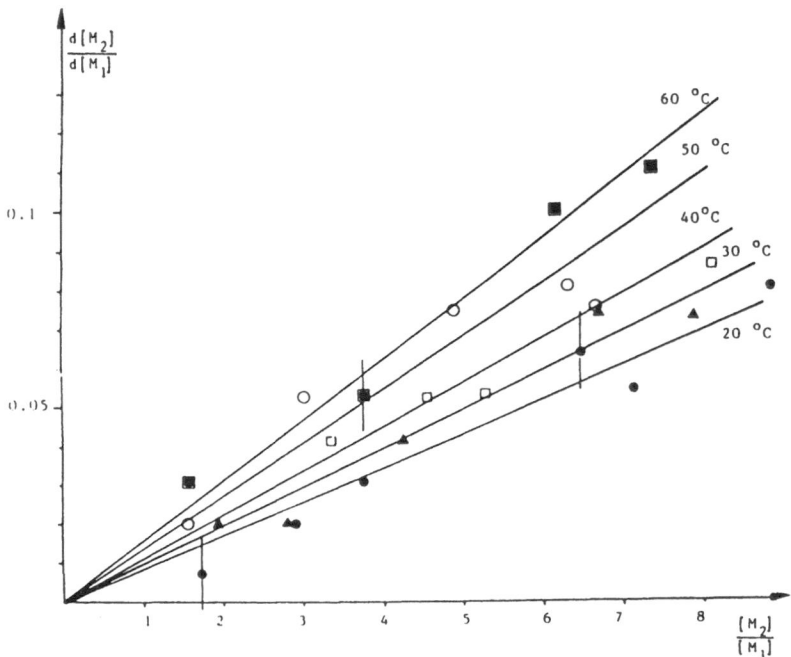

Ethene / 1-Octene copolymerization diagram in dependence on temperature:
● 20 °C, ▲ 30 °C, □ 40 °C, ○ 50 °C, ▣ 60 °C.

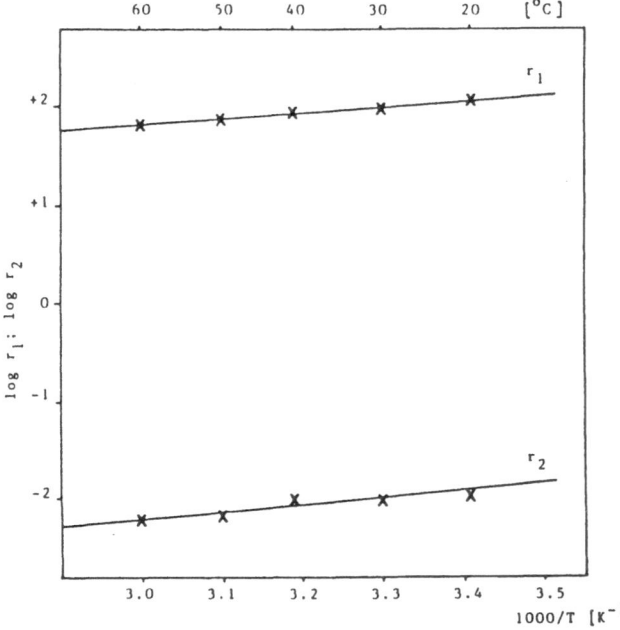

Arrhenius plot of r_1 and r_2 of the binary copolymerization:
ethene (1) and 1-octene (2).

Figure 4

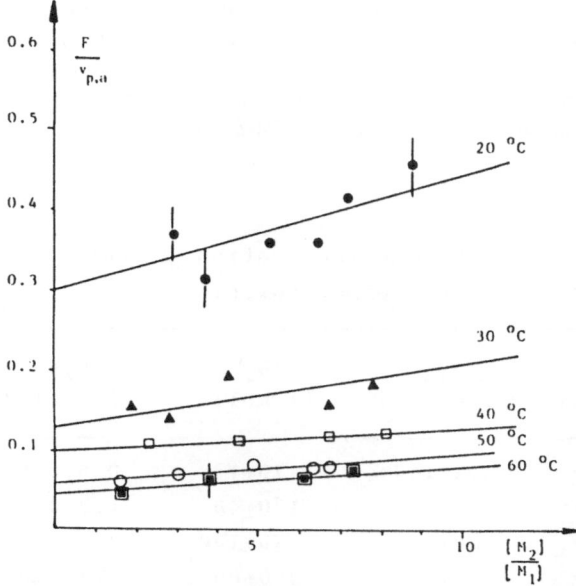

Plot of copolymerization data according to equations (14) and (15)
● 20 °C; ▲30 °C; □40 °C; ○ 50 °C and ■ 60 °C.

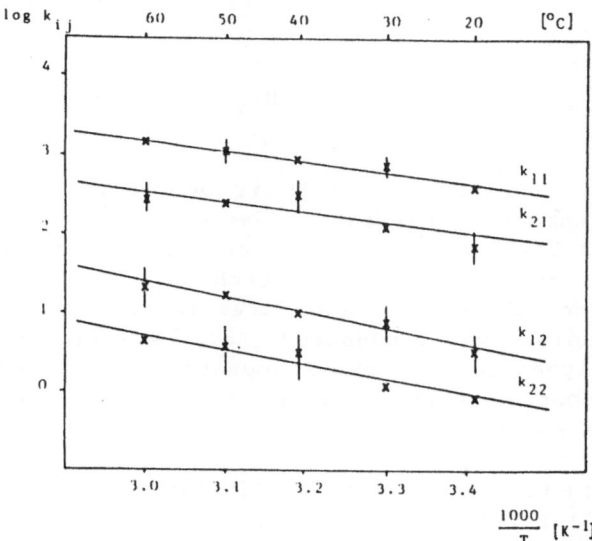

Arrhenius plot of the four rate constants of binary copolymerization:
ethene (1) and 1-octene (2)

$E_{11} = 25 \pm 3$ kJ/mol

$E_{12} = 37 \pm 12$ kJ/mol

$E_{21} = 30 \pm 10$ kJ/mol

$E_{22} = 37 \pm 12$ kJ/mol

Figure 5

Fig. 5. These data show that again the rate constants are different by orders of magnitude and that the activation energies differ in a reasonable mode.

Finally, in Table 3 are collected all of the rate constants in dependence on temperature: all rate constants increase with increasing temperatures.

Table 3. Rate constants of the copolymerization process of ethene (1) and 1-octene (2) for different temperatures.

T [°c]	k_{11} [1/mol·s]	k_{12} [1/mol·s]	k_{21} [1/mol·s]	k_{22} [1/mol·]
20	400±100	3.3±1.7	70±15	0.8±0.5
30	750±200	7.6±4.0	120±40	1.2±0.7
40	900±200	10±5	320±100	3.2±1.8
50	1200±300	17±8	250±80	3.7±2.0
60	1400±400	22±11	280±90	4.5±2.5

Ethene homopolymerization (20°C)

k_p = 193 1/mol·s

E_A = 45 kJ/mol

Ziegler Catalyst System MgH_2 / $TiCl_4$ / $AlEt_3$

In this context, a very interesting result is demonstrated in the lower part of Table 3: comparing the rate constants of ethene homo-polymerization (k_p = 193 1/mol·s[10]) with the one of ethene/1-octene copolymerization (k_{11} = 400 1/mol·s), we reach an increase by a factor of 2. In other words, octene also accelerates the ethylene polymerization. Previously, we recognized this effect qualita-tively[10]: in ethene polymerization, small amounts of added propene or 1-butene led to a clear enhancement of the overall activity compared to the standard polymerization.

In 1986, Tait[11] reported similar effects qualitatively for 1-hexene and 1-decene, Kaminsky[12] for 1-hexene and Duranel[13] for 1-propene. Tait is discussing the role of the α-olefin as a coordination- or adsorption phenomenon whereas Duranel believes that this phenomenon refers to the heterogeneity of the active centers.

Table 4 compiles all the determined enthalpy and entropy data of this copolymerization.

Table 4. Enthalpy and entropy values of the copolymerization
of ethene (1) and 1-octene (2).

reaction	enthalpy [kJ/mol]	entropy[*] [J/Kmol]
C_2H_4 (g) $\longrightarrow$ 1/n $(C_2H_4)_n$ (c)	-108	-173
C_2H_4 (g) $\longrightarrow$ 1/n $(C_2H_4)_n$ (a)	-101	-155
$K-M_1$ + M_1 $\longrightarrow$ KM_1	23 ± 3	-116 ± 12
$K-M_1$ + M_2 $\longrightarrow$ KM_2	35 ± 12	-115 ± 11
C_8H_{16} (1) $\longrightarrow$ 1/n $(C_8H_{16})_n$ (a)	-80[**]	-110[**]
$K-M_2$ + M_2 $\longrightarrow$ KM_2	35 ± 12	-127 ± 13
$K-M_2$ + M_1 $\longrightarrow$ KM_1	28 ± 10	-114 ± 11

(a) amorphous solid polymer

(c) crystalline solid polymer

[*] standard conditions for monomers:

concentration of 1 mol/l in solution and 1 bar for a gas

[**] values calculated from literature data for 1-butene and 1-hexene

Ziegler Catalyst System MgH_2 / $TiCl_4$ / $AlEt_3$.

The activation parameters surprisingly show that all of the four
insertion reactions are accompanied by the same high decrease in
activation entropy, i.e. there is a steric control in the reaction
of the active centers with the two monomers, but there is no
difference between the ethene and the 1-octene molecule. Table 4
shows on the contrary that the rate determining differences between
the ethene and the α-olefin reaction are caused through the
activation enthalpies. These values for the 1-octene insertion are
found to be about 10 kJ/mol higher than for the ethene insertion.
Or, may we say in other words, the activation of the double bond of
the α-olefin needs more energy than the activation of the double
bond of the ethene molecule? To explain these facts is not an easy
task and, at this point, other principal questions arise such as
whether the instrument used here for the analysis of the activation
parameters is at all sharp enough. For instance in equation (1),
$[M_2]/[M_1]$ is the monomer ratio in the reaction solution, as
determined by gas chromatography and we do not know the true monomer
ratio [1-octene] / [ethene] on, or around the active centers. It may
be assumed that the adsorption behavior of ethene and the longer
α-olefin molecule differ a lot from one another. Furthermore, the
fractionation of blockcopolymers, which were produced with this
catalyst system, led to the result that also homopolymers were
formed of both monomers besides from the true blockcopolymers. This
leads to the conclusion that different and multiple active sites are
positioned on the catalyst. All these results combined teach us that

the mutual interactions of monomer, comonomer, and the multiple active site in Ziegler catalyst systemes are of a very complex nature, and that we need to perform more precise experiments.

EXPERIMENTAL PART

Materials: Toluene was stirred over Na/K-alloy and distilled under argon. Triethylaluminium (TEA) and $TiCl_4$ were commercially obtained and used after distillation. Argon was used after passing through a heated BTS-column and two molecular sieve 3 Å columns. Ethene was used after passing through a molecular sieve 3 Å column. α-olefins were used after distillation.

Preparation of MgH_2-supported Ti catalyst: The catalysts were prepared by the procedure described previously[6].

Copolymerization procedure: The copolymerization of ethylene with α-olefins was carried out in a 250 ml glass reactor (Fa. Buechi, Uster) with toluene as solvent. The catalyst suspension and the co-monomer were placed in the reactor followed by ethene at a pressure of 2 bar. Inside of the reactor a small glass vessel was fixed, containing the TEA. The polymerization was initiated by flowing the TEA into the reaction mixture. The rate of ethene polymerization was followed by a flow measurement detector. The reaction was stopped by adding methanol.

Terpolymerization procedure: The terpolymerization of ethene with two α-olefins was carried out in a similar way as the copolymeri-zation procedure.

Analytical procedure: ^{13}-NMR spectra[14] were obtained at 120°C with a 75.5 MHz Bruker WM-300 Fourier Transform spectrometer. Measure-ments were taken of a sample solution of ca. 6% (w/w) in 1,2,4-tri-chlorbenzene in a sampling tube of 10 mm, with 1,1,2,2-tetrachlor-ethane-d_2 as an internal lock. A pulse width of 45° and a typical number of transients of 8000 to 9000, were used.

REFERENCES

1) Cooper, W., "Kinetics of Polymerization Initiated by Ziegler-Natta and Related Catalysts" in: "Comprehensive Chemical Kinetics 15, Non Radical Polymerization" Bamford, C.H., Tipper, C.F.H., ed., Elsevier Scientific Publ. Co., Amsterdam 1976, p. 133

2) Böhm, L.L., Makromol. Chem. 182, 3291 (1981)

3) Böhm, L.L., J. Appl. Polym. Sci. 29, 297 (1984)

4) Kissin, Y.V., Adv. Polymer Sci. 15, 91 (1974)

5) Kissin, Y.V., Beach, D.L., J. Polym. Sci., Polym. Chem. Ed. 22, 333 (1984)

6) Kinkelin, E., Fink, G., Bogdanović, Makromol. Chem., Rapid Commun. 7, 85 (1986)

7) Mayo, F.R., Lewis, F.M., J. A. Chem. Soc. 66. 1594 (1944)

8) Seppälä, J.V., J. Appl. Pol. Sci. 31, 657 (1986)

9) Ojala, T.A., Ph.D. Thesis, University of Düsseldorf, 1987

10) Kinkelin, E., Ph.D. Thesis, University of Düsseldorf, 1985

11) Tait, P.J.T., Downs, G.W., Akinbami, A.A., International
    Symposium Transition Metal Catalyzed Polymerization,
    The University of Akron, Ohio, June 1986, p. 38

12) Spitz, R., Duranel, L. Masson, P., Darricades-Llanro, M.F.,
    Guyot, A., International Symposium Transition Metal
    Catalyzed Polymerization, The University of Akron, Ohio,
    June 1986, p. 37

13) Kaminsky, W. in : "Catalytic Polymerization of olefins"
    ed. by Keii, T., Soga, K., Elsevier, Amsterdam, 1986, p. 293

14) ^{13}C-NMR Nomenclature: Carman, C.J. and Wilkes, C.E., Rubber
    Chem. Technol. 44, 781 (1971)

Copolymerization of ethylene and 1-olefins with Ziegler catalysts under high pressure

G. Luft, H. Grünig and R. Mehner

Institut für Chemische Technologie, Technische Hochschule Darmstadt, Petersenstraße 20, D-6100 Darmstadt, FRG

ABSTRACT

After the polymerization of ethylene with Ziegler-type catalysts has been investigated under high pressure the performance of the catalyst was tested in the copolymerization with 1-olefins. The heterogenous titanium catalyst was used together with tetra-iso-butyl dialuminoxane as cocatalyst. The polymerization experiments were performed batchwise in an autoclave reactor of 100 ml capacity at pressures of 1000 and 1500 bar, temperatures between 150 and 230°C and a residence time of 2 minutes. 1-butene, 1-hexene, 1-octene and 1-decene were the comonomers. Their concentration in the feed was increased up to 20 mol %.

From the results of the tests the conversion and the productivity were determined. The polymers yielded were analyzed for their composition. Besides density and average molecular weight the molecular weight distribution was measured. Furthermore the reactivity ratios were determined from amount of comonomer incorporated into the polymer and the composition of the feed.

INTRODUCTION

The objectives for the use of Ziegler-type catalysts in the olefin polymerization under high pressures are greater effeciency of the catalyst, high rate of polymerization, the variation of the product properties in a wide range and the elimination of large amounts of solvent.

In earlier articles the synthesis of waxes and of high molecular polyethylene with magnesium/chloride-supported titanium catalyst and different aluminium alkyl catalysts under high pressures and high temperatures was described.

The objectives of this work were to test the ability of the same catalyst for copolymerization of 1-olefins and to investigate the influence of the incorporation of comonomers on density and molar mass of the polymer.

EXPERIMENTAL

The polymerization experiments were performed batchwise in the pilot plant shown in Fig. 1. Its heart is an inductively heated autoclave reactor with a capacity of 110 ml. The reactor is designed for a maximum pressure of 2500 bar and a temperature of up to 300°C. It was equipped with a fast running, magnetically driven agitator.

W. Kaminsky and H. Sinn (Eds.)
Transition Metals and Organometallics as
Catalysts for Olefin Polymerization
© Springer-Verlag Berlin Heidelberg 1988

Fig. 1. Pilot plant

First the liquid comonomers were fed into the reactor, then ethylene
was added by a 2-stage membrane compressor and the reaction was started
by injection of the catalyst slurry. After a residence time of 2 minutes the reaction was stopped by releasing the pressure.

The catalyst system consisted of $TiCl_3$ supported on $MgCl_3$ ball-milled to
5 μm and suspended in a hydrocarbon (Isopar L). As cocatalyst, tetra-
isobutyl dialuminoxane (TIBAO) was used in a molar ratio of Al/Ti = 50.

RESULTS

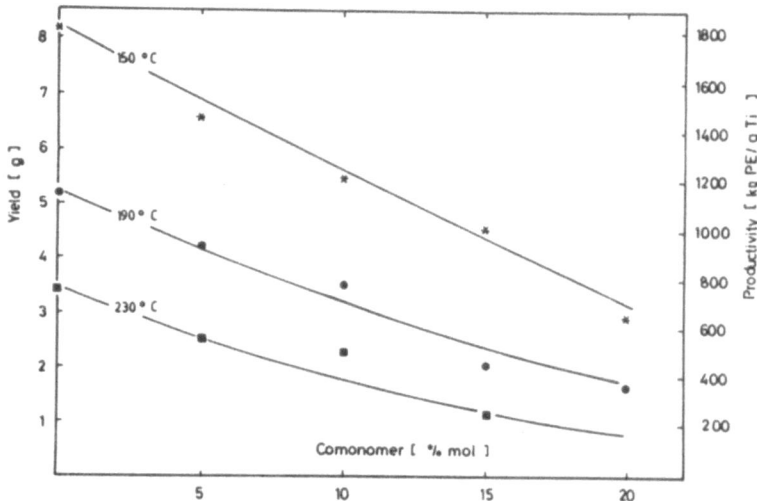

Fig. 2. Influence of the comonomer on yield and productivity.
        Comonomer 1-butene, pressure 1500 bar

First the influence of the comonomers on yield and productivity was
determined. As shown in Fig. 2 yield and productivity decreased significantly with increasing content of 1-butene in the feed. When 20
mole % of butene were added only half of the yield and productivity
were obtained as in the homopolymerization experiments. Yield and productivity decreased more steeply with increasing concentration of comonomer in the feed when 1-decene or 1-hexene were used.

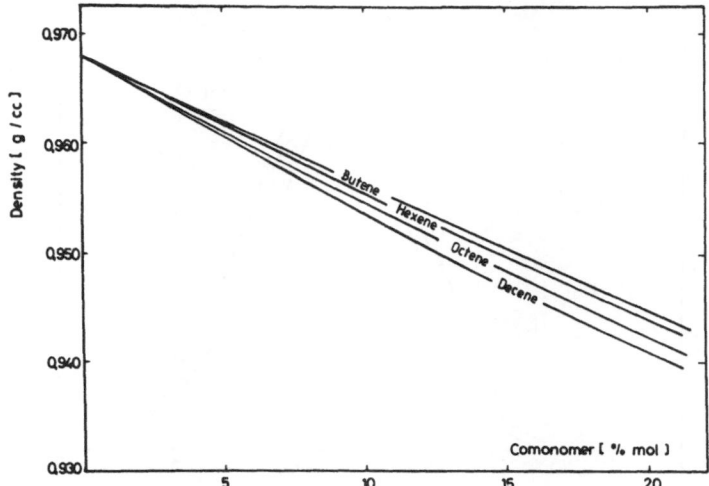

Fig. 3. Influence of the comonomer on the density. Pressure
1500 bar, temperature 190°C

The density of the polymer decreased with increasing amount of comonomer in the feed (Fig. 3). The effect was greater for higher olefins
like 1-decene and 1-octene because of the higher rate of their incorporation.

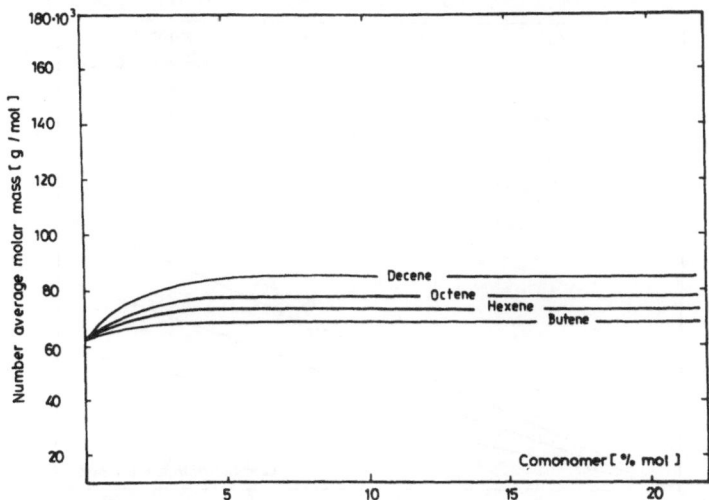

Fig. 4. Number average molar mass. Reactor pressure 1500 bar,
temperature 190°C

The number average molecular weight of the copolymers was found slightly higher than $\bar{M}_n$ of the homopolymers produced under the same conditions of pressure and temperature (Fig. 4). Again the effect was greater for higher olefins. At comonomer concentrations beyond 5 mol % the molar mass was no longer dependent of the amount of the comonomer in the feed gas.

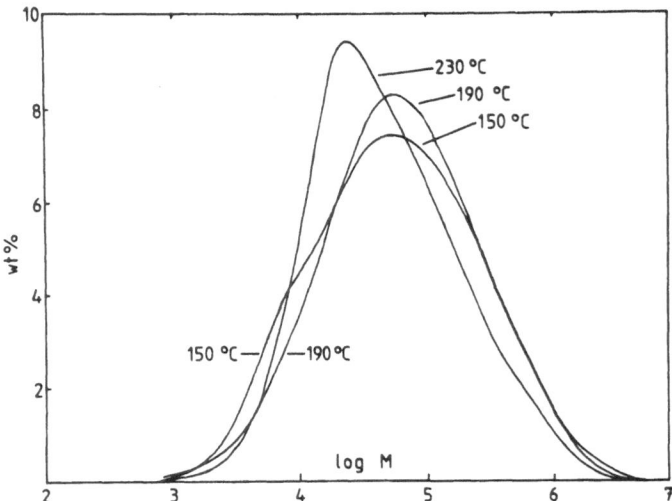

Fig. 5. Influence of the polymerization temperature on the molar mass distribution. Reactor pressure 1500 bar, comonomer 1-butene, comonomer in the feed 10 mol %

Figure 5 shows the influence of the reaction temperature on the molecular weight distribution of a copolymer produced with 10 mol % of 1-butene in the feed. When the reactor temperature was 150°C, the distribution curve exhibits a shoulder at lower masses, which disappears at higher

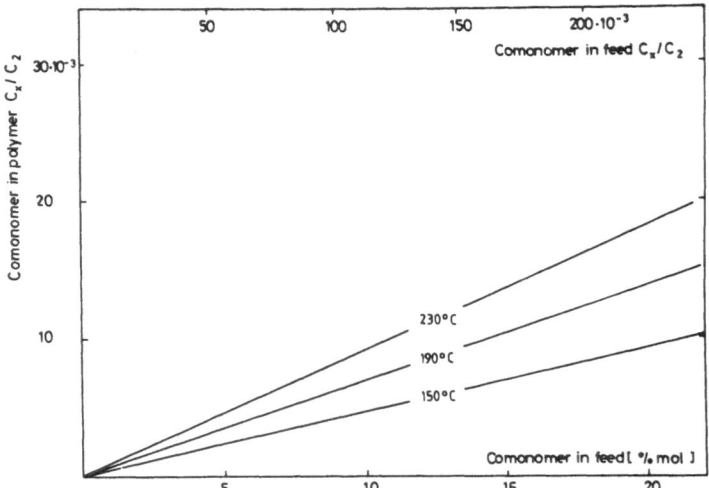

Fig. 6. Incorporation of different comonomers. Pressure 1500 bar, temperature 190°C

temperatures. The polydispersity determined from the ratio of $\bar{M}_w$ to $\bar{M}_n$ decreased with increasing polymerization temperature.

The incorporation of different 1-olefins into the polymer is presented in Fig. 6. In this diagram the amount of comonomer incorporated into the polymer is plotted versus its concentration in the feed. The experiments were performed at 1500 bar and 190°C. With increasing comonomer concentration in the feed the amount of comonomer incorporated increased. The rate of incorporation increased also with increasing chain length of the comonomer.

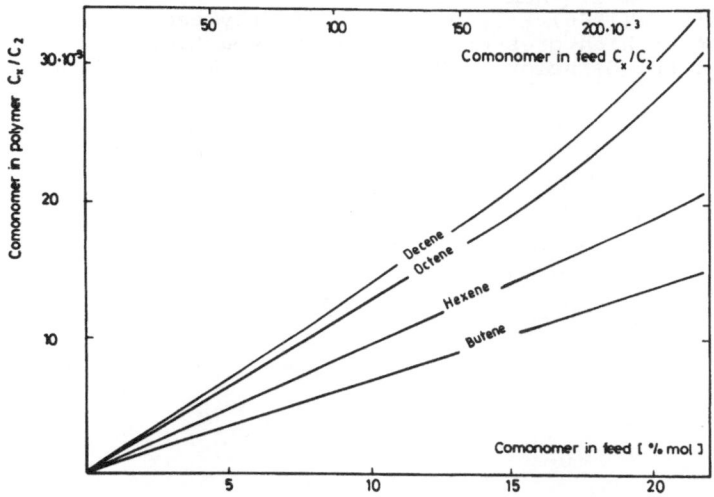

Fig. 7. Influence of the temperature on the incorporation. Comonomer 1-butene, pressure 1500 bar

The last figure (Fig. 7) shows the influence of the temperature on the incorporation of 1-butene at 1500 bar. When the reactor temperature was increased from 150 to 230°C the rate of incorporation of 1-butene increased. The same results were obtained also with other 1-olefins. A lower amount of comonomer was incorporated at lower reactor pressures of 1000 and 500 bar.

Table 1. Reactivity ratios of the system ethylene (1) and 1-butene (2)

Pressure (bar)	Temperature (°C)	Reactivity ratios		
		r_1	r_2	$r_1 \cdot r_2$
1500	150	22.7	0.10	2.270
1500	190	14.3	0.08	1.144
1500	230	9.9	0.04	0.396

From the amount of comonomer incorporated and its concentration in the feed the reactivity ratios were determined. The data are listed in Table 1 for the system ethylene (1) and 1-butene (2).

The reactivity ratios obtained at 1500 bar are comparable to those for low pressures. r_1 values are markedly greater than 1 decreasing steeply with increasing temperature. The values for r_2 are about 3 decades smaller and practically independent of the temperature.

REFERENCES

Grünig HG, Luft G (1986) Synthese von wachsartigen Polyethylenen durch Reaktion von Ethylen mit Aluminiumalkylen unter Hochdruck. Die Angew. Makromol. Chemie 142:161-169

Grünig HG, Luft G (1986) Polymerization of ethylene with Ziegler catalysts under high pressure. In Polymer Reaction Engineering, Ed. Reichert KH, Geiseler W, Hüthig und Wepf Verlag, Basel, p 293-297

On the Activation of Supported Zirconium Organic Catalysts in the Polymerization of Ethylene

J. Ulbricht, J. Giesemann and A. Leistner

"Carl Schorlemmer" Technical University
Department of Chemistry, DDR-4200 Merseburg, GDR

ABSTRACT

Former results point to a deactivation of the active centres of catalysts from tetrabenzyl zirconium and alumina by a reaction with the free hydroxyl groups remaining on the catalyst surface. By pretreatment of the catalyst with n-butyl lithium the attempt was made to drive back this deactivation reaction. Three methods of catalyst modification were studied. Addition of an excess of n-butyl lithium at the beginning of the polymerization effects an increase in the maximum polymerization rate and a suppression of the deactivation to a great extent. A similar result was obtained with catalysts on silica-alumina carriers.

INTRODUCTION

By reaction of tetrabenzyl zirconium (Zrbzl$_4$) with γ-alumina, supported catalysts for the polymerization of ethylene are obtained. During the supporting reaction Zrbzl$_4$ is fixed onto the carrier by splitting off toluene as a result of the reaction between the hydroxyl groups on the surface of the support and the ligands of the zirconium component. But an interaction between the Lewis acid - Lewis base pair sites on the support surface and the zirconium compound is also discussed. The opinion of the authors who studied these catalysts differs as to the composition and the structure of the surface compounds so formed (1,2,3).

Unfortunately, only a small part of these surface compounds is transformed in polymerization-active centres (4). Because of this small concentration most methods for studying the structure are not sensitive enough to differentiate between active and inactive centres. Consequently, there is very little detailed structural information available for any supported complex. Some authors emphasize that no definite proof of any structure for such an active centre has yet been reported (5,6). Therefore the study of supported catalysts is mainly restricted to kinetical studies, because the polymer formed during the polymerization is a result of the reactivity of these active centres.

The polymerization reactivity of the catalysts mentioned above depends on the pretreatment, the loading and the porosity of the carriers (7,8,9,10). The kinetic profile of these catalysts is characterized by a gradual increase in the rate at the beginning of the polymerization, followed by a period of constant polymerization and finally a decline in the rate.

W. Kaminsky and H. Sinn (Eds.)
Transition Metals and Organometallics as
Catalysts for Olefin Polymerization
© Springer-Verlag Berlin Heidelberg 1988

DEACTIVATION OF THE CATALYSTS

During the polymerization the catalysts lose their activity relatively quickly. From our former studies it follows that the decline is caused by a decrease in the concentration and the reactivity of the propagation centres (4). These centres are deactivated by a chemical reaction and not by restriction of the monomer diffusion to the active sites, due to the inclusion of the catalyst particles into the polymer formed. The decline in the polymerization rate also occurs if the monomer feed is interrupted and no polymer is formed during the deactivation phase, as recently shown (11).

Catalysts prepared from the same support, pretreated at different calcinating temperatures, and impregnated with the same quantity of $Zrbzl_4$ exhibit different deactivation rates. Catalysts from supports with a lower calcinating temperature have higher deactivation rates because of their higher content of free hydroxyl groups remaining on the surface. Catalysts from oxide carriers normally possess free hydroxyl groups since it is impossible to remove all these groups by reaction with metal organic compounds, as we, and other authors established (3).

The decline in the polymerization rate follows a first order law. We found a functional relationship between the rate constant of deactivation and the concentration of the free hydroxyl groups on the catalyst (11). No exact linear relationship was established between these two quantities because of a certain difference in the reactivity of the hydroxyl groups and also of the active centres on the catalyst surface. From this result it must be concluded that during polymerization the propagation centres of the catalysts are deactivated by a reaction with the free hydroxyl groups remaining on the catalyst surface or with their protons.

MODIFICATION OF ALUMINA-SUPPORTED CATALYSTS WITH BUTYL LITHIUM

If the assumption about the deactivation reaction mentioned above is correct then by blocking a part of these hydroxyl groups it should be possible to drive back or suppress the deactivation. Consequently, we tried to block these groups via a reaction with n-butyl lithium. Three methods of catalyst modification were studied:

1. Modification of the support with n-butyl lithium before impregnation with $Zrbzl_4$.
2. Modification of the impregnated catalyst with n-butyl lithium.
3. Addition of n-butyl lithium into the polymerization mixture.

In line with the first method, the non-impregnated carrier containing a certain concentration of free hydroxyl groups on its surface is treated with n-butyl lithium. The metal alkyl does not only react with the free hydroxyl groups by deprotonation but also with the coordinatively unsaturated aluminium cations and oxygen anions in the support surface, which act as Lewis acid and Lewis base centres. The latter reaction can be interpreted as a concerted reaction, in which the latent carbanion of the n-butyl lithium attacks the Lewis acid centre while the electrophilic metal atom attacks the Lewis base centre. In this way butyl groups are attached to the support surface. Table 1 shows the result of the reaction of n-butyl lithium at a molar ratio butyl lithium / hydroxyl groups = 0.5 and 1.0 in a mixture of cyclo-hexane and pentane with alumina calcinated at 600 °C.

Table 1.   Modification of γ-Al$_2$O$_3$ with n-butyl-Li

$\dfrac{BuLi}{OH}$ (mole ratio)	content of butyl groups a) $(\dfrac{mmol}{9Al_2O_3})$	unreacted OH groups $(\dfrac{mmol}{9Al_2O_3})$
0	0	0.74
0.5	0.30	0.67 b)
1.0	0.48	0.48 b)

a) butyl groups of the modified support from determination of the
   butane formed by reaction with H$_2$O
b) calculated

With increasing treatment with n-butyl lithium the concentration of
the fixed butyl groups increases and the free hydroxyl groups decreases.
Because of the low concentration of the n-butyl lithium added the
modified support still contains unreacted hydroxyl groups which are
able to interact with Zrbzl$_4$. The catalysts so modified have a maximum
polymerization rate which is three times higher than before but there
is no change in the deactivation (Fig. 1). Contrary to our expectations,
this modification method does not influence the deactivation of the
catalyst but enhances the polymerization maximum.

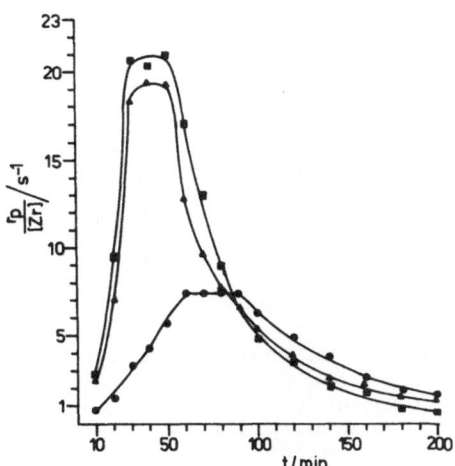

Fig. 1.   Effect of n-butyl-Li on the overall polymerization rate r$_p$ with
the Al$_2$O$_3$/BuLi-supported Zr(CH$_2$C$_6$H$_5$)$_4$ catalyst

ethylene = 0.490 MPa; (content of Zr)$\cdot$10^4 in mol Zr / g supported
catalyst = ($\bullet$): 2.4, ($\blacktriangle$): 2.1, ($\blacksquare$): 2.1; mole ratio n$_{BuLi}$/n$_{Zr}$ = ($\bullet$): 0,
($\blacktriangle$): 1.8, ($\blacksquare$): 3.3; thermal pretreatment of the support: 600 $^\circ$C;
solvent: toluene; polymerization temperature = 80 $^\circ$C

In the second modification method the butyl lithium reacts directly
with the supported catalyst. A catalyst with 0.2 mmol Zrbzl$_4$/g catalyst
was treated with n-butyl lithium dissolved in a cyclohexane/pentane

mixture at a molar ratio of butyl lithium/Zrbzl$_4$ = 1 - 5. The chemical analysis shows the formation of butyl groups on the catalyst and the transformation of hydroxyl groups; however, the ratio of the benzyl groups to the zirconium content was unchanged. With increasing concentration of the n-butyl lithium related to the zirconium content of the catalyst the maximum polymerization rate and the deactivation rate decrease (Fig. 2). An analogous decline of the polymerization rate with increasing concentration of the metal organic compound was established with catalysts impregnated with Zrbzl$_4$ once a special value of the loading was exceeded.

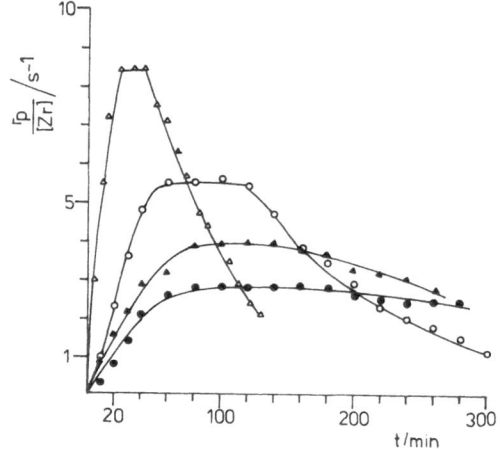

Fig. 2.   Dependence of the overall polymerization rate r$_p$ with the Al$_2$O$_3$/Zr(CH$_2$C$_6$H$_5$)$_4$/n-butyl-Li catalyst on the mole ratio BuLi/Zr ethylene = 0.686 MPa; (content of Zr)$\cdot$10^4 in mol Zr/g supported catalyst = ($\bullet$), ($\blacktriangle$), (O), ($\triangle$): 2.0; mole ratio n$_{BuLi}$/n$_{Zr}$ = ($\triangle$): 0, (O): 1, ($\blacktriangle$): 3, ($\bullet$): 5; the other experimental conditions as in Fig. 1

The third method of modification studied was the addition of the n-butyl lithium into the reaction mixture at the beginning of the polymerization. This method allows the addition of a great excess of the n-butyl lithium in relation to the concentration of the reactive groups on the catalyst. The polymerization rate-time curves in Fig. 3 show not only an increase in the polymerization rate but also extensive suppression of the deactivation. The small decline of the rate after the maximum at the highest n-butyl lithium concentrations is not caused by a chemical reaction, since after interruption of the monomer feed at different reaction times up to 64 hours, the polymerization is continued at the same rate as before. Table 2 illustrates the influence of the n-butyl lithium added into the polymerization mixture on the concentration and the reactivity of the polymerization centres of the catalysts. The catalyst efficiency rises from 10 to 15 %, however, the average propagation rate constant is four times greater than without n-butyl lithium, which points to the generation of propagation centres which are now more reactive.

If the n-butyl lithium is added at a later moment and not at the beginning of the polymerization then the maximum polymerization rate does not increase to such a high value (Fig. 4). The later the moment of addition, the lower the increase of the rate, and the catalyst deactivation disappears.

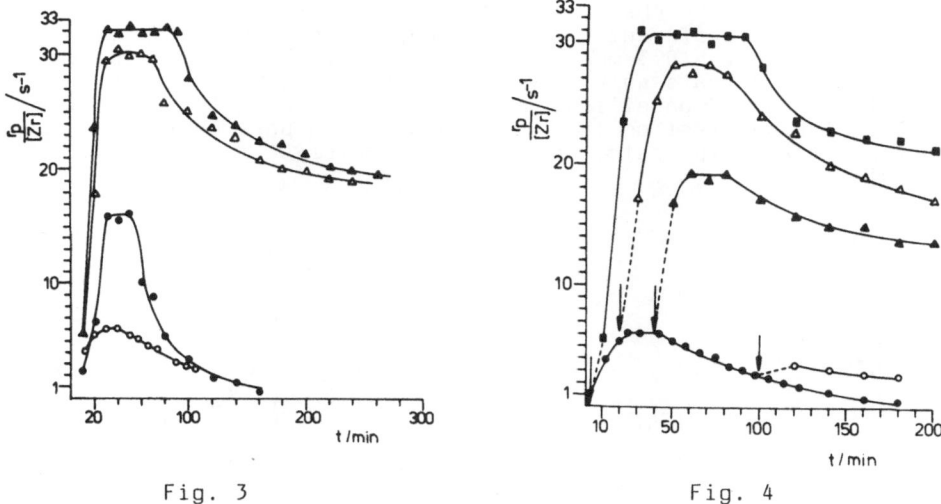

Fig. 3   Fig. 4

Fig. 3.   Effect of n-butyl-Li on the overall polymerization rate r_p with the Al_2O_3-supported $Zr(CH_2C_6H_5)_4$ catalyst

(content of $Zr) \cdot 10^4$ in mol Zr/g supported catalyst = (O), (●), (△), (▲): 2.1; mole ratio n_{BuLi}/n_{Zr} = (O): 0, (●): 10, (△): 40, (▲): 240; the other experimental conditions as in Fig. 1

Fig. 4.   Time-dependence of the overall polymerization rate r_p with the Al_2O_3-supported $Zr(CH_2C_6H_5)_4$ catalyst on the addition of n-butyl-Li into the reaction mixture

mole ratio n_{BuLi}/n_{Zr} = (●): 0, (■), (△), (▲), (O): 100; arrows indicate the moment of n-butyl-Li addition; the other experimental conditions as in Fig. 3

Table 2.   Catalyst efficiency f and average propagation rate constant $\bar{k}_p$ of the Al_2O_3-supported $Zr(benzyl)_4$ catalyzed ethylene polymerization in toluene at 80 °C

loading of the support $(\frac{mol\ Zr \cdot 10^4}{g\ Al_2O_3})$	cocatalyst	$\frac{n_{cocatalyst}}{n_{catalyst}}$ (mole ratio)	f $(\frac{mol \cdot 10^2}{mol\ Zr})$	$\bar{k}_p$ $(\frac{1}{mol \cdot s})$
2.1	-	-	10.3	289
2.3	n-C_4H_9Li	40	15	1240

MODIFICATION OF SILICA-ALUMINA SUPPORTED CATALYSTS WITH BUTYL LITHIUM

The activation effect of n-butyl lithium is not restricted to catalysts from alumina carriers. We also established an analogous effect on silica-alumina carriers. The activity of supported catalysts depends to a great extent on the chemical nature of the carrier used. Fig. 5 shows the influence of different carriers thermally pretreated at 400 °C and supported with 0.2 - 0.3 mmol $ZrbzI_4$/g carrier. The maximum polymerization rate of catalysts supported on alumina is much higher than that of catalysts supported on silica, which is described in the literature. Moreover, the catalyst on the silica-alumina support exhibits an even higher polymerization rate. However, this catalyst loses its activity very quickly.

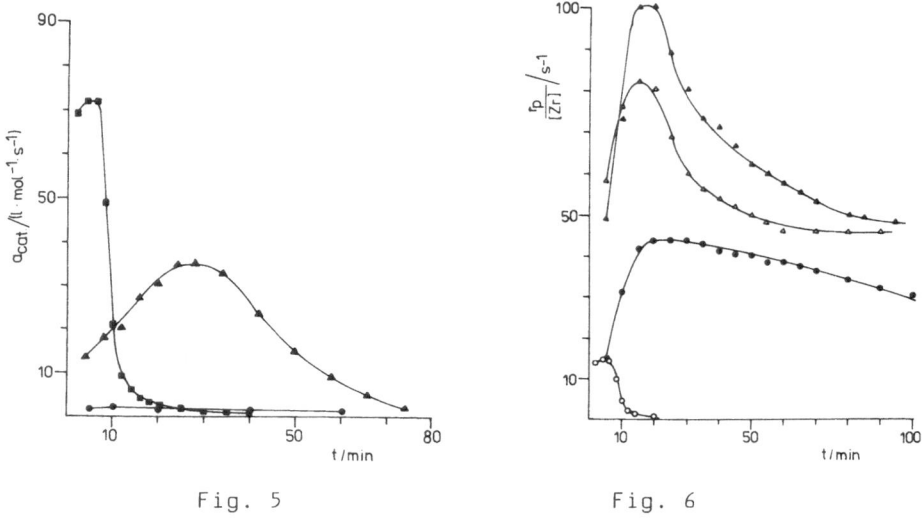

Fig. 5   Fig. 6

Fig. 5.   Influence of the type of the support on the catalytic activity a_{cat}

support = (●): SiO_2, (▲): Al_2O_3, (■): silica-alumina (78 wt % Al_2O_3); (loading of the support)$\cdot10^4$ in mol Zr/g support = (●): 3.2, (▲): 2.2, (■): 2.5; ethylene = (●): 0.882 MPa, (▲): 0.686 MPa, (■): 0.588 MPa; thermal pretreatment of the supports: 400 °C; the other experimental conditions as in Fig. 1

Fig. 6.   Variation of the overall polymerization rate r_p for the silica-alumina (78 wt % Al_2O_3)-supported $Zr(CH_2C_6H_5)_4$ catalyst as a function of n-butyl-Li as cocatalyst

ethylene = 0.588 MPa; (content of Zr)$\cdot10^4$ in mol Zr/g support = 2.4; mole ratio n_{BuLi}/n_{Zr} = (O): 0, (●): 25, (△): 50, (▲): 100; thermal pretreatment of the support: 400 °C; the other experimental conditions as in Fig. 1

In order to drive back this rapid deactivation reaction the addition of n-butyl lithium to the polymerization with catalysts on silica-alumina carriers was studied. Fig. 6 shows a similar effect found in the poly-merization with catalysts on alumina supports. The maximum polymerization rate increases and the deactivation is indeed forced back. A catalyst is formed, which is more active than catalysts on pure alumina modified by n-butyl lithium.

The similarity of the kinetic profile of catalysts from silica-alumina carriers rich in alumina (78 wt % Al_2O_3) in comparison with pure alumina carriers points to similar active sites in both cases. $ZrbzI_4$ seems to be favourably anchored onto hydroxyl groups bound with alumina. From NMR studies it follows that hydroxyl groups on silica-alumina with more than 75 wt % alumina are mainly attached onto aluminium atoms, i.e. the concentration of hydroxyl groups on aluminium and silicium are 2.4 and 0.2 hydroxyl groups / nm^2 (12). On the other hand, because of the higher polymerization rate, it must be assumed that the formation of the active sites is influenced by the higher Brönsted and Lewis acidity of the silica-alumina carriers.

REFERENCES

(1) Ballard DGH (1973) Adv Catal 33: 263
(2) Zakharov VA, Yermakov YI (1979) Catal Rev-Sci Eng 19: 67
(3) Firment LE (1982) J Catal 77: 491, (1983) J Catal 82: 196
(4) Ulbricht J, Giesemann J (1986) Makromol Chem Makromol Symp 3: 345
(5) Evans HE, Weinberg WH (1980) J Amer Chem Soc 102: 2548
(6) Pino P, Mülhaupt R (1980) Angew Chem 92: 869
(7) Sommer H, Giesemann J, Ulbricht J (1984) Acta Polym 35: 253
(8) Ernst E, Giesemann J, Ulbricht J (1986) Acta Polym 37: 28
(9) Sommer H, Giesemann J, Ulbricht J (1984) Acta Polym 35: 369
(10) Ulbricht J, Giesemann J, Ernst E (1986) Polimery 31: 374
(11) Ernst E, Giesemann J, Mellerowicz T, Ulbricht J (1986) Acta Polym 37: 67
(12) Schreiber LB, Vaughan RW (1975) J Catal 40: 226

The influence of crystal water on the performance of a Ziegler-Natta catalyst in propylene polymerization

Thomas Garoff*, Eero Iiskola and Pekka Sormunen

Neste Oy, Technology Centre
SF-06850 Kulloo, Finland

INTRODUCTION

Ziegler-Natta catalysts are the most common catalysts when producing polypropylene. The polymerization process is disturbed even by minute amounts of water. Great care is therefor taken to avoid moisture both when producing the catalyst and when performing the polymerization.

The catalyst is prepared by activating $MgCl_2$ with alcohol. The active structure of the crystals is achieved during the titanation of the $MgCl_2$ adduct in $TiCl_4$. Water, if present, reacts also in this stage with $TiCl_4$. In this way the titanation stage is acting as a moisture lock. Moisture if present in small amounts as crystal water in the support material do thus not disturb the proceeding polymerization process.

There are several synthesis routs described in the patent literature were $MgCl_2$. $6H_2O$ have been used as support material. The catalysts have usually enough activity in polyethylene polymerization but show less or no activity in polypropylene polymerization (1 - 13). Great efforts have been done to bring the hydrated support material to a dry state either by calcination to MgO or by careful drying in HCl atmosphere or by extensive chlorination with chlorinating chemicals as $SOCl_2$, $SOCl_3$, $HSiCl_3$ followed by normal titanation (14 - 19).

Since there have been no systematic study of the influence of crystal water in the support material our attempt has been to investigate this influence both on the properties of the catalysts and the produced polymer materials.

EXPERIMENTAL

A series of $MgCl_2$. $x(EtOH)$. $(6-x)(H_2O)$ support materials was prepared. This was done by veighing 0.1 mol portions of anhydrous $MgCl_2$ in inert conditions. The salt was placed in an 1 l reactor. A slurry was made with 300 ml of heptane. Appropriate amounts of ethanol and water was added droppvise while stirring the slurry. The slurry was then heated just as much to bring the salt-ethanol-water complex to melt. Then the solution was allowed to cool, letting the $MgCl_2$-EtOH-H_2O complex recrystallize. After wash the composition of the dry support material was checked by elementary analyzes of Mg and Cl and by analyzing the ethanol and water content of the support material.

---

* To whom all correspondence should be addressed

W. Kaminsky and H. Sinn (Eds.)
Transition Metals and Organometallics as
Catalysts for Olefin Polymerization
© Springer-Verlag Berlin Heidelberg 1988

Ziegler-Natta catalysts were prepared out of these support materials by treating, them twice with 300 ml of boiling $TiCl_4$. An internal donor was added during the first titanation. After titanation the catalyst were washed with heptane and dried.

The Ti, Mg and Al contents in catalysts and support materials were determined by AAS.

The chloride content was determined by potentiometric titration and the ethanol contents of the support materials were measured by a Micromat gas chromatograph.

The water content of the support material was measured by Karl-Fischer titanation. The material was dissolved in absolute alcohol and the moisture determinated using a Mitsubishi Moisturemeter.

In the Table 1 the support materials are listed together with their state of crystallinity.

Table 1. Codes and composition of the prepared support materials.

Code	Formula	Apperance
MS-132	$MgCl_2 * 6EtOH$	Crystalline
PW-19	$MgCl_2 * 5EtOH * H_2O$	Crystalline
PW-20	$MgCl_2 * 4EtOH * 2H_2O$	Partly crystalline
PW-26	$MgCl_2 * 3.5EtHO * 2.5H_2O$	Partly crystalline
PW-22	$MgCl_2 * 2EtOH * 4H_2O$	Amorphous
PW-23	$MgCl_2 * EtOH * 5H_2O$	Amorphous
PW-24	$MgCl_2 * 6H_2O$	Crystalline

The Mg, Cl and Ti percentages were determined from the catalysts. The X-ray diffraction spectra were also taken for each support material and each catalyst.

The powder X-ray diffraction patterns were recorded employing an automated Philips PW 1700 wide angle diffractometer, a Xe-proportional counter and a graphite monochromator in the diffracted beam. Cu-K-alpha-radiation was used at 40 kV and 40 mA. The step-scanning program was used in recording the X-ray diffractograms in the 2.0 - 62.0 degrees 2 theta range, accumulating the counts every 0.1 degrees 2 theta step for 2 s giving a total measuring time of 1200 s per sample.

POLYMERIZATION

Catalyst activity was examined by polymerization of propylene in heptane slurry. The catalyst solutions were prepared in the following way: triethylaluminium was mixed with external donor. 20 ml of this solution was mixed with 50 ml of heptane and 30 - 40 mg of the catalyst (Al/Ti=200). These solutions were added to the reactor.

Polymerization was carried out at 70 °C and at 10 bar pressure (0.3 bar H_2 and 9.7 bar propylene). Polymerization was maintained for 3 hours, after which the polymer was filtrated and dried. The filtrate was also dried and weighed in order to measure the evaporation residue.

CHARACTERIZATION OF THE POLYMER

Several analysis were carried out: bulk density, isotacticity, melt index, molecular weight and particle size distribution. Bulk density was made according to ASTM D 1895-69. Melt index was measured using a Göttfert MP-E according to ASTM 1238, temperature used was 230 °C and weight 2.16 kg.

The molecular weight of the polymer was determined by GPC (Waters 150C ALC/GPC). Particle size distribution was measured with sieveseries (DIN 4188).

RESULTS AND DISCUSSION

The results of the elementary analyses of the catalyst are shown in the pillar-diagram in Figure 1 and also in Table 2. The outstanding trend in the catalyst series is the increase in the Ti content. Figure 2 shows the trend. There is an drastic increase in the Ti percentage as soon as the amount of crystal water exceeds *$3H_2O$. This is mostly do to the difficulty to elute the reaction products from the catalyst. If only ethanol is present in the support material the main reaction is as follows:

$$MgCl_2 . EtOH + TiCl_4 \rightleftharpoons MgCl_2 . TiCl_3(EtO) + HCl \qquad (1)$$

With excess of $TiCl_4$ present the reaction product $TiCl_3(EtOH)$ is easily eluted away, leaving the activated amorphous $MgCl_2$ open for $TiCl_4$ to coordinate.

Table 2. Composition of the catalysts.

Code	Mg %	Cl %	Ti %
MS-132	14.0	50.2	2.5
PW-19	14.6	51.0	4.2
PW-20	17.7	58.0	4.0
PW-26	16.5	54.0	4.9
PW-22	6.7	39.0	17.0
PW-23	13.6	52.0	9.0
PW-24	6.8	32.0	19.0

COMPOSITION OF THE CATALYSTS

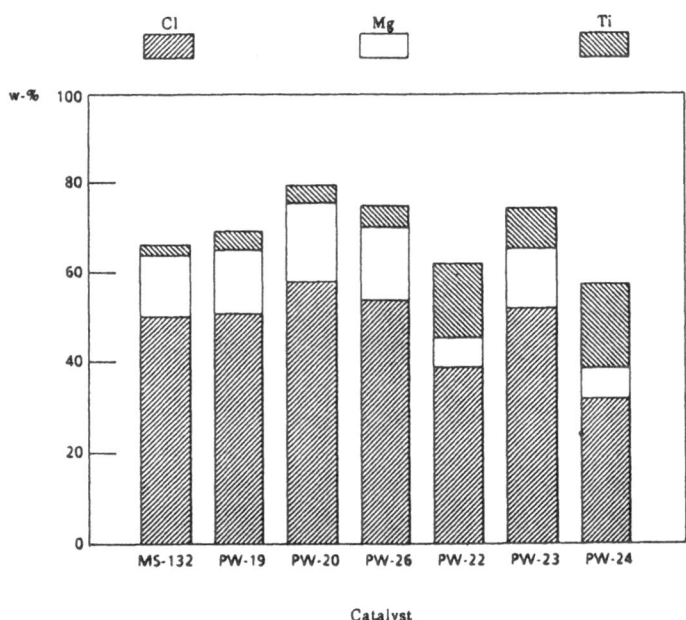

Fig. 1. The Mg, Ti and Cl content of the catalyst.

If water is present as crystal water, then the main reaction is:

$$MgCl_2 + H_2O + 2TiCl_4 \rightleftharpoons MgCl_2 + Cl_3Ti - O - TiCl_3 + 2HCl \qquad (2)$$

The reaction product $Cl_3Ti - O - TiCl_3$ is far less soluble in excess of $TiCl_4$. Thus the $MgCl_2$ crystals are only partly opened for $TiCl_4$ to coordinate, and partly filled up with the reaction product.

ACTIVITY

$TiCl_4$ that is coordinated to the active sites in the $MgCl_2$ crystals are the active centers polymerizing propylene.

Reaction products like $Cl_3Ti-OEt$ and $Cl_3Ti - O - TiCl_3$ are inactive. These products are not only increasing the weight of the catalysts, but also blocking the active centers in the activated support material. So it is not surprising to find a correlation between an increasing amount of crystal water in the support material, increasing amount of Ti in the catalyst and a decreasing activity. This relationship is clearly shown by Figures 2 and 3.

There is a drastic increase in the Ti % of the catalysts when more than three crystal water molecules per $MgCl_2$-molecule have been present in the original support material.

The decrease in activity (in Figure 2) expressed as kg PP/g cat. is almost linear, dropping from the maximum of 10 kg PP/g cat. when no water has been present to 1 kg PP/g cat. when all the crystal ethanols have been replaced by crystal water.

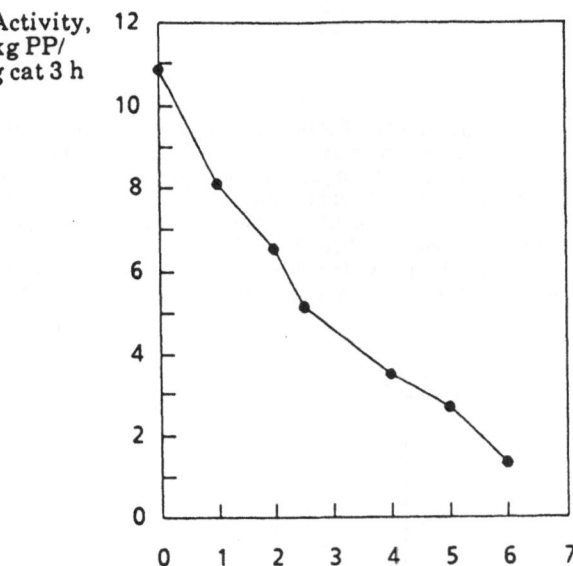

Amount of crystalwater in original support material (x)

$MgCl_2 \cdot xH_2O$

Fig. 2. The activity (kg PP/g cat. 3 h) of the catalyst as a function of the amount of crystal water in the used support material.

In table 3 the polymerization results are listed: the amount of catalyst used (m_{cat}) the amount of polymer gained (m_{pol}), the evaporation residue and the activities.

Table 3. Polymerization results

Code	$m_{cat.}$ mg	$m_{pol.}$ g	Evaporation residue g	%	Activity kg PP/g cat.	Activity 10^3 kg PP/g Ti
MS-132	34.4	373.4	7.6	2.0	10.85	434
PW-19	47.3	383.6	15.6	3.9	8.11	193
PW-20	36.7	240.1	10.2	4.1	6.54	163
PW-26	39.9	205.3	11.2	5.2	5.15	105
PW-22	39.4	137.7	6.2	4.3	3.49	21
PW-23	41.3	110.9	5.9	5.1	2.69	30
PW-24	36.5	48.0	5.8	10.8	1.32	7

Activity expressed as kg PP/g Ti gives an even clearer picture of the situation.
The activity is dropping drastically from over 400 when no water is present to about
100 when half of the ethanol have been replaced by water. When preparing the catalyst
from a support material containing 4, 5 or 6 crystal water molecules the activity
dropps to a level of 10 - 20 kg PP/g Ti. It is thus obvious that inactive Ti is
present in the catalyst.

MOLECULAR WEIGHT DISTRIBUTION

The molecular weight was determined of the polymers. Figure 3 shows the result. The
molecular weight results are clearly forming two groups, one consisting of results
close to 300.000 g/mol and one group of results close to 100.000 g/mol. Again the
support material containing equal amounts of crystal water and crystal ethanol seem
to be the deviding border. If less water than ethanol is present in the support
material the resulting molecular weight is higher 250.000 - 300.000 g/mol and when
more water is present in the support material the molecular weight in the produced
polymer drops to 100.000 g/mol. The numerical results are listed in Table 4.

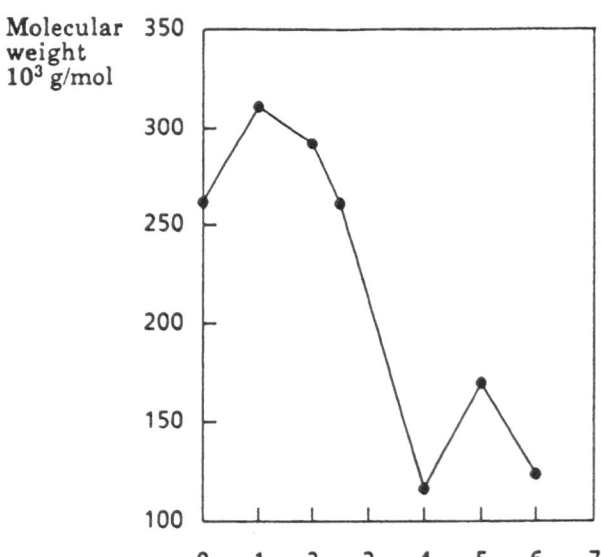

Amount of crystalwater in
original support material (x)
$MgCl_2 \cdot xH_2O$

Fig. 3. Molecular weight of the produced polypropylene as a function of the amount
of crystal water in the used support material.

Table 4. Properties of the polymer

Code	Bulk density g/cm^3	Melt index g/10 min	Molecular weight 10^3*g/mol	Isotacticity %	Isotacticity index %
MS-132	0.40	33.29	262	99.4	97.4
TG-95	0.35	18.96		98.3	
PW-19	0.29	9.69	311	97.4	93.6
PW-20	0.23	1.99	291	96.7	92.8
PW-26	0.33	30.39	261	97.2	92.2
PW-22	0.37	32.95	116	96.8	92.6
PW-23	0.38	55.14	170	98.3	93.3
PW-24	0.36	55.33	124	98.3	87.7

The drop in molecular weight is also shown as an increase in the melt index. This is shown in Figure 4. There is a first area where the melt index is below 10 g/10 min corresponding to the higher molecular weight results. When more crystal water is present in the support material there is a corresponding increase in the melt index, up to 50 - 60 g/10 min if 5 to 6 ethanol molecules have been replaced by crystal water. The results are also listed in table 4.

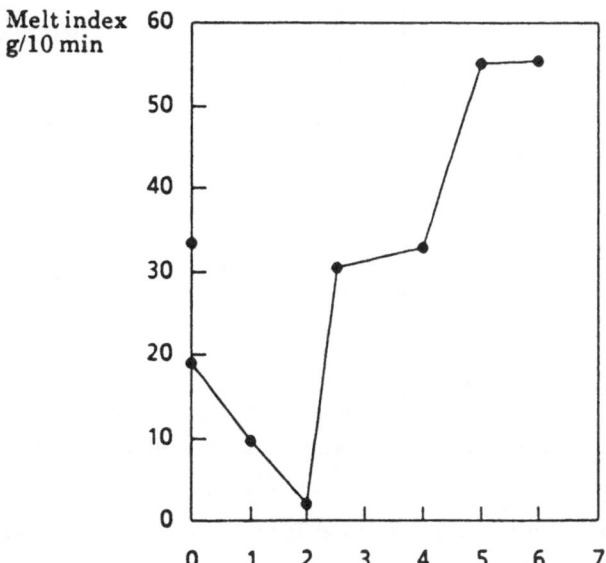

Fig. 4. The melt index of the produced polypropylene as a function of the amount of crystal water in the used support material.

BULK DENSITY

No particular efforts had been made to influence the particle size distribution. Thus the isolated effect of an increasing amount of crystal water in the support material was possible to investigate. The bulk densities were low throughout the whole series of investigation. There is an avaridge result of 0.35 g/cm^3 whit a clear drop in the results when $MgCl_2$. H_2O . 5(EtOH) and $MgCl_2$. 2(H_2O) . 4(EtOH) were used as support material. The results are listed in Table 4.

ISOTACTICITY, EVAPORATION RESIDUE AND ISOTACTICITY INDEX

The isotacticity for the resulting polymers were all in the region of 97 - 98 % which can be regarded as satisfying. The isotacticity is however just telling the result of the polymerization as it appears in the polymer. The evaporation residue shows how much atactic and short chained isotactic polypropylene have dissolved in the solution during the polymerization. The resulting curve is first slowly rising from the ideal values of <2 % to 4 - 5 % when 4 - 5 crystal water molecules are present 'in the support material. If $MgCl_2$. 6(H_2O) was used there was a drastic increase in the evaporation residue to over 10 %. Figure 5 is showing the isotacticity index as a function of the crystal water present in the support material. The total isotacticity is first dropping from the ideal value of 97 % when no water have been involved to a index level of 92 - 93 %. The results are stabilizing here for almost the hole investigation series until the index drastically falls down to 87 % when only crystal water have been present in the support material ($MgCl_2$. 6(H_2O)).

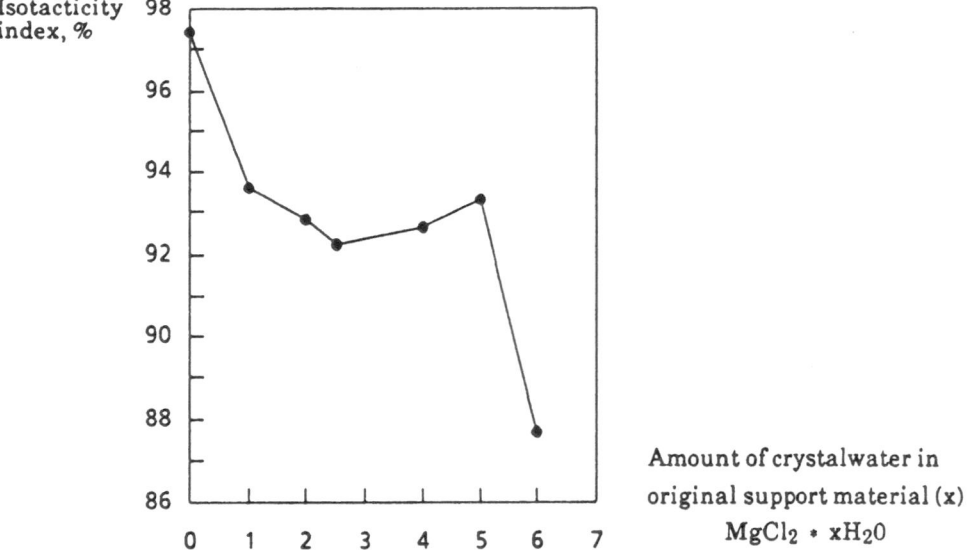

Fig. 5. Isotacticity index of the produced polymer as a function of the amount of crystal water in the used support material.

The polymer samples were also investigated by NMR. The results are listed in table 5.

Table 5. The isotactic, syndiotactic and atactic precentages in the polymer samples.

Code	Isotactic %	Syndiotactic %	Atactic %
MS-132	92.5	4.8	2.7
PW-19	90.6	5.6	3.8
PW-20	86.2	8.5	5.3
PW-26	91.6	5.2	3.2
PW-22	90.2	6.6	3.2
PW-23	92.4	5.0	2.6
PW-24	93.8	4.4	1.8

PARTICLE SIZE DISTRIBUTION OF THE POLYMER

The results in polymer particle size distribution shows the same behavior as the molecular weight distribution. If the crystal water content was below 50 % there was a main particle size distribution around 1 mm, if more water have been involved in the support material the particle size distribution shifted drastically towards finer material with a maximum around 0.18 mm. The distribution also spread out on a wider size region. Figure 11 is showing a typical particle size distribution for the first group, and Figure 12 for the second.

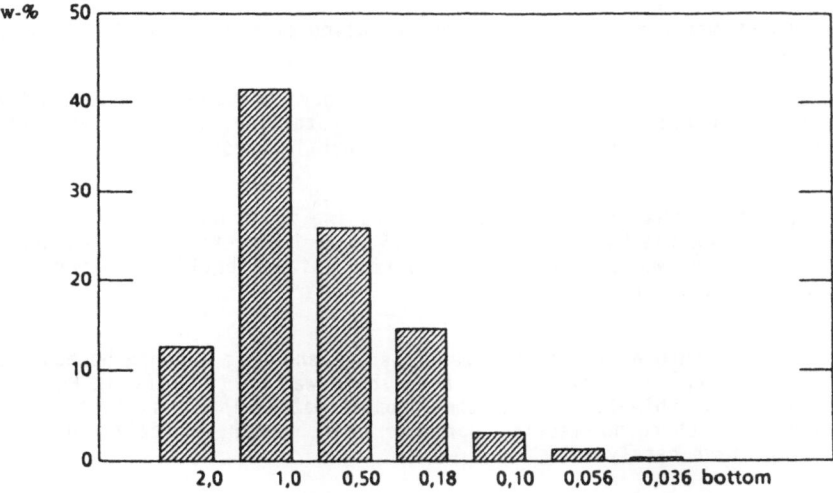

Fig. 6a.

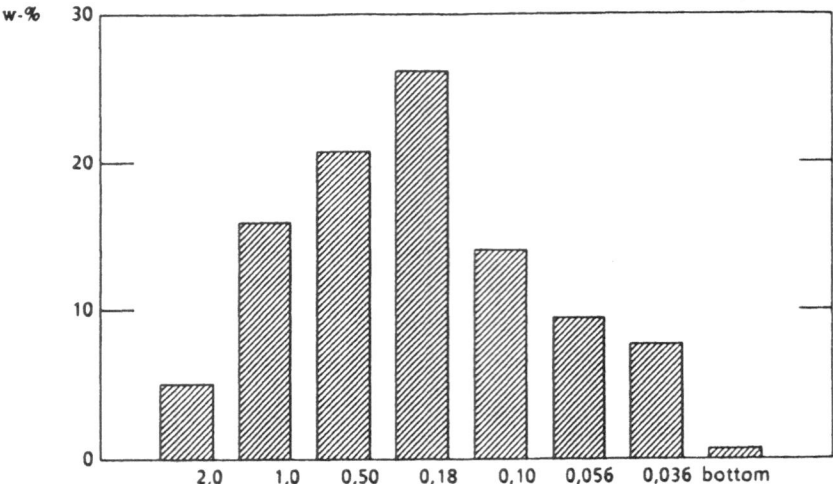

Fig. 6b. Particle size distribution for the resulting polymers produced with the PW-19 (above) and the PW-24 (below) catalysts, d (mm).

X-RAY DIFFRACTION PATTERNS

The structure of the catalysts and the support materials were investigated by X-ray diffraction (XRD). At first the support materials with different crystal water/crystal ethanol ratio, ranging from 0:6 to 6:0, were characterized. We were interested in

changes in the support material structure after replacing crystal ethanol molecules with water molecules.

It was not possible to get any diffractograms of four and five crystal water containing support materials because they were not crystalline. Also two and half and three crystal water containing products were only partly crystalline.

When replaceing crystal ethanol with water, increasing disorder was noticed as increasing peak number. On the $MgCl_2$. $2.5 H_2O$. $3.5 EtOH$ diffractogram the disorder had gone so far that only two main peaks were distinguistable besides an amorphous hump with very small crystalline peaks.

The peaks in the $MgCl_2$. $6H_2O$ diffractogram were sharper and more intense because of a tighter lattice. The peak at 9 degrees 2 theta angle was not present in $MgCl_2$. $6H_2O$ but clearly distinquisable in all the other support materials. There is a strong correlation between the changing lattice parameters in support material and the activity of the respective catalysts.

Catalysts prepared out of support material containing up to 2 crystal water had X-ray patterns that resembled each other (Figure 7). Differences in activity could be explained with variations in crystal size calculated from the peak at 50 degrees 2 theta angle.

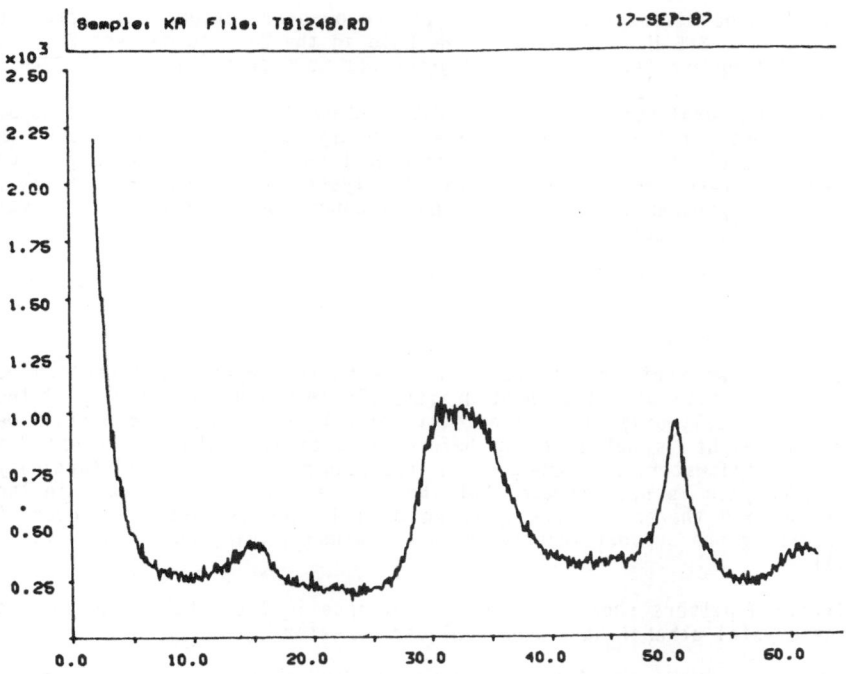

Fig. 7. X-ray diffraction pattern for catalyst prepared at out of $MgCl_2 . 6(EtOH)$.

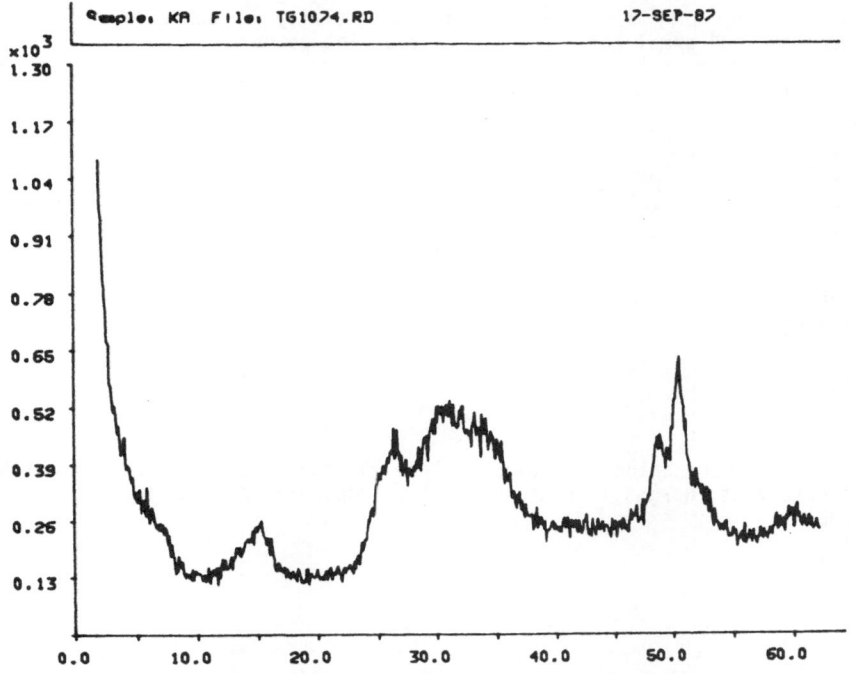

Fig. 8. X-ray diffraction pattern for catalyst prepared ut of $MgCl_2 . 6(H_2O)$.

Essential changes in the diffraction pattern appeared on diffractograms of catalyst prepared out of $MgCl_2$. 2.5 H_2O . 3.5 EtOH. The halo in the 25 - 40 degrees 2 theta angle range was split up and the peak at 15 degrees was more crystalline in nature.

varying degree of structural collapse could be detected on the difraction patterns of catalyst made of support material containing 4 to 6 crystal water. The pattern in Figure 8 is a good example of all these diffraction patterns. The important magnesium dichloride micro structure was absent in these catalysts indicating low activity. Two additional peaks appeared at 27° and 48° which cannot be detected in a normal $MgCl_2$ * $TiCl_4$ Ziegler-Natta catalyst.

CONCLUSIONS

This investigation shows that crystal water in the $MgCl_2$ support material can be present up to the mol ratio of 1:1 without drastically lowering the activity of the catalysts in propylene polymerization. An activity of 8 kg PP/g cat/3 h can still be reached. Molecular weight is not dropping before 50 % of the ethanol content has been replaced by crystal water, the same goes for the mean particle size. Isotacticity of the resulting polymer is not effected but there is an constant increase in the evaporation residue and thereby a corresponding drop in the isotacticity index. A drastic drop occurs if the support material have contained nothing but crystal water ($MgCl_2$. 6(H_2O)).

The X-ray diffraction patters show some kind of collapse in the catalyst as soon as over 50 % of the crystal ethanol have been replaced by water.

To avoid significant drop in catalytic activity the $MgCl_2$ support material should contain not more than one crystal water molecule per $MgCl_2$ support molecule.

If H_2O is used as the only activating agent low activity is achieved, molecular weight is around 100.000 g/mol and the polymer is a fine powder.

REFERENCES

1. US 4,520,121
2. GB 1-546912
3. US 4,473,660
4. GB 1291522
5. S 7204267-4
6. SF 57422
7. GB 1539 175
8. DE 26 43143 C2
9. EP 0 015 099
10. US 4,218,339
11. EP 0045 975 US
12. EP 0045 976 US
13. EP 0 086 473
14. Nirisen, ø, Polymerization of polypropylene on magnesium chloride supported Ziegler-Natta catalysts, Norges Tekniske Høgskole, Trondheim, anhandling
15. GB 2018 788 A
16. GB 1536 171
17. GB 2000 514 A
18. S 7713028-4
19. SF 57422

INDUSTRIAL ASPECTS OF THE PRODUCTION OF CATALYSTS FOR ETHYLENE POLYMERIZATION

T. Dall'Occo, U. Zucchini, I. Cuffiani
Dutral S.p.A. Ausimont Compo N.V.
Centro Ricerche "G. Natta", Ferrara, Italy

INTRODUCTION

High activity Ziegler-Natta catalysts, based on activated magnesium chloride, for the low pressure polymerization of ethylene, have been thoroughly studied in order to find significant rules to drive catalyst synthesis and forecast its performance. The catalyst is actually the key not only to controlling the main polymer properties but also to improving economics and achieving the full potential of the most advanced polymerization processes.

For slurry and gas phase polymerization processes updated catalysts should fulfill a broad range of requirements as shown in Table 1.

Table 1. Main requirements of industrial catalysts for the polymerization of ethylene.

STORAGE AND FEEDING	POLYMERIZATION PROCESS	POLYMER STRUCTURE
- no storage and aging problems	- high productivity - proper kinetic - no reactor fouling	Control of: - molecular weight - molecular weight distribution
- easy feeding to the reactor	- control of morphology average particle size and bulk density of the polymer particles	- density - chain branchings and unsaturations
- low sensitivity to the "poisons"		- wax production

As is well known, whereas polymer structure (hence properties) mainly depends on the chemical composition of the catalyst, the kinetic behaviour of the polymerization reaction and the parallel growth of the polymer particle are instead dependent on the physical structure of the catalyst as well. Thus a paramount aspect is to control the building and shaping of the catalyst in every detail of its synthesis.

Dutral produces industrial catalysts and is highly skilled at controlling average particle size (APS) and particle size distribution (PSD) of the catalyst.

Typical performances of these catalysts are shown together with the influence of pressure and polymerization time on the replication phenomenon. Kinetic behaviour in dependence on catalyst APS is also highlighted.

W. Kaminsky and H. Sinn (Eds.)
Transition Metals and Organometallics as
Catalysts for Olefin Polymerization
© Springer-Verlag Berlin Heidelberg 1988

EXPERIMENTAL

Ethylene polymerization tests were carried out on lab scale in slurry
at the following conditions: reactor, 2.5 dm^3; hexane, 1.0 dm^3; cocatalyst Ali-Bu$_3$, 1.0 g; catalyst, 0.015 g; temperature, 85°C; ethylene
partial pressure, 0.63 MPa; hydrogen partial pressure, 0.47 MPa; time
3 hours.

Kinetic data were recorded according to the rate of ethylene consumption. APS and PSD of catalysts and polymers were determined by MALVERN
2600c microphotosizer instrument.

PERFORMANCE OF THE CATALYSTS

It is quite clear that in the design of industrial catalysts no one preparation method is able to completely control at will both the chemical
composition (i.e. amount of transition metal, ligands, etc.) and the
physical structure (i.e. size, shape, texture, etc.) of the catalyst.
We have found a method which allows us to obtain - by carefully changing
the chemico-physical parameters of the synthesis - a broad range of
high yield catalysts characterized by different shape, size and capability to control polymer MWD as shown in Table 2.

In order to improve the control of the polymerization kinetic and of
the polymer particle morphology we have selected - inter alia - proper
titanium/magnesium ratios. This brings about a high titanium content
in the catalysts. Nevertheless, high polymer productivities in respect
to the catalysts are still obtained so that, in industrial practice,
the de-ashing step is unnecessary (chlorine residues in the polymer
less than 20-30 ppm).

From Table 2 one can note that all the catalysts yield polymers with
high bulk density and narrow PSD independently of catalyst APS. This
can be clearly seen in Fig. 1.

The extraordinary ability of these catalysts to control shape and size
of the polymer particles (by the well known replication phenomenon[1]
is also shown by the microphotographs in Figures 2 through 6.

A replication factor (i.e. polymer size to catalyst size relationship)
of about 24 has been found (Tab. 3) for all the catalysts having similar chemico-physical structure. This value is in good agreement with
the calculated one assuming a spherical shape of the particle, thus
confirming the ability of the catalysts to control the replication
phenomenon.

KINETIC BEHAVIOUR

As far as kinetic profiles is concerned it has been found that catalysts
of similar chemical composition (GF1B/M, GF1B, GF1E) and physical texture, but of different APS (respectively 5, 15, 43 µm) practically
show the same "hybrid" - type rate curve as defined by Floyd[2] (Fig.8).
(A more evident decay and a slightly higher activity is shown by the
catalyst with the smallest APS). Thus with these catalysts there is
no important effect of particle size in polymerization.

TABLE 2. Result of ethylene slurry polymerization test[a] with commercial and development grades Dutral catalyst.

Grade	C A T A L Y S T					H I G H D E N S I T Y P O L Y E T H Y L E N E						
	Ti	Specific Surface Area	Porosity	A.P.S.	Span[b]	Yield	$MI_{2.16}$	$MI_{21}/MI_{2.16}$	Bulk Density	Flowa-bility	A.P.S.	Span[b]
	(wt%)	(m^2/g)	(cm^3/g)	(µm)		(Kg/gcat)	(g/10')		(g/cm^3)	(s)	(µm)	
GF1 B/M*	13.2	6.5	0.05	4.6	1.1	14.4	2.6	24.2	0.52	23	113	0.6
GF1 B/S**	13.5	8.6	0.08	7.8	0.8	12.3	2.2	24.7	0.38	25	187	0.7
GF1 B***	13.5	< 1	<0.01	15.2	0.8	12.7	3.7	24.3	0.40	18	391	0.8
GF1 E***	13.9	14.1	0.08	42.7	1.4	11.2	3.4	24.9	0.39	18	921	1.3
GF1 C***	17.0	38.7	0.2	14.0	1.2	25.0[c]	0.15	55.0	0.34	22	490	1.1
GF1 D**	22.0	---	---	43.1	1.4	14.0[c]	0.15	45.0	0.32	23	1000	1.3

*Experimental grade; **Development grade; *** Commercial grade;
a) Polymerization conditions: see experimental part;
b) Span = $(D_{90}-D_{10})/D_{50}$;
c) $P_{ethylene}$ 0.7 MPa, $P_{hydrogen}$ 0.4 MPa, Temperature 75°C.

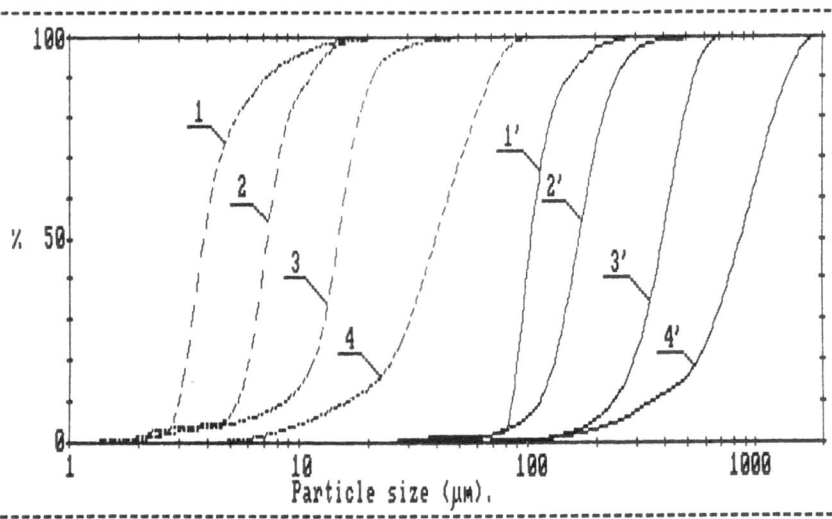

Fig. 1. Particle size distribution of catalysts 1-4 and related polymers 1'-4'.
1 - GF1 B/M; 2 - GF1 B/S; 3 - GF1 B; 4 - GF1 E.

TABLE 3. Comparison between found and calculated catalyst replication
factor.

CATALYST	REPLICATION FACTOR	
	$Rf^{a)}$	$Rc^{b)}$
GF1 B/M	24.6	27.3
GF1 B/S	24.0	25.9
GF1 B	24.2	26.2
GF1 E	21.6	25.1

a) Found replication factor: $Rf = APS_{pol}/APS_{cat}$

b) Calculated replication factor: $Rc = \sqrt[3]{Y \cdot \dfrac{\delta cat}{\delta pol}}$

where:

Y = yield (g pol/g cat)

δcat = catalyst density (g/cm^3)

δpol = polymer density (g/cm^3)

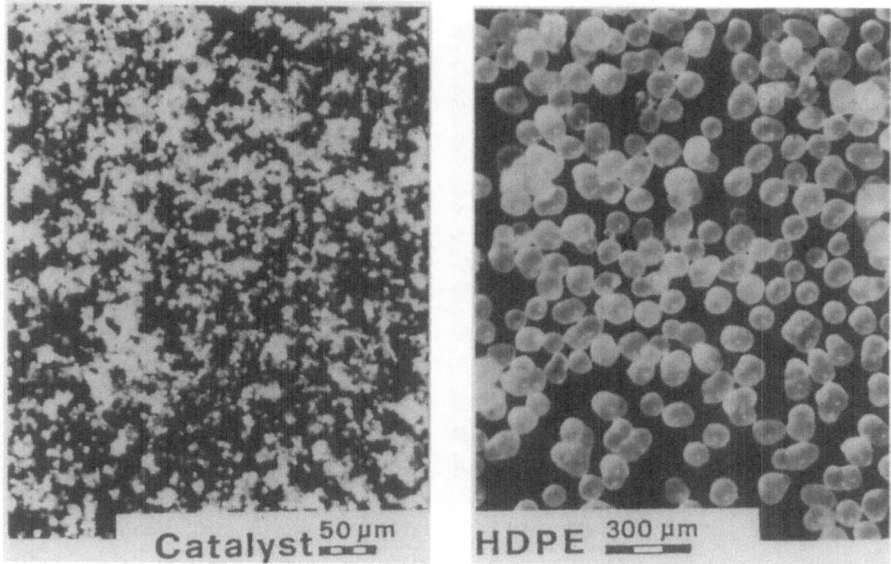

Fig. 2. Optical microphotograph of GF1 B/M catalyst particles and related polymer particles.

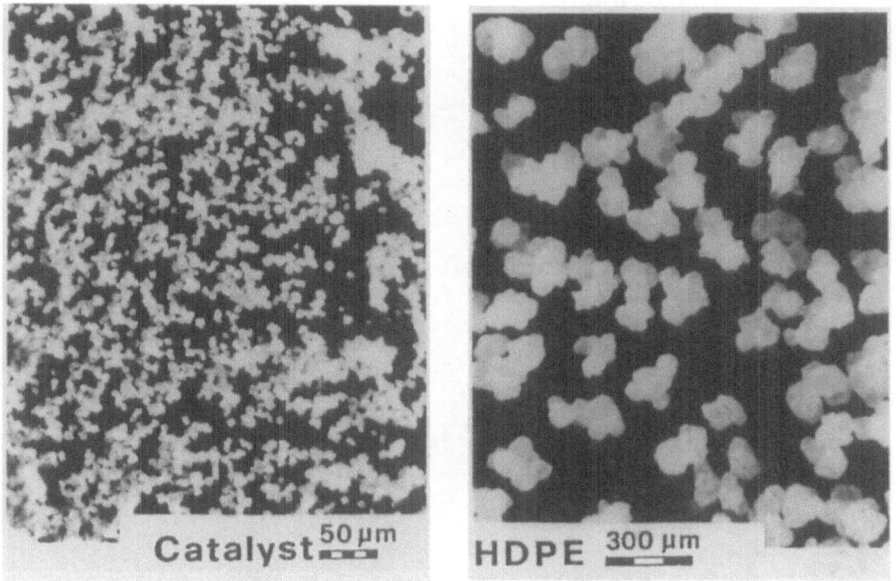

Fig. 3. Optical microphotograph of GF1 B/S catalyst particles and related polymer particles.

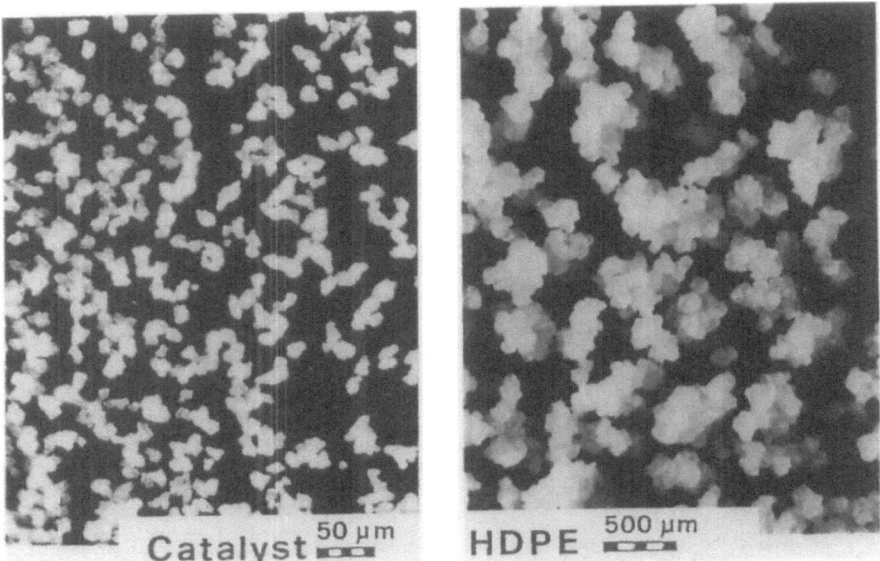

Fig. 4. Optical microphotograph of GF1 B catalyst particles and related polymer particles.

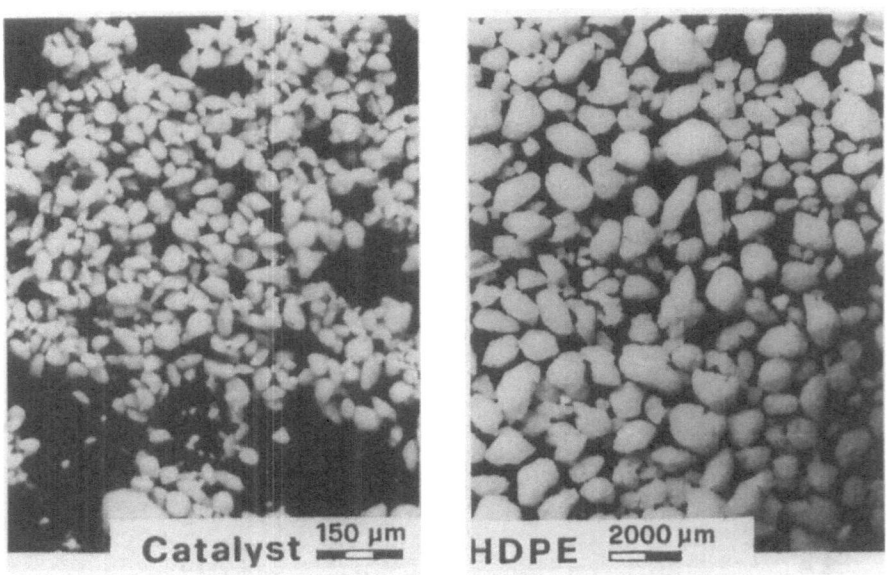

Fig. 5. Optical microphotograph of GF1 E catalyst particles and related polymer particles.

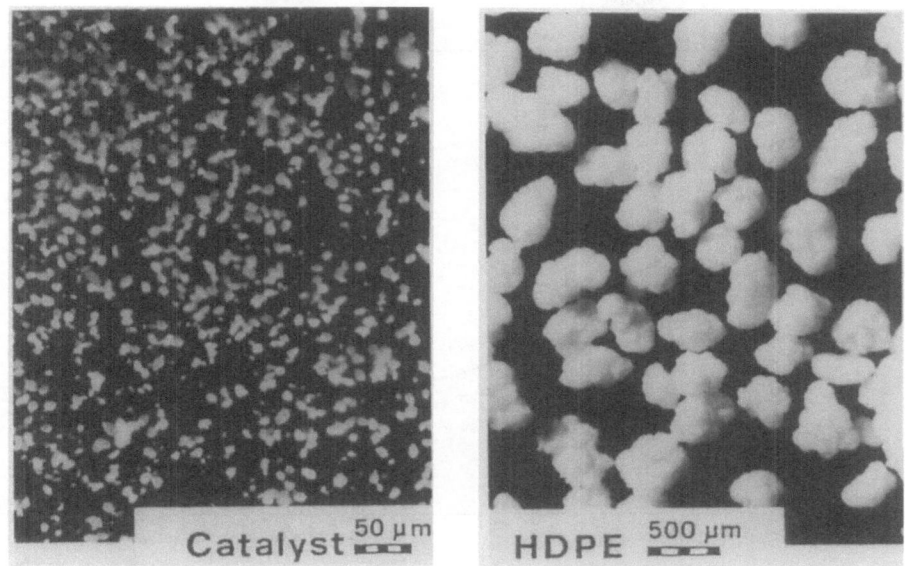

Fig. 6. Optical microphotograph of GF1 C catalyst particles and related polymer particles.

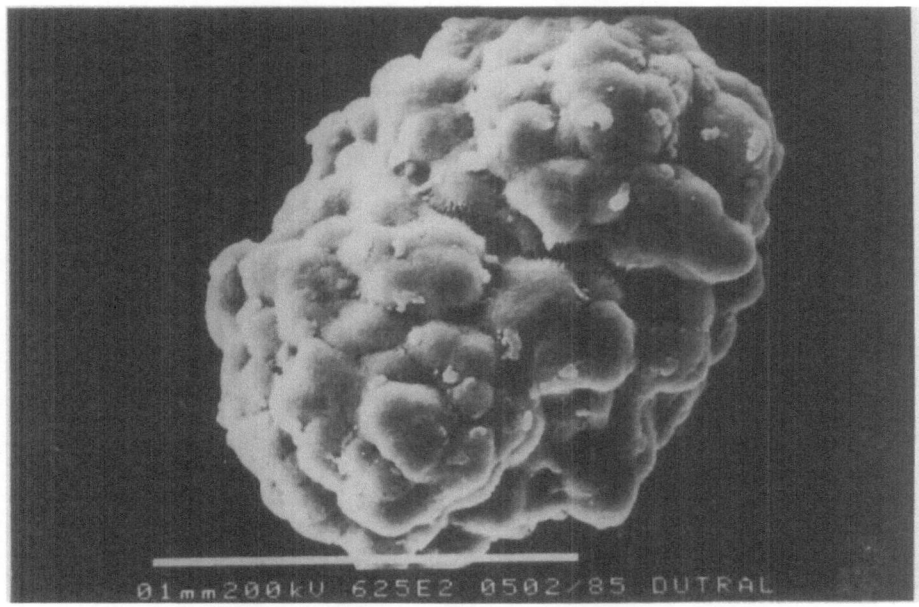

Fig. 7. S.E.M. picture of a polymer particle from GF1 B catalyst (625 X).

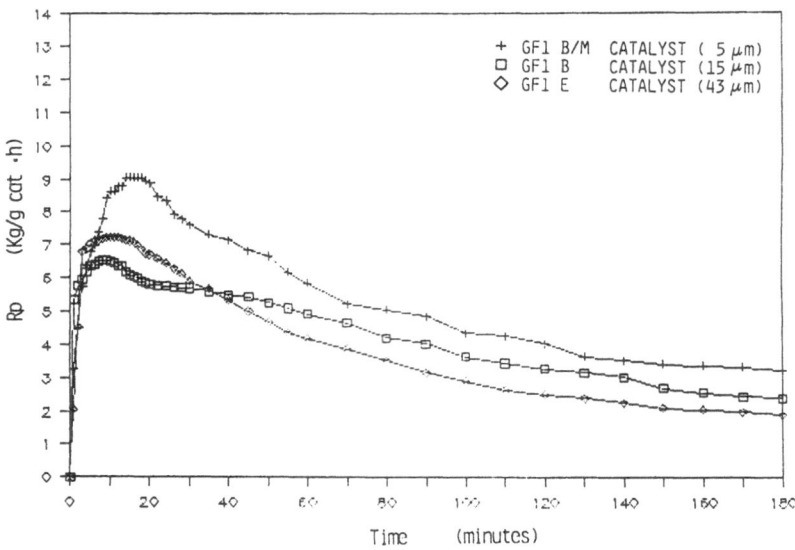

Fig. 8. Polymerization rate Rp as a function of the reaction time for catalysts with different sizes.

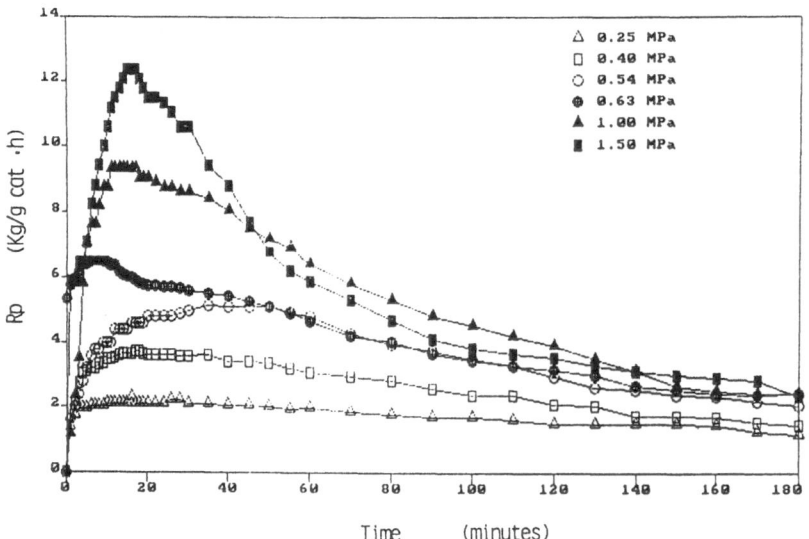

Fig. 9. Polymerization rate Rp as a function of the reaction time at different ethylene partial pressures with GF1 B catalyst.

If we examine in greater detail the behaviour of the GF1 B catalyst we can recognize (Fig. 9) the typical influence of monomer pressure on the rate-time profile as already pointed out for other catalysts by Tait [3] and the appearance of a pronounced decay at higher pressure.

As a matter of fact, outstanding characteristics of this catalyst are: kinetic stability, regular shape and compactness of the related polymer particles (Fig. 7) which allow high bulk density in a broad range of polymerization conditions. Moreover, neither kinetic profile nor PSD and replication phenomenon are affected by either chemical (cocatalyst precontact) or physical (ultrasound treatment) modifications of the catalyst. This is well shown also in the case of a controlled prepolymerization by which it is possible to increase catalyst APS (compare curves 3 and 4 in Fig. 10) without modifying polymer PSD (Fig. 10) and kinetic behaviour (Fig. 11).

Significantly, in order to bring about substantial changes in the kinetic behaviour it is necessary to deeply destroy the physical structure, hence the texture, of the catalyst particle. For instance by milling the catalyst or by synthesizing it without control of the morphology one obtains catalyst APS respectively of 4 and 8 μm, as shown in curves 1 and 2 in Fig. 10. In this case the related kinetic curves change from "hybrid" type to "decay" type profile (Fig. 12). This again confirms that catalyst kinetic behaviour is controlled not only by chemical but also by certain physical factors (i.e. break up, etc.).

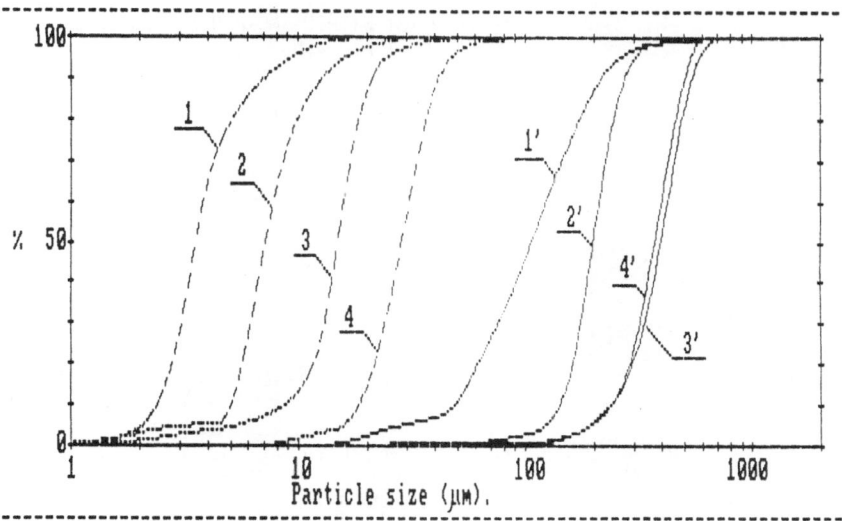

Fig. 10. Modification of GF1 B catalyst: particle size distribution of catalyst 1-4 and related polymer 1'-4'. 1-ground catalyst; 2-catalyst obtained without control of the synthesis parameters; 3-catalyst as such; 4-prepolymerized catalyst (3g/g cat.).

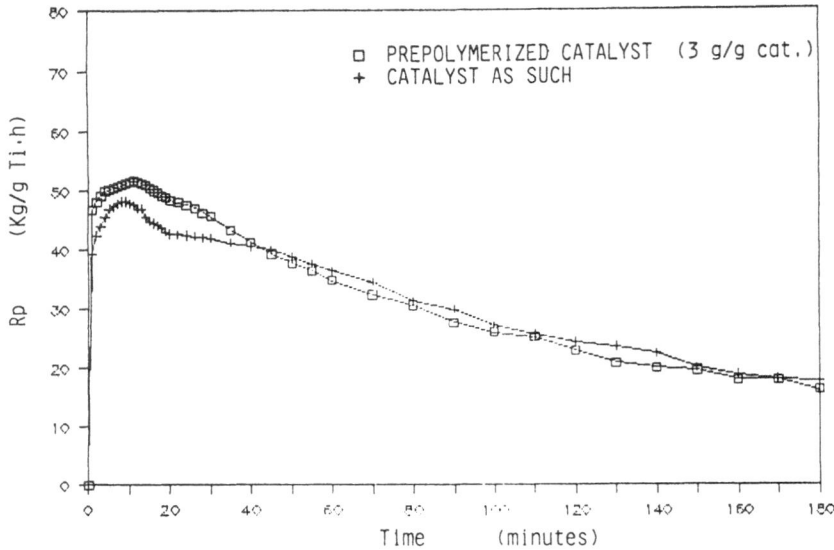

Fig. 11. Effect of the prepolymerization on the kinetic behaviour of the GF1 B catalyst.

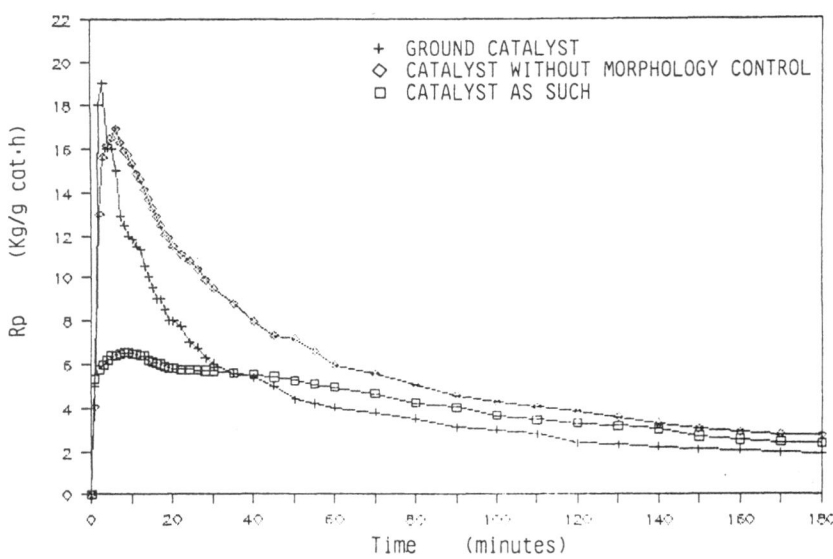

Fig. 12. Effect of GF1 B catalyst modification on the kinetic behaviour.

POLYMER PARTICLE GROWTH

The dependence of polymer particle growth on the polymerization time is governed by the replication phenomenon. This is clearly shown, for instance, by the GF1 B catalyst as depicted in Fig. 13 where polymer APS is compared with the yield. This behaviour should be consistent with a growth model predicting a rapid increase of polymer APS during the early stages of polymerization. It is interesting to note that found and calculated replication factors are in very good agreement (Tab. 4). The real occurence of the replication phenomenon is also confirmed by the polymer PSD measured at different times as shown in Fig. 14.

Finally, it is worth saying that this mechanism works well over the whole industrial range of monomer pressures (Fig. 15).

In conclusion a broad specturm of polymer APS with controlled size can be covered by changing, in a concerted way, pressure or polymerization time as shown in Fig. 16 for the GF1 B catalyst. Similar behaviours are peculiar also of the other catalysts.

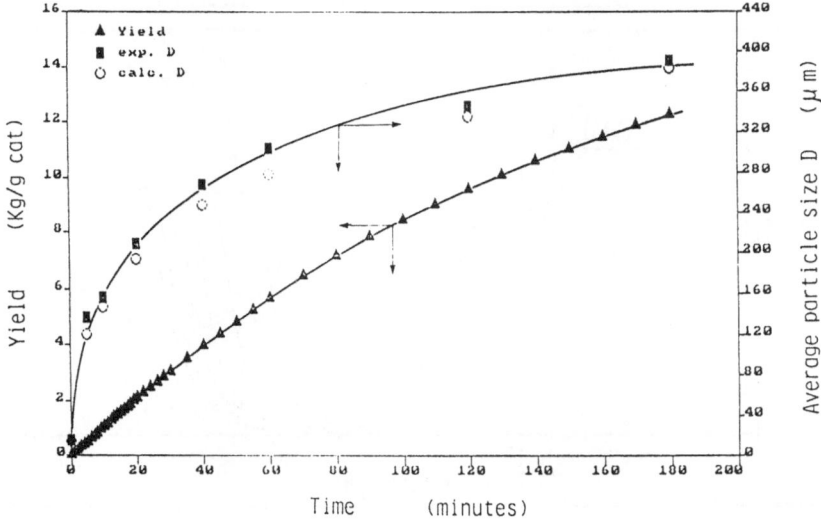

Fig. 13. Average polymer particle size D and yield as a function of reaction time with GF1 B catalyst.

TABLE 4. Comparison between found and calculated dimension of the polymer particles at different polymerization times (GF1 B catalyst).

Time	D_e	D_c	$\Delta_1 = D_e - D_c$	Replication factor R_f	R_c	$\Delta_2 = R_f - R_c$
(min)	(μm)	(μm)	(μm)			
5	137	123	14	9	8	1
10	156	151	5	10	10	-
20	209	199	10	14	13	1
40	268	253	15	18	17	1
60	304	284	20	20	19	1
120	345	344	1	22	23	-1
180	392	392	-	24	26	-2

R_f = found replication factor

R_c = calculated replication factor

D_e = experimental average polymer particle size

D_c = calculated size according to [4]: $D_c = 2 \cdot \sqrt[3]{\dfrac{3}{4\pi\delta N} \cdot \int_0^t R_p \, dt}$

where: δ = polymer density; N = number particles; R_p = polymerization rate; t = polymerization time.

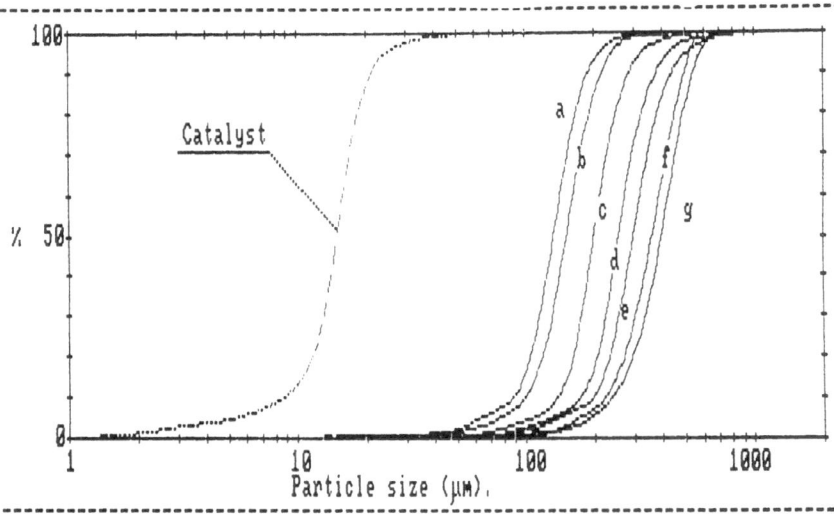

Fig. 14. Particle size distribution as a funciton of reaction time with GF1 B catalyst after: a- 5'; b- 10'; c- 20'; d- 40'; e- 60'; f- 120'; g- 180'.

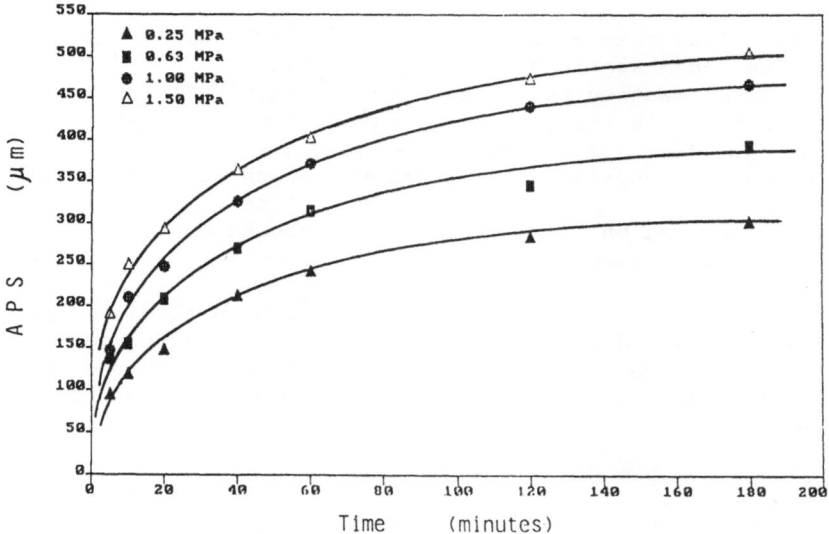

Fig. 15. Average polymer particle size as a function of reaction time at different ethylene partial pressures with GF1 B catalyst.

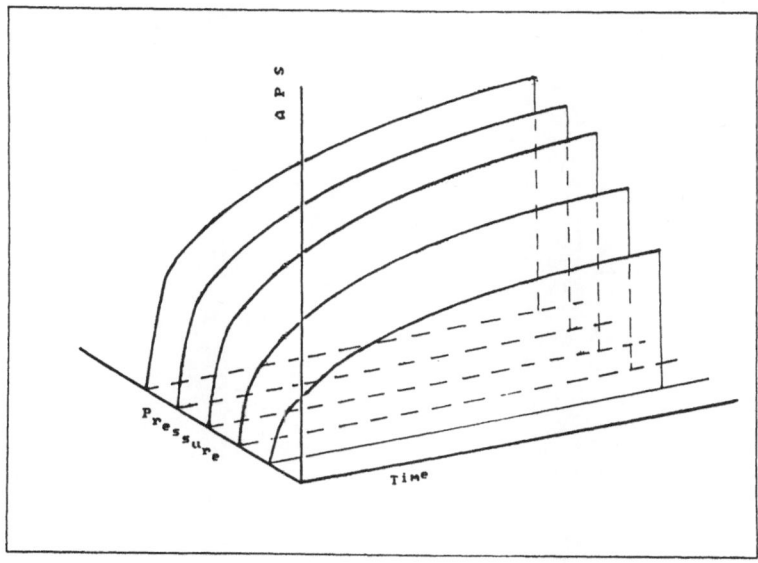

Fig. 16. Overall view of the influence of reaction time and pressure on the polymer particle growth with GF1 B catalyst.

CONCLUSION

For some industrial catalysts of similar chemical composition but of
different size and shape it has been shown that:
- ethylene polymerization replication factor is practically the same
- kinetic behaviour is independent of catalyst particle size
For the same catalyst it has been found that:
- only considerable physical modifications can change the kinetic profile
- replication phenomenon is working and it depends on monomer pressure
and polymerization time.

Thus with these catalysts able to control polymer particle morphology
(size, shape and bulk density) it is possible to enhance reactor
throughput, to improve existing polymerization processes and, for the
sake of energy savings, even to eliminate the pelletizing step in polymer manufacture.

REFERENCES

1. J. Boor, Jr., "Ziegler Natta catalyst and polymerization" Academic
   Press, New York, 1979, p. 180.
2. S. Floyd, T. Heiskanen, T.W. Taylor, G.E. Mann, W.H. Ray, J. Appl.
   Polym. Sci., 33 1021 (1987).
3. P.J.T. Tait, Catalytic polymerization of olefins (Stud. in surf.
   sci. and cat. 25), T. Keil and K. Soga Eds., Elsevier 1986, p. 305
   and references therein.
4. F.J. Karol, Catal. Rev. Sci. Eng. 26 557 (1984).

The Application of X-Ray Absorption Spectroscopy to Study the Titanium Environment on $MgCl_2$ Supported Ziegler Natta Catalysts

P J V Jones[‡] and R J Oldman[*]

Research & Technology Department, ICI Chemicals and Polymers Group, Wilton, Cleveland[‡] and The Heath, Runcorn, Cheshire,[*] UK

INTRODUCTION

As a problem in structural chemistry the aim to characterise the catalytic site on $MgCl_2$ based $TiCl_4$ Ziegler Natta catalysts presents substantial difficulties. Problems centre on the instability of Ti complexes in air, and the consequent need for in-situ studies, and the dilution and dispersion of active material. The potential of x-ray absorption spectroscopy (XAS) opposite some of these difficulties makes it well suited to characterise catalyst structure. The technique uses fine structure at core level absorption edges to derive information on chemical state and near neighbour bonding round elements of specific interest. At the absorption edge information comes from transitions which occur to the valence band. For higher x-ray energies absorption fine structure (EXAFS/XANES) results from back scattering of the photo-electron by neighbouring atoms. Since XAS is concerned with local order information can be derived from amorphous systems independent of dispersion. Also, using x-rays, material is accessible in pores, at interfaces or in-situ under working environments including liquids. Taking advantage of these possibilities, using XAS at the Daresbury Synchrotron Radiation Source, we have been able to study the Ti environment in $MgCl_2$ supported $TiCl_4$ Ziegler Natta catalysts made by milling and precipitation. The results of the first stages of this work are presented in this paper.

XAS MEASUREMENTS AT DARESBURY SYNCHROTRON RADIATION SOURCE

The spectra reported here were recorded using the new double crystal, focusing monochromator on station 8.1 at Daresbury SRS designed for high resolution and high intensity. This is the most advanced XAS monochromator in the world and yielded correspondingly good quality spectra. For many of the samples examined Ti is diluted in a relatively strongly absorbing matrix containing Cl, making fluorescence detection essential. Also it would be extremely difficult to prepare suitably thin samples ($\sim 20\mu m$) for transmission x-ray absorption measurements. Fluorescence XAS (FLEXAFS) effectively measures the number of core holes created by the absorption process, in this case via Ti K_α fluorescence (I_f). I_f/I_0 is linearly dependent on absorption coefficient (1) for dilute samples or thin layers in the conventional 45°/45° geometry (Fig 1). Departure from these conditions effectively leads to a reduction in EXAFS amplitudes and uncertainty in quantificatio of co-ordination numbers. In order to correct for this a computer program ABSCAL was written to calculate the departure from ideal behaviour for any sample. Ti K_α radiation is relatively soft and therefore in order

W. Kaminsky and H. Sinn (Eds.)
Transition Metals and Organometallics as
Catalysts for Olefin Polymerization
© Springer-Verlag Berlin Heidelberg 1988

to minimise air path only a single scintillation photomultiplier detector was used as close to the sample as possible in the standard 45°/45° configuration.

After preparation air sensitive samples in sealed vessels were transferred to an ultra pure atmosphere VAC HE series Dri Lab for loading into small cells with a Mylar window designed for FLEXAFS (Fig 1). Complete sealing of the cell was achieved with a smear of silicone grease between the window and front of the cell body and between the insert and rear stopper. Sample cells were loaded individually and examined immediately by XAS.

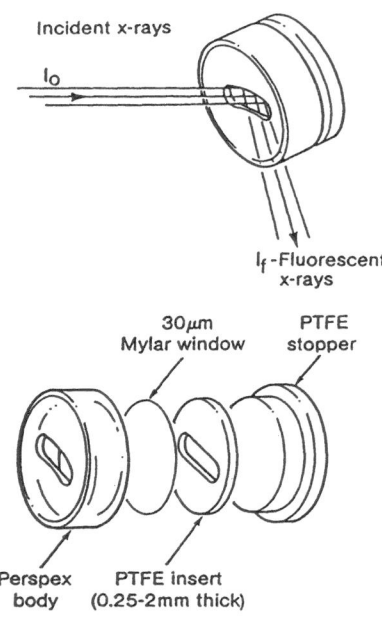

Fig 1. Fluorescence cell for EXAFS

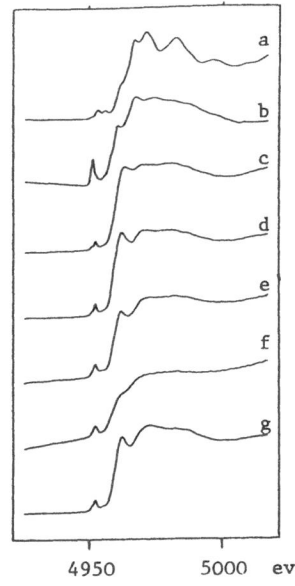

Fig 2. Ti K edge XAS for (a) TiO_2 (b) $TiCl_4$ (c) $EB_2 \; TiCl_4$ (d) $(EB_2TiCl_4)_2$ (e) catalyst B (f) alkylated catalyst B (g) catalyst A

RESULTS

Absorption Edge Position

Absorption edge position depends roughly on oxidation state due to reduced screening of nuclear charge with valency and is modified by the number and identity of surrounding ligands. Work on Ti in various minerals and oxides gives ∿2ev splitting between Ti^{III} and Ti^{IV} (2). Table 1 shows edge positions (Fig 2) for a variety of standard materials, ester complexes and catalysts. Edge positions were defined by taking the first derivative of the data to identify an inflection point or point of steepest ascent on the edge. The situation is complicated because of the difficulty in separating the true edge from the first 'shape' resonance in the XANES caused by multiple scattering of the photoelectron by surrounding ligands (3). However Table 1 shows a very clear trend in line with reference 2 and indicates that the dominant Ti species in the catalyst is Ti^{IV}.

Table 1. X-Ray Absorption Edge Data

	Pre-Edge Features (ev)			Edge Position (ev)
Ti foil		4953.6		4958.4
30% TiCl$_2$/MgCl$_2$		4956.3		4961.4
γ TiCl$_3$ (H)		4956.6		4962.6
γ TiCl$_3$ (AA)		4956.5		4962.7
TiCl$_4$		4956.4		4964.2
TiO$_2$	4955.7	4958.4	4961.2	4967.8
EB$_2$TiCl$_4$	4955.9	4957.3	4960.4	4965.7
DNHPTiCl$_4$	4956.1	4957.2	4960.8	4965.8
DIBPTiCl$_4$	4956.3	4957.9	4961.1	4965.8
(EBTiCl$_4$)$_2$	4955.8	4957.5		4964.4
EA$_4$MgCl$_2$TiCl$_4$	4955.9	4957.4		4965.0
Catalyst A	4955.9	4957.6		4965.0
Catalyst B		4957.5		4964.3
Alkylated Catalyst B		4957.6		4964.4

Pre-Edge Features

In XAS, transitions at an absorption edge occur from a core level to
the valence band. For the Ti K edge this is a transition from the 1s
level to a valence band dominated by d electrons, which is forbidden,
and hence pre-edge spectral features are generally weak. They are
however observed (Fig 2) and carry information on the nature of the
valence orbitals. For example in tetrahedral TiCl$_4$ the valence orbitals
have more 'p' like character and therefore the pre-edge transition
should be more allowed (3). The edge spectrum for TiCl$_4$ in Fig 2 does
indeed show a relatively intense feature at 4956.4 ev confirming the
dramatic change in valence orbitals in switching from octahedral to
tetrahedral co-ordination. This represents an extreme but work on Ti in
minerals and oxides, rutile solid solutions and alkoxides (2,4)
correlates the form and intensity of pre-edge features with symmetry at
the Ti site and increasing octahedral distortion and mixing of
orbitals. We would therefore expect to be able to differentiate bridged
and monomeric Ti ester complexes on this basis. Figure 2 shows edge
spectra for rutile and two TiCl$_4$ ester complexes, monomeric and dimeric
ethyl benzoate, which represent substantial octahedral distortion.
Both show a tendency towards a single pre-edge transition but this is
accentuated for the bridged complex (EBTiCl$_4$)$_2$. Figure 2 also shows
similar data for the catalysts and indicates at least a badly distorted
octahedron and most probably a bridged structure.

Extended X-ray Absorption Fine Structure (EXAFS)

In order to carry out quantitative structural interpretation of EXAFS
data, and in particular to calculate x-ray fine structure from trial
structures for comparison with experiment using program EXCURVE at
Daresbury, it is necessary to develop phase shifts using standard
materials to theoretically describe the backscattering of the emitted
photoelectron by the surrounding structure. Taking into account
simplicity and stability of materials, the ethylacetate complex
$EA_4MgCl_2TiCl_4$, TiO_2 (rutile) and Ti foil were chosen for this purpose.
The derived phase shifts were tested against literature data for the Ti
halides as shown in Table 2. The halides show excellent agreement
between experiment and theory and confirm the ability of the technique
to distinguish between different kinds of sites, eg tetrahedral
($TiCl_4$ r=2.19Å) and octahedral ($TiCl_3$ r=2.46Å). No difference in
structure was detected for the two samples of γ-$TiCl_3$, Stauffer AA and H.

Table 2. EXAFS Results For Standard Materials

	Element	Co-ordination Number	Bond Distance(Å)*	Debye-Waller Term (Å²)≠
Ti	Ti	12	2.89 (2.90)	0.013
TiO_2	O	4	1.94 (1.95)	0.007
		2	1.97 (1.98)	0.007
	Ti	2	2.98 (2.96)	0.007
	O	4	3.53 (3.49)	0.007
		4	3.56 (3.56)	0.007
	Ti	8	3.56 (3.57)	0.010
$EA_4MgCl_2TiCl_4$	Cl	1	2.27 (2.26)	0.007
		1	2.29 (2.28)	0.007
		1	2.31 (2.31)	0.007
		1	2.31 (2.33)	0.007
		1	2.44 (2.47)	0.010
		1	2.52 (2.49)	0.011
$TiCl_4$	Cl	4	2.17 (2.19)	0.008
γ-$TiCl_3$	Cl	6	2.44 (2.46)	0.016
30% $TiCl_2$/$MgCl_2$	Cl	6	2.54 (2.50)	0.015

*   Literature values are shown in parenthesis.
≠   The Debye-Waller term, $2\sigma^2$, represents thermal and static
     disorder.

As an additional test Table 3 shows the structural data derived by
EXAFS for various ester complexes compared to literature data where
available (5,6). Most importantly Table 3 compares x-ray absorption
results for the two ethyl benzoate complexes which show excellent
agreement with literature values. Coupled with the pre-edge data, from
the EXAFS models we can conclude that the 'unknown' isobutylphthalate
complex and n-heptylphthalate complex are monomeric. Attempts to
fit a bridged model to these resulted in an unreasonably high value for
the Debye-Waller disorder term for the bridging atoms, ie the computer
fitting routine was attempting to reject them.

Table 3.   EXAFS Results for $TiCl_4$/Ester Complexes and Catalysts

	Element	Co-ordination Number	Bond Distance(Å)*	Debye-Waller Term (Å²)≠
$(EBTiCl_4)_2$	O	1	2.05 (2.05)	0.003
	Cl	1	2.20 (2.21)	0.004
		2	2.23 (2.22)	0.004
		1	2.53 (2.46)	0.005
		1	2.54 (2.52)	0.006
EB_2TiCl_4	O	1	2.09 (2.05)	0.003
		1	2.09 (2.07)	0.003
	Cl	1	2.21 (2.22)	0.003
		1	2.22 (2.23)	0.003
		1	2.27 (2.29)	0.003
		1	2.29 (2.30)	0.005
$DIBPTiCl_4$	O	2	2.08	0.003
	Cl	4	2.26	0.003
$DNHPTiCl_4$	O	2	2.08	0.003
	Cl	4	2.27	0.003
Catalyst A	O	1	2.04	0.013
	Cl	3	2.20	0.020
		2	2.52	0.022

*   Literature values are shown in parenthesis.
≠   The Debye-Waller term, $2\sigma^2$, represents thermal and static
     disorder.

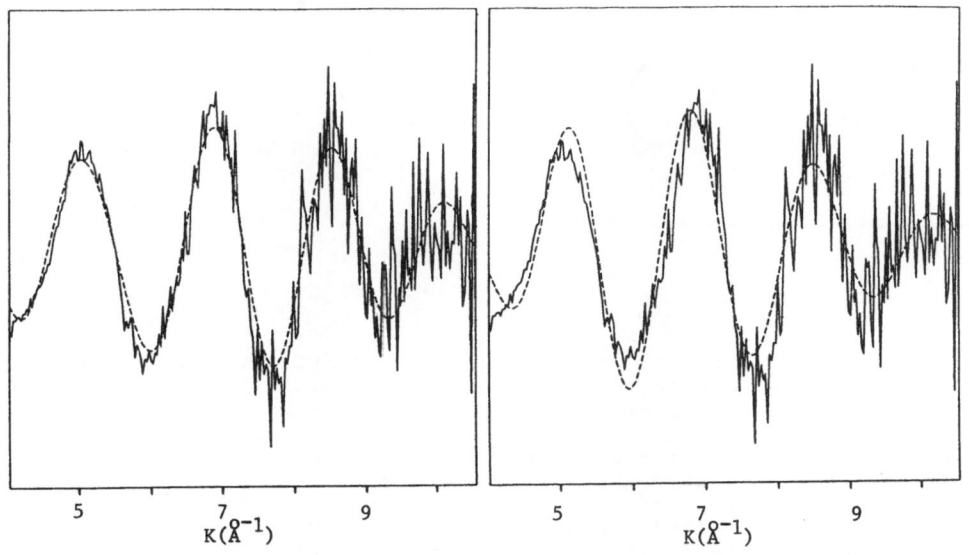

Fig 3.   Comparison between theoretical and experimental K^3 weighted
EXAFS at the Ti K edge for catalyst A (a) double Cl bridge to $MgCl_2$
with a terminal ester group (b) free monomeric ester complex.

Results for catalyst A in Table 3 and Fig 3a for the best fit to experimental EXAFS indicate that Ti is bound to $MgCl_2$ as a double bridge structure (ie 2 Ti-Cl bonds at ∿ 2.52Å). This implies a monomeric Ti unit. A slightly better fit is obtained if it is assumed that one of the terminal positions is occupied by an ester group (ie 3 Ti-Cl at 2.20Å and one Ti-O at 2.04Å) as in Fig 4. These distances, which relate directly to the packing around Ti, are consistent with formation of a Ti^{IV} complex as predicted by absorption edge position in Table 1. A worse fit to the experimental EXAFS is obtained assuming a free monomeric complex (Fig 3b). For completeness a number of other suggested models were tried, eg a quadruple bridge complex involving 4 Ti-Cl bonds at ∿ 2.5Å, ie a dimeric Ti unit on the (100) cut of $MgCl_2$ or a monomeric unit on the (110) cut (7). Also suggested in reference 7 is a triple bridge possibility for a monomeric Ti unit on the (100) $MgCl_2$ cut. None of these approaches gave satisfactory fits to experimental EXAFS.

By comparing pre-edge data and edge positions it is reasonable to conclude that in catalyst B also Ti is bound to $MgCl_2$ via a double Cl bridge. During the alkylation process the structure of catalyst B changes but edge position data suggests that oxidation state remains as Ti^{IV}. Current data quality does not allow us to take the structural analysis any further.

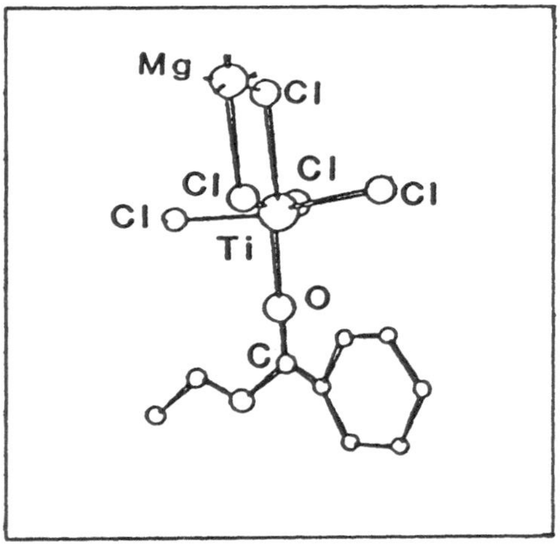

Fig 4. Structural model for the environment round Ti in Catalyst A

DISCUSSION

Work so far has concluded that Ti is bound to $MgCl_2$ via a double Cl bridge as a Ti^{IV} complex. There is some indication that the ester, forms part of that complex. Cleavage of $MgCl_2$ provides an abundance of suitable sites for such a complex. What we cannot say at present is whether this complex is responsible for all or any of the catalytic activity. It is probable that particular edge or step sites are necessary for the appropriate activated geometry necessary for alkylation and sustained chain growth. However washing experiments do confirm that all the Ti is firmly bound, ie we are observing a complex with $MgCl_2$ and not merely a 'free' complex. The stereochemical implications of the above interpretation are being assessed using molecular graphics techniques.

REFERENCES

1   Jaklevic J, Kirby JA, Klein MP, Robertson AS, Brown GS,
     Eisenberger P (1977) Solid State Comm 23:679-682

2   Waychunas GA (1986) J Physique C8 47:841-844

3   Bianconi A (1981) EXAFS for Inorganic Systems, Proc Daresbury Study
     Weekend, SERC, p 13

4   Poumellec B , Marucco JF, Touzelin B (1986) J Physique C8 47:709-712

5   Bart JCJ, Bassi IW, Calcaterra M, Albizzati E, Giannini U,
     Parodi S (1981) Z Anorg Allg Chem 482:121-132

6   Guolin G, Youchang X, Yougi T (1984) Scientia Sinica 27:1-11

7   Busico V, Corradini P, De Martino L, Proto A, Albizzati E (1986)
     Makromol Chem 187:1115-1124

HIGHLY ACTIVE SUPPORTED PROPYLENE POLYMERIZATION CATALYSTS PREPARED BY ACTIVATION OF SUPPORTS DERIVED FROM PRECOMPLEXED MAGNESIUM ALKYLS

Karayannis NM, Johnson BV*, Hoppin CR, and Khelghatian HM

Amoco Chemical Company, P.O. Box 400, Naperville, IL 60566, USA

New highly active supported propylene polymerization catalysts are prepared by activating supports derived from Mg dialkyl complexes with either ethyl 2,4,6-trimethylbenzoate or di-n-butyl n-butylphosphonate. These catalysts realize yields of 16-17 kg/g cat in 2h slurry polymerizations at 70°C, 10.2 atm, with 0.8-1.5% boiling hexane extractables, 0.40-0.41 g/ml polymer powder bulk density, and narrow polymer particle size distribution, which renders them suitable for use in both slurry and solventless propylene polymerization processes.

INTRODUCTION

Within the framework of our research on supported propylene polymerization catalysts,(1) we became interested in exploring the possibility of preparing well-performing catalysts from supports derived from Mg dialkyls. Accordingly, research in this direction was undertaken, and our intitial work involved the preparation of catalyst supports by reacting MgRR' (R = n-butyl; R' = n-butyl or n-octyl) with $SiCl_4$. Activation of these supports by treatment with $TiCl_4$ and diisobutyl phthalate (DIBP) resulted, however, in catalysts with good morphology, but poor performance characteristics and low polymer powder bulk density. During subsequent work we discovered that use of a complex of MgRR' with either a sterically hindered ester of benzoic or 2,6-disubstituted benzoic acid (2) or di-n-butyl n-butylphosphonate (DBBP) (3) as the starting material for support precipitation results in the production of catalysts with high activity and stereospecificity, excellent morphology, and satisfactory polypropylene powder bulk density. The results of our research leading to the development of these catalysts are described in some detail in this communication.

RESULTS AND DISCUSSION

Our initial studies on suitable precomplexing agents for Mg dialkyl involved the use of a variety of esters of benzoic and substituted benzoic acids. This work revealed that only sterically hindered esters are suitable for this application, as depicted from the slurry polymerization data of Table 1. Steric hindrance in the vicinity of the coordinating C=O oxygen of the ester delays reactions between metal alkyls and esters, leading to the formation of metal alkoxides and alkylation and reduction products of the ester moiety.(4) An infrared study of the reactivities of a variety of esters with Mg dialkyls confirmed the effects of steric hindrance in delaying the reactions between the components of the MgR_2.ester complex (R=n-butyl). The results of this study, given in Table 2, indicate that complex formation between the Mg dialkyl and the ester occurs immediately upon mixing solutions of these compounds, as demonstrated by the shift of the $\nu_{C=O}$ mode of the free ester toward lower wave numbers.(5) Rough estimates of the relative stabilities

* Present Address: 3M Center, St. Paul, MN 55144, USA

W. Kaminsky and H. Sinn (Eds.)
Transition Metals and Organometallics as
Catalysts for Olefin Polymerization
© Springer-Verlag Berlin Heidelberg 1988

of these complexes ($t_{1/2}$) were obtained by assuming approximate second-order reaction rates and using the equations shown in Footnote a of Table 2. From the combined information of the data of Tables 1 and 2, it became clear that good catalyst performance is realized only with ethyl esters of 2,6-dialkylsubstituted benzoic acids (ethyl 2,6-dimethyl-(EDMB) and 2,4,6-trimethyl-(ETMB) benzoates) or isopropyl benzoate (IPB). Esters of 2,6-dichloro- or 2,6-bis(trifluoromethyl)-benzoic acids produced catalysts with relatively poor performance, despite the high stability of their complexes with MgR_2. The presence of Cl- or CF_3-substituents on the aromatic ring has apparently adverse effects on the performance characteristics of the resulting catalysts. Regarding polymer powder bulk densities, EDMB and ETMB gave satisfactory results, but use of IPB as precomplexing agent resulted consistently in polypropylene powders with unacceptably low bulk densities. In view of its somewhat better overall performance, ETMB was selected as the precomplexing agent of choice for optimization work.

The fact that ETMB and EDMB are very costly rare chemicals not readily available in commercial quantities dictated the need for further studies aimed at the development of systems involving readily commercially available and relatively inexpensive substitutes for these esters. Numerous Lewis bases, including organophosphoryl esters, organic ethers, and amines were evaluated, and some of our results are included in Table 1. As may be seen from these data, DBBP was found to be an excellent substitute for ETMB. In contrast, the less sterically hindered triethyl phosphate produced poorer catalysts, while use of triisopropyl phosphite led to inactive catalysts, indicating that P-ligands are probably not suitable for this application. Among various ethers evaluated, 2,2,5,5-tetramethyltetrahydrofuran produced catalysts with satisfactory bulk density. Optimization of catalyst preparations using this precomplexing agent might lead to performance improvements. Finally, 2,4-di-tert-butylpyridine produced the best-performing catalysts among various amines evaluated. Unfortunately this compound is about as rare and expensive as ETMB. Our supplementary studies led to the selection of DBBP, a low-cost chemical available commercially in bulk quantities, as precomplexing agent with potential commercial applicability. Interaction of DBBP with MgR_2 leads to complex formation, as manifested by $\nu_{P=O}$ shifts

from 1255 in free DBBP to a doublet at 1235 and 1211 cm^{-1} (DBBP acts as an O-ligand, binding through the P=O oxygen).(6)

Optimization studies using MgRR'.L complexes (L=ETMB or DBBP) led to the following catalyst preparation scheme (L to Mg molar ratios employed: 1.5 for ETMB, 0.4 for DBBP):

$$MgRR' \xrightarrow[\text{Room temp}]{L} MgRR' \cdot L \xrightarrow[\substack{45°C \\ 16h}]{SiCl_4} Solid \xrightarrow[\substack{Chlorobenzene \\ 130°C, 1.5h}]{TiCl_4,\ DIBP} Catalyst\ (1)$$

The solid product of the reaction between MgRR'.L and $SiCl_4$ contains still coordinated L, as shown by the appearance of $\nu_{C=O}$ at below 1680 (L=ETMB) or $\nu_{P=O}$ at below 1240 (L=DBBP) cm^{-1}. During the activation of the solid support with $TiCl_4$ and DIBP, most of the precomplexing agent is displaced by DIBP, as shown by the analytical data in Table 3. However, small amounts of the precomplexing agent remain in the final catalytic complex. The presence of Si in

catalysts made with L=DBBP is presumably due to polymeric residues
(e.g., $Si(C_4H_9PO_3)_2$) produced by reaction between DBBP and $SiCl_4$.(6)
The new catalysts routinely exhibit slurry polymerization yields of
16-17 kg/g cat with 0.8-1.5% boiling hexane extractables, polymer
powder bulk densities of 0.40-0.41 g/ml and narrow particle size
distribution. The catalysts consist of clusters of spherical
particles, which are significantly larger when L=ETMB, and produce
polypropylene powders with the same morphology (Figure 1),
containing less than 1% coarse (>850 microns) and 0.4% fine (<150
microns) particles (Table 3). These features render the new
catalysts suitable for use in all the commercial propylene
polymerization processes, i.e., slurry, bulk, and gas-phase.
Systems involving the new catalysts with $(C_2H_5)_3Al$ and
diphenyldimethoxysilane as cocatalyst exhibit essentially linear
polymerization kinetics, as shown in Figure 2.

In conclusion, the present work describes a novel method of
improving the performance characteristics of supported propylene
polymerization catalysts produced by activation of supports derived
from Mg dialkyls. The use of Mg dialkyl complexes with suitable
Lewis bases instead of uncomplexed MgRR' improves the catalyst
activity, stereospecificity, and bulk density to levels comparable
to those of commercial catalysts. Moreover, the excellent
morphology of the new catalysts makes them suitable for use in all
the propylene polymerization processes. Use of ETMB as
precomplexing agent leads to the best-performing catalysts, but
DBBP, which yields catalysts with equally high activity and slightly
lower stereospecificity, is preferable in view of its low cost and
ready commercial availability.

EXPERIMENTAL

The following Mg dialkyls were used in this work: MAGALA 7.5-E
(composition $[(n-C_4H_9)_2Mg]_{7.5} \cdot (C_2H_5)_3Al)$ solution in hexane,
containing 16.3 wt % Mg and 0.25 wt % Al (Texas Alkyls) and BOMAG-D
(composition $(n-C_4H_9)_{1.5} (n-C_8H_{17})_{0.5} Mg)$ solution in heptane,
containing 8% diethyl ether (Schering). A typical optimized
catalyst preparation is as follows (all operations conducted in a
dry, deoxygenated nitrogen atmosphere): 150 ml MAGALA 7.5-E
solution and 11.6 ml ETMB or 9 ml DBBP are combined at ambient
temperature ($MgR_2.L$ remains in solution). The solution of the
complex is then added dropwise to a 1-liter flask equipped with a
condenser and a mechanical stirrer containing 150 ml $SiCl_4$, and the
resulting mixture is allowed to stir at 45°C for 16 hours. The
precipitate formed (yield 11.4 g) is separated by filtration and
thoroughly washed with predried hexane; 7.5 g of the solid is
suspended in 280-ml chlorobenzene in the 1-liter flask, 100 ml $TiCl_4$
is added, and then 4 ml DIBP is added dropwise. The resulting
suspension is stirred at 130°C for 1.5 hours, and subsequently the
supernatant is decanted and the solid is thoroughly washed with
toluene and later with hexane, and finally dried under a current of
dry, deoxygenated nitrogen. The final catalyst is a tan or
beige-colored solid and is obtained in yields ranging between 3-5g.
Batch slurry propylene polymerization evaluations were performed in
2-liter Parr reactors by using techniques previously described.(7)
Conditions: 2-hour polymerizations at 70°C, 10.2 atm using 10-15 mg
catalyst and $(C_2H_5)_3Al-(C_6H_5)_2Si(OCH_3)_2$ as cocatalyst (Al/Si/Ti
molar ratios = 200/20/1), 650 ml hexane and 0.006 mole H_2. Yields
were determined from the Mg content of the dried polymer powder.
Boiling hexane extractables reported are the sum of the hexane
solubles and extractables calculated as described elsewhere.(7)
Bulk densities were obtained by weighing 100 ml of untapped polymer

powder. Catalyst particle size distributions were determined on light mineral oil suspensions of the catalyst using a Cilas 715 granulometer, and polymer powder particle size distributions by sieve analysis. Melt flow rates of the slurry polymer ranged between 2-8g/10 min, and its polymolecularity index (Mw/Mn) was in the 5-7 range. Infrared studies of solutions of the MgR_2.ester complexes were performed between NaCl windows using a Perkin-Elmer 683 spectrophotometer.

REFERENCES

(1)   Karayannis NM, Skryantz JS (to Standard Oil Company, Indiana) U.S. Patent 4,277,370 (1981); Johnson BV, Karayannis NM, Skryantz JS (to Standard Oil Company, Ind), U.S. Patent 4,526,882 (1985); Arzoumanidis GG, Lee SS, (to Amoco Corporation) U.S. Patent 4,540,679 (1985).
(2)   Johnson BV, Karayannis NM, Hoppin CR, Ornellas L (to Standard Oil Company, Indiana), U.S. Patent 4,581,342 (1986).
(3)   Karayannis NM, Skryantz JS, Johnson BV (to Amoco Corporation) U.S. Patent 4,657,882 (1987).
(4)   Johnson BV, Abstr. Intern. Chem. Congress of Pacific Basin Socs., Honolulu, Hawaii, December 16-21, 1984; No. 09P19.
(5)   Driessen WL, Groeneveld WL, Van der Wey FW, Recl. Trav. Chim. Pays-Bas, <u>89</u>, 353 (1970).
(6)   Karayannis NM, Mikulski CM, Pytlewski LL, Inorg Chim. Acta Rev., <u>5</u>, 69 (1971).
(7)   Karayannis NM, Lee SS, Makromol. Chem., <u>183</u> 1171 (1982); <u>184</u>, 2275 (1983).

TABLE 1

2h Propylene Slurry Polymerizations Using Catalysts
Prepared with MgRR'·L as Starting Material[a]

L	Yield kg/g cat	Boiling Hexane Extractables wt %	Polymer Powder Bulk Density g/ml
None (MgRR' used)	1-6	5-20	0.22-0.31
Ethyl Benzoate	6.1	4.2	0.36
Isopropyl Benzoate	16.0	0.8	0.32
Ethyl-2,6-Dimethyl Benzoate	13.5	1.5	0.38
Ethyl-2,4,6-Trimethyl Benzoate	16.4	0.9	0.41
Ethyl-2,3,5,6-Tetramethyl Benzoate	8.5	2.3	0.30
Ethyl-2,6-Dichlorobenzoate	5.0	1.3	0.21
Ethyl-2,6-Bis(trifluoromethyl) Benzoate	4.2	3.6	0.36
Di-n-butyl n-Butylphosphonate	14-16.6	1.5	0.40
Triethyl Phosphate	6.8	2.3	0.32
Triisopropyl Phosphite	Negligible		
Diisoamyl Ether	2.4	6.7	0.33
2,2,5,5-Tetramethyltetrahydrofuran	6.2	3.2	0.41
2,2,6,6-Tetramethylpiperidine	2.0	10.7	0.24
2-Methyl-5-ethylpyridine	5.9	1.9	n.d.
2,4-Di-tert-butylpyridine	11.1	0.7	n.d.

a Polymerization conditions described in experimental section.
n.d. = not determined.

TABLE 2

Relative Stability of MgR$_2$·Ester Complexes

Ester	Free Ester CO cm^{-1}	MgR$_2$·Ester Complex CO cm^{-1}	$t_{1/2}$ minutes[a]
Ethyl Benzoate	1730	b	<0.5
Ethyl p-Anisate	1724	b	<1.0
tert-Butyl Benzoate	1723	1663	3
Ethyl o-Toluate	1726	b	<1.0
Ethyl Pivalate	1737	1679	5
Ethyl 2,6-Dimethoxybenzoate	1737[c]	1657[c]	5
Ethyl 2,6-Dichlorobenzoate	1753	1694	54
Ethyl 2,6-Bis(trifluoromethyl) benzoate	1757	1698	d
Ethyl 2,6-Dimethylbenzoate	1736	1679	920
Ethyl 2,4,6-Trimethylbenzoate	1732	1675	1200
Diisobutylphthalate	1732	b	<1.0

All spectra recorded in hexane unless noted.
a $t_{1/2} = 1/k_2 a$ where $k_2 = x/a(a-x)t$ (x determined from IR spectra
at time t); a = molar concentration of MgR$_2$·ester complex;
x = $_{CO}$ absorbance reduction at time t.
b Complex lifetime was too short to permit measurement
($t_{1/2}$ estimated from color change).
c Recorded in toluene.
d Only partially complexed. Complex appears to be extremely stable.
Note: The relative stability of the MgR$_2$ complex with isopropyl
benzoate was not determined. It is assumed that it is
comparable to that of the tert-butyl benzoate complex, in
view of the comparable stabilities of the complexes of these
two esters with triethyl aluminum.(4)

TABLE 3

Typical Analytical and Particle Size Distribution (PSD)
Data for Catalysts Prepared From MgRR'·L Complexes
and Slurry Polypropylenes Made From These Catalysts

	L = ETMB	L = DBBP
Analysis, wt %		
Ti	1.7	1.8
Mg	19.4	19.2
Cl	59.7	58.2
DIBP	10.9	10.5
ETMB	0.4	--
P	--	0.4
Si	0	0.8
Catalyst PSD, μ		
D_{10}%	2.5-7.0	2.5
D_{50}%	18.0-27.0	13.0
D_{90}%	40.0-49.0	18.0
Polymer PSD, wt %		
>850 μ	0.9	0.3
850-425 μ	22.0	1.0
425-250 μ	74.2	81.4
250-180 μ	2.4	16.4
180-150 μ	0.2	0.7
<150 μ	0.3	0.2

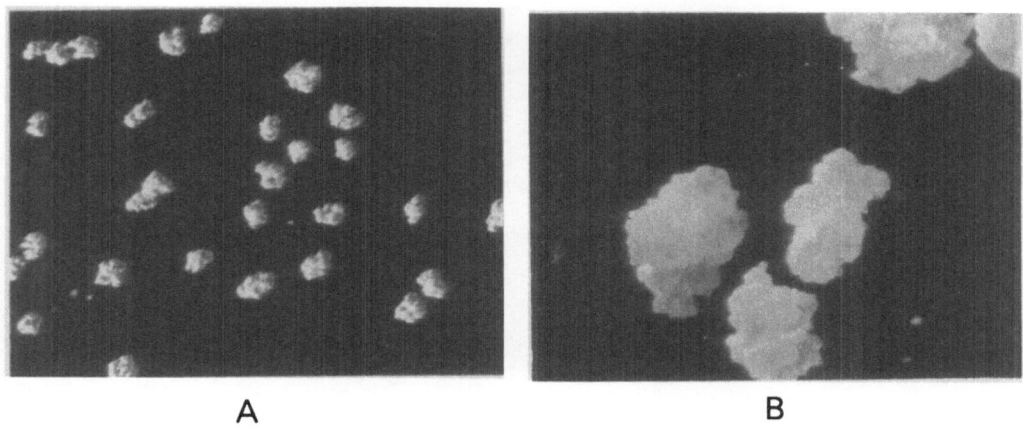

Figure 1

Photomicrographs of A. Catalyst Made Using L = etmb (x250);
B. Polypropylene Produced With the Same Catalyst (x50)

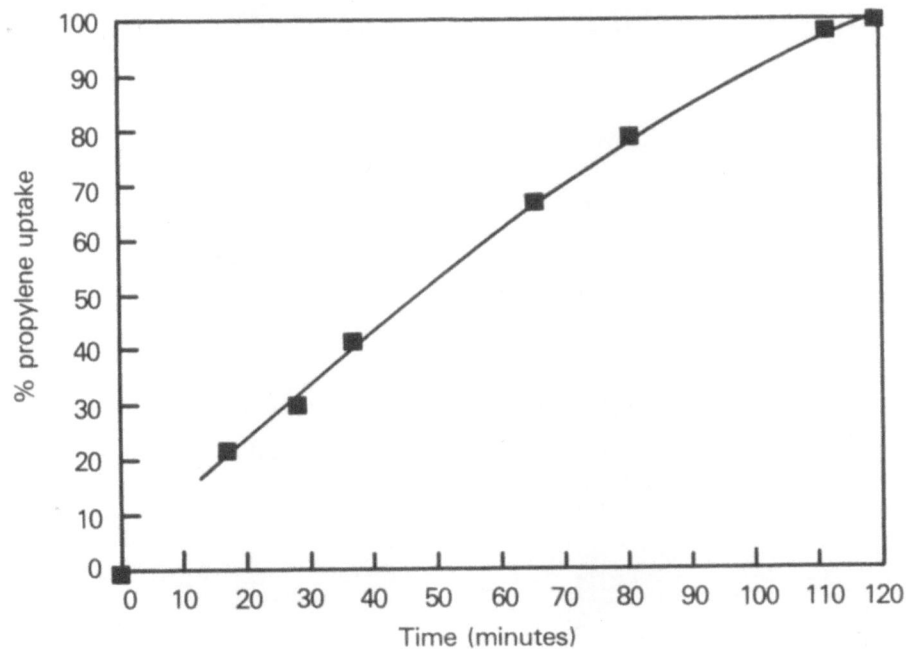

Figure 2

Propylene Uptake vs. Time During Slurry Polymerization Using a
Catalyst Made With L = etmb Under the Conditions of Table I

SYNTHESIS OF METAL ORGANIC ZIRCONIUM CATALYSTS CONTAINING *OR* LIGANDS FOR ETHYLENE POLYMERIZATION

Krzysztof Szczegot

Institute of Chemistry, Pedagogical University, Opole (Poland)

ABSTRACT

Metal organic zirconium catalysts containing *OR* ligands for ethylene polymerization were obtained by the reaction of alkoxy derivatives of zirconium $Zr(OC_nH_{2n+1})_mCl_{4-m'}$ $n = 3,4,5,8$, $m = 4$ or 2 with organoaluminium compounds of the type R_kAlCl_{3-k} (R - alkyl, $k = 1 - 3$). The conditions of these catalysts formation as well as their catalytic activity in low-pressure ethylene polymerization are presented.

INTRODUCTION

The original Ziegler-Natta catalysts for olefin polymerization were based upon transition metal halides, mainly titanium and vanadium and were activated by alkylation with aluminium alkyls. Numerous publications have appeared covering a variety of compositions and procedures for the preparation of such catalysts. One of the ways for modification of conventional Ziegler-Natta catalysts is the addition of a third component, often Lewis base, to the catalytic system. Another way is a change of the ligands in dicomponent catalytic system. In particular, an effect of alkoxy ligands (*OR*) in such system (i.e. alkoxy derivative of transition metal + organoaluminium compound) seems to be very interesting. Both steric and inductive effects of *OR* ligands can play an important part in the catalyst. An effect of the number and nature of alkoxy or aryloxy ligands in titanium-aluminium catalysts on the polymerization of ethylene was widely studied [1-7].

Zirconium metal organic catalysts are a new stage in Ziegler-Natta catalysis. In this paper is described the synthesis of metal organic zirconium-aluminium catalysts containing *OR* ligands.

EXPERIMENTAL

The catalysts were obtained by the reaction of alkoxy derivatives of zirconium with organoaluminium compounds. The following alkoxides of zirconium were prepared:

$$Zr(OC_nH_{2n+1})_mCl_{4-m'} \quad n = 3,4,5,6 \text{ or } 8, \quad m = 2 \text{ or } 4.$$

The alkoxides were prepared by alcohol interchange from the isopropoxide of zirconium accordingly to [8]. Most of the required operations involved highly moisture- and air-sensitive materials and were carried out under an inert gas atmosphere (argon). The catalysts were handled in vacuum atmosphere drybox under

W. Kaminsky and H. Sinn (Eds.)
Transition Metals and Organometallics as
Catalysts for Olefin Polymerization
© Springer-Verlag Berlin Heidelberg 1988

argon. Aluminium alkyls were delivered by FLUKA AG and were used without further purification. Polymerization reactions were carried out at 50°C in thermostated 1 dm^3 steel autoclave under pressure 0.5 MPa. Hexane was used as a solvent for the polymerization process. Hexane and other solvents were purified with standard methods. Molecular weight of polyethylene ($\bar{M}_\eta$) was estimated by method described in [9].

RESULTS AND DISCUSSION

An influence of Al:Zr molar ratio on the reaction of zirconium compounds with organoaluminium agent is sagnificant in consideration of further activity of zirconium-aluminium complexes in low-pressure ethylene polymerization. Tetraalkoxy derivatives of zirconium react only with Et$_3$Al or Et$_2$AlCl (Tables 1 and 2). Especially reaction with Et$_3$Al is effective, independently on the kind of zirconium compound and Al:Zr molar ratio. Sesquiethylaluminium chloride is ineffective reacting agent, even in different solvents and temperatures (Table 3). Dialkoxy derivatives of zirconium easily react with organoaluminium compounds used (Table 4). Preliminary polymerization of ethylene has shown that the catalysts based on Zr(OR)$_4$ and organoaluminium compounds are inactive in low-pressure ethylene polymerization. On the other hand, dialkoxy derivatives of zirconium form active complexes with Et$_3$Al$_2$Cl$_3$ and Et$_2$AlCl as cocatalysts (Table 5). Both yield of PE and molecular weight of PE are a function of Al:Zr molar ratio in the catalyst, see Tables 6 and 7 respectively. Unexpectedly, similar relationships are observed when catalyst concentration is increased (catalyst activity – catalyst concentration) (Tables 8 and 9). Yield of PE and $\bar{M}_\eta$ are not a function of the length of OR ligand (Table 10). Zirconium-aluminium catalysts based on Zr(OR)$_2$Cl$_2$ show high stability as is presented in Fig. 1.

The observed lack of catalytic activity of Zr-Al complexes based on Zr(OR)$_4$ can be explained with steric and inductive effects of OR ligands. The contrast between catalytic properties of the catalysts based on the alkoxides of zirconium on the one hand and those of titanium alkoxides on the other, invites some comment. It may well be that this behaviour is connected with the atomic radius of the central atom. Thus zirconium require more effective shielding than the smaller titanium. It is notheworthly that in combination with oxygen the maximum covalency of titanium is 6 whereas zirconium can exhibit a covalency of 8 [8].

The alternative possibility cannot be ignored that the abnormal behaviour of some zirconium catalysts is due to the intermolecular bonding $0 \longrightarrow Zr$ being stronger for zirconium than for titanium. Another explanation is also considered that the +I inductive effect of alkyl (R) group in OR ligand could be responsible for the striking properties of zirconium-aluminium complexes based on Zr(OR)$_4$. For example, the structure R-O$^+$=Zr$^-$ would be stabilised by +I effect of R group.

Table 1. Effect of Al:Zr molar ratio on the reaction of $Zr(OR)_4$ with organoaluminium compound. Temperature 50°C, concentration of zirconium compound $[Zr] = 0.1$ mol.dm^{-3}, hexane.

No	Zirconium compound	Et_3Al		Et_2AlCl		$Et_3Al_2Cl_3$	
		Al:Zr	reaction*	Al:Zr	reaction*	Al:Zr	reaction*
1	$Zr(OC_3H_7)_4$	1:1	+	2:1	−	4:1	−
2	$Zr(OC_3H_7)_4$	2:1	+	4:1	+	10:1	−
3	$Zr(OC_3H_7)_4$	6:1	++	8:1	++	15:1	−
4	$Zr(OC_3H_7)_4$	10:1	++	12:1	++	20:1	−
5	$Zr[OCH(CH_2CH_3)_2]_4$	2:1	+	2:1	−	10:1	−
6	$Zr[OCH(CH_2CH_3)_2]_4$	6:1	++	6:1	+	20:1	−
7	$Zr[OCH(CH_2CH_3)_2]_4$	10:1	++	10:1	+	30:1	−
8	$Zr[OCH(CH_2CH_3)_2]_4$	15:1	++	15:1	+	50:1	−

*
++ - after 5 minutes
+ - after 15 minutes
− - reaction does not take place.

Table 2. Effect of both organoaluminium and zirconium compounds on the reaction between them. Temperature 50°C, hexane, $[Zr] = 0.1$ mol.dm^{-3}, Al:Zr molar ratio = 6:1

No	Zirconium compound	Organoaluminium compound		
		Et_3Al	Et_2AlCl	$Et_3Al_2Cl_3$
1	$Zr(OC_4H_9)_4$	++	+	−
2	$Zr(OC_6H_{11})_4$	++	+	−
3	$Zr(OC_8H_{17})_4$	++	+	−
4	$Zr[OCH(CH_3)_2]_4$	++	+	−
5	$Zr[OCH(CH_3)CH_2CH_2CH_3]_4$	++	+	−
6	$Zr[OC(CH_3)_2CH_2CH_3]_4$	++	+	−
7	$Zr[OCH(CH_2CH_3)_2]_4$	++	+	−
8	$Zr[OCH(C_6H_5)CH_3]_4$	++	−	−

Table 3. Effect of activation temperature and a kind of solvent on the reaction of zirconium compound with sesquiethylaluminium chloride. $[Zr] = 0.1$ mol·dm^{-3}

No	Zirconium compound	Al:Zr molar ratio	Temperature [°C]	Solvent	Reaction
1	$Zr(OC_4H_9)_4$	6:1	40	hexane	-
2	$Zr(OC_4H_9)_4$	6:1	64	hexane	-
3	$Zr(OC_4H_9)_4$	6:1	81	benzene	-
4	$Zr(OC_4H_9)_4$	6:1	110	toluene	+
5	$Zr[OC(CH_3)_2CH_2CH_3]_4$	10:1	40	hexane	-
6	$Zr[OC(CH_3)_2CH_2CH_3]_4$	10:1	64	hexane	-
7	$Zr[OC(CH_3)_2CH_2CH_3]_4$	10:1	81	benzene	-
8	$Zr[OC(CH_3)_2CH_2CH_3]_4$	10:1	110	toluene	+

Table 4. Effect of a kind of both organoaluminium compound and solvent on the reaction of $Zr(OR)_2Cl_2$ with organoaluminium compounds. Temperature 50°C, Al:Zr molar ratio = 6:1

No	Zirconium compound	Solvent	Organoaluminium compound		
			Et_3Al	Et_2Al	$Et_3Al_2Cl_3$
1	$Zr(OC_4H_9)_2Cl_2$	hexane	++	+	+
2	$Zr(OC_4H_9)_2Cl_2$	benzene	++	+	+
3	$Zr(OC_4H_9)_2Cl_2$	toluene	++	+	+
4	$Zr(OC_8H_{17})_2Cl_2$	hexane	++	+	+
5	$Zr(OC_8H_{17})_2Cl_2$	benzene	++	+	+
6	$Zr(OC_8H_{17})_2Cl_2$	toluene	++	+	+

Table 5. Effect of organoaluminium compound on a yield of PE. Catalyst: $[Zr(OC_8H_{17})_2Cl_2$ + organoaluminium compound]. Polymerization time 1 h, $[Zr]$ = 1.5 $mmol.dm^{-3}$

Organo-aluminium compound	Al:Zr molar ratio	Yield of PE [g]	Catalyst activity [gPE/gZr.h]	Molecular weight of PE $\bar{M}\eta$
$Et_3Al_2Cl_3$	6:1	97	709	5 042 200
Et_2AlCl	4:1	21	153	4 070 300
Et_3Al	4:1	-	-	-

Table 6. Yield of PE as a function of Al:Zr molar ratio. Catalyst: $[Zr(OC_8H_{17})_2Cl_2$ + $Et_3Al_2Cl_3]$. Polymerization time 1 h.

$[Zr]$ $[mmol/dm^3]$	Al:Zr molar ratio	Yield of PE [g]	Catalyst activity [gPE/gZr.h]	Molecular weight of PE $\bar{M}\eta$
2	2:1	20	110	1 396 400
2	4:1	78	427	3 845 900
2	6:1	211	1 151	9 013 100
2	10:1	102	559	6 645 100

Table 7. Yield and $\bar{M}\eta$ of PE as a function of Al:Zr molar ratio. Catalyst: $[Zr(OC_6H_{13})_2Cl_2$ + $Et_3Al_2Cl_3]$. Polymerization time 1 h.

$[Zr]$ $mmol.dm^{-3}$	Al:Zr molar ratio	Yield of PE [g]	Catalyst activity [gPE/gZr.h]	Molecular weight of PE $\bar{M}\eta$
2	2:1	22	121	1 325 400
2	4:1	74	406	3 174 100
2	6:1	60	629	7 966 100

Table 8. Effect of zirconium compound concentration on a yield and $\bar{M}\eta$ of PE. Catalyst: $\left[Zr(OC_5H_{11})_2Cl_2 + Et_3Al_2Cl_3\right]$. Polymerization time 1 h.

$[Zr]$ $\left[mmol.dm^{-3}\right]$	Al:Zr molar ratio	Yield of PE $[g]$	Catalyst activity $\left[gPE/gZr.h\right]$	Molecular weight of PE $\bar{M}\eta$
0.50	8:1	8	175	751 580
0.75	8:1	10	146	2 350 300
1.00	8:1	127	1 392	2 554 900

Table 9. Effect of catalyst concentration on a yield and molecular weight of polyethylene. Catalyst: $\left[Zr(OC_8H_{17})_2Cl_2 + Et_3Al_2Cl_3\right]$. Polymerization time 1 h.

$[Zr]$ $\left[mmol.dm^{-3}\right]$	Al:Zr molar ratio	Yield of PE $[g]$	Catalyst activity $\left[gPE/gZr.h\right]$	Molecular weight of PE $\bar{M}\eta$
1.0	6:1	41	449	2 589 700
1.5	6:1	97	708	5 042 200
2.0	6:1	211	1 156	9 013 100

Table 10. Yield of PE and molecular weight of PE as a function of *OR* ligand length. Catalyst: $\left[Zr(OR)_2 + Et_3Al_2Cl_3\right]$. Polymerization time 1 h. $[Zr]$ = 2 mmol.dm^{-3}.

Zirconium compound	Al:Zr molar ratio	Yield of PE $[g]$	Catalyst activity $\left[gPE/gZr.h\right]$	Molecular weight of PE $\bar{M}\eta$
$Zr(OC_4H_9)_2Cl_2$	4:1	96	526	3 090 300
$Zr(OC_6H_{13})_2Cl_2$	4:1	74	406	3 174 100
$Zr(OC_8H_{17})_2Cl_2$	4:1	78	426	3 845 900

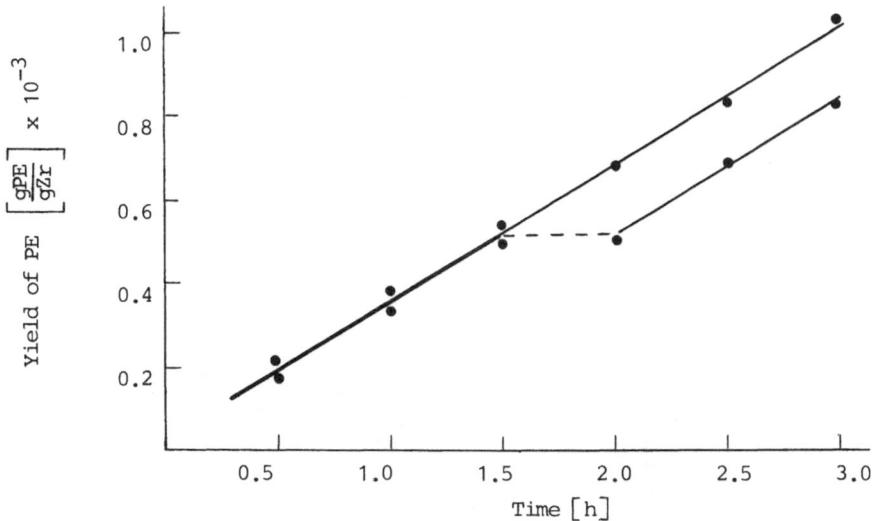

Fig. 1. Yield of ethylene polymerization vs. time. Catalyst: $[Zr(OC_8H_{17})_2Cl_2$ + $Et_3Al_2Cl_3]$, activation time 15 min. $[Zr]$ = 0.5 mmol·dm^{-3}, Al:Zr = 6:1, Hexane. $[------]$ - lack of ethylene.

As a result a positive charge of Zr in the catalytic centre of Zr-Al complex is reduced and coordination of ethylene molecule to this centre is impossible.

Catalytic activity of zirconium-aluminium complexes in low-pressure ethylene polymerization (based on $Zr(OR)_2Cl_2$) is of the same order as for conventional Ziegler-Natta catalysts. However molecular weight of polyethylene is much higher.

REFERENCES

[1] Nowakowska M, Rybczyk M, Uhniat M (1971) Polimery 16: 500

[2] Nowakowska M, Maciejewska H (1976) Polimery 21: 103

[3] Nowakowska M, Czaja K, Makowski M, Marcinkiewicz K, Majzner L (1978) Vysokomol Soed A 20: 2243

[4] Czaja K, Nowakowska M, Makowski M (1978) Polimery 23: 299

[5] Szczegot K, Nowakowska M, Krueger A (1981) Vysokomol Soed A 23: 2551

[6] Czaja K, Szczegot K, Nowakowska M (1983) Przemysł Chem 62: 690

[7] Szczegot K, Nowakowska M, Iwański L, (1984) Vysokomol Soed A 26: 625

[8] Bradley DC, Mehrotra RC, Swanwick JD, Wardlaw W (1953) J Am Chem Soc 2025

[9] Elliott JH, Horowitz KH, Hoodock T (1970) J Appl Polym Sci 14: 2947

3. Homogeneous Catalysts for Olefin Polymerization

Synthesis and Characterisation of Chiral Metallocene Cocatalysts for Stereospecific α-Olefin Polymerisations

H.H. Brintzinger

Fakultät für Chemie, Universität Konstanz,
D-7750 Konstanz, F.R.G.

INTRODUCTION

Homogeneous ethylene polymerisation catalysts containing $(C_5H_5)_2TiCl_2$ and various aluminum alkyls – first described by Natta, Pino and co-workers (1) – were being actively studied about fifteen to twenty years ago by a number of research groups (2-8). Similar reaction systems can also lead to nitrogen-fixing $(C_5H_5)_2Ti$-alkyl complexes, the identification of which was, at that time, the aim of much of my research, then at the University of Michigan. Questions related to polymerisation catalysis by these titanocene derivatives became a topic of increased personal interest to me, when Professor Overberger joined the Department there in 1968, who had then just completed a series of studies on the $(C_5H_5)_2TiCl_2/AlEt_2Cl$-induced polymerisation of styrene (9). These reactions – whatever their detailed mechanism – raised the question whether an isotactic polymer of styrene or another vinyl derivative might not be produced if some chiral organometallic molecule acts as a stereospecific polymerisation catalyst. These ideas gradually developed into our first attempts to prepare complexes of this kind and then into a formal DFG project to synthesize chiral "ansa-metallocenes": These sandwich molecules were to contain two suitably substituted ring ligands connected by a chelate bridge, so as to retain them in a well-defined "stereorigid" chiral molecular shape. The fixation of ligand geometry by a chelate bridge had been found to be beneficial for other types of stereoselective homogeneous catalysts (10); it was our hope that it would also facilitate the interpretation of any stereoselectivity that might be observed with these compounds.

Our project "Chirale ansa-Metallocene" emerged from a special program on "Homogenkatalyse", sponsored by DFG from 1973 to 1978. The work of Professor Kaminsky on alumoxane-activated zirconocene catalysts (11), which he reported during this DFG program, substantially encouraged us in our efforts to synthesize ansa-metallocenes of potential use in α-olefin polymerisation catalysis.

SYNTHESIS OF RACEMIC ansa-METALLOCENES

From our initial attempts to synthesize chiral ansa-metallocenes (12) it soon became clear that the formation of substitutional isomers and of metal-ring linkage diastereomers caused more serious problems than anticipated. The synthesis of substitutionally uniform bis-cyclopenta-dienyl ligand molecules proved to be achievable, rather straight-forwardly, by a number of approaches – by use of electronic (13) or of steric (14,15) effects to induce regioselective substitution or by the reductive coupling of appropriately substituted fulvenes (14,16).

W. Kaminsky and H. Sinn (Eds.)
Transition Metals and Organometallics as
Catalysts for Olefin Polymerization
© Springer-Verlag Berlin Heidelberg 1988

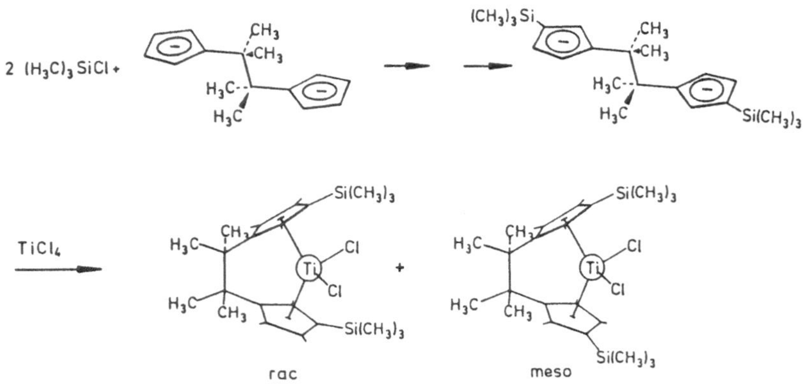

Fig. 1. Regioselective formation of a bisubstituted ansa-titanocene by use of steric effects (14)

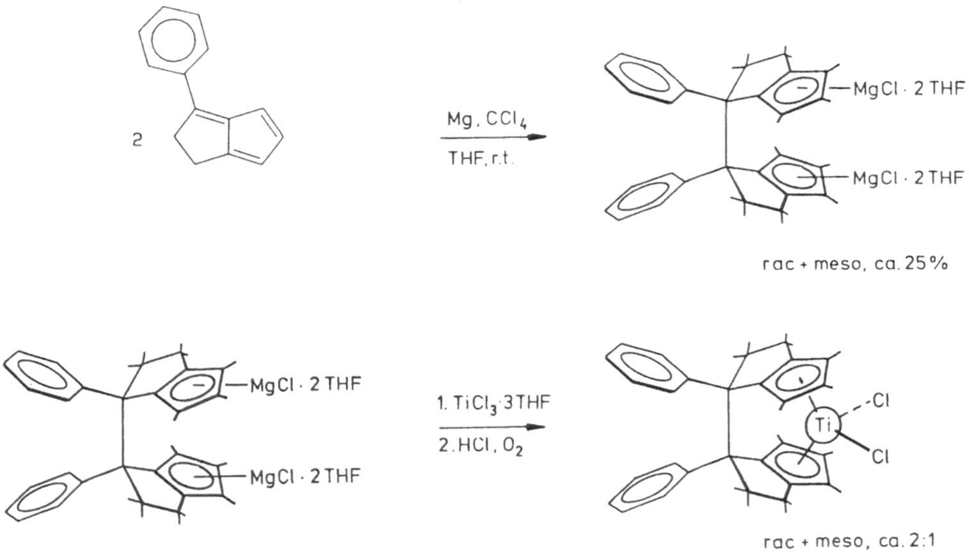

Fig. 2. Formation of a chiral ansa-titanocene with configurationally stable ligand framework by reductive coupling of a dihydropentalene derivative (16)

The concurrent formation of meso diastereomers along with the desired racemic product, however, still is a major obstacle for an efficient synthesis of metallocenes containing various bridging and terminal substituents. Despite considerable variations in type and conditions of the metal-ligand linkage reaction, comparable amounts of meso and racemic products were invariably obtained even with bulky trimethyl-silyl or t-butyl ring substituents (14,15), which might be assumed to avoid each other in an axially symmetric, racemic product geometry.

Apparently, the configuration of the chiral intermediate formed by attachment of the metal to one of the two interconnected, substituted cyclopentadienyl rings induces only negligible enantiofacial selectivity for the consecutive reaction step in which the chelate ring is closed by attachment of the metal to the second ring ligand. Which factors might be put to work to increase the enantiofacial selectivity of this ligand exchange reaction is not clear at present.

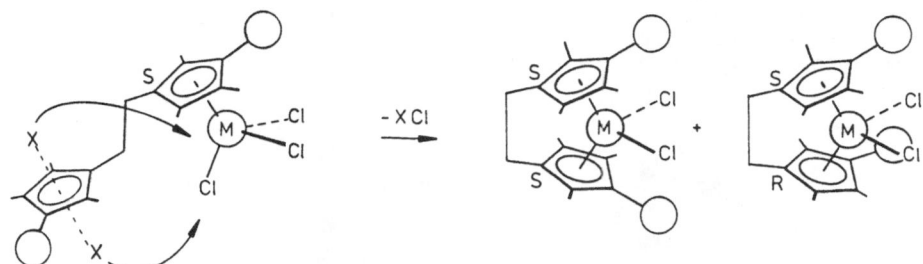

Fig. 3. Model for the chelate-ring closure step in the formation of a bisubstituted ansa-metallocene: Attachment of the metal to the re or si face of the second ring is kinetically controlled by the arbitrary enantiofacial position of the cationic leaving group (Li, MgCl, R_3 Sn)

In one particular case, this problem was completely and efficiently solved by utilizing bis-indenyl ethane as a ligand molecule (13). Reaction of its dilithium salt with ZrCl₄ and subsequent hydrogenation yielded exclusively the desired racemic diastereomer of $C_2 H_4$-(tetra-hydroindenyl)₂ZrCl₂. A meso/racemate mixture of the titanium analogue is obtained from an analogous reaction with TiCl₄ or TiCl₃, but this product mixture is completely converted to the racemic diastereomer by a photoinduced rearrangement. Other chirally bisubstituted ansa-metallocenes also show a photoinduced meso-racemat conversion; in these cases, however, both diastereomers were present in about equal concentrations in the final, photostationary state (15). Apparently, steric or electronic interactions of yet unknown origin steer the metal-ring linkage formation reaction in the direction of the racemic isomer in the case of these ethylene-bridged bis-indenyl and bis-tetrahydroindenyl complexes. These complexes were mainly utilized, so far, to study the homogeneous catalysis of isotactic α-olefin polymer-isation in the presence of methylalumoxane (17-20), since a separation of the meso/racemate mixture obtained in other cases - though feasible by chromatography or fractional crystallisation - proved cumbersome.

ENANTIOMER SEPARATION AND CHARACTERISATION

In order to provide optically pure complexes - rather than racemates - for further studies on the stereochemical origins (20) of the high degree of isotacticity observed with these α-olefin polymerisation catalysts (18), we have investigated possibilities to obtain their separate enantiomers. The stereochemical designation of these ansa-metallocene enantiomers is based on an adaption by Schlögl (21) of the Cahn-Ingold-Prelog rules (22): In a metallocene derivative, the metal is viewed as being σ-bonded to each of the ring carbon atoms.

Designation of the configuration at the bridge-head carbon atom – for
which its metal neighbour has highest priority, followed in that order
by its substituted and its unsubstituted ring neighbours and, finally,
the C atom of the interannular bridge – is sufficient for an unambiguous description of the sense of helicity of the complex molecule.

The S-configurated titanium complex was obtained via a stereoselective
exchange of its chloride ligands against an S-binaphthol ligand (13).
More widely applicable is a method based on conversion of the dichloro
titanium and zirconium complexes into diastereomeric O-acetyl mandelic
acid derivatives, which are cleanly separable by crystallisation, and
from which the dichloro derivatives are easily regenerated (23).

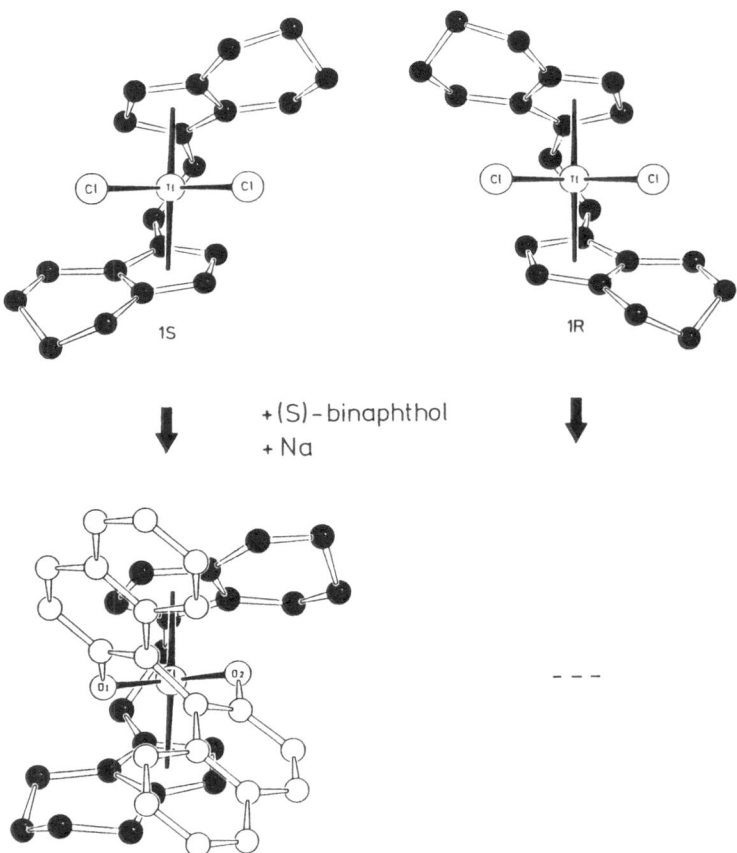

Fig. 4. Stereoselective formation of an S-binaphthol complex from the
S-configurated ethylene bis(tetrahydroindenyl) titanium dichloride (13)

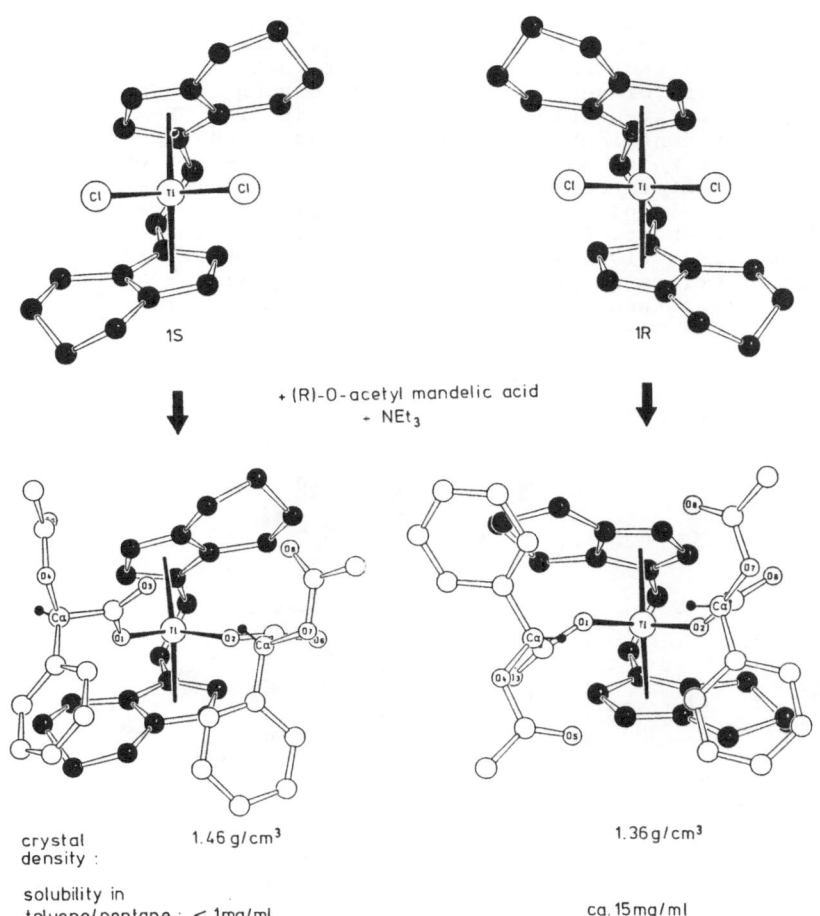

crystal
density :   1.46 g/cm³   1.36 g/cm³

solubility in
toluene/pentane : < 1mg/ml   ca. 15mg/ml

Fig. 5. Formation of diastereomeric O-acetyl-mandelic acid derivatives
from racemic ethylene bis(tetrahydroindenyl) titanium dichloride (23)

The crystal structures of the diastereomeric O-acetyl-R-mandelic acid
derivatives revealed, to our surprise, that different conformations of
the interannular bridge, and hence of the substituted ansa-metallocene
framework as a whole, are adopted in these diastereomer pairs : In the
crystalline diastereomer containing the 1S-configurated ansa-metallocene, the ethylene bridge has an M conformation (λ in the Corey-Bailar
notation (24)) which places the tetramethylene substituents in their
normal "forward" position. The diastereomer with 1R configuration however also contains the ethylene bridge in its M (i.e. λ) conformation.
This R-λ geometry, which places the tetramethylene substituents in a
"backward" position, has not been observed in any other derivative;
its adoption in this instance is undoubtedly caused by the necessity
to allow for reasonably close crystal packing while avoiding repulsive

interactions with the bulky mandelate groups. From a comparison of the R-δ structure (computer-generated by mirroring its S-λ counterpart) with the unusual R-λ conformer it is apparent that the ring-substituent atoms 4 and 5 are closer to the in-plane oxygen ligand atoms in the latter; these unfavourable interactions in the R-λ (and S-δ) conformer probably contribute to their reduced tendency to undergo crystallisation. To which degree the less-favourable conformer of each enantiomer is present in equilibrium in solutions of these compounds, or of intermediates involved in α-olefin polymerisation catalysis remains unknown at present. At any rate, conversion to these diastereomeric mandelic acid derivatives provides a convenient means to monitor the degree of enantiomeric purity on a microanalytical scale by ¹H-NMR spectrocopy, for instance in the course of a catalytic reaction (23).

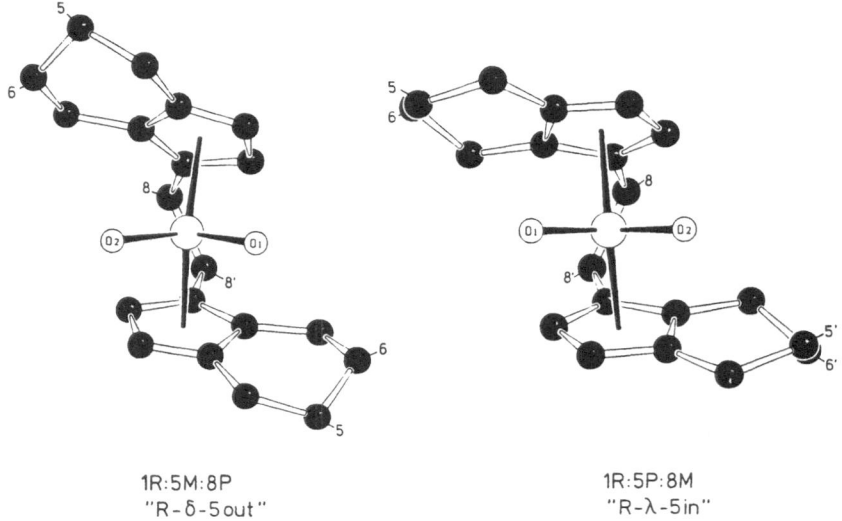

1R:5M:8P
"R-δ-5out"

1R:5P:8M
"R-λ-5in"

Fig. 6. Different conformers of an R-configurated ethylene bis(tetrahydroindenyl) metal complex, derived from the structures in Fig. 5

CONCLUSIONS

Further efforts are now to be directed at a systematic variation of substituents and bridging groups in these chiral ansa-metallocenes in order to delineate optimal spatial requirements for the stereoselective catalysis of α-olefin polymerisation as well as of other catalytic C-C bond formation reactions. Substitution with functional groups other than hydrocarbon residues might be desirable in order to stabilize the crucial – presumably cationic (25,26) – intermediates essential for α-olefin polymerisation and to facilitate their stereospecific transformation reactions. Prerequisite for studies of this kind is a possibilty to induce selective formation of racemic rather than meso-configurated ansa-metallocenes, i.e to gain a means of control on the stereochemistry of the chelate ring formation reaction.

ACKNOWLEDGEMENTS

Financial support of these studies by Deutsche Forschungsgemeinschaft, by Fonds der Chemischen Industrie and by funds of the University of Konstanz is gratefully acknowledged.

REFERENCES

1 Natta G, Pino P, Mazzanti G, Giannini U (1957) J Amer Chem Soc
   79:2975
2 Breslow D S, Newburg N R (1957) J Amer Chem Soc 79:5072; Breslow DS
   Newburg NR (1959) J Amer Chem Soc 81:81; Long WP, Breslow DS
   (1960) J Amer Chem Soc 82:1953; Long WP, Breslow DS (1975) Ann
   Chem 1975:463
3 Chien JCW (1959) J Amer Chem Soc 81:86; Chien JCW, Hsieh JTT (1975)
   in Chien JCW (ed) Coordination Polymerisation, Acad Press, New
   York, p 305
4 Zefirova AK, Tikhomirova NN, Shilov AE (1960) Dokl Akad Nauk SSSR
   132:1082; Zefirova AK, Shilov AE (1961) Dokl Akad Nauk SSSR 136:
   599; Stepovik LP, Shilova AK, Shilov AE (1963) Dokl Akad Nauk
   SSSR 148:122; Dyachkovskii FS, Shilova AK, Shilov AE (1967) J
   Polym Sci C16:2333; Borodko YG, Kyashina EF, Panov VB, Shilov AE
   (1973) Kinet Katal 14:255
5 Belov G P, Kuznetsov V I, Solovyeva T I, Chirkov N M, Ivanchev S S
   (1970) Makromol Chem 140:213; Belov G P, Belova V N, Raspopov L
   N, Kissin Y V, Brikenshtein A, Chirkov A M (1972) Polym J 3:681
6 Henrici-Olive G, Olive S (1967) Angew Chem 79:764; (1968) Kolloid Z
   228:43; (1969) Makromol Chem 121:70; (1969) Advan Polym Sci 6:421;
   (1969) J Organomet Chem 16:339; (1969) Polymerisation, Katalyse-
   Kinetik-Mechanismen, Verlag Chemie, Weinheim; (1971) Angew Chem
   83:782
7 Reichert KH, Schötter E (1968) Z Phys Chem 57:74; Reichert KH,
   Meyer K (1969) Kolloid-Z 232:711; Meyer K, Reichert KH (1970)
   Angew Makromol Chem 12:175; Reichert KH (1970) Angew Makromol
   Chem 13:177; Reichert KH, Meyer K (1973) Makromol Chem 169:163
8 Fink G, Schnell D (1974) Angew Makromol Chem 39:131; Fink G,
   Rottler R, Schnell D, Zoller W (1976) J Appl Polym Sci 20:2779;
   Fink G, Rottler R (1981) Angew Makromol Chem 94:25; Fink G,
   Rottler R, Kreiter CG (1981) Angew Makromol Chem 96:1; Fink G,
   Zoller W (1981) Makromol Chem 182:3265
9 Overberger CG, Diachkovsky FS, Jarovitzky PA (1964) J Polym Sci 2:
   4113; Overberger CG,Jarovitzky PA (1965) J Polym Sci 3:1483
10 Fryzuk MD, Bosnich B (1978) J Amer Chem Soc 100:5491; Bosnich B,
   Fryzuk MD (1981) Top Inorg Organomet Stereochem 12:119
11 Kaminsky W, Vollmer HJ, Heins E, Sinn H (1974) Makromol Chem 175:
   433; Kaminsky W, Kopf J, Thirase G (1974) Ann Chem 1974:1531;
   Kaminsky W, Sinn H (1975) Ann Chem 1975:424; Kaminsky W, Vollmer
   HJ (1975) Ann Chem 1975:438; Kaminsky W, Kopf J, Sinn H, Vollmer
   HJ (1976) Angew Chem 88:688; Andresen A, Cordes HG, Herwig J,
   Kaminsky W, Merck A, Mottweiler R, Pein J, Sinn H, Vollmer HJ
   (1976) Angew Chem 88:689; Sinn H, Kaminsky W, Vollmer HJ, Woldt
   R (1980) Angew Chem 92:396; Kaminsky W, Miri M, Sinn H, Woldt R
   (1983) Makromol Chem Rapid Comm 4:417
12 Schnutenhaus H, Brintzinger HH (1979) Angew Chem 91:837
13 Wild FRWP, Zsolnai L, Huttner G, Brintzinger HH (1982) J Organomet
   Chem 232:233; Wild FRWP, Wasiucionek M, Huttner G, Brintzinger HH
   (1985) 288:63

14 Gutmann S, Hund U, Brintzinger HH (1987) unpublished results

15 Wiesenfeldt H, Barsties E, Brintzinger HH (1987) unpublished results

16 Burger P, Brintzinger HH (1987) unpublished results

17 Ewen JA (1984) J Amer Chem Soc 106:6355

18 Kaminsky W, Külper K, Brintzinger HH, Wild FRWP (1985) Angew Chem 97:507; Kaminsky W (1986) Angew Makromol Chem 145:149

19 Longo P, Grassi A, Pellecchia C, Zambelli A (1987) Macromolecules 20:1015

20 Pino P, Cioni P, Wei J (1987) J Amer Chem Soc, in print

21 Schlögl K (1966) Fortschr Chem Forsch 6:479; (1971) IUPAC Bulletin 24:413

22 Cahn RS, Ingold C, Prelog V (1966) Angew Chem 78:413.

23 Schäfer A, Karl E, Zsolnai L, Huttner G, Brintzinger HH (1987) J Organomet Chem 328:87

24 Corey EJ, Bailar Jr JC (1959) J Amer Chem Soc 81:2620

25 Eisch JJ, Piotrowski AM, Brownstein SK, Gabe EJ, Lee FL (1985) J Amer Chem Soc 107:7219

26 Jordan RF, Bajgur CS, Willet R, Scott B (1986) J Amer Chem Soc 108: 7410; Jordan RF, Lapointe RE, Bajgur CS, Echols SF, Willet R (1987) J Amer Chem Soc 109:4111

Some New Results on Methyl-Aluminoxane

H. Sinn, J. Bliemeister, D. Clausnitzer, L. Tikwe, H. Winter,
O. Zarncke

Institute for Technical and Macromolecular Chemistry, University
of Hamburg, Bundesstr. 45, D-2000 Hamburg 13, FRG

Aluminoxanes are formed by controlled hydrolysis of aluminumalkyles. The typical structural element is an oxygen atom joining two
aluminum atoms that still bear alkyl groups. The simplest representative of the aluminoxanes is the μ-oxo-bisdialkylaluminium or
tetraalkyl-dialuminum-oxide respectively. However, its preparation
and characterization especially with methyl groups has not been
reported yet.

Generally the reaction between water and aluminumalkyles is very
exothermal, accompanied with flames and yields aluminum oxide. For
the careful, controlled and incomplete hydrolysis of aluminumalkyles several methods have been developed. These methods aim for
small amounts of product for catalytical investigations by directly
mixing both reactants, alkyl and water. The methods differ in the
carrier that is used to dilute the water to slow down the reaction
rate. Vandenberg (1), one of the first describing the preparation
of aluminoxanes, reacts aluminumalkyles and water in benzene.
Sakharovskaya (4) et al. use a nitrogen stream as carrier for the
water vapor used in the reaction. Storr (3) et al. developed the
condensation method, where water vapor is condensed into a cooled
solution of aluminumtriethyl in benzene. The molecular sieve method
(6, 7) was developed in our institute. In this case water adsorbed
on a molecular sieve serves as a water source for the reaction.

Direct methods are quite hazardous as a runaway reaction can occur
very easy. Due to the weak physical bonds between the water and the
carrier, the water is given off quite readily so that controlled
reaction conditions are difficult to achieve.

However, the water needed for partial hydrolysis can also be introduced into the reactor in a chemically bonded form as salt hydrates. In this case the crystal water is given off more slowly so that
a contolled reaction can be guaranteed. These considerations led to
the crystal water method (10, 11, 12).

Other methods of preparation aim at producing the Al-O-Al group
with appropriately chosen organoaluminum compounds like dialkylaluminum chloride and lithiumdialkyl aluminate (13, 14, 15). The
reactions between alkoxyaluminum dichlorides and methylaluminum
dichloride (16) or between methoxyaluminum compounds
$(Me_x(OMe)AlCl_{2-x}; x=0,1,2)$ and methylaluminum compounds
$(Me_yAlCl_{3-y}; y=1,2,3)$ in the presence of triethylaluminum have also
been shown to yield aluminoxanes (17).
Another way to prepare aluminoxanes is the reaction of PbO with
aluminumtrialkyles (18).

Other procedures use solvents that form complexes with the reactants and the products of the reaction (e. g. (19)). These methods
are of minor interest in this context. A very detailed study on the
mechanism of aluminoxan formation in these solvents was done by

W. Kaminsky and H. Sinn (Eds.)
Transition Metals and Organometallics as
Catalysts for Olefin Polymerization
© Springer-Verlag Berlin Heidelberg 1988

Boleslawsky et al. (20).

The latest improvement to the crystal water method was made by
Kaminsky and Hähnsen (21,22). Partially dehydrated aluminum sulfate
is used as a water source. However, the dehydration process is very
uneconomic and insoluble aluminoxanes produced during the hydrolysis cannot be separated from the residual aluminum sulfate. A
com-plete mass balance of the reaction is nearly impossible.

In connection with Ziegler-Natta polymerisation attention was drawn
to aluminoxanes on the discovery that small amounts of water added
to the wellknown Breslow-system Cp₂TiCl₂/Al(CH₃)₃ (23) gave a dramatic increase in activity and facilitated the regulation of the
molecular weight of the formed polyethylene even in systems free of
halogens (24). Especially powerful systems were obtained by the use
of Cp₂ZrR₂ as transition metal component (25). Thus

- activities up to 25*10⁶ (g PE/g Zr*h)

- transition times less then 5*10⁻⁵ s

- polymer formation of 15,000 macromolecules / Zr-atom*h

were observed.

As this homogenous catalyst sytem can be also used for α-olefine
polymerisation, it was possible to apply the principle postulated
in 1958 (26)

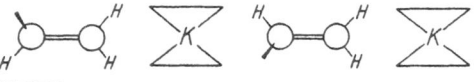

A 891.1

Abb. 1. Bei symmetrischem Komplex K sind die beiden entgegen-
gesetzten Anlagerungspositionen des Olefins thermodynamisch
gleichwertig

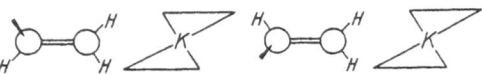

A 891.2

Abb. 2. Bei unsymmetrischem Komplex K sind die beiden ent-
gegengesetzten Anlagerungspositionen des Olefins thermodynamisch
ungleichwertig, d. h. eine Anlagerungsposition ist bevorzugt.

Reproduction from (26)

to use chiral transition metal compounds for producing stereoregu-
lated poly-α-olefines in homogenous phase (27).

It is still not known why maximum activity with respect to zirconium requires such a large excess of aluminoxane, with typical
ratios of 100 to 10000 Al-atoms/Zr-atom. This must depend on structural elements of the aluminoxanes though it is as yet unclear how.
This is obviously connected to the fact that aluminoxane structure
is still vague. It is known however that oligomeric aluminoxanes
are needed for catalytic activity and that methylaluminoxanes are
much more effective in catalysis than other aluminoxanes.

Several structures have been proposed for aluminoxanes: rings (28), clathrates (29) and linear or branched chains. The stucture of the aluminoxane apparently changes with reaction conditions. The clarification of these structural inconsistencies calls for a method for reproducible methylaluminoxane production.

As aluminoxanes to this point have been difficult to obtain in large quantities, neither could their structure be elucidated nor could other fields of application be developed. Therefore our aim was to develop methods for the production of aluminoxanes that can be applied on an industrial scale. Readily available aluminoxanes will enable structural elucidation, a decrease in cost for aluminoxane based catalysts, the development of mixed metaloxanes with defined aluminum content, the use of aluminoxanes or mixed metaloxanes as catalyst support and the production of special ceramics.

In new efforts to react aluminum alkyles - especially trimethylaluminum - with water directly in an inert solvent like toluene under controlled conditions, we made some unexpected observations:

1) If ice is immersed in trimethylaluminum solution at low temperatures, more or less gas evolution depending on the temperature takes place on the bare surface. This increases the surface temperature with respect to the bulk temperature and the solution temperature. Depending on the reaction temperature the reaction rate either increases until it cannot be controlled or the rate diminishes until no more gas is evolved and the surface grows dull.

2) The inactive surface may be reactivated by scrubbing the ice with a piece of stainless steel wire. On the newly formed bare surface gas evolution or - at higher temperatures - small gas eruptions take place. However, the gas evolution decreases rapidly. The scrubbing can be repeated several times at reaction temperatures between 200 and 250 K.

3) If a piece of ice is crushed in trimethylaluminum solution after the surface reaction ceased, gas is evolved only on the freshly formed surface.

4) If very small pieces of ice are used, the gas bubbles adhering to the ice drive the pieces to the surface of the solution to form a layer of foam. In this layer the ice is coated with insoluble aluminoxane that contains less than one methyl group per aluminum atom.

From these observation we derived the following conclusions:

a) It is neccessary to create a defined ice surface in a solution of aluminumalkyl and keep it reactive by continuously cleaning the surface.

b) The velocity of solution flowing past reactive surface must be high enough to carry formed aluminoxane away from the ice to avoid further hydrolysis and formation of insoluble products.

c) Either the formation of small ice particles that can
float to the surface of the solution must be prevented or
the circulation of the solution must be properly directed in
order to tear gas and solid apart.

d) To avoid formation of insoluble aluminoxane it is also
useful to work with an excess of aluminumalkyl with respect
to available water at the time of reaction.

e) Gas evolution on the surface must not prevent the
solution from reaching the surface and transporting soluble
reaction products away from the surface. This can be controlled by adjusting either surface or ice temperature. For
this reason the ice temperature should be adjustable independently of solution temperature. This may be done within
limits by putting the ice on a cooled plate within the solution. The plate is cooled by a separate cooling circuit,
independent of the one used for cooling the solution (see
figure 3 in method 4).

These conclusions have indicated the way to several working methods. They are described in the experimental section. These methods
avoid the formation of insoluble material almost quantitatively,
especially if an excess of trimethylaluminum with respect to reacting water is used. Residual aluminumalkyl and solvent can be used
directly for the next reaction so that in a series of reactions
both water (introduced into the reactor as ice) and aluminumalkyl
can be reacted completely. The reaction yields more than 90% soluble aluminoxane. The rest is insoluble in aromatic solvents. No
other by-products except methane naturally are formed.

The reactor described in method 4 has a constant reactive surface
of permanently temperature controlled ice. Using this method the
reaction rate and its dependence on aluminumalkyl concentration and
temperature can be measured. We found

$dNCH_4/dT = k(T)*S*[Al]$

with

$dNCH_4/dT$ rate of methane evolution in moles / hour

S surface area in cm²

$[Al]$ trimethylaluminum concentration in moles / liter

$k(T) = k_o*exp\,(E_A/RT)$

$k_o = 1.225*10^{17}$ [1/cm²h]

$E_A = -84$ KJ   energy of activation

$R = 8.314$ Jmol⁻¹K⁻¹   gas constant

T absolute temperature

These values are still uncertain with respect to k_o, because the
temperature difference could be measured precisely whereas the base
temperature has an error of ± 1.5°C.

Aluminoxanes prepared in the usual fashion by distilling off the
excess aluminumalkyl and the solvent under reduced pressure still
contain a small amount of free aluminumalkyl. The ratio of Al-atoms

in aluminoxane to Al-atoms in this amount of free alkyl is about 20.

In the case of trimethylaluminum this remaining alkyl may be further removed by dissolving the isolated raw aluminoxane in cumene or by using cumene as a solvent for the preparation and distilling off the cumene with a very efficient rectification column. As cumene has a higher boiling point than trimethylaluminum the alkyl is enriched in the first fractions .

However, these fractions contain less alkyl than the corresponding fractions resulting from a distillation of an equally concentrated solution of trimethylaluminum in cumene. This indicates an associative bond between trimethylaluminum and aluminoxane. This supposition is in accordance with ^{1}HNMR-spectra, although these are not sufficiently informative.

The molecular mass of 1130 ± 30 was measured kryoscopically in benzene. Another colligative method for molecular mass determination is ebullioscopy. Tetrahydrofurane is known to be a strong donor, therefore associates of aluminoxanes or between aluminoxanes and aluminumalkyl should be cleaved to yield tetrahydrofuranates in this solvent. Ebullioscopic measurements in tetrahydrofurane gave a molecular mass of 360 ± 40.

The CH_3/Al-ratio was measured by complete hydrolysis, gasvolumetric determination of CH_4 and complexometric titration of aluminum to be

$$CH_3/Al = 1.59.$$

Elementary analysis shows the H/C-ratio to be nearly 3 (2.98) as should be expected for a compound containing methyl groups only. The CH_3/Al-ratio computed from this method gives

$$CH_3/Al = 1.587.$$

Considering the oxygen value the analysed aluminoxane has to consist of two linear or branched associated molecules of 7-8 aluminoxane units. Bonded to this adduct are three molecules of trimethylaluminum. On dissolution in tetrahydrofurane apparently tetrahydrofuranates are formed as can be seen from the evolved heat. However, instead of forming 5 particles as expected, only 3 can be measured by the colligative methods. Therefore we postulate - at least as a working hypothesis - that the aluminoxane molecule is strongly associated to one molecule of trimethylaluminum so that they are not dissociated by tetrahydrofurane. Only one molecule of trimethylaluminum yields the tetrahydrofuranate, the same amount of trimethylaluminum that can be separated by distillation with cumene. At this time these suppositions are quite daring, but correspond to experimental evidence.

Further observations can be reported only cursory:

Methyl groups in trimethylaluminum and methylaluminoxane react with dimethylgallium hydroxide under methane evolution. Obviously a ga-O-al structure is formed where ga and al indicate a single valency of Ga and Al respectively. Nearly all but not all methyl groups react in this way, even if an excess of dimethylgallium hydroxide is used. If trimethylaluminum is added to this reaction mixture, it reacts with the excess dimethylgallium hydroxide under methane evolution. On light heating all gallium may be distilled off as galliumtrimethyl. After distillative removal of the excess aluminum-trimethyl a white residue remains containing less than 0,02 Ga atoms per Al atom. It has been shown by hydrolysis to be an aluminoxane.

Tetraisobutyldialuminoxane reacts with a repeatedly added excess of trimethylaluminum by giving off aluminumtriisobutyl that is distilled out of the reaction vessel together with the excess of aluminumtrimethyl. The residue is an oligomeric aluminoxane.

Apparently structural rearrangements of aluminoxanes are possible even at room temperature.

It is possible to perform a "dry" hydrolysis of methylaluminoxanes in a fluidized bed reactor using argon partially saturated with water vapor for fluidization of the powdered aluminoxane. The reaction is slow enough to be interruptable at any degree of "hydrolysis". The resulting compounds containing more or less CH_3-groups respectively, are insoluble in hydrocarbons and open a way to organometallic adsorbents and special ceramics.

Thermolysis of methylaluminoxanes yields yellow products at 300°C and brownish red products at 500°C. The yellow products react with water and diluted acid under gas evolution, whereas no reaction can be detected with the brown products.

The described preparative access to aluminoxanes will be used to pursue the reported observations and elucidate the structures.

EXPERIMENTAL

Method 1

The reactor is a temperature controlled glass autoclave (made by Büchi or SFS) equipped with a magnetically coupled stirrer, temperature sensor, gas connections and openings for filling and emptying. The stirrer consists of a common impeller and a cutter taken from a kitchen mixer.

After purging with argon, the reactor is filled with 360 ml toluene and 40 ml (~ 400 mMol) trimethylaluminum in an argon stream.

When the reactor has been cooled to -80°C 3.76g (≈ 200 mMol) ice of -80°C is added. The autoclave is now sealed, the exhaust valve connected with the gas measuring unit is opened, and the stirrer is switched on. Stirring rate is increased until the ice is crushed. Following a strong initial gas evolution, the volume flow rate stabilizes at 1 to 2 l/h. As soon as the gas evolution decreases, stirring rate and reaction temperature are increased. The diagram shows that flow rates up to 3 l/h are safely handled under these conditions.

When 8 - 9 l gas (at room temperature) have been evolved, the thermostate is shut down, and the reaction mixture is heated to room temperature. The overall gas evolution is 9.5 - 10 l at room temperature. This is approximately equal to the volume calculated for the complete reaction of 200 mMol water with an excess of trimethylaluminum.

The reaction product, an almost colourless solution, is filtered through a glass filter (G4, Schott) and the solvent and excess alkyl are distilled off in vacuo. The 16 - 17 g product is a foamy glasslike material that can be crushed to a white powder. The substance spontaneously ignites on contact with air, is soluble in aromatic hydrocarbons like benzene, toluene or cumene, less soluble in methylcyclopentane and nearly insoluble in alkanes.

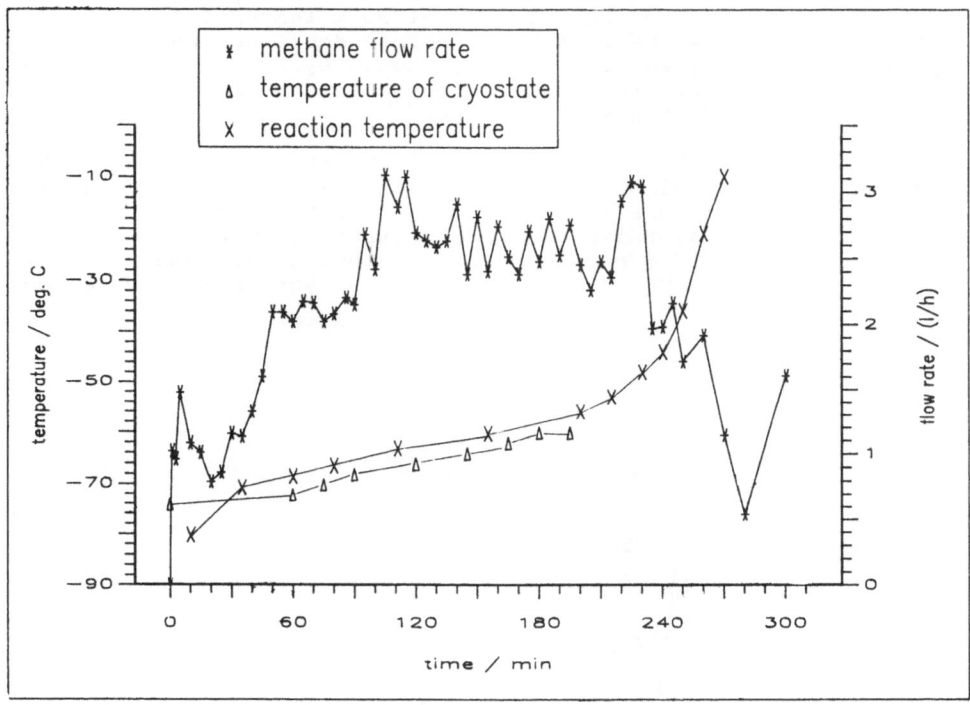

Fig. 1   Thermostate and autoclave temperatures and rate of gas evolution as a function of time

Method 2

If no kryostate is available for cooling, but only a dry ice/-ethanol bath, the reaction rate may be regulated by the stirring speed.

In this case the autoclave used in method 1 may either be immersed in the cooling bath or the cooling jacket is filled with the cooling fluid. In this way temperatures down to -60°C can be reached.

With a little practice the reactor can be filled in such a way that a concentrated solution of trimethylaluminum at the bottom of the autoclave is covered with a layer of almost pure solvent. Only after the ice has been added, the reactor has been sealed and the exhaust valve has been opened the stirrer should be switched on at low speed. Immediately, the reaction starts with a barely manageble flow rate of 5 l/h methane - in a reaction of 40 ml trimethylaluminum with 3.74 g ice in 360 ml toluene. Sometimes the frothing causes a part of the reaction mixture to leave the reactor. For this reason, the exhaust valve should be a ball valve at a suitable large diameter and should be connected to a cooled receiver.

As soon as the gas evolution slows down, stirring speed is increased until the ice particles are crushed and their surface is
eroded. Stirring speed is increased in intervals so that frothing
remains manageable, i.e. the flow rate lies between 3 and 4 1/h. If
stirring speed cannot be increased further to increase gas evolution and about 8 1 methane have been evolved, the coolant may be
taken away to complete the reaction at room temperature within
another 3 1/2 hours. Caution! If the rate decreases while less than
8 1 of gas are evolved, the reactor should by no means be heated to
room temperature, but the reaction must continue at low temperature
until the said amount has been evolved.

The reacted solution is pressed out of the reactor with argon and
filtered through a glass filter (G4, Schott). The 2-3 g residue on
the filter is insoluble in aromatic hydrocarbons and upon hydrolysis gives a CH_3/Al-ratio $\leq$ 1. The filtrate is treated according
to method 1 to yield 12g of a soluble methylaluminoxane. The diagram represents an experiment where cooling was on for three hours
at -68°C.

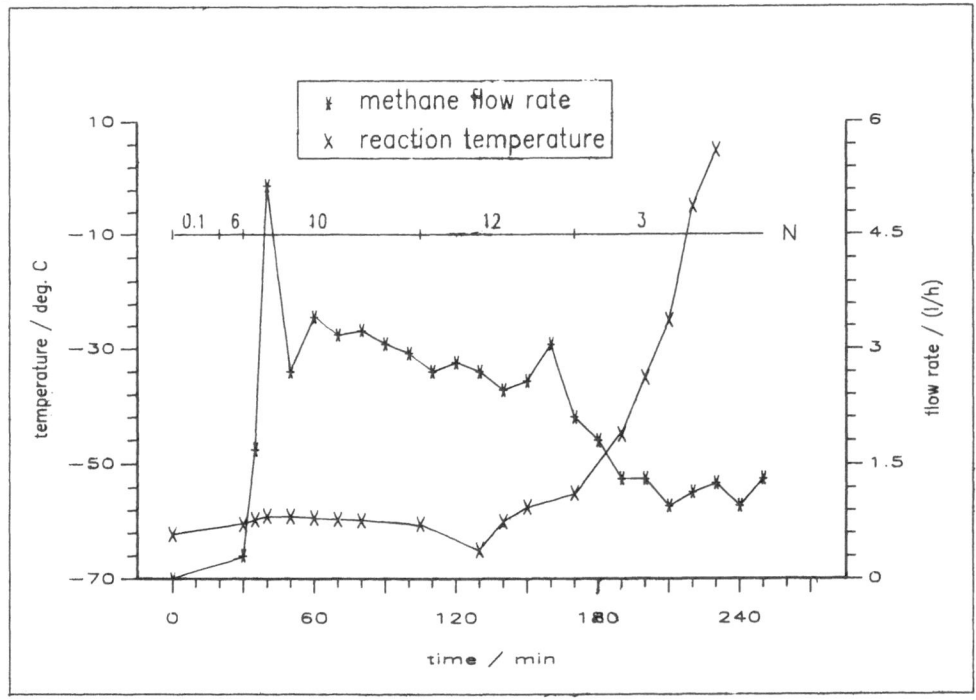

Fig. 2   Flow rate of methane and reaction temperature of the ex-
          periment described above. The rate was regulated by
          changing the stirring speed (N is given in 100 rounds
          per minute, i.e. N = 12 corresponds to 1200 min^{-1})

Method 3

A 1 1 three necked flask is filled with 360 ml cumene dried over
Na/K-alloy and 40 ml trimethylaluminum. The flask is equipped with
a KPG-stirrer, a thermostated dropping funnel and a stopcock connected to the vakuum line and the gas measuring device. The reac-

tion temperature (-40 °C) is controlled by immersing the flask in a cooling bath of dry ice/ethanol.

In a seperate vessel an ice suspension is prepared by slowly adding 4 ml water to 50 ml cumene cooled to -30°C that is stirred with an Ultraturrax-stirrer. The resulting thin and flat ice crystals are heavier than cumene at temperatures higher than -30°C and float at lower temperatures. The suspension is transferred to the dropping funnel that is cooled at -20°C and added in small portions to the alkyl solution. The reaction starts immediately evolving methane and producing a finely structured foam apart from small amounts of a white substance (insoluble aluminoxane).

If foaming is too strong or the flow rate exceeds 0.3 l/min, the reaction vessel is cooled by a short immersion in the coolant.

Within one hour all the water has been added and 10 l gas have been evolved. At this time the reaction temperature should be about -20°C. During the next hour the reactor is heated to room temperature. Only small amounts of methane are evolved at this stage, a few residual ice crystals that were covered with insoluble aluminoxane burst with little eruptions.

Even though these eruptions are quite spectacular, the reaction is safe after 9-10 l gas have been evolved and only severe carelessness could lead to violent reactions.

The heated and slowly stirred reaction mixture becomes clear and bright while some insoluble material (aluminoxane or aluminum oxide) settles to the bottom. The product mixture is filtered through a glass filter and the filtrate worked up for soluble aluminoxane. The insoluble residue amounts to 3-6 g and decomposes violently in air.

The filtrate yields 13 to 16 g raw product that is soluble in toluene and shows the usual activity in homogenous Ziegler-catalysis. It still contains a small amount of free trimethylaluminum.

The diluted alkyl solution distilled from the raw product is used for the next hydrolysis after its concentration has been adjusted to 40 ml trimethylaluminum in 400 ml solution.

Note: The amount of insoluble aluminoxane is roughly proportional to the reaction rate, especially in violent reactions when large bubbles are formed. Because the foam layer can not be cooled efficiently and contains very little alkyl thus causing a local excess of water, presumably the insoluble aluminoxane is produced within the foam layer from ice crystals adhering to gas bubbles .

Method 4

A cylindrical 3 l-autoclave 80 mm in diameter is fitted with a cooling jacket and closed with a flange at the top and the bottom respectively. From each flange a magnetically coupled shaft projects into the reactor. The free volume of the magnetically coupled stirrer is filled with cumene that can enter the reactor only by diffusing through bearings and gaskets. Inside the reactor a doughnut shaped plate is mounted horizontally on the lower flange so that the plate surrounds the vertical stirring shaft without hindering its rotation.

When the reactor has been purged with argon the depression of the plate is filled with water that is frozen by switching on the plate's cooling system. Now the scraper that is driven by the bottom shaft is lowered onto the ice. Because the scraper can move freely up and down, it is pressed onto the surface by its own weight slowly milling away the surface layer from the ice as the shaft rotates.

The upper shaft drives the stirrer that can be operated independently of the scraping unit to mix the reactor content.

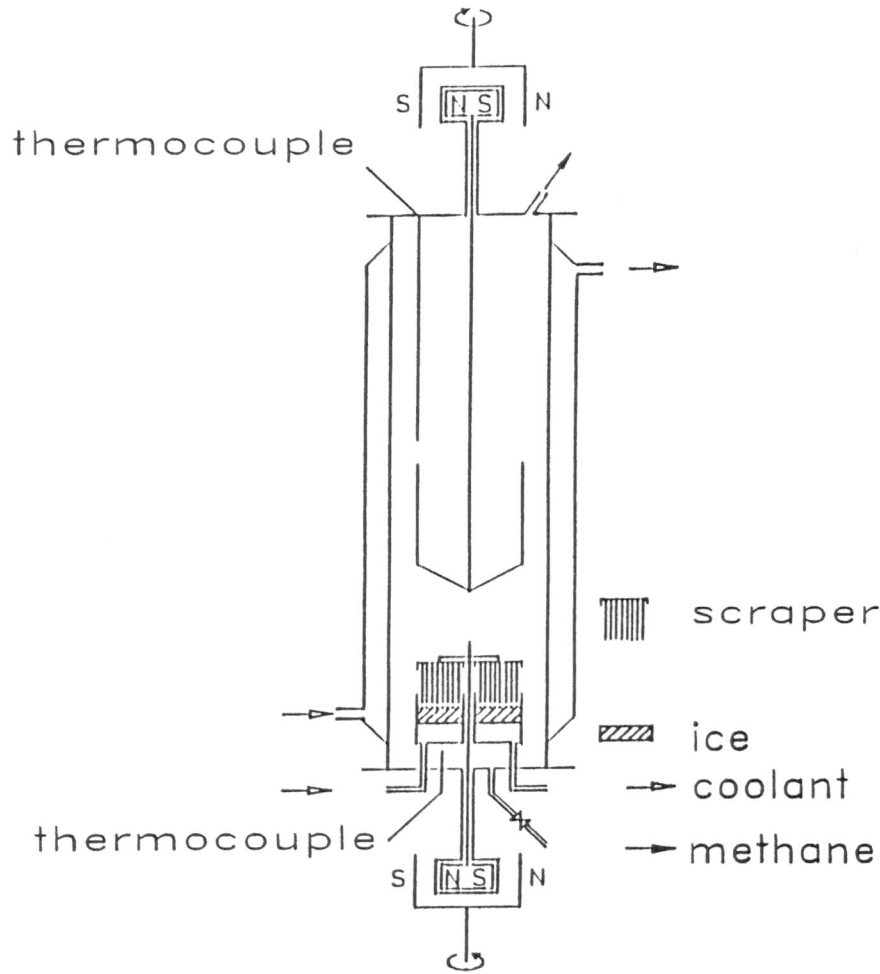

Fig. 3 The Reactor Used in Method 4

The plate is filled with 7.5g water and cooled to 205-210 K. The surface area (41 cm²) of the ice thus created, remains constant during the reaction. When the water is frozen, 800 ml cooled toluene or cumene is added creating a solvent column of 10 cm above the plate and completely covering the stirrer.

When the solvent has been cooled to 210 K by letting coolant flow through the cooling jacked 80 ml (800 mMol) trimethylaluminum are added. At these temperatures the alkyl is only partially soluble, the rest forms a layer of crystals on top of the solvent.

When the pressure remains constant, the autoclave is connected to a receiver, which in turn is connected to a mercury valve and a gasmeter. After fitting the reactor with a thermocouple, firstly the scraper is set in motion, then also the stirrer to suspend the trimethylaluminum throughout the solution.

Tiny gas bubbles are evolved near the ice. By reducing the coolant flow rate, temperature is increased to 220-230 K and methane flow rate to 12 l/h. After 3 hours 18 l gas (at room temperature) have been evolved so the reactor can be heated to room temperature during the next hour, giving another 2 l of methane. Only a few flakes of insoluble material are formed in the reaction. They are filtered off with a glass filter (G4). The solution is worked up in the usual way for aluminoxane (27 g). The trimethylaluminum solution that was distilled off is made up to the starting concentration and can be used for the next reaction.

The produced soluble aluminoxane still contains traces of trimethylaluminum.

Analytical data was calculated for:

$(CH_3)_2Al-(OAlCH_3)_5-OAl(CH_3)_2*(CH_3)_2Al-(OAlCH_3)_5-OAl(CH_3)_2*3Al(CH_3)_3$

	calculated	found
Molar mass (kryoscop., benz.)	1114	1129 ±30
Molar mass (ebulliosc., THF)	352	360 ±40
Decomposition analysis		
(CH_4)/Al	1.56	1.59
Titration AL %	43.6	43.2
Elementary analysis (done by Pascher Labs)		
Al %	43.6	43.0 ±0.2
C % (with addition of V_2O_5)	30.2	30.45±0.03
O %	18.7	18.8
H %	7.5	7.62±0.05
H/C	3.0	2.98

REFERENCES

1) Vandenberg EI (1960) J Polymer Sci 47:4
2) Korneev NN, Popov AF, Larikov EJ, Zhigach AF, Sakharovskaya GB (1964) J Gen Chem USSR 34:3425
3) Storr A, Jones K, Laubengayer AW (1968) J Am Chem Soc 90:3173
4) Sakharovskaya GB, Korneev NN, Popov AF, Snegova SZ, Zhigaih AF, Sobolewskii HV, Patent: Brit. 1, 319, 746 (Ch. C08g) 6. Jun. 1973, Appl. 6621/71 11. Mar. 1971
5) Aoyagi H, Vada T, Tadokoro Y, Horikivi S, Patent:Jpn. Kokai Tokyo Koho 79, 64, 600 (Cl. C08G 79/10) 24 May. 1979 Appl. 77/131, 909 1. Nov. 1977
6) Herwig J (1979) Ph. D. Thesis, University of Hamburg
7) Vollmer HJ (1980) Ph. D. Thesis, University of Hamburg
8) Wolinska A (1982) J Organomet Chem 234:1-6
9) Boleslawski M, Serwatowski J (1983) J Organomet Chem 254:159-66
10) Razuvajev GA, Sangolov JA, Nelkenbaum JJ, Minsker KS (1975) Jzw Akad Nauk SSR Ser Chim 19:2547
11) Hähnsen H (1980) Diploma Thesis, University of Hamburg
12) Rafikov SR, Minsker KS, Sangalov JA, Nelkenbaum JJ, Patent:S.U. 56684, C. A. 87 (1977)
13) Tani H, Araki T, Oguni N, Aoyoagi T, Ueyama N (1966) J Pol Sci Pol Lett (B) 4:97
14) Tani H, Araki T, Oguni N, Aoyagi T, Ueyama N (1967) J Am Chem Soc 89:173
15) Ueyama N, Araki T, Tani H (1973) Inorg Chem 12:2218
16) Kosinska W, Zardecka K, Kunicki A, Boleslawski M, Pasynkiewicz S (1978) J Organomet Chem 153:281
17) Kosinska W, Kunicki A, Boleslawski M, Pasynkiewicz S (1978) J Organomet Chem 161:289-297
18) Boleslawski M, Pasynkiewicz S (1972) J Organomet Chem 43:81

19) Pasynkiewicz S, Sadownik A, Kunicki A (1977) J Organomet Chem 124:265-269
20) Boleslawski M, Serwatowski J (1983) J Organomet Chem 255:269-78
21) Hähnsen H (1984) Ph. D. Thesis, University of Hamburg
22) Kaminsky W, Hähnsen H, DE/02.11.82/DE 3240383, Fw. Hoechst Oligomeric aluminoxanes
23) Breslow DS, Newburg NR a) (1957) J Am Chem Soc 79:5072 b) (1959) J Am Chem Soc 81:81
24) Andresen A, Cordes H-G, Herwig J, Kaminsky W, Merck A, Mottweiler R, Pein J, Sinn H, Vollmer H-J (1976) Angew Chem 88:689
25) Kaminsky W, Miri M, Sinn H, Wold R (1983) Makromol Chem, Rapid Commun 4:417-21
26) Patat F, Sinn H (1958) Angew Chem 70:496
27) Kaminsky W, Külper K, Brintzinger HH, Wild FRWP (1985) Angew Chem 97:507
28) Sinn H, Kaminsky W (1980) Adv Organomet Chem 18
29) Lasserre S, Derouault J (1983) Nouv J Chem 7:659

The authors are indepted to the Fonds Chemie for material support and to the Max-Buchner-Stiftung for a grant.

Asymmetric Hydrooligomerization of Propylene

P. Pino, P. Cioni, M. Galimberti, J. Wei, N. Piccolrovazzi

Swiss Federal Institute of Technology, Institut für Polymere, Universitätstrasse 6, 8092 Zürich, Switzerland

ABSTRACT

Propylene was polymerized in the presence of hydrogen using $(-)(R)$ $(EBTHI)Zr(CH_3)_2$ (Ia) or $(+)(S)(EBTHI)ZrCl_2$ (Ib) as catalyst precursor and $(-Al(CH_3)-O-)_n$ as co-catalyst. The polymers were fractionated by solvent extractions and oligomeric fractions were isolated having number average molecular weight between 310 and 2040 and between 445 and 3150 respectively. All the fractions were optically active, the optical rotation being positive using (Ia) and negative using (Ib) as catalyst precursors. Oligomers and polymers have a highly isotactic structure as shown by ^1H and ^{13}C NMR. End groups determination shows that all chains contain a n-propyl group, the other chain end being in the case of (Ia) either an isobutyl- or a n-butyl group. The results give new relevant informations on the origin of regio- and stereospecificity in the polymerization of propylene.

INTRODUCTION

The discovery of the ethylene-bis-tetrahydroindenyl $ZrX_2/[-Al(CH_3)$ $-O-]_n$ (X=Cl, $-CH_3$)(Ib and Ia) catalytic system which polymerizes propylene to isotactic polypropylene [1,2] and the possibility of separating a single antipode of the above Zirconium complex,[3] has offered for the first time the possibility to prepare well defined chiral non racemic catalysts for the stereospecific polymerization of α-olefins. Unfortunately only a vanishing small optical activity is foreseen in solution [4] for high molecular weight isotactic polypropylene; however for propylene oligomers, if the end groups are, as usual, different a measurable optical activity is expected.[4] The chemical structure, the sign and the value of the optical rotation of oligomers can yield very interesting information on the origin of stereospecificity and regiospecificity in the polymerization of propylene with the above catalytic system.[5]

Using the catalytic system (Ib) oligomers can be easiliy obtained[1] carrying out the reaction between 60°C and 80°C; however (Table 1) stereoregularity also seems to decrease with temperature, and therefore in this case oligomerization yields less precise stereochemical informations.

W. Kaminsky and H. Sinn (Eds.)
Transition Metals and Organometallics as
Catalysts for Olefin Polymerization
© Springer-Verlag Berlin Heidelberg 1988

Table 1

Oligomerization of propylene with the catalytic system Ib at different temperatures

Temperature [a] ($^\circ$C)	Productivity ($kgPP/molZrxhxmolC_3H_6xL^{-1}$)	$\bar{M}_\eta$	mmmm [b] %
0	41	83'000	86
30	1499	6'500	72
60	2735	4'800	37
80	n.d.	n.d.	18

a) of the polymerization experiment; b) from ^{13}C NMR

With the catalytic system Cp_2 Zr $Cl_2/[-Al(CH_3)-O-)]_n$ it is known[6] that in the polymerization of ethylene lower molecular weight polymers are obtained operating in the presence of hydrogen. Therefore catalysts of the type (Ib) were used at 0°C in order to obtain highly stereoregular propylene oligomers, carrying out the polymerization in the presence of 1-3 atm. of hydrogen (Scheme 1). Substantial amounts of hydrogen were consumed during the reaction and a mixture of oligomers were obtained which was fractionated by solvent extraction at room-temperature or at the boiling temperature of the solvents[7] (Table 2).

Scheme 1

$$nC_3H_6 + H_2 \xrightarrow{\text{Cat}} H-(C_3H_6)_n-H$$

Table 2

Fractionation of the polymers obtained using $(-)(EBTHI)Zr (CH_3)_2/[-Al(CH_3)-O-]_n$ (Ia)[5] or $(+)(EBTHI)ZrCl_2/$[b]$(Al(CH_3)-O-)_n$ (Ib) as catalytic system.

Catalyst precursor: (Ia) Fraction	%	$\bar{M}_n$	Mp $^\circ$C	Fraction	%	$\bar{M}_n$	Mp; $^\circ$C
B	1.0	310	n.d.	F	8.5	420	35.9
C	6.0	380	-50	G	5.6	840	76.1
D	8.1	490	20	H	4.6	965	96.9
E	4.6	670	62.1	I	51.0	2040	138.0
				J	10.1	a)	150.9

Catalyst precursor:(Ib) Fraction	%	$\bar{M}_n$	Mp $^\circ$C	Fraction	%	$\bar{M}_n$	Mp $^\circ$C
C'	4.1	445	n.d.	F'	4.4	600	68.0
D'	2.6	585	46.1	G'	4.3	1400	91.3
				H'	7.6	1990	107.1
E'	3.6	900	75.0	I'	63.4	3150	140.2
				J'	10.0	c)	152.2

a) $\bar{M}_n = 17'000$ b) $[\alpha]_{436}^{25} = + 424$; the product was kindly supplied to us by Prof. H.H. Brintzinger, University of Konstanz (FRG) and fully characterized in our laboratory. c) $\bar{M}_n = 20'000$.
B: methanol soluble (25°C); C.C': methanol insoluble, acetone soluble (25°C); D,D': acetone insoluble, ethyl acetate soluble (25°C); E,E': ethyl acetate insoluble, diethyl ether soluble (25°C);F,F': boiling methanol soluble; G,G': boiling methanol insoluble, acetone soluble; H,H': boiling acetone insoluble, boiling diethyl ether soluble; I,I': boiling diethyl ether insoluble, boiling heptane soluble; J,J': boiling heptane insoluble.

Structural characterization of the oligomers

Structural characterization of the oligomers fractions was carried out using I.R., NMR, vapour pressure osmometry and, for the low molecular weight fractions (up to C_{30}), using Gas-Chromatography and Mass-Spectrometry.

Gas-Chromatography coupled with Mass-Spectrometry of the oligomers soluble in toluene at room-temperature obtained with the catalytic systems Ia, after distillation of the solvent, (Fig. 1), shows the presence of couples of compounds having similar retention time and identical molecular weight, corresponding to the hydrooligomers of propylene $((C_3H_6)_n+2H)$.

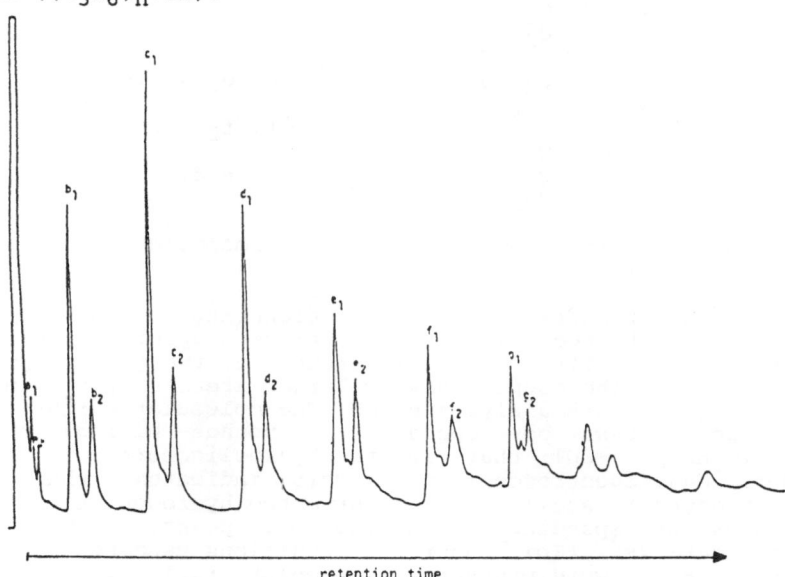

Fig. 1 Gas-Chromatographic analysis of the propylene low molecular weight hydrooligomers soluble in toluene. Column: carbowax 20 M, 4m Temperature 110°C-220°C.

In the Table 3 it is shown that the molar ratio between the two isomeric paraffines corresponding to each couple of peaks, with the exception of the tetramers, have a value of about 2.5, the isomer with the lower retention time prevailing.

Table 3

G.C.-M.S. analysis of the low molecular weight oligomers obtained with the catalytic system (Ia).

	M^+/e	Formula	Isomeric ratio *
a_1	170	$C_{12}H_{26}$	$a_1/a_2 = 1.4$
a_2	170	$C_{12}H_{26}$	
b_1	212	$C_{15}H_{32}$	$b_1/b_2 = 2.4$
b_2	212	$C_{15}H_{32}$	
c_1	254	$C_{18}H_{38}$	$c_1/c_2 = 2.6$
c_2	254	$C_{18}H_{38}$	
d_1	296	$C_{21}H_{44}$	$d_1/d_2 = 2.6$
d_2	296	$C_{21}H_{44}$	
e_1	338	$C_{24}H_{50}$	$e_1/e_2 = 2.6$
e_2	338	$C_{24}H_{50}$	
f_1	380	$C_{27}H_{56}$	$f_1/f_2 = 2.5$
f_2	380	$C_{27}H_{56}$	
g_1	422	$C_{30}H_{62}$	n.d.
g_2	422	$C_{30}H_{62}$	

* moles of isomer with lower retention time/moles of isomer with higher retention time

Surprisingly the results obtained fractionating in the same way the hydrooligomers obtained with the catalytic species (Ib) yielded different results. First of all groups of two pair of peaks were detected (Fig. 2). The couples having larger area (e.g. b_1 and b_3, c_1 and c_3 in Fig. 2) were analyzed by MS. The molecular weight indicates that the larger peaks correspond to compounds having the general formula $((C_3H_6)_n +2H)$ that is to hydrooligomers of propylene. Furthermore Mass-Spectroscopic analysis indicated that the two compounds present in smaller amounts have two hydrogen atoms less than the compounds corresponding to the two main peaks. It was concluded that with catalyst (Ib) under the conditions used non hydrogenated oligomers $((C_3H_6)_n)$ were present. The molar ratio between the two paraffines corresponding to the two main peaks of each quartet (Fig. 2) was lower than that found using the catalytic system Ia, (Fig.1) the isomer with higher retention time being prevailing (Table 4).

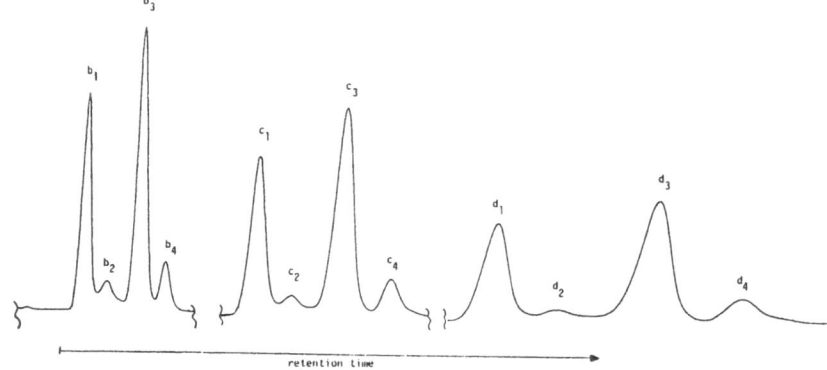

Fig. 2 Gas-Chromatographic analysis of the propylene low molecular weight hydrooligomers soluble in toluene. Column: carbowax 20 M, 4m Temperature 100°-200°C.

Table 4

G.C.-M.S. analysis of the oligomers and hydrooligomers obtained with the catalytic system (Ib)

	M^+/e	Formula	Isomeric ratio *
a_1	212	$C_{15} H_{32}$	
a_2	210	$C_{15} H_{30}$	
a_3	212	$C_{15} H_{32}$	$a_1/a_3 = 0,62$
a_4	210	$C_{15} H_{30}$	
b_1	254	$C_{18} H_{38}$	
b_2	252	$C_{18} H_{36}$	
b_3	252	$C_{18} H_{38}$	$b_1/b_3 = 0,66$
b_4	252	$C_{18} H_{36}$	
c_1	296	$C_{21} H_{44}$	
c_2	294	$C_{21} H_{42}$	
c_3	296	$C_{21} H_{44}$	$c_1/c_3 = 0,62$
c_4	294	$C_{21} H_{42}$	
d_1	338	$C_{24} H_{50}$	
d_2	336	$C_{24} H_{48}$	
d_3	338	$C_{24} H_{50}$	$d_1/d_3 = 0,62$
d_4	336	$C_{24} H_{48}$	
e_1	380	$C_{27} H_{56}$	
e_2	378	$C_{27} H_{54}$	
e_3	380	$C_{27} H_{56}$	$e_1/e_3 = 0,62$
e_4	378	$C_{27} H_{54}$	

* moles of isomer with higher retention time/moles of isomer with lower retention time

Finally a small quantity of the hydrogenated trimers and tetramers obtained with the catalytic system (Ia) were separated from toluene by careful rectification.
Hydro-trimers were analized with G.C.-MS. and their structures were identified by comparison of their NMR ^{13}C and ^{1}H spectra with that of authentic samples.[8] The isomer with M^+/e 128 and lower retention time was identified as 2,4- dimethylheptane (II), the isomer with higher retention time was identified as 4-methyloctane (III). Similarly for the tetramers[8] the isomer with lower retention time was identified as u-[9] 2,4,6-trimethylnonane (IV) and that one with higher retention time was identified as u- 4,6-dimethyldecane (V).

Table 5

NMR spectra of isotactic and syndiotactic polypropylenes as well as of some low molecular weight models

Compound	Chem. Shift of the Signals corresponding to CH_2 groups between two successive tertiary carbon atoms ppm	P/Q	
		Found	Calc.
High molecular weight isotactic polypropylene	1.36 [10)] 0.9 [10)]	4	4
High molecular weight syndiotactic poly-propylene	1.07[10)]	1.5	1.5
(4S: 6R)-2,4,6,8-tetra-methylnonane	1.11 [12),13)] 0.92 [12),13)]	n.d.	7
(4S: 6S) or (4R: 6R)-tetramethylnonane	1.06 [11),13)]	n.d.	3
(4S : 6R or 4R : 6S)-2,4,6-trimethylnonane	1.27 1.05	2.7	2.8

P= Intensity of the signals below 0.98 ppm.; Q= Intensity of the signals between 0.98 and 1.45 ppm.

As shown in table 5 for the 2,4,6-trimethylnonane obtained in the hydrooligomerization the ratio between the intensity of the signals below 0,98 p.p.m. and between 0,98 and 1.45 p.p.m. is much nearer to 2.8, as expected for the u diastereoisomer, than to 1.87 as expected for the l diastereoisomer.
Taking into account that no head to head, tail to tail enchainments are present in the oligomers prepared with the catalyst (Ia) as shown by the substantial absence of absorption at 752 cm^{-1} in the I.R. spectrum and that the same terminal groups (n-propyl, isobutyl and n-butyl) are present also in the higher oligomers, the ^{1}H NMR data (Table 6) indicate that all the oligomers fractions have a highly isotactic structure.

Table 6

^{1}H NMR spectra of the hydrooligomers fractions obtained with the catalytic system (Ia)[a]

Fract. $\bar{M}_n$	Calculated				Experimentally found	
	Isotactic Structure		Syndiotactic Structure			
	P/Q[b]	P/T[b]	P/Q[b]	P/T[b]	P/Q[b]	P/T[b]
C 380	2.92	4.69	1.5	3.78	2.6	4.6
D 490	3.11	4.52	1.5	3.58	2.8	4.5
F 420	2.99	4.62	1.5	3.69	2.9	4.6
G 840	3.42	4.28	1.5	3.32	3.0	4.2
H 965	3.49	4.24	1.5	3.28	3.1	3.9
I 2040	3.74	4.11	1.5	3.13	2.9	3.6

a) Taking for the isomeric composition the value calculated from the gas-chromatographic data for the fractions from $C_{15}H_{32}$ to $C_{27}H_{56}$.
b) For the meaning of P and Q see table 5; T=intensity of the signals above 1.45 p.p.m.

The identification of the hydro-trimers and -tetramers and the quantitative ^{13}C NMR investigation of the end groups in the fraction F (Table 6) indicate that n-propyl, n-butyl and isobutyl groups are the only end groups. Furthermore the concentration of the n-propyl groups corresponds to the sum of the concentrations of the n-butyl and isobutyl groups.
On this basis the general formula VI and VII has been assigned to the hydrooligomers.

Prevailing absolute configuration of the oligomers

As the catalytic systems used have been prepared from chiral non racemic Zirconocene complexes a prevalence of either (VIa) or (VIb) and of either (VIIa) or (VIIb) is expected, that is the oligomers should show optical rotation.
Indeed with the catalytic system (Ia) prepared from (-)(R)(EBTHI)Zr $(CH_3)_2$ hydrooligomers with positive optical rotation have been obtained. On the contrary with the catalyst (Ib) prepared from a (+)(S)(EBTHI)ZrCl$_2$ oligomers with negative optical rotation are formed.
By comparison with the literature data[14],[15] (Table 7) a prevailing (S) configuration has been assigned to the trimers prepared with (Ia) that is they have prevailingly the structure VIa (n=1) and VIIa (n=1). Although the optical rotation measurements for the hydrogenated trimers were not very precise due to the small amount of product available and to the presence of impurities (toluene, hydrogenated tetramers) an optical purity of at least 70% could be estimated for VIa (n=1). The literature values of the trimers as well as of other compounds with similar structure are compared with the experimental values and with the values calculated according to Brewster[16] in Table 7.

Table 7

Optical rotation of the hydrooligomers obtained with the catalytic system (Ia) and of related paraffines.

Compound	$\{[\Phi]_D^{25}\}_{max}$ (literature)	$[\Phi]_D^{25}$ (exp.)	$[\Phi]_D^{20}$ (calculated)[16]
(S)2,4-dimethylheptane	+14.9 [14]	+11.2	+ 10
(S)4-methyloctane	+1.67 [15]	+ 2.9	+ 2
(4R:6S)2,4,6-trimethyl-nonane	+20.7*[8]	+18.4	+ 15
(4S:6R)4,6-dimethyl-decane	+3.5* [8]	+ 2.2	+ 3

* at 20°C

As usual in the case of optically active paraffines Brewster calculations predict correctly not only the sign of the optical rotation but also, with good approximation, the value of the optical rotation.[16]
For the mixture of higher hydrooligomers, the values of the molar optical rotation can be estimated according to Brewster for (VI) and (VII) taking as molecular weight the number average molecular weight of the fractions. In Table 8 the values calculated according to Brewster for mixtures containing 71% of (VI) (see Table 3) are given for the oligomers produced with catalytic system (Ia). For the oligomers produced with the catalyst (Ib), the values of the optical rotation have been calculated assuming 38% of (VI) (see Table 4) and neglecting the contribution to the rotation given by the olefinic compounds.

Table 8

Optical rotation of the mixture of hydrogenated oligomers obtained with the catalytic systems (Ia) and (Ib)

Oligomer Fraction	$\bar{M}_n$	Catalytic System	Exp. Value $[\Phi]_D^{25}$	Calculated[16] Value $[\Phi]_D^{20}$	% of VI	Prevailing absol.conf.[a]
B	310	Ia	+15.9	+16.67	72	(S)
C	380	Ia	+14.63	+18.00	71	(S)
D	490	Ia	+14.52	+19.20	71	(S)
E	670	Ia	+13.64	+19.74	71	(S)
C'	445	Ib	-10.07	-12.37	38	(R)
D'	585	Ib	- 9.16	-12.96	38	(R)
E'	900	Ib	- 7.24	-13.68	38	(R)

a) of the first asymmetric carbon atom following the n-propyl end group.

The sign of the rotation experimentally measured leaves no doubts about the prevailing absolute configuration of the asymmetric carbon atoms in the isotactic hydrooligomers obtained with the catalysts (Ia) and (Ib) under the only assumption that no configurational inversion is present in the oligomers chain. The configuration of the asymmetric carbon atom following the n-propyl group is (S) in the case of (Ia) and (R) in the case of (Ib). The lower absolute value of the optical rotation obtained with the catalyst (Ib) has been attributed, tentatively, to the larger percentage of (VII) present in the mixtures (as shown by the Gas-Chromatography (Fig. 2) of the lower oligomers and by ^{13}C NMR) in comparison to the mixtures obtained with the catalytic system (Ia) (Fig.1). However this low value of the optical rotation could be also attributed at least in part to the presence in the products prepared with the catalyst (Ib) of significant amounts of olefins and to a possible lower asymmetric induction connected with the use of the catalytic system (Ib) instead of (Ia).

Concerning the optical purity of the hydrogenated oligomers as no experimental values for higher oligomers are available, only an order of magnitude can be estimated from Table 10. In fact there is no doubt that for higher molecular weigth paraffines, the approximations introduced in the conformational analysis according to Brewster,[16] could cause remarkable discrepancies between experimental and calculated values.

Discussion

The hydrooligomerization of propylene has shown the possibility to synthesize the two series of paraffines (VI) and (VII) with a moderate optical purity. Particularly when no significant amount of olefins is present, as in the case of the catalyst (Ia), the single members of the series can be easily separated by Gas-Chromatography.

The good agreement between the maximum optical rotation of VIa(n=1) and VIIa(n=1) already known in the literature and the value calculated according to Brewster, indicates that for both the series VI and VII Brewster calculation can be safely used to evaluate sign and order of magnitude of maximum value of optical rotation of the hydrooligomers fractions.

The substantially constant order of magnitude of molar optical rotation in solution or in the melt, considering the molar optical rotation calculated according to Brewster, indicates that the molar ratio between the isomers (VIa) and (VIIa) remains constant independently from the molecular weight and corresponds to the value obtained from the G.C. of the mixture of low molecular weight isomers.

Under the above assumption it appears that, as expected,[4] the length of the isotactic chain does not influence the molar optical rotation of the hydrooligomers which should be experimentally measurable at the D line up to a number average molecular weight of about 3000.

More interesting are the indications arising from the propylene hydrooligomerization on mechanism of stereospecific polymerization (Table 9) which have been already discussed elsewhere.[5],[17]

Table 9

Catalyst Precursor	(−)(R)(EBTHI) Zr(CH$_3$)$_2$	(+)(S)(EBTHI) ZrCl$_2$
(iso-C$_4$H$_9$)/n-C$_4$H$_9$	2.5 [a]	0.62 [a]
Optical rotation Sign	(+)	(−)
Prevailing configuration of the asymmetric carbon atom following the n-propyl group in the hydrooligomers chain	(S)	(R)

(a) from Gas-Chromatographic analysis of the low hydrooligomers.

The asymmetric hydrooligomerization seems to proceed with a "like selectivity"[18] (lk). In view of the very high regio- and stereo-specificity of the process and of the high optical purity of the catalyst precursors, which however are not chemically pure, a higher enantiomeric excess would be expected.

Possible origins of the relatively low optical purity could be a racemization of the Zirconium complex during the interaction with the aluminoxanes and/or the presence of two conformers (λ and δ) in the zirconium containing catalyst precursors.[19] These factors are at the present under investigation.

The remarkable difference in the ratio between (VI) and (VII) using as catalyst (Ia) or (Ib) was unexpected. In fact if the catalytic system corresponds to a tight ion pair[20], as shown in the scheme 2, and if regio- and stereoselectivity is controlled by the zirconium moiety similar results should be espected.

Scheme 2

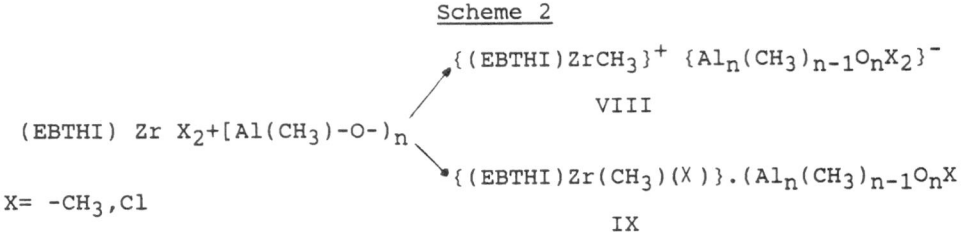

$$\{(EBTHI)ZrCH_3\}^+ \ \{Al_n(CH_3)_{n-1}O_nX_2\}^-$$

VIII

$$(EBTHI)\ Zr\ X_2 + [Al(CH_3)-O-]_n$$

$$\{(EBTHI)Zr(CH_3)(X)\}.(Al_n(CH_3)_{n-1}O_nX)$$

$$X = -CH_3, Cl$$

IX

A catalytic system containing the Zr moiety IX having different structure depending on the nature of X does not seem probable because of the severe steric crowding around the Zr atom which would render very difficult the approach of the monomer to the growing chain. At the present we think that regioselectivity and stereospecificity are controlled by very peculiar steric features of the ion couple including VIII. In any case our remarks are restricted to the reaction conditions used and changes in monomer concentration, temperature, purity of the catalyst's precursors can certainly give further interesting indications on the propylene polymerization with this type of catalytic systems.

REFERENCES

1) Kaminsky W.,Külper K., Brintzinger H.H., Wild F.R.W.P.(1985) Angew. Chem. Int. Ed. Engl. 24:507.
2) Ewen J.A.(1984) J. Am. Chem. Soc. 106:6355.
3) Wild R.W.P.F., Wasiucionek M., Huttner G., Brintzinger H.H. (1985) J. Organometallic Chem. 288:63.
4) Pino P., (1966) Adv. Polym.Sci 4; 393.
5) Pino P., Cioni P., Wei J. (1987) J. Am. Chem. Soc. 109:6189.
6) Kaminsky W., Lüker H. (1984) Makromol. Chem. Rapid, Commun. 5:225.
7) Natta G., Pino P., Mazzanti G. (1957) Gazz. Chim. Ital. 87:528.
8) Hydrotrimers and Hydrotetramers were prepared in our laboratory by P.Prada. Experimental details will be published elswhere.
9) Prelog V., Helmchen G., (1982) Angew. Chem. 94: 614.
10)Ferguson R.C. (1967) Trans. N.Y. Acad. Sci. 29: 495.
11)Pucci S, Pino P., Strino E. (1968) Gazz. Chim. Ital. 98:421.
12)Pino P., Pucci S., Benedetti E., Bucci P. (1965) J. Am. Chem. Soc. 87:3263.
13)Suter U. (1973) Diss. 5133 ETH-Zürich.
14)Levene P.A., Marker R.E. (1931) J. Biol. Chem. 92:455.
15)Levene P.A., Marker R.E. (1931) J. Biol. Chem. 91:761.
16)Brewster J.H. (1959) J. Am. Chem. Soc. 81:5475.

17) Pino P., paper presented at the Rolduc Meeting 2, Rolduc Abbey, Limburg, The Netherlands (1987).
18) Seebach D., Prelog V. (1982) Angew. Chem. Int. Ed. Engl. 21:654.
19) Schäfer A., Karl E., Zsolnai L., Huttner G., Brintzinger H.H. (1987) J. Organomet. Chem., 328:87.
20) Eisch J.J., Pitrowski A.M., Brownstein S.K., Gabe E.G., Lee F.L. (1985) J. Am. Chem. Soc. 107:7219.

Propylene Polymerizations with Group 4 Metallocene/Alumoxane Systems.

John A. Ewen, Luc Haspeslagh, M. J. Elder, Jerry L. Atwood[*], Hongming Zhang[*] and H. N. Cheng[±]

Fina Oil and Chemical Company, Box 1200, Deer Park, Texas 77536
[*]Department of Chemistry, University of Alabama, Tuscaloosa, AL 35487
[±]Research Center, Hercules Incorporated, Wilmington, Delaware 19894

INTRODUCTION

The existing Ti and Zr metallocene/methylalumoxane (MAO) isospecific propylene polymerization catalyst systems are unsuitable at conventional polymerization temperatures. The most significant problems have been that Ti(IV) complexes are reduced to inactive species (1,2) and that the Zr analogues produce wax (3-7).

In this contribution we summarize three aspects of the group 4 chiral metallocene/alumoxane system that influence the polymerization results: The C-5 ring substituents (CpR), the bridge linking the C-5 ring and the transition metal (6-8).

Isospecific propylene polymerizations catalyzed by metallocenes have been extended to include hafnium (6,7). The important finding is that the polypropylene molecular weights obtained with rac-ethylenebis(1-indenyl)Hf(IV) dichloride and rac-ethylenebis(4,5,6,7-tetrahydro-1-indeny)Hf(IV) dichloride are an order of magnitude higher than those produced by the Zr analogues.

The CpR groups and the indenyl ligand bridges influence the molecular weights, polymerization rates and catalyst stereospecificities significantly. The ligand steric effects are more pronounced for the smaller transition elements (Ti > Hf > Zr).

^{13}C-NMR spectroscopy with pentad resolution has been used to measure the stereo- and regio-specificities of three zirconocenes (9). The steric imperfections in the isotactic fraction and the atactic polymer are proportional to the polymerization temperature and can be profoundly influenced by Cp ligand effects.

EXPERIMENTAL

Synthetic procedures and characterization of the metallocenes, polymer molecular weight determination, ^{13}C NMR characterization and DSC procedures have been described (6-8). The metallocene dichlorides were recrystallized from dry toluene no longer than three weeks prior to a polymerization and the Hf metallocenes were synthesized from 99.99% HfCl$_4$ purchased from Research Organic Research Inorganic Chemical Company (Roc Ric). The cocatalysts were prepared by one of the following procedures. A: Methylalumoxane (MAO; FW = 1100) toluene solution (0.56 M in Al) from 24 h hydrolysis at 25°C of 400 ml of 2.0 M trimethylaluminum (TMA) with 45.6 g of Al$_2$SO$_4$.16H$_2$O (DE 3 240 383). B: Solid MAO obtained by vacuum evaporation of residual TMA and the solvent from A at 25°C into a cold

W. Kaminsky and H. Sinn (Eds.)
Transition Metals and Organometallics as
Catalysts for Olefin Polymerization
© Springer-Verlag Berlin Heidelberg 1988

trap containing n-butanol. C: TMA/MAO solutions obtained by $CuSO_4 \cdot 5H_2O$ hydrolysis of TMA (1). Cocatalyst C and bulk polymerizations were employed to prepare the samples analyzed by ^{13}C-NMR spectroscopy. The metallocenes are more sensitive to water than the most highly active heterogeneous systems. Toluene was distilled under nitrogen from sodium/benzophenone. Polymerization grade propylene was dried to < 1 ppm water with 3A molecular sieves, deoxygenated with BASF R3-11 catalyst and freed from COS and arsine with 40 lbs of Calsicat E-315 (PbO). The polymerization procedures were as follows: (I) 35 ml of toluene containing the metallocene and MAO were added to a Magnedrive Packless Zipperclave already containing propylene at the reaction temperature. (II) I, with toluene added to the reactor prior to the propylene. The zirconium produced polymer samples were recrystallized from 1 wt-% xylene solutions for DSC analyses.

RESULTS

Polymerizations and Polymer Molecular Weights. The polymerization conditions developed for hafnium are displayed in Table 1.

Table 1. Polymerization conditions and yields for the chiral hafnium complexes.

Complex (mg)	MAO (g)	Pol. T, °C	Yield, g	$10^{-3} \cdot \overline{M}_w$	$\overline{M}_w/\overline{M}_n$
Et[Ind]$_2$HfCl$_2$ (0.17)[a]	B (0.28)	50	9	>724	2.2
Et[Ind]$_2$HfCl$_2$ (1.7)[b]	B (0.28)	50	76	361	2.4
Et[Ind]$_2$HfCl$_2$ (3.4)[c]	B (0.56)	50	230	304	2.3
Et[IndH$_4$]$_2$HfCl$_2$ (1.48)[d]	A (0.38)	80	95	42	2.4
Et[IndH$_4$]$_2$HfCl$_2$ (1.76)[e]	B (0.28)	70	25	76	2.3
Et[IndH$_4$]$_2$HfCl$_2$ (1.41)[f]	B (0.28)	50	25	150	2.2

Polymerization procedure and (propylene/toluene) in liters:
[a] II (0.66/0.5) [b] II (1/0.5) [c] II (2/1) [d] I (1.2/0) [e] II (2/1)
[f] II (0.63/0.5)

Molecular weight distributions close to 2 have been obtained with complexes synthesized from 99.99% pure HfCl$_4$. 3 liters of a swollen mass containing as little as 25 g of polymer is formed in polymerizations containing 1 liter of toluene. The dry samples are compressed sheets resembling plaster-of-paris for yields of 200-400 g in a 4 liter reactor. A fine powder is obtained when the molecular weights are decreased ten fold with hydrogen. The cross-links are consistent with trapped entanglements due to the long chain lengths.

The examples in Table 1 imply an inverse dependence of molecular weight on the Et[Ind]$_2$HfCl$_2$ concentration. A smaller but proportional dependency on monomer has been observed with this particular species.

The tabulated rates and molecular weights are higher for $Et[Ind]_2HfCl_2$ than for $Et[IndH_4]_2HfCl_2$; as they are for zirconium (3-5). This ligand effect is consistent with the steric environments implied from the hafnium metallocene crystal structures exhibited in Fig. 1 (6).

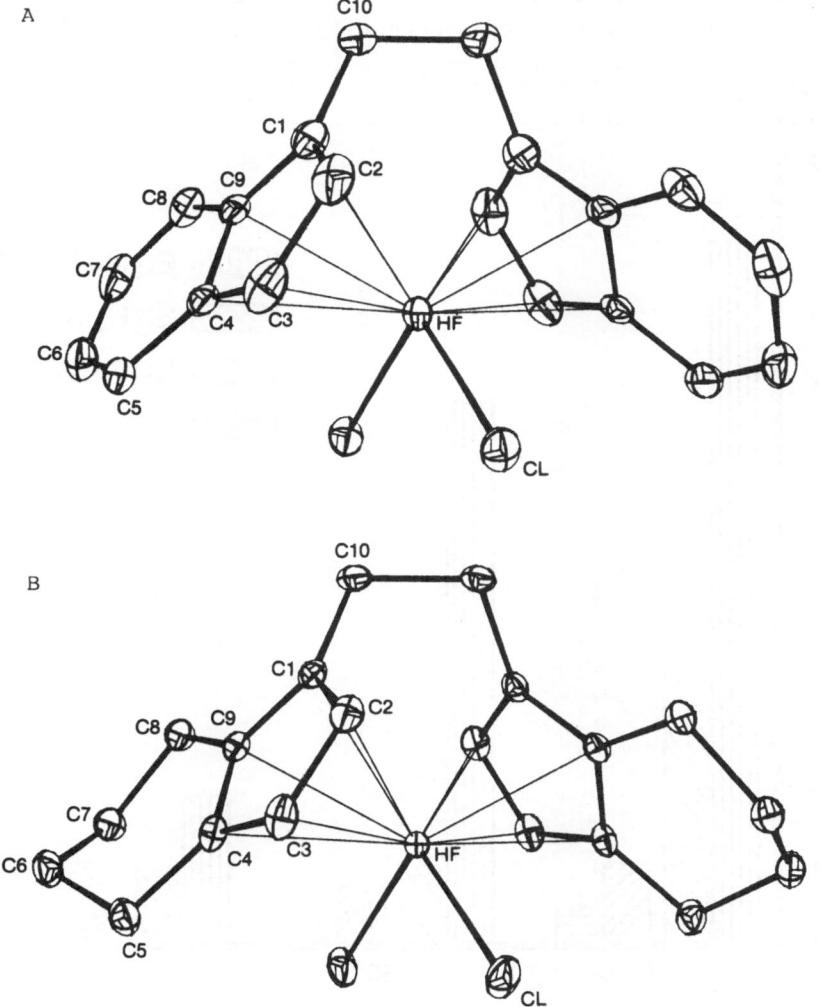

Fig. 1. Molecular structures and atom numbering schemes of the 1-R-enantiomers of (a) rac-$Et[Ind]_2HfCl_2$ and (b) rac-$Et[IndH_4]_2HfCl_2$.

The indenyl derivative's Hf-ring centroid distance is 0.02 angstroms
longer and the unit cell is 82 cubic angstroms smaller than in the case of
the more basic and bulkier tetrahydroindenyl ligand. The rac-
Et[IndH$_4$]$_2$HfCl$_2$ Hf-ring centroid distance and the Hf-Cl length are 0.02
and 0.04 angstroms shorter, respectively, than the corresponding distances
in rac-Et[IndH$_4$]$_2$ZrCl$_2$ due to the lanthanide contraction (6).

The steric requirements of Et[IndH$_4$]$_2$MCl$_2$ compared to Et[Ind]$_2$MCl$_2$ is more
pronounced for the smaller Ti analogues (9). Milligrams of Et[Ind]$_2$TiCl$_2$
have provided gram quantities of polypropylene at -60°C whereas
Et[IndH$_4$]$_2$TiCl$_2$ is almost as inactive as Et[IndH$_4$]$_2$MCl$_2$ and Et[Ind]$_2$MCl$_2$
(M = Zr, Hf) are at this temperature (1,2).

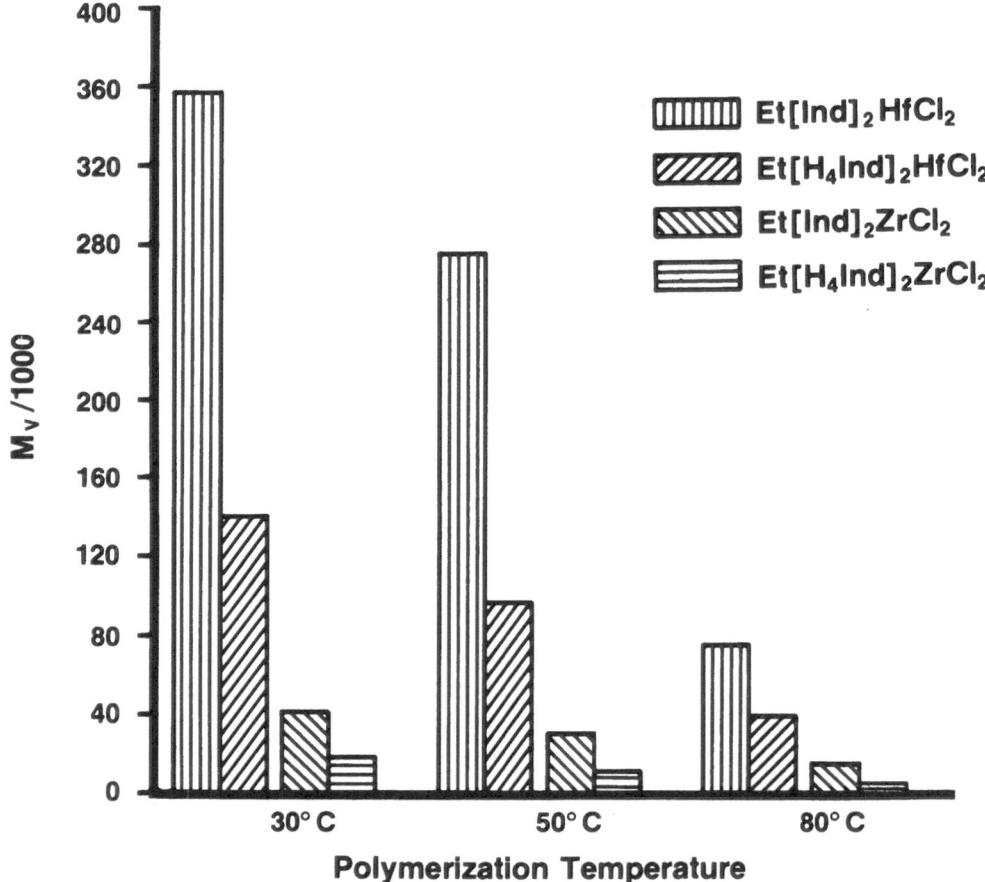

Fig. 2. Typical temperature dependencies of polypropylene molecular
weights with Hf and Zr.

Table 2. Optimum polymerization conditions and yields for zirconium complexes.

Complex (mg)	MAO (g)	Pol. T, °C	Yield, g	$10^{-3} \cdot \overline{M}_v$
rac-Et[3-MeInd]$_2$ZrCl$_2$ (1.42)[a]	B (0.28)	50	33	31
rac-Me$_2$Si[Ind]$_2$ZrCl$_2$ (1.40)[a]	B (0.28)	50	164	52
rac-Et[Ind]$_2$ZrCl$_2$ (1.39)[b]	B (0.28)	80	355	14
rac-Et[IndH$_4$]$_2$ZrCl$_2$ (1.39)[b]	A (0.38)	80	469	7

Polymerization procedure and (propylene/toluene) in liters: [a]II (1/0.5) [b]I (1.2/0)

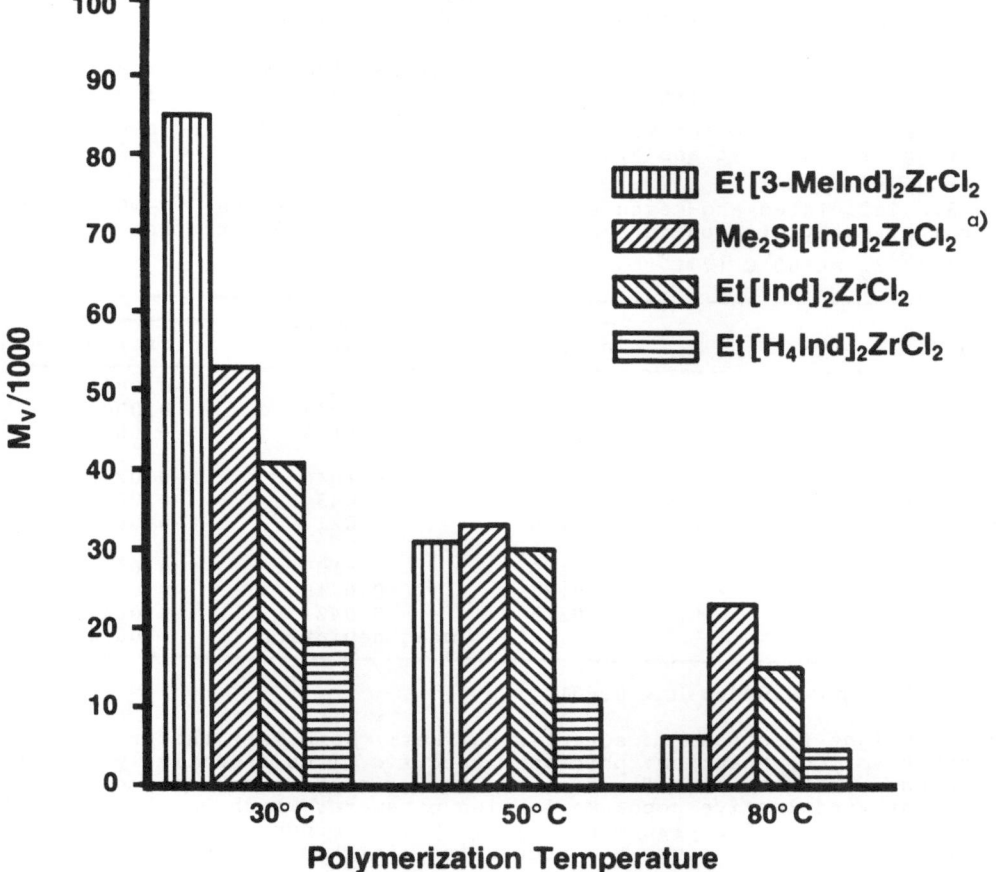

Fig. 3. Polypropylene molecular weight temperature dependencies with Zr. [a]Using MAO C in bulk; higher MWs were obtained with MAO B (Table 2).

The conditions at which the most favorable yields have been obtained with four zirconium complexes are indicated in Table 2. Rows 2 and 4 of both Table 1 (Hf) and Table 2 (Zr) imply faster Zr insertion rates. The Zr bulk polymerization rates are an order of magnitude higher than in earlier slurry experiments (3). The higher activation energies in bulk are consistent with a low solubility of MAO in propylene much below $70^{\circ}C$ (7).

The influence of the ligands on polypropylene molecular weights are shown in Fig. 3. The molecular weight-temperature profiles differ significantly with more severe alterations in the structures of the cyclopentadienyl ligands. The degree of polymerization obtained with rac-Et[3-MeInd]$_2$ZrCl$_2$ at $30^{\circ}C$ is surprisingly high for a zirconium complex. The sharper polymerization temperature dependence of the molecular weights for rac-Et[3-MeInd]$_2$ZrCl$_2$ is in accord unusually large differences between both the enthalpies and entropies of activation for insertion vs. termination.

Catalyst Stereospecificities. The chiral zirconocene catalysts' stereo-and regio-specificities and the composition of the polypropylenes produced by them at different polymerization temperatures were calculated from ^{13}C-NMR spectra of the methyl pentad region of the 'as-polymerized' and xylene insoluble samples. The polymers were produced in bulk polymerizations to avoid toluene fractionations and sample inhomogeneities (7).

Table 3 lists the pentad band intensities of the ^{13}C-NMR spectrum for polypropylene obtained with rac-Et[Ind]$_2$ZrCl$_2$ at $80^{\circ}C$. The calculated intensities were estimated with Doi's two parameter model for a mixture consisting of isotactic and atactic polymer (10).

Table 3. Calculated and measured band intensities for the ^{13}C-NMR spectrum of methyl region for polymer mixture obtained with rac-Et[Ind]$_2$ZrCl$_2$ at $80^{\circ}C$ (7).[a]

pentad	measured	band intensities calculated	difference
mmmm	0.6807	0.6810	−0.0003
mmmr	0.1049	0.1047	0.0002
rmmr	0.0175	0.0108	0.0067
mmrr	0.0803	0.1046	−0.0243
mrmm + rmrr	0.0404	0.0434	−0.0030
mrmr	0.0140	0.0217	−0.0077
rrrr	0.0110	0.0107	0.0003
rrrm	0.0136	0.0216	−0.0080
mrrm	0.0375	0.0421	−0.0046
		mean deviation	0.0061

[a] $w = 0.8536$; $\beta = 0.0446$; $\sigma = 0.5014$

w is the weight fraction of isotactic polymer; β and σ are the probability parameters for the enantiomorphic and aspecific sites rspectively. Kissin has proposed that the mole fraction of steric inversions be calculated from the stereoselective copolymerization parameter which can, in turn, be calculated from the %-m placements (11). This procedure takes into account that NMR spectra count two successive mistakes as stereoregular placements. However, these double catalyst errors are negligible for practical purposes in highly isotactic polymers and β = rr = r/2 is approximately equivalent to the mole fraction of steric inversions.

95.5 % of the methine carbon atoms of the isotactic fraction have the same relative configuration. The meso placements are connected by single units of the opposite handedness. The ...mmmmmmrrmmmmmm... microstructure of the polymers obtained with rac-Et[Ind]$_2$MCl$_2$ (M = T, Zr, Hf) is a consequence of the enantiomorphic site control mechanism (1,12).

14.6 wt-% of the sample obtained with rac-Et[Ind]$_2$ZrCl$_2$ is atactic polymer produced by an achiral species resulting from catalyst thermal instabilty, reactions with impurities or MAO and/or (less likely) isomerization to the meso stereoisomer. The fraction of atactic polymer decreases with decreasing polymerization temperature. The same trends were observed in the ^{13}C-NMR spectra for rac-Et[IndH$_4$]$_2$ZrCl$_2$ and rac-Me$_2$Si[Ind]$_2$ZrCl$_2$ between 20-80°C. The lowest amount of atactic polymer found in these experiments was 5.9 wt-% along with 1.19 mole-% steric inversions in the isotactic fraction with more thermally stable rac-Et[IndH$_4$]$_2$ZrCl$_2$at 20°C. The polymer compositions obtained with rac-Et[Ind]$_2$ZrCl$_2$ and rac-Et[IndH$_4$]$_2$ZrCl$_2$ are therefore, suprisingly (3), closely analogous to those made with a mixture of meso and rac enantiomers of Et[Ind]$_2$TiCl$_2$ (1).

It is also noted that the 'as-polymerized' samples contain about 4% regio-irregularities which are mostly stereoregular (meso). It is concluded that most of the head-to-head and tail-to-tail enchainments result from "2-1" insertions with the chiral species. Regio-irregularites are unimportant in low temperature chain-end controlled Ti catalyzed polymerizations (1) and for heterogeneous catalysts (13).

Table 4 summarizes the ^{13}C-NMR determined mole fractions of inverted configurations in the isotactic fractions of polymers obtained with metallocene catalysts. The steric imperfections increase with increasing polymerization temperature and are influenced by the ligands attached to the transition metals. The ligand steric effect on stereo-specificity is greatest for the smaller and least stereospecific Ti complexes (1,2).

Table 4. ^{13}C-NMR determined stereo-irregularities of the isotactic fractions and DSC melting points of the 'as-polymerized' samples of polypropylenes obtained metallocenes (7).

Complex	Pol. T, °C	m. pt., °C	Inversions, mole-%
rac-Et[Ind]$_2$TiCl$_2$	-60	94	11.01
rac-Et[IndH$_4$]$_2$TiCl$_2$	0	99	9.38
rac-Et[Ind]$_2$ZrCl$_2$	80	126	4.68
rac-Me$_2$Si[Ind]$_2$ZrCl$_2$	80	134	3.13
rac-Et[Ind]$_2$ZrCl$_2$	50	135	2.57
rac-Me$_2$Si[Ind]$_2$ZrCl$_2$	50	144	1.55
rac-Et[IndH$_4$]$_2$ZrCl$_2$	20	147	1.18

The data Table 3 yield an empirical linear correlation between the mole fractions of steric inversions in the isotactic fraction and the DSC melting points for the low melting polypropylenes:

$$\text{Mole fraction of inversions} = -0.500[\ \log_{10}(T_m)\ ]\ +\ 1.09 \tag{1}$$

where T_m is in degrees centigrade. The DSC melting points are very sharp, well defined and clearly correlate with structural defects in the polymer chains. The exponential relationship results in the melting points being very sensitive to small differences in the imperfections of highly isotactic materials.

The DSC melting points for polymers produced at 50°C with differing ligands and transition metals are listed in Table 5. The conclusions are the same with DSC data on higher molecular weight samples formed at lower polymerization temperatures.

Table 5. DSC analyses of polypropylenes obtained with zirconium and hafnium metallocenes at 50°C.

Complex	m. pt., $^{\circ}$C	Complex	m. pt., $^{\circ}$C
rac-Me$_2$Si[Ind]$_2$ZrCl$_2$	144	rac-Et[3-MeInd]$_2$ZrCl$_2$	none
rac-Et[IndH$_4$]$_2$HfCl$_2$	141	rac-Et[IndH$_4$]$_2$ZrCl$_2$	137
rac-Et[Ind]$_2$HfCl$_2$	136	rac-Et[Ind]$_2$ZrCl$_2$	135

The stereorigid dimethylsilyl bridged species with only 1 atom hinging the two indene ligands is more stereospecific than in the case of the more flexible ethylene bridged indenyl analogue (2 independent hinges). The almost complete absence of stereospecificity with rac-Et[3-MeInd]$_2$ZrCl$_2$ (28 mole-% inversions at 50°C) is in accord with opposing steric effects from the C5 methyl substituents and from the C6 ring attached to the other Cp ligand (7).

The magnitude of the ligand steric effect on stereo-regulation for the indenyl vs. the tetrahydroindenyl ligands is larger for the smaller Hf complexes. The smaller Hf molecules are slightly more stereospecific than the Zr analogues. The stereospecificities of these 4 complexes increase with the order of decreasing M-C5 ring centroid distances.

CONCLUSIONS

Ligand and transition metal steric and electronic effects exert a strong influence on the propylene polymerizations. The emerging trends in reactivity and stereospecifcity can be qualitatively rationalized to some extent on the basis of differences in the chemistries of the transition metals, the metallocene crystal structures and the basicities of the ligands.

Trends in stereospecificities can be understood in terms of differences in the crystal structures of the complexes and ligand stereo-rigidities. The magnitude of the ligand steric effects on polymerization rates and the polymer stereoregularity generally decrease in the order of increasing transition metal size: Ti >> Hf > Zr. There are significant ligand steric and electronic effects on polypropylene molecular weights. In the absence of overbearing ligand steric constraints the insertion rates increase (Ti >> Zr > Hf) according to decreasing M-C σ-bond strengths (14).

The reasons for the much higher ratio of k_p/k_t for Hf relative to Zr have not been sorted out. The same order of magnitude for the polymerization rates with the two transition metals indicate that this remarkable difference is predominantly due to a smaller k_t for Hf. Suggested relevant, fundamental differences between Hf and Zr include the M-C σ-bond strengths, the Lewis acidities of the transition metals and their ionic radii (14).

REFERENCES

1. Ewen JA (1984) J Am Chem Soc 106:6355-6364
2. Ewen JA (1986) Ligand Effects on Metallocene Catalyzed Ziegler-Natta Polymerizations. In: Keii T and Soga K (eds) Catalytic Polymerization of Olefins, Elsevier, Amsterdam Oxford New York Tokyo, pp 271-292
3. Kaminsky W, Kulper K, Brintzinger HH, Wild FW (1985) Angew Chem 97:507-508
4. Kaminsky W, Kulper K, Niedoba S (1986) Makromol Chem 187:378-387
5. Kaminsky W (1986) Preparation of Special Polyolefins from Soluble Zirconium Compounds with Aluminoxane as Cocatalyst. In: Keii T and Soga K (eds) Catalytic Polymerization of Olefins, Elsevier, Amsterdam Oxford New York Tokyo, pp 293-304
6. Ewen JA, Haspeslagh L, Atwood JL, Zhang H (1987) J Am Chem Soc (in press)
7. Ewen JA, Haspeslagh L, Elder MJ, Cheng HN (1987) Macromolecules (in press)
8. Ewen JA, Haspeslagh L, Atwood JL, Zhang HM (submitted)
9. Wild FRWP, Zsolnai L, Huttner G, Brintzinger HH (1982) J Organomet Chem 232:233-247
10. Zhu SN, Yang XZ, Chujo R, Doi Y (1983) Polymer J 15:859-868
11. Kissin YV (1985) Isospecific Polymerizations of Olefins. Springer-Verlag, New York Berlin Heidelberg Tokyo p 259
12. Sheldon RA, Fueno T, Tsuntsugu T, Furukawa J (1965) J Polym Sci Part B 3:23.
13. Doi Y, Suzuki E, Keii T (1981) Makromol Chem Rapid Commun 2:293-297
14. Cardin DJ, Lappert MF, Raston CL (1986) Chemistry of Organo-Zirconium and -Hafnium Compounds. John Wiley and Sons, New York p 16

Isotactic Polymerization of Olefins with Homogeneous Zirconium Catalysts

W. Kaminsky, A. Bark, R. Spiehl, N. Möller-Lindenhof, S. Niedoba

Institute for Technical and Macromolecular Chemistry, University of Hamburg, Bundesstr. 45, D-2000 Hamburg 13

INTRODUCTION

Titanocene, zirconocene and hafnocene, for example biscyclopenta-dienyl- and bisindenyl transition metal complexes form together with alumoxane highly active Ziegler-Natta catalysts (1-3). A key substance responsible for the high activity is methylalumoxane. Whereas zirconocenes in combination with aluminumtrialkyls give only minor polymerization activities, exchange of the cocatalyst leads to activities greater by a factor of 100,000. In the case of ethylene polymerization with the system Cp_2ZrCl_2 and methylalumoxane values of 40,000,000 gPE/g $Zr \cdot h$ can be obtained (Tab. 1).

Table 1 Polymerization Activity of the Bis(cyclopentadienyl)zirco-niumdichloride/methylalumoxane Catalyst Applied to Ethylene in 330 ml Toluene

Activity (95°C, 8 bar)	$= 39,8 \cdot 10^6$ g PE/g $Zr \cdot h$
[Zirconocene]	$= 6,2 \cdot 10^{-8}$ mol/l
[Alumoxane] (M = 1200)	$= 7,1 \cdot 10^{-4}$ mol/l
molecular weight of the obtained polyethylene	78 000
degree of polymerization	2 800
macromolecules per Zr atom in h	46 000
rate of growth of one macromolecule	0,087 s
turnover time	$3,1 \cdot 10^{-5}$ s

As shown by Tait (4), every zirconium atom forms under this condition an active complex and produces about 20 000 polymer chains in one hour. The turnover time of $2 \cdot 10^{-5}$ sec is of the same value as that of many enzymes.

Table 2 gives an overview about the efficiency of the zirconocene/alumoxane catalyst.

The analogous titanium and hafnium compounds form active catalysts too. Especially at temperatures above 50 °C the zirconium catalyst is more stable and active than the titanium or hafnium systems. Of the cocatalysts, methylalumoxane is much more effective than the ethyl-alumoxane or isobutylalumoxane (9).

W. Kaminsky and H. Sinn (Eds.)
Transition Metals and Organometallics as
Catalysts for Olefin Polymerization
© Springer-Verlag Berlin Heidelberg 1988

<u>Table 2</u> Efficiency of the Zirconocene/Alumoxane Catalyst

	Literature:
1. Ethylene - Homopolymerization	
high activity ($40 \cdot 10^{6}$ g PE/g Zr·h	
M_w/M_n = 2, highly linear	3
2. Ethylene - Copolymerization	
random distribution, LLDPE	
comonomers: propene, butene, hexene,	
diolefins	5
3. EPDM - Elastomers	
low transition metal concentration in the	
polymer	6
4. Propene Polymerization to Atactic Polymers	7
5. Propene Polymerization to Highly Isotactic	
Polymers	
high activity	7
6. Propene Polymerization to Stereoblock Polymers	8
7. Cyclopentene Polymerization to Isotactic Polymers	*
8. Copolymerization of Long Chained α-Olefins	
(Iso-butene, 4-Methylpentene)	*
9. Oligomerization to Optically Active Hydrocarbons	*
10. Polymerization in the Presence of Filling	
Materials	3

    * = this paper

The molecular weight of the polyethylene can be varied over a wide range between 10 000 and 2 000 000 by changing both the polymerization temperature between 20 and 100°C and the zirconium concentration. In addition, the molecular weight can be influenced by addition of hydrogen (10). In contrast to most heterogeneous catalysts, only traces of hydrogen were needed to lower the molecular weight of the polymer.

The catalyst can easily produce co- and terpolymers. The density of the polymer can be lowered by using hexene-1 or butene-1 as an additional monomer in copolymerizations. Incorporation of 1 mol-% hexene-1 yields a polymer which, when in the molten state, has a density lower than 0,95 g/cm^3. With 5 mol-% incorporated the density is 0,92 g/cm^3 and decreases to 0.89 g/cm^3 for 9 mol-%, so that LLDPE can be obtained.

Using dienes in addition to propene and ethylene leads to EPDM-elastomers (6).

Since alumoxane is rather stable towards carbon hydrates like starch and cellulose the homogeneous catalysts can be equally distributed on the surface. Under these conditions ethylene addition leads to natural substances equally surrounded by polyolefins.

ISOTACTIC POLYMERIZATION OF PROPENE

Bis(cyclopentadienyl)zirconium or titanium IV-compounds produce solely atactic polypropylene in polymerizations of propene. It can be freely chosen to work either in solution or in liquid monomer. At low temperatures mean molecular weights ($M\eta$) of the atactic polypropylenes reach values of more than 500 000.

On the other hand it is possible to make highly isotactic polypropylene with stereorigid chiral zirconocenes (11,13).

By use of the racemic mixture of ethylene bis(indenyl)zirconiumdichloride or ethylene bis(tetrahydroindenyl)zirconiumdichloride together with the cocatalyst methylalumoxane in polymerizations of propene highly isotactic polypropylenes can be produced (Table 3).

Table 3   Polymerization of 70 ml Propene with $7 \cdot 10^{-7}$ mol/lit rac-Et(Ind)$_2$ZrCl$_2$ (I) or $8,4 \cdot 10^{-6}$ mol/lit rac-Et(Ind H$_4$)$_2$ZrCl$_2$ (II) and $1,6 \cdot 10^{-2}$ mol Al/l methylalumoxane in 330 ml toluene

Monomer	Catalyst	T/°C	Time/ min.	Yield/ g	Activity/ kgPP/mol Zr·h
Propene	I	15	620	3,6	1 500
Propene	I	21	500	31	16 000
Propene	I	35	120	26,6	43 000
Propene	II	-10	270	4,5	300
Propene	II	15	170	26,7	2 900
Propene	II	20	120	31,3	4 750
Propene	II	60	90	38,7	7 700

The analogous titanium compounds lead to polypropylene of lower isotacticity (12). Molecular mass distribution M_w/M_n has typical values between 1.9 and 2.6 which are extraordinarily narrow. NMR-spectroscopic investigations show that the isotacticity index is in the range of 99%.

COPOLYMERIZATION OF LONG-CHAIN α-OLEFINES

The copolymerization of 4-methyl-1-pentene can be achieved with the chiral catalyst Et(Ind)$_2$ZrCl$_2$ as well as with Cp$_2$ZrCl$_2$ (Table 4).

Herewith the activities display to be independent of the given molar ratio of 4-methyl-1-pentene and ethylene. They are in the range of approximately 3 000 g copolymer/mol zr·h. The maximum of the polymerization rates decrease with increasing comonomer concentration for both catalyst systems. The homopolymerization of ethylene with the chiral catalyst happens to be of a lower activity by a factor of

Table 4 Copolymerization of 4-methylpentene-1 (M) and ethylene (E)
with two different zirconocenes and methylalumoxane
[Zr]: $2 \cdot 10^{-6}$ mol/l, [Al]: $2,1 \cdot 20^{-2}$ mol/l, ethene pressure:
2 bar, solvent toluene, volume: 250 ml, temperature: 30°C

Transition Metal Compound	M/E-Ratio in Polymerization Solution	Activity g polymer/ mol Zr·s	Mol-% M in the Polymer	Mη
Cp_2ZrCl_2	0	2 809		490 000
	3,45	3 367	2,8	190 000
	6,65	3 282	3,5	160 000
	10,56	1 960	5,2	160 000
$Et(Ind)_2ZrCl_2$	0	940		114 000
	3,45	3 546	6,1	41 000
	6,65	3 355	10,9	26 000
	10,56	3 356	14,6	20 000

three compared with the activity of the copolymerization. The mean
molecular mass is for the application of the $Et(Ind)_2ZrCl_2$ catalyst
smaller than by engagement of Cp_2ZrCl_2.

Calculated from the ^{13}C-nmr-spectroscopic data the 4-methyl-1-pentene
amounts lie in the range of 0,82 mol-% incorporation for 14,63 mol-%
given in the starting mixture (Figure 1).

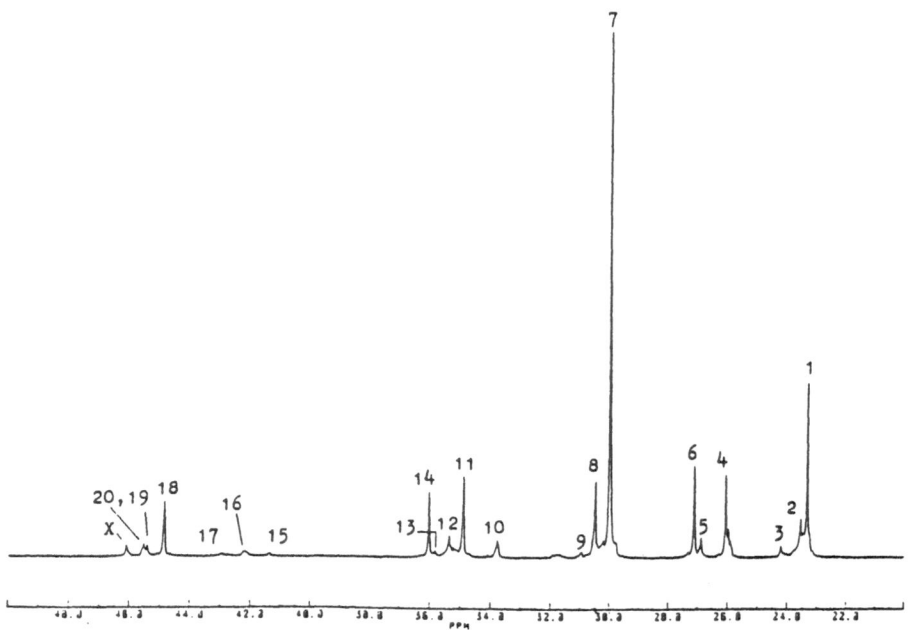

Fig. 1 ^{13}C-NMR (75 MHz) Spectrum of a Ethylene/4-Methyl-1-Pentene
(14,6 mol-%) Copolymer. Numbers of Peaks see Table 5

Table 5 conveys chemical shifts of the copolymers as they have been
found compared to those calculated based on Lindeman and Adams (14)
method.

Table 5 Calculated and measured chemical shifts for an ethylene (E)/
4-methyl-1-pentene (M) copolymer with 14,6 mol-% M

$$\delta \quad \gamma \quad \beta \quad \alpha \qquad \alpha \quad \beta \quad \gamma \quad \delta$$
$$- CH_2 - CH_2 - CH_2 - CH_2 - CH - CH_2 - CH_2 - CH_2 - CH_2 -$$
$$CH_2 \ (sc)$$
$$CH \ (sc)$$
$$H_3C \qquad CH_3$$

Peak No.	Typ	Sequency	Chemical Shift Calculated	Measured
1	CH_3	EM*E	22.62	23.25
2	CH_3	PM*E	22.62	23.45
	CH_3	MM*M	22.62	
3		ME*M	25.08	24.15
4	CH(sc)	MM*M	25.92	25.98
		MM*E	25.92	
		EM*E	25.92	
5	$\beta\delta$	MME*E	27.52	26.82
6	$\beta\delta$	EME*E	27.52	27.05
7	$\delta\delta$	(EE*E)	29.96	29.88
8	$\gamma\delta$	ME*EE	30.21	30.40
9	$\gamma\gamma$	ME*EM	30.46	30.89
10	CH	MM*E	32.91	33.77
11	$\alpha\delta$	EM*EE	34.72	34.82
12	$\alpha\delta$	MM*EE	34.97	35.29
13	$\alpha\gamma$	MME*M	35.22	35.76
14	CH	EM*E	34.98	35.97
15	$\alpha\alpha$	EM*ME	39.48	41.31
16	$\alpha\alpha$	MM*ME	39.73	42.16
17	$\alpha\alpha$	MM*MM	39.98	42.95
18	CH_2(sc)	EM*E	43.86	44.71
19	CH_2(sc)	MM*E	44.11	45.38
20	CH_2(sc)	MM*M	44.36	45.52

From this in analogy to Kimura (14) the following triad distributions
can be deduced (Table 6).

Table 6   Triads of Some Ethylene/4-Methyl-1-Pentene Copolymers in
mol-%

[M]	[MMM]	[MME] + [EMM]	[MEM]	[EME]	[EEM] [MEE]	+ [EEE]
2.59				2.57	5.22	92.21
5.73			0.67	3.98	10.88	84.47
6.10	0.30	0.53	0.40	4.70	14.21	79.86
10.86	0.35	1.30	1.08	9.30	17.80	70.17
14.40	0.74	2.54	2.50	11.38	23.43	59.41
14.63	0.97	2.36	2.25	11.54	24.39	58.49

Via correlation of the monomer ratios M_{MP} in the starting polymerization mixture with those in the produced copolymer the copolymerization parameters r_1 and r_2 can be obtained. Determination after
Fineman-Ross leads to the following values for $Et(Ind)_2ZrCl_2$:
$r_1 = 50.00$, $r_2 = 0.004$.

In the case of Cp_2ZrCl_2 under the same conditions the r-parameter
happens to be $r_1 = 102.7$. The amount of incorporations depends on
both the applied comonomer concentration and on the catalyst concentration either. The system $Et(Ind)_2ZrCl_2$ is also capable to copolymerize 2-methyl-propene with ethylene (Table 7).

Table 7   Copolymerization of 2-Methylpropene (MP) and Ethylene (E)
with $Et(Ind)_2ZrCl_2$ and Methylalumoxane $(2,1 \cdot 10^{-2}$mol Al/l),
Solvent Toluene, Volume 250 ml

Zr (mol/l)	Ethylene (bar)	MP (mol)	Polymerization Temperature (°C)	Time (h)	Yield (g)	MP in the copolymer (mol-%)	Mη
10^{-7}	4	10	50	0,75	2,62	< 1	93 000
$5 \cdot 10^{-8}$	4	40	50	2,0	2,41	< 1	120 000
$2 \cdot 10^{-6}$	2	70	30	3,2	2,16	< 1	101 000
$2 \cdot 10^{-6}$	2	80	30	95	0,64	1,36	19 000

The incorporation of 2-methyl-propene in the polymer is smaller than
in the case of 4-methyl-1-pentene. The maximum amount of incorporation
was 2.8 mol-%. In the case of Cp_2ZrCl_2 the amount of incorporation
lies below the limit of detection.

CYCLOPENTENE POLYMERIZATION TO ISOTACTIC POLYMERS

The chiral catalyst allows the polymerization of cycloalkenes like
cyclopentene or cycloheptene to isotactic polycycloalkenes. No ring
opening occurs. In the literature there are some examples of the
polymerization of cycloalkenes with Ziegler-Natta catalysts but always a ring opening has to be stated which lies in the range of 20 to
30%. This provides the polymers with elastomeric features. In contrast here to the polymers produced with the homogeneous catalyst are
highly crystalline and are obtained in high yields. Table 8 shows the
reaction conditions.

<u>Table 8</u> Reaction Conditions for the Polymerization of Cyclopentene
with the Chiral Catalyst Et(Ind)$_2$ZrCl$_2$ Methylalumoxane in a
Stirred Vessel

Temperature	30 °C
Et(Ind)$_2$ZrCl$_2$	0,21 mg (5·10^{-7} mol)
Methylalumoxan	200 mg (1,67·10^{-4} mol)
Solvent	250 ml toluene
Monomer	100 ml cyclopentene
Reaction Time	90 h
Yield	13,6 g
Activity	2 200 g Polycyclopentene/g Zr·h

Isotactic polycyclopentene is insoluble in common hydrocarbons. Its
density is 1.104 g/cm^3 (23°C).

In Fig. 2 there is depicted the IR-spectrum of isotactic polycyclopentene.

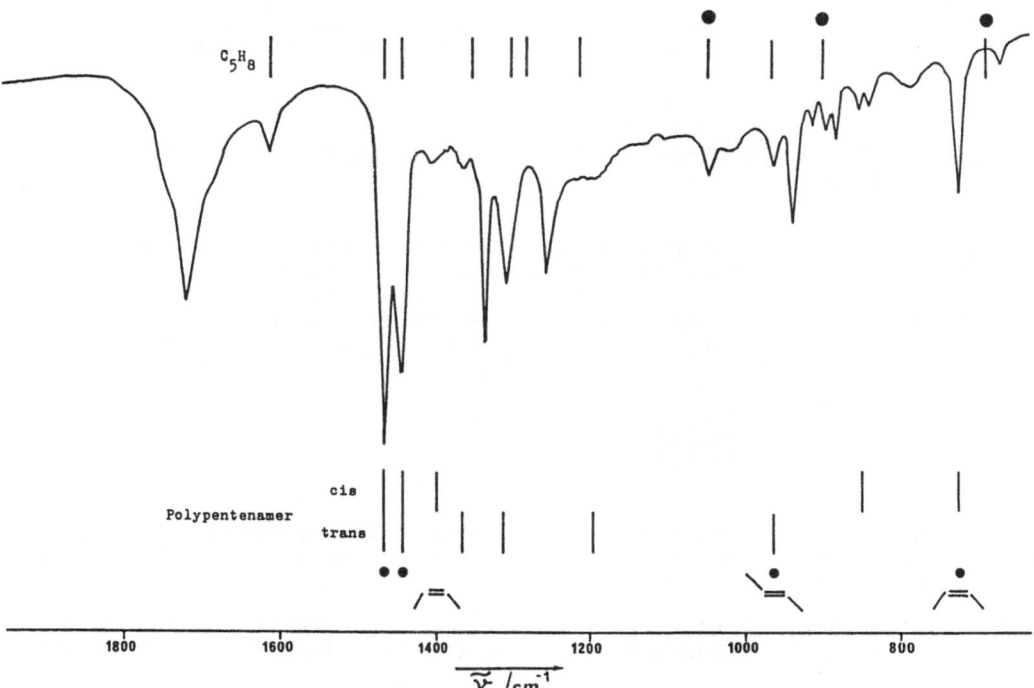

Fig. 2 IR-Spectrum of Isotactic Polycyclopentene, KBr-molding

It becomes obvious that not all bands of cis- and trans-polypenten-
amers can be detected. More information can be gained from inspection
of the Debye-Scherrer diagramme (Fig. 3).

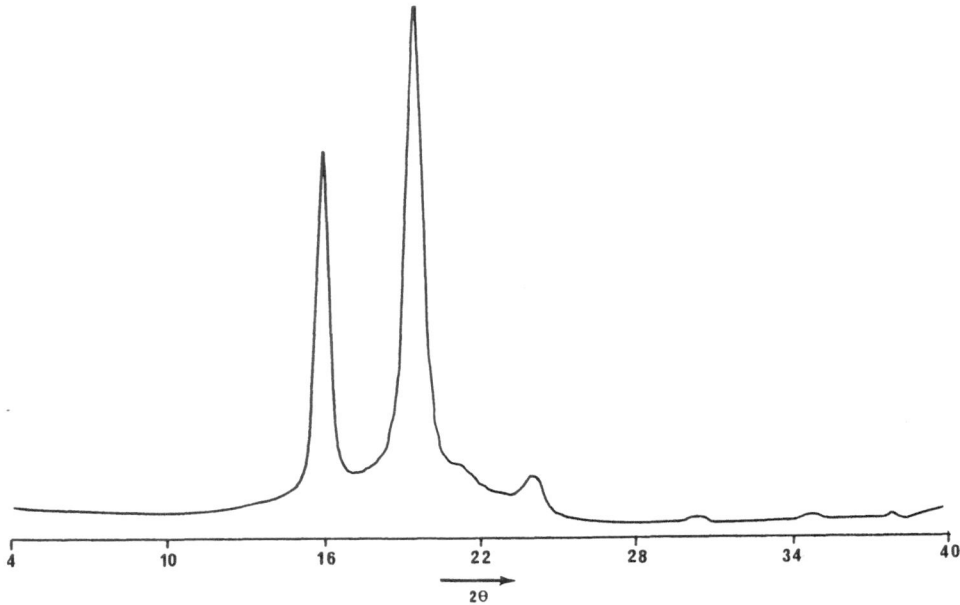

Fig. 3 Debye-Scherrer-X-ray Spectrum (Cu-K-α) of Isotactic Polycyclopentene

In comparison to usual powder diagrammes here three unusually sharp
lines can be seen. This is an indication for high crystallinity
which is the most striking point of difference from the wholly
amorphous polypentenamers. The observed lines lead to calculations
of the lattice distances, which are found to be for the three X-ray
deflection maxima:

2 θ	d/A
15.975	5.55
19.454	4.56
23.932	3.72

More information is provided by CPMAS-NMR (cross polarization magic
angle spinning) spectroscopy, which has to be engaged because of the
insolubility of the polymer (Fig.4).

In consideration of the analytical methods engaged the following
structures can be deduced for isotactic polycyclopentene (structures
see next page):

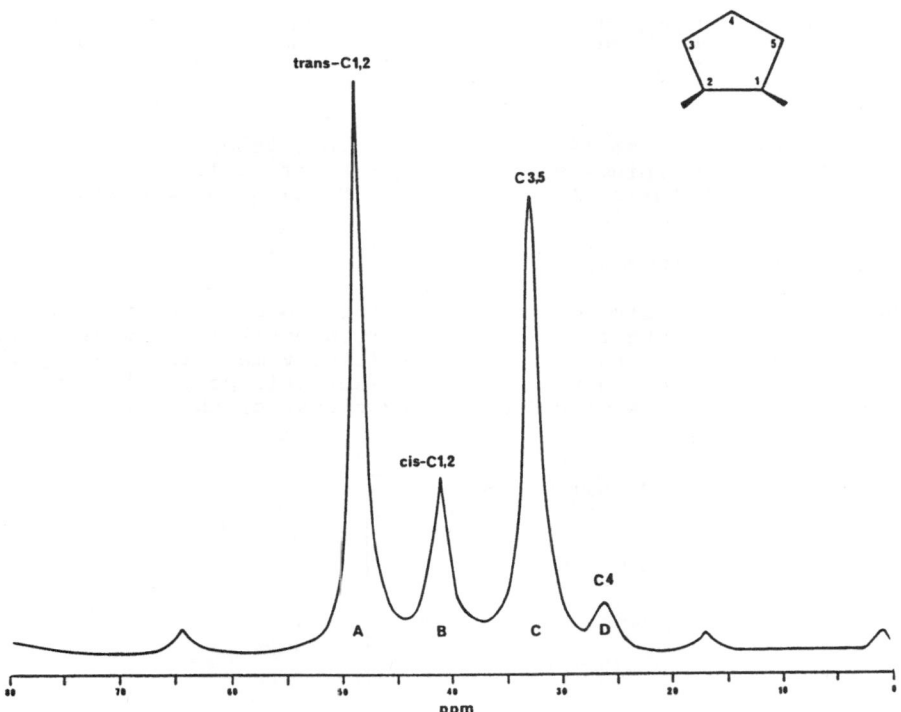

Structure of Isotactic Polycyclopentene

Fig. 4 Solid State ^{13}C-NMR Spectrum of Isotactic Polycyclopentene

A peculiar characteristic of the isotactic polycyclopentene is its high melting point which probably exceeds 300 °C. The dsc-diagramme (Fig. 5) displays in its first heating curve an exothermic conversion which starts at 130 °C and ends up not beyond 330 °C. The specific enthalpy for this phase conversion lies in the range of 0.5 - 0.8 kJ/g, depending on each single sample. Neither in the cooling down process nor in the second heating curve there occurs no conversion any more. From this it can be concluded that the process in mind is not a reversible process like melting, but rather an irreversible reaction (decomposition, oxidation). Whereas at the beginning of the thermal treatment the sample was white afterwards it was coloured yellow and showed a brittle consistence.

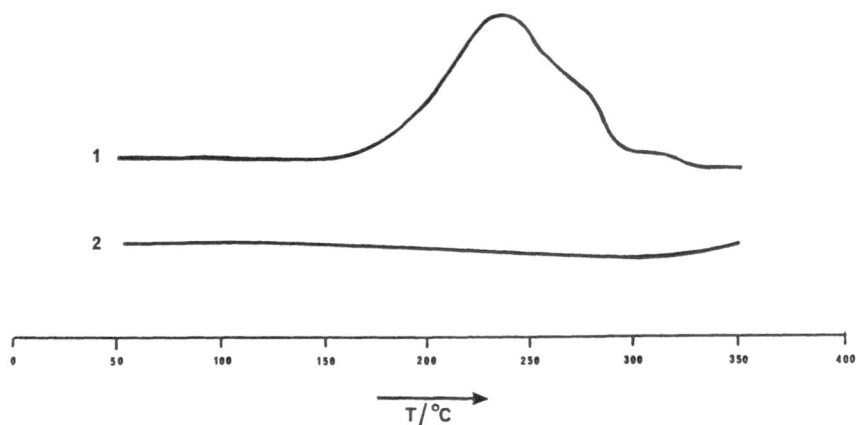

Fig. 5   DSC-Diagramme of Isotactic Polycyclopentene,
$\Delta T/\Delta t$ = 50 k/min, Curve 1: Original Material,
Curve 2: Material After the First Melting and Cooling

OPTICALLY ACTIVE OLIGOMERS

By the help of optically active binaphtol it is possible to separate the racemic mixture ethylenebis(tetrahydroindenyl)zirconiumdichloride into the pure enantiomers (15). For the first time this opens up the opportunity to produce only one kind of isotactic polyolefins by Ziegler-Natta catalysis provided the end groups of the polyolefin chain are different:

For long polymer chains in solution there results no optical rotation. Only if it becomes possible to produce during the polymerization exclusively right or left handed helical structures of the polyolefins these will be optically active.

The situation is different if the polymerization grade is low. Then the terminal carbon atoms are chiral by the influence of the nearby different end groups. So it is possible to synthesize optically active oligomers. This is shown by asymmetric hydrooligomerization of propene by Pino (16). Using these optically active enantiomers for the oligomerization of butene yields optically active alkanes and alkenes (Tab. 9).

Table 9   Purity and Optical Rotation of the Di- and Trimers of 1-Butene Oligomerization with S-Et(IndH4)$_2$ ZrCl$_2$/Methylalumoxane

Oligomer	Compound	Purity (%)	α_{589}^{25}	α_{365}^{25}
Dimeres	3-methylheptane	97	- 8,3	- 23,4
Trimeres	3-methyl-5 ethylnonane	94	-34,1	- 102,6

The optical purity of these products is in the range of 70 - 80 %.

The special use of chiral Ziegler-Natta catalysts opens up totally new possibilities in the field of olefin polymerization and oligo-merization.

*

We thank the Deutsche Forschungsgemeinschaft and the Hoechst Aktien-gesellschat for sponsoring this research.

LITERATURE

1   Sinn H, Kaminsky W, Vollmer H-J, Woldt R   (1980) Angew Chem Int Ed Engl 19:390
2   Sinn H, Kaminsky W   (1980) Adv Organomet Chem 18:99
3   Kaminsky W (1983) Transition Metal Catalyzed Polymerization, in: Quirk RP (ed), MMI Press, Symposium Series Vol 4, Harwood Academic Press Publ, New York, p 225
4   Tait P, This Proceedings Book
5   Kaminsky W, Schlobohm M (1986) Makromol Chem Macromol Symp 4:103
6   Kaminsky W, Miri M (1985)   J Polym Sci Polym Chem Ed 23:2151
7   Kaminsky W (1986) Angew Makromol Chem 145/146:149
8   Kaminsky W, Buschermöhle M (1987) in: Recent Advances in Mechanistic and Synthetic Aspects of Polymerization, Fontanille M, Guyot A (ed), Reidel Publ Comp Series C, Dordrecht, 215:503
9   Kaminsky W, Miri M, Sinn H, Woldt R (1983) Makromol Chem Rapid Commun 4:464
10 Kaminsky W, Lüker H (1984) Makromol Chem Rapid Commun 5:225
11 Wild FRWP, Zsolnai L, Huttner G, Brintzinger HH (1982) J Organo-met Chem 232:233
12 Ewen JA (1984) J Am Chem Soc 106:6355
13 Kaminsky W, Külper K, Brintzinger HH, Wild FRWP (1985) Angew Chem 97:507
14 Lindeman LP, Adams JQ (1971) Anal Chem 43:1245
    Kimura K, Yuasa S, Maru Y (1984) Polymer 25:441
15 Wild FRWP, Zsolnai L, Huttner G, Brintzinger HH (1985) J Organo-met Chem 288:63
16 Pino P, Cioni P, Galimberti M, Wei J, Piccolrovazzi N, This book

Copolymerization of Ethene and α-Olefins with a Chiral Zirconocene/Aluminoxane Catalyst

H. Drögemüller, K. Heiland, and W. Kaminsky

Institute of Technical and Macromolecular Chemistry, University of Hamburg, Bundesstr. 45. D-2000 Hamburg 13

INTRODUCTION

Recent works with the soluble, chiral Ziegler-Natta catalyst system composed of rac-ethylene-bis(1,1'-indenyl)zirconiumdichloride as the catalyst component and methylaluminoxane as the cocatalyst component (Fig. 1) in the field of the homopolymerization of propene (1, 2) and other α-olefins (3) showed interesting behaviour. So we were interested to use this system also for the copolymerization of ethene with α-olefins like propene and butene-1.

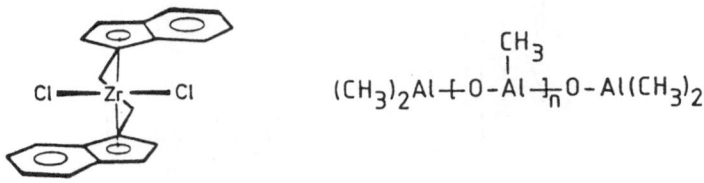

catalyst   cocatalyst

Fig. 1 Soluble Ziegler-Natta catalyst system (4) composed of rac-ethylene-bis(1,1'-indenyl)zirconiumdichloride/methylaluminoxane

The copolymerizations of ethene with propene are carried out in a laboratory scale bubble column reactor. This reactor typ was preferred because of the very high heat and mass transfer rates and the easy kinetic measurement. Points of interest are the reactivity ratios r_1 and r_2, their dependence on the polymerization temperature, the kinetic behaviour of the copolymerizations and the product properties as a function of the experimental conditions.

THE BUBBLE COLUMN REACTOR AND THE PRINCIPLE OF THE MEASUREMENTS

Fig. 2 shows the flowsheet of the plant.

The bubble column reactor consists of a 2 liter glass tube fitted with a perforated plate. It contains the solvent for the catalyst components and the monomers. The solvent is toluene. The monomers are pumped in a circuit by a compressor and bubbled through the solution. Consumed monomer is replaced from a pressure regulator. The composition of the two monomers in the inlet stream of the reactor is held constant by an automatic gas blending device. The gas blending device also measures the dosed quantity of the monomers. The plant is fitted

W. Kaminsky and H. Sinn (Eds.)
Transition Metals and Organometallics as
Catalysts for Olefin Polymerization
© Springer-Verlag Berlin Heidelberg 1988

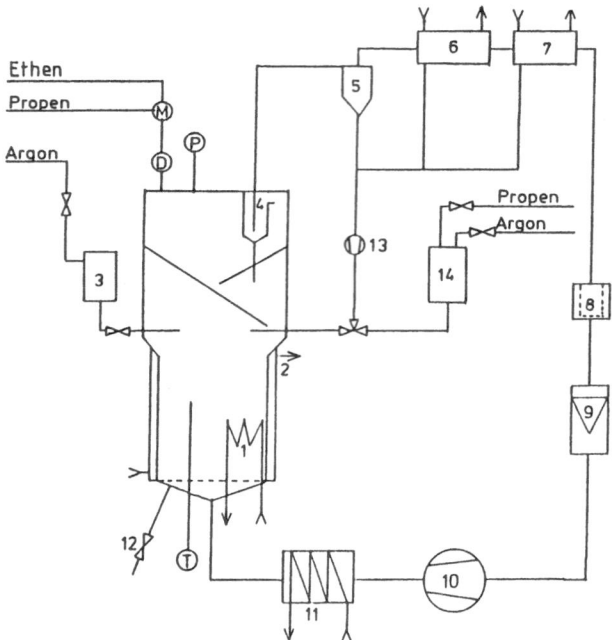

Fig. 2   Flowsheet of the Plant
1) cooling coil, 2) heating jacket, 3) dosing container,
4, 5) cyclones, 6, 7) condensers, 8) filter, 9) flowmeter,
10) compressor, 11) heat exchanger, 13) outlet valve, 13)
solvent pump, 14) solvent container, M) gas blending device,
P) pressure measurement, T) thermocouple

with an efficient temperature and pressure regulation. Reachable constancy: $\Delta T \leq 0,3\ °C$, $\Delta P \leq 0,05$ bar.

The monomers are purified by passing the inlet stream through two columns with copper catalyst and molecular sieves 10 A. For the calculation of the reactivity ratios r_1 and r_2 of ethene and propene the linearized copolymerization equation is used.

$$(f-1) \cdot F/f = r_1 \cdot F/f - r_2$$

$$f = [dM_1]/[dM_2] \quad F = [M_1]/[M_2]$$

Eq. 1   Fineman-Ross equation

To receive the reactivity ratios the composition of the monomers in the fluid phase F and in the copolymer f has to be determined.

The experiments are carried out in the following way: The fluid phase in the reactor is saturated with any composition ratio of the monomers. After the injection of the catalyst the polymerization begins. The pressure regulator replaces the consumed monomer by a mixture of the

monomers with the constant composition that is desired in the copolymer. If the fluid phase was saturated with the 'false' composition F*
the comonomer composition in the fluid phase would change in the
direction of the searched for value F. After 3-4 experiments this
value is found. In the last experiment the fluid phase is saturated
with the right composition. After the injection of the catalyst a
constant composition of the monomers in the fluid phase is obtained
during the run.

The value of f is known because it is identical with the composition
of the monomers in the inlet stream of the reactor. In this way it is
possible to get the reactivity ratios without analyzing the copolymerization products. The composition of the monomers in the fluid
phase is controlled by GC-measurements of the gas phase.

RESULTS OF THE ETHENE/PROPENE COPOLYMERIZATION

Fig. 3 shows the copolymerization diagram of ethene/propene for the
temperatures of 25°C and 50°C and Fig. 4 the Fineman-Ross plot for
50°C.

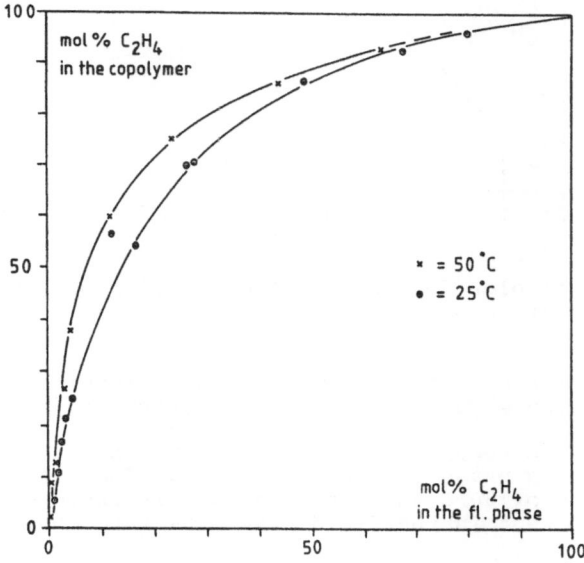

Fig. 3   Copolymerization diagram of ethene/propene for the tempera-
         tures of 25°C and 50°C

The reactivity ratios of the ethene/propene copolymerizations were
determined to r_1= 6,61, r_2= 0,06 at the temperature of 50°C and
r_1= 6,26, r_2= 0,11 at the temperature of 25°C. Their products are
$r_1 \cdot r_2$= 0,40 (50°C) and $r_1 \cdot r_2$= 0,69 (25°C).

In comparison with the data of other soluble Ziegler-Natta catalyst
systems like $Cp_2Ti(CH_3)_2$/methylaluminoxane (5), (r_1=18,6, r_2=0,032,
$r_1 \cdot r_2$=0,60) and $Cp_2Zr(CH_3)_2$/methylaluminoxane (r_1=20,1, r_2=0,015,
$r_1 \cdot r_2$=0,30), the r_1 value of this system is about three times lower,

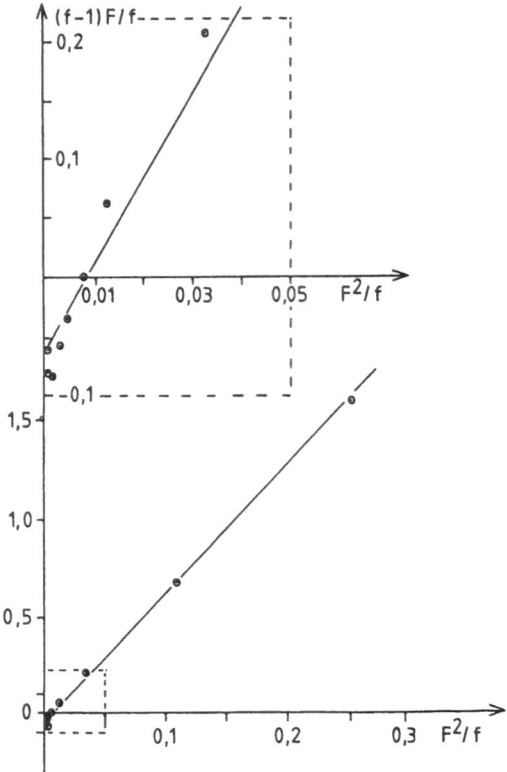

Fig. 4   Fineman-Ross plot for the copolymerization of ethene/propene at a temperature of 50°C

the r_2 value about three times higher. The same results are found in comparison with soluble Vanadium based catalysts (6).

The dependence of the average rate konstants k_p upon the monomer composition in the fluid phase is shown in Fig. 5. The shape of the curve for the temperature of 50°C shows a behaviour as expected. With increasing ethene content in the fluid phase the rate constants increase.

The weight-average molecular weights of the copolymers as a function of the copolymer composition is shown in Fig. 6. The molecular weights were measured by GPC. With increasing ethene content in the copolymer the molecular weights increase. The molecular weight distribution M_w/M_n by GPC is about 2,5 to 4,5.

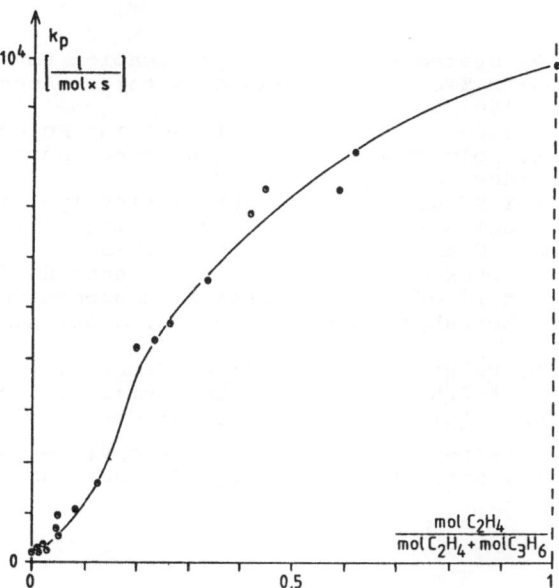

Fig. 5 Dependence of the average rate constants k_p upon the comonomer content in the fluid phase at a temperature of 50°C

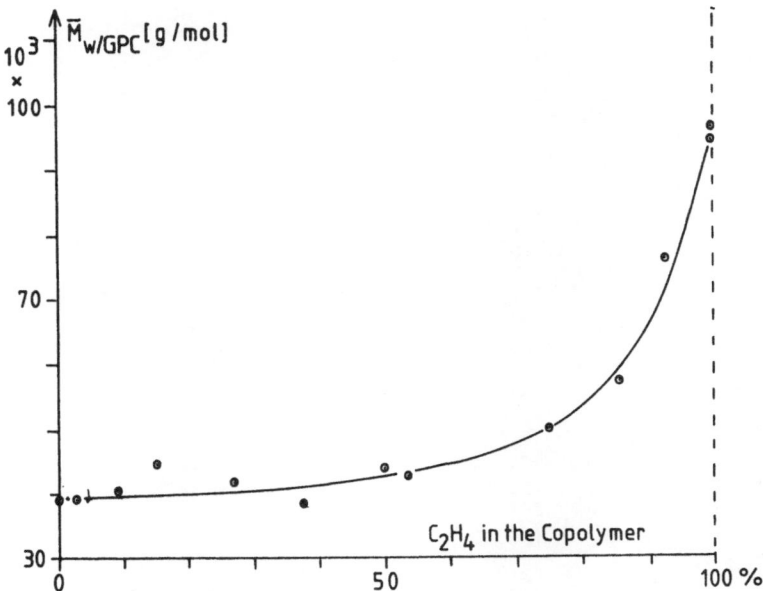

Fig. 6 Dependence of the weight average molecular weights upon the copolymer composition at a temperature of 50°C

REFERENCES

1 Kaminsky W (1986) Stereoselektive Polymerisation von Olefinen mit homogenen, chiralen Ziegler-Natta-Katalysatoren. Angew Makromol Chem145/146: 149-160
2 Drögemüller H, Niedoba S, Kaminsky W (1986) in: Reichert KH, Geiseler W (eds), Polymer Reaction Engineering. Hüthig & Wepf, Basel-Heidelberg-New York, p 299
3 Kaminsky W, Külper K, Niedoba S (1986) Olefinpolymerization with highly active soluble Zirconium compounds using Aluminoxan as co-catalyst. Makromol Chem, Makromol Symp 3:377-387
4 Wild FRWP, Wasiacionek M, Huttner G, Brintzinger HH (1985) Synthesis and Crystal Structure of a Chiral Ansa-Zirconocene Derivate with Ethylene Bridged Tetrahydroindenyl Ligands, J Organomet Chem 288: 63-67
5 Busico V, Mevo L, Palumbo G, Zambelli A, Tancredi T (1983) Preliminary Results of Ethylene/Propene Copolymerization in the Presence of $Cp_2Ti(CH_3)_2Al(CH_3)_3/H_2O$, Makromol Chem 1984:2193-2198
6 Cozewith C, Ver Strate G (1971) Ethylene-Propylene Copolymers. Reactivity Ratios, Evaluation, and Significance. Macromolecules 4:482-489

Newer Aspects of Active Centre Determination in Ziegler-Natta Polymerization using ^{14}CO Radio-Tagging.

P.J.T. Tait

Department of Chemistry, UMIST, Manchester M60 1QD, U.K.

INTRODUCTION

The determination of the number of active centres in Ziegler Natta polymerization continues to be of increasing interest to researchers in this field of study. Since the pioneering work of Natta (1,2) well in excess of one hundred papers containing values for active centre concentrations have appeared in the scientific literature. Many of these papers have been concerned directly with the development and applications of appropriate methods for active centre determination to both first and subsequent generations of Ziegler-Natta catalysts. Indeed recent developments of high activity catalyst systems have done much to focus interest on this area of discovery since a knowledge of the number of actual propagating centres is a prerequisite for the determination of values of the rate constant of chain propagation. A knowledge of these quantities is essential in any informed programme for catalyst characterization and development.

MODELS AND METHODS

Unfortunately this area of research has not yet reached the stage where there are any direct methods for the determination of active centre concentrations and this situation is likely to remain so until appropriate spectroscopic methods can be developed. A major problem in the development of potentially useful methods continues to be our lack of understanding of the exact nature of the active centres in Ziegler-Natta polymerizations. At the present time all methods are indirect and so there exists the need to relate the parameters which have been measured to the concentrations of active centres actually present in the polymerization systems. Since the measured parameters and concentrations of active centres are bridged by means of appropriate models the accuracy of active centre determinations as currently carried out depends largely on the validity of the models which have been adopted.

The use of models in polymerization catalysis is widespread and is of great value to our understanding of these complex chemical reactions. It is important

W. Kaminsky and H. Sinn (Eds.)
Transition Metals and Organometallics as
Catalysts for Olefin Polymerization
© Springer-Verlag Berlin Heidelberg 1988

to recognise that the models used may be of several different types.

(a) Models of active centres - essential for an understanding of the mechanism and the various steps in the polymerization sequence. It is often assumed that the active centres are fixed geometric sites, invariant with respect to time. However, this is too simplistic a model to account for the kinetic behaviour of many polymerization systems. A model with a more dynamic character is often more appropriate. Additionally, as will be demonstrated later in this paper, the value of the propagation rate constant for a particular polymerization systemn may vary with the time of polymerization.

(b) Kinetic models - essential for a quantitative description of the polymerization reaction and for linking the parameters measured to the concentration of active centres.

(c) Morphological models - increasingly necessary for describing the disintegration of catalyst particles and to account for any diffusion limitations arising from the precipitation of solid polymer in the near vicinity of active centres.

A number of different methods for active centre determination have been successfully developed and these have been reviewed by various authors (3-8). In attempting to reconcile values for active centre concentrations obtained using different methods it is important to recognise which parameters have been measured and how these can be related to the active centre concentrations. Details for the more frequently used methods are listed in Table 1.

Table 1. More frequently used methods for active centre determination.

	Method	Parameters
1	Molecular Weight Data	M_n values as a function of time.
2	Quenching with Tritiated Alcohols	Tritium content of polymer produced as a function of time
3	Radio-Tagging with ^{14}CO	^{14}CO content of polymer produced for selected contact times.
4	Catalyst inhibitors (allene, CO, CO_2, CS_2 etc)	Amount of poison adsorbed and corresponding decrease in rate of polymerization.

In all cases it is necessary to relate the parameters measured to the concentrations of active centres and this will of necessity involve the use of an appropriate model with the introduction of various limiting approximations which will vary with the method selected.

^{14}CO RADIO-TAGGING

TAGGING SEQUENCE AND CONTACT TIME

The use of carbon monoxide to tag polymer chains in Ziegler-Natta
polymerization reactions has its origins in research carried out jointly in the
research laboratories of the Hercules Powder Co in the USA and ICI in the UK.
The establishment of the method however owes much to the detailed studies of
Yermakov et al (5) who suggested that the reaction sequence on treatment of a
polymerization system with ^{14}CO could be described as follows:

Coordination

$$L_x Ti\text{-}CH_2\text{-}CH_2{\sim}P + {}^{14}CO \longrightarrow \overset{\overset{\displaystyle ^{14}CO}{\downarrow}}{L_x Ti}\text{-}CH_2\text{-}CH_2{\sim}P \qquad (1)$$

Insertion

$$\overset{\overset{\displaystyle ^{14}CO}{\downarrow}}{L_x Ti}\text{-}CH_2\text{-}CH_2{\sim}P \longrightarrow L_x Ti\text{-}\underset{\overset{\|}{O}}{{}^{14}C}\text{-}CH_2\text{-}CH_2{\sim}P \qquad (2)$$

Quenching

$$L_x Ti\text{-}\underset{\overset{\|}{O}}{{}^{14}C}\text{-}CH_2\text{-}CH_2{\sim}P \xrightarrow{ROH/H^+} H\text{-}\underset{\overset{\|}{O}}{{}^{14}C}\text{-}CH_2\text{-}CH_2{\sim}P \qquad (3)$$

The laboratory use of the technique is comparatively simple. An excess of ^{14}CO
(preferably over the Al) is used. The radio-tagged polymer is then isolated,
thoroughly decontaminated, and its content determined either by combustion
analysis, or, as in our laboratories, by gel scintillation counting. In
relating the measured radioactive content of the polymer C_i, to the propagating
centres C_p^*, the following assumptions are made.
(a) The carbon monoxide inserts into all active transition-metal carbon bonds
[via reaction sequence (1) → (2)] during the selected contact time between the
^{14}CO and the polymerization system, t_c.
(b) The carbon monoxide is inserted only into active transition-metal carbon
bonds, thus obviating complications arising from chain transfer.
(c) Multiple insertions do not occur.

Under conditions where the above model is valid:

$$GA/a = C_i = C_p^* \qquad (4)$$

where G = polymer yield/g (mol Ti)$^{-1}$; A = polymer activity/dpm g^{-1} and a =

specific activity of $^{14}CO/dpm \, mol^{-1}$.

The increase in the radioactivity of the isolated polymer with the time of
contact between the carbon monoxide and the polymerization system observed by
Yermakov and Zakharov (5,9) has been attributed to a slow copolymerization of
^{14}CO with monomer. However, it has been demonstrated (6,10) that for the
polymerization of propylene using δ-TiCl$_3$0.33 AlCl$_3$ - AlEt$_2$Cl catalyst systems
this increase of polymer radioactivity with increase in t_c takes place even in
the absence of monomer; an observation which has also been confirmed (11,12)
for highly active MgCl$_2$/donor$_1$/TiCl$_4$ - Al(i-Bu)$_3$/donor$_2$ catalyst systems, as
shown in Fig.1

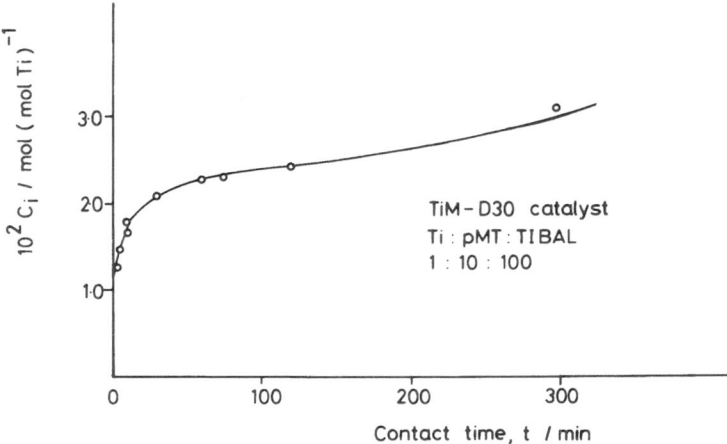

Fig.1. Plot of C_i versus contact times. Propylene polymerization at 1 atm and
60 °C using MgCl$_2$/EB/TiCl$_4$ - Al(i-Bu)$_3$/pMT high activity catalyst.
Ti : pMT : TIBAL = 1: 10 : 100. (EB = ethyl benzoate; pMT = p-methyl toluate)

The equilibrium nature of the initial reaction between L$_x$Ti-CH$_2$-CH$_2$ P and ^{14}CO
has been established by Ajayi et al (13,14) who have also demonstrated that
this reaction can be described in terms of a Langmuir-Hinshelwood adsorption
isotherm, as has been established also by Lesna and Mejzlik (15).
The major issue of debate over the years has been the rate and extent of
insertion of the complexed CO molecule into the metal-carbon bond of the
growing polymer chain (16,17), and whether the value of $\overset{\bullet}{C_i}$ obtained from short
contact times or that obtained from longer contact times should be regarded as
being directly related to the number of active centres. Uncertainty concerning
this situation remains the greatest drawback to the acceptance of the method
which otherwise offers distinct advantages due to its ease of usage.

POLYMER DECONTAMINATION

It should be noted that the final step in reaction sequence (3) is a quenching reaction which requires the use of acidified alcohol. It is useful for this reason to distinguish between this method which combines both tagging and quenching reactions and other methods which involve only tagging reactions, and also between this method and quenching reactions where the tag is introduced in the actual quenching reaction, e.g., quenching using tritiated alcohols. To illustrate the magnitude of the effects which can be introduced should acid not be used in the final quenching values of C_i and k_p for propylene polymerization in the presence and in the absence of acid are listed in Table 2 (11,12).

Table 2. Values of C_i and k_p for propylene polymerization obtained for polymer isolated in the presence and absence of acid.

	Polymerization Time/min	R_p(max)/ gPP(mmol Ti h atm)$^{-1}$	$10^2 C_i$/ mol(molTi)$^{-1}$	k_p/dm^3 s^{-1} mol^{-1}
Acid Present	12	584	3.5	454
Acid Absent	12	638	9.2	190

High activity catalyst containing 2.5% weight Ti
Temperature = 60 °C. Pressure = 1 atm
For acid present: 250 cm^3 EC180; polymerization system treated with MeOH(3X) containing dilute H$_2$SO$_4$ (5 cm^3).
Contact time = 10 min.

The importance of using thorough decontamination procedures has been highlighted by Bukatov et al (18) and must be regarded as essential for any proper application of the technique. The use of appropriate blank experiments is required to establish the method for any particular polymerization system.

USE OF ^{14}CO RADIO-TAGGING IN ALUMINOXANE SYSTEMS

The use of soluble Kaminsky type catalysts affords a useful method for investigating whether short or long contact times should be used in relating C_i to C_p^*. Investigations have been carried out at UMIST using a variety of zirconocene compounds, $(RC_5H_4)_2ZrCl_2$ where R = H, Me, n-Pr, i-Pr and t-Bu, using methylaluminoxane as the cocatalyst for the polymerization of ethylene in toluene (19,20). All polymerizations show very stable rate-time profiles except for R=t-Bu. Typical rate-time profiles are shown in Fig.2.

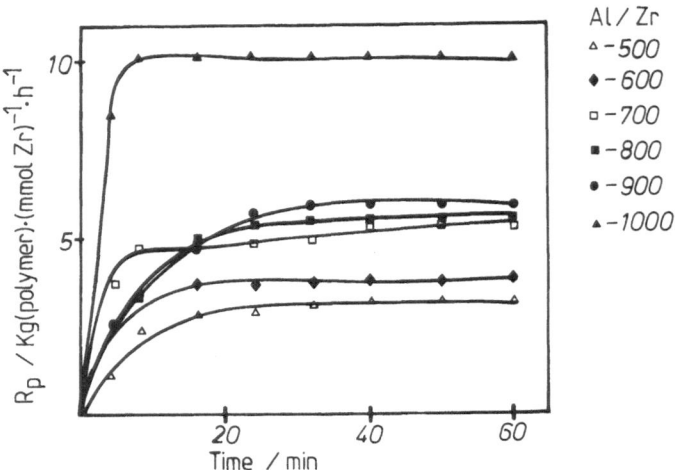

Fig.2. Plots of R_p versus time for the polymerization of ethylene using $(CH_3C_5H_4)_2ZrCl_2$/methylaluminoxane in toluene at 60 °C. $[Zr] = 0.024$ mmol dm^{-3}. Pressure = 1 atm.

These polymerizations using $(RC_5H_4)ZrCl_2$ - methylaluminoxane catalyst systems are characterised in general by:

(a) Very high activities when these are expressed in terms of the amount of polymer produced per mole of transition metal.

(b) Rates of polymerization which increase continuously with increase in Al : Zr ratio, even beyond 1000 : 1.

(c) The need for the use of very high ratios of Al : Zr to obtain favourable activities.

(d) Short settling periods followed by long periods of constant overall rates of polymerization.

The variation of C_i with contact time in the presence of monomer was investigated and typical results are shown in Fig.3.

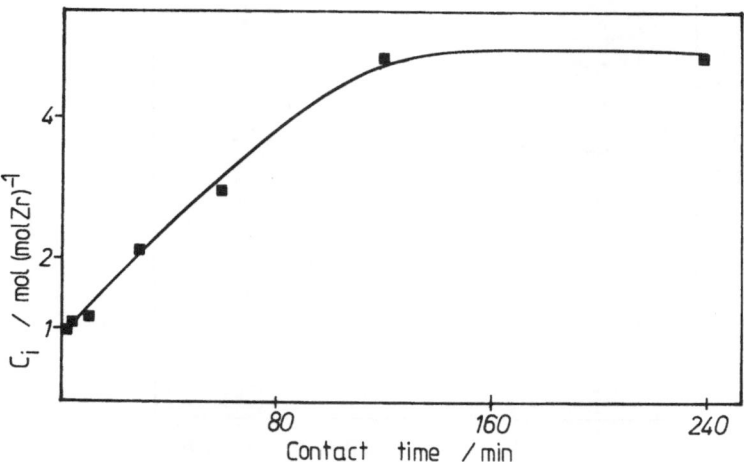

Fig.3. Variation of C_i with contact time for the polymerization of ethylene using the catalyst system $(C_5H_5)_2ZrCl_2$/methylaluminoxane in toluene at 60 °C. $[Zr] = 0.024$ mmol dm^{-3}; $[Al] : [Zr] = 1000 : 1$. Pressure = 1 atm; Polymerization time = 10 min. $[CO]/[Zr] = 23$.

It is evident that for this catalyst system C_i increases with t_c as observed previously for δ-TiCl$_3$0.33 AlCl$_3$ systems (6,10,21) and now in this paper for MgCl$_2$/donor/TiCl$_4$ catalyst systems (11,12). However at short contact times C_i is very nearly equal to the total Zr present in the catalyst, indicating that nearly 100% of the Zr is involved in active centres, and also indicating that as short contact times as are possible, whilst still allowing complete insertion, should be used in the determination of C_p^*. This conclusion is reached on the assumption that no more than one chain can grow at the same time on a given Zr atom.

These results are considered to be significant although it has to be accepted that the conclusions reached from a study of a homogeneous system may not apply to heterogeneous Ziegler-Natta catalyst systems. Nevertheless the similarities between the plots of C_i versus t_c for both homogeneous and heterogeneous catalysts are very striking, encouraging the belief that an extrapolation may be justified.

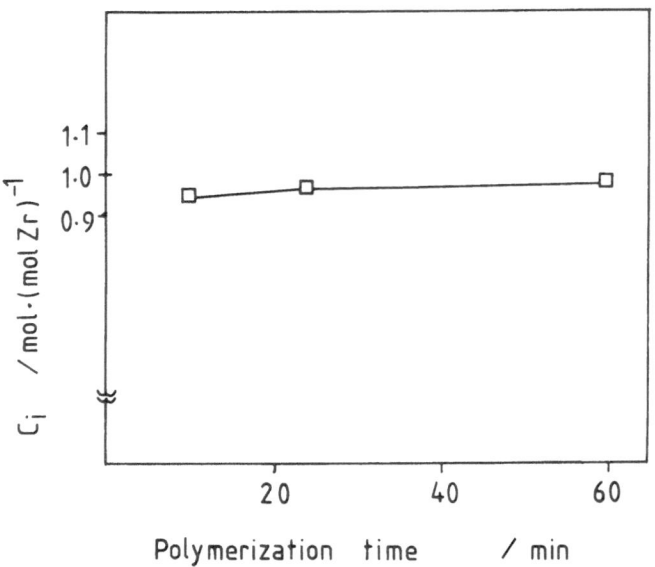

Fig.4. Plot of C_i versus time for the polymerization of ethylene using the catalyst system $(CH_3C_5H_4)_2ZrCl_2$/methylaluminoxane in toluene at 60 °C. $[Zr]$ = 0.024 mmol dm^{-3}; $[Al]$: $[Zr]$ = 1000 :1. Pressure = 1 atm.

Fig.4 shows that the value of C_p^* remains extremely constant within a given polymerization reaction; an observation which is consistent with the rate-time plots shown in Fig.2. The corresponding behaviour of k_p with time is shown in Fig.5.

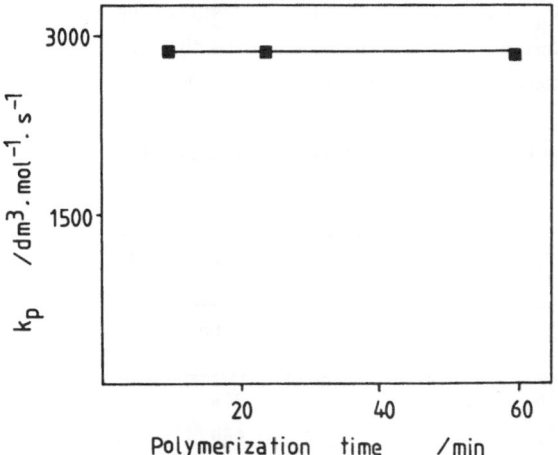

Fig.5. Plot of k_p versus time for the polymerization of ethylene using the catalyst system $(CH_3C_5H_4)_2ZrCl_2$/methylaluminoxane at 60 °C. $[Zr]$ = 0.024 mmol dm^{-3}; $[Al]:[Zr]$ = 1000 :1. Pressure = 1 atm.

The increase in C_i with t_c is not very easy to rationalise. Copolymerization of ethylene and ^{14}CO might be expected to take place (5), but as can be seen from an inspection of Fig.4 the incorporation of ^{14}C reaches a steady value while some 80% of the ^{14}CO is still unreacted.   The possibility certainly exists that alkyls and acyls may be exchanged in a scrambling reaction such as:

$$[Zr] - {}^{14}CO - CH_2 - CH_2 \sim P + Al_n(CH_3)_nO_n$$

$$[Zr] - CH_3 + Al_n(CH_3)_{n-1}O_n(-{}^{14}CO-CH_2 - CH_2 \sim P) \qquad (5)$$

Further research using comparative methods of active centre determination is required before unambiguous conclusions can be reached.

ADVANTAGES OF ^{14}CO RADIO-TAGGING
Determination of Concentrations of Centres Producing Isotactic and Atactic
Polymer.
One distinct advantage of the use of ^{14}CO radio-tagging is that its use allows a ready differentiation between centres producing heptane insoluble and those producing heptane soluble polymer. Fractionation of the polymer produced into fractions which are insoluble and soluble in boiling n-heptane is easily achieved, and to a first approximation this separation allows estimation of the concentrations of centres producing either isotactic or atactic polymer

respectively. Values of C_p^* and k_p at different polymerization temperatures for the polymerization of propylene using a high activity $MgCl_2/EB/TiCl_4$ catalyst with $AlEt_3$ as cocatalyst are listed in Table 3 (23-25.)

Table 3. Values of C_p^* and k_p for propylene polymerization at different temperatures.

Temp/°C	$10^2 C_p^*/mol(mol\ Ti)^{-1}$			$k_p/dm^3\ mol^{-1}\ s^{-1}$		
	Total	insol	Sol	Total	insol	Sol
40	7.14	3.44	3.70	369	703	59
50	8.03	4.93	3.10	669	1034	89
60	10.30	6.68	3.61	858	1256	122
70	11.31	7.47	3.84	1341	1952	163

Notes: (a) C_p^*(total) was determined using the unextracted polymer.
  (b) C_p^*(insol) was determined using the insoluble fraction from polymer extracted with boiling n-heptane for 24 h.
  (c) C_p^*(sol)= C_p^* (total) - C_p^* (insol)
  (d) Polymerization time before injection of ^{14}CO was 2 min.
  (e) Contact time = 4 min.

It is evident that only a small percentage of the total titanium is active (7.1 - 11.3% mol/mol and that this percentage increases with increase in temperature. However while C_p(insol) and C_p(sol) have about the same value at 40 °C the value for C_p(insol) increases with temperature in the range 40 - 70 °C while the value for C_p(sol) remains about the same. The values for k_p for this catalyst are much higher than for δ-$TiCl_3$0.33 $AlCl_3$ - $AlEt_2Cl$ or Solvay & Cie type catalysts (21,22). It is significant that k_p(insol) is much higher than k_p(sol), reflecting the differences in the transition states (23).

Determination of Concentrations of Active Centres as a Function of Polymerization Time

Another advantage of the use of ^{14}CO radio-tagging is that it allows the variation of C_p^* as a function of time within polymerizations to be investigated. In the case of the polymerization of ethylene using $((RC_5H_4)_2ZrCl_2/$aluminoxane catalyst systems a constant value of C_p^* was observed, apart from perhaps during the initial settling periods. However this situation is not the case for all polymerization systems. In the polymerization of propylene using $MgCl_2/DIOP/TiCl_4$ - $Al(i-Bu)_3$/triphenylsiloxane catalyst systems C_p^* is found to increase with the time of polymerization whilst the corresponding value of k_p decreases with time, as is shown in Fig.6 (26). (DIOP = diisooctylphthalate.)

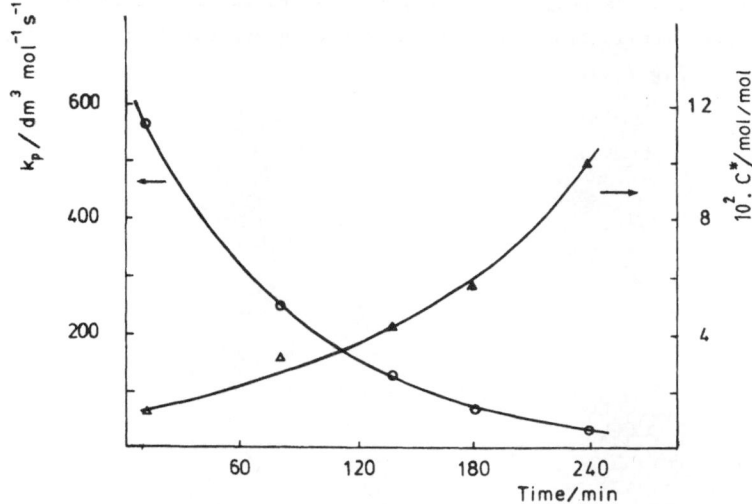

Fig.6. Plots of C_p^* and k_p versus time for the polymerization of propylene using a $MgCl_2$/DIOP/$TiCl_4$ - Al(i-Bu)$_3$/triphenylsiloxane catalyst system at 60 °C. Al : Ti = 180 : 1. Pressure = 1 atm. Contact time = 20 min.

These results are consistent with those of Giannini (27) who has reported previously a decrease in the value of k_p with time, and with those of other workers (28,29) who report very high values of k_p for the initial stages of polymerization.

The above observations may be explained in terms of the morphology of the catalyst particles as the polymerization proceeds. The $MgCl_2$ supporting matrix is made up of primary crystallites and has been prepared in such a way that the particles are spherical. However these particles are very fragile and break easily. Rapid alkylation may be expected to take place on and within the outer regions of the catalyst particles with the formation of highly active polymerization centres. The rate of diffusion of aluminium alkyl into a catalyst particle is relatively low, being much lower than that of propylene. As the polymerization proceeds the catalyst matrix fractures gradually from the outside, and as the zone of alkylation spreads inwards towards the centre of the particle the number of active centres increases. Centres formed in the outer regions of a particle where the concentration of aluminium alkyl is high are expected to have a high activity but to be unstable. Centres formed further within a particle where the concentration of aluminium alkyl is lower are regarded as having a lower activity but a higher stability. Diffusion limitations of monomer may also be important. The catalyst particle as such does not disintegrate and completely fall apart but is held together by precipitated polymer. As polymerization proceeds the catalyst particle increases in size, replicating the shape of the original catalyst particle, as is illustrated in Fig.7(a).

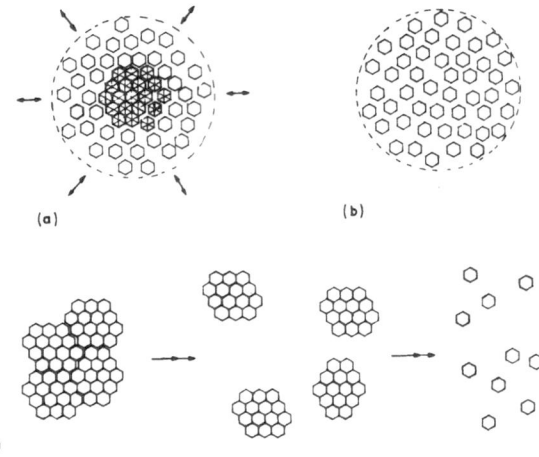

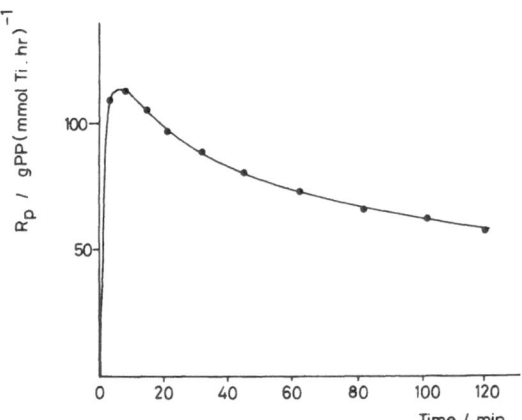

Fig.7. Diagramatic representation of catalyst particles during polymerization.

The primary crystallites become separated as the particle grows in size giving rise to the familiar multigrain model (30). These phenomena are believed to account for the typical rate-time profiles shown by $MgCl_2/DIOP/TiCl_4$ - $Al(i-Bu)_3$/catalyst systems in the polymerization of propylene (Fig.8).

Fig.8. Typical rate-time profile for the polymerization of propylene using a $MgCl_2/DIOP/TiCl_4$ - $Al(i-Bu)_3$/triphenylsiloxane catalyst at 70 °C. Al : Ti = 180 : 1. Pressure = 1 atm.

The most efficient use of titanium will occur for this type of catalyst when the situation shown in 7(b) is achieved.

When ball-milled $MgCl_2/EB/TiCl_4$ catalysts (EB = ethyl benzoate) are used higher activities and higher C_p^* values can normally be achieved (31). This situation is due to the greater breakdown of the $MgCl_2$ particles during ball-milling giving rise to higher concentrations of active centres which are highly active but deactivate quickly due to the high surrounding concentration of aluminium alkyl. The situation is shown schematically in Fig 7(c). To achieve the highest activity it is necessary to move to the far right of this diagram.

USE OF ^{14}CO RADIOTAGGING IN COPOLYMERIZATION EXPERIMENTS

Some of the most puzzling phenomena to emerge from kinetic studies on the copolymerization of ethylene and α-olefins are the rate enhancement effects which have been observed for a number of catalyst systems (32-36). Copolymerization studies of ethylene and 4-methylpentene-1 have been carried out at UMIST (37,38) using the catalyst systems:

(a) $\delta\text{-TiCl}_3 0.33AlCl_3 - Al(i\text{-}Bu)_3$
(b) $MgCl_2/EB/TiCl_4 - Al(i\text{-}Bu)_3$
(c) $MgCl_2/DIBP/TiCl_4 - Al(i\text{-}Bu)_3$ (DIBP = diisobutylphthalate)

The kinetic features of copolymerization experiments are found to be very sensitive to the order of addition of the components of the polymerization system as is shown in Fig.9.

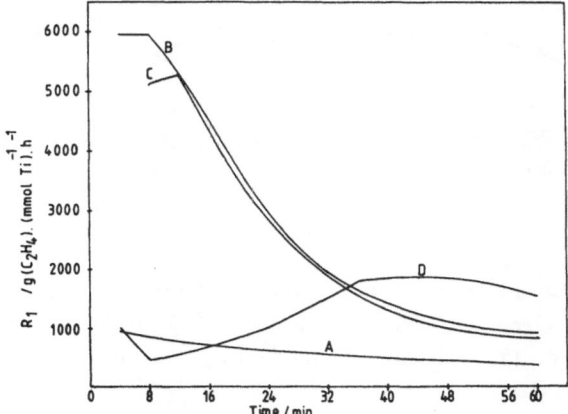

Fig.9. Effects of order of addition of components on rate-time profiles for copolymerization experiments at 60 °C. Catalyst system: $MgCl_2/EB/TiCl_4$ - $Al(i\text{-}Bu)_3$.

A   Homopolymerization:EC180/ethylene/cocatalyst/catalyst

B   EC180/ethylene/4-MP-1/cocatalyst/catalyst

C   EC180/cocatalyst/catalyst/4-MP-1 for 4 min then ethylene admitted

D   EC180/ethylene/cocatalyst/catalyst for 4 min then 4-MP-1 added

     (EC180 is a mixture of isomers of pentamethylheptane, boiling in the range
     170-180 °C)

In order to obtain high values for the rate of ethylene polymerization (R_1) it
is necessary for the ethylene and the 4-MP-1 to be added together followed by
catalyst and cocatalyst.

The following order of addition of the components of the polymerization system
was therefore used: EC180/ethylene/α-olefin/cocatalyst/catalyst

Representative rate-time profiles for the three catalyst systems using
varying concentrations of 4-MP-1 are shown in Fig 9-11. It is apparent that
significant rate enhancement takes place and that this depends on the
concentration of 4-MP-1 present.

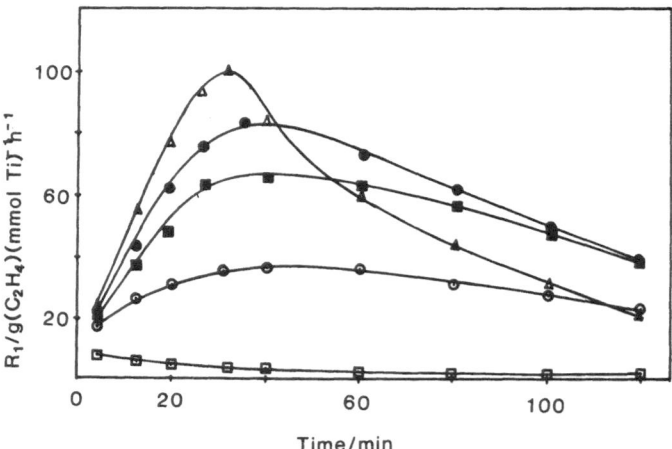

Fig.10. Typical rate time plots for the homopolymerization of ethylene and the
copolymerization of ethylene and 4-MP-1 using a δ-TiCl$_3$0.33 AlCl$_3$-Al(i-Bu)$_3$
catalyst system at 60 °C. [Ti] = 1.62 mmol dm^{-3}; [Al] = 3.30 mmol dm^{-3}.
Pressure = 1 atm.

□ = homopolymerization of ethylene;   O = 0.16 mol dm^{-3} 4-MP-1;

■ = 0.24 mol dm^{-3} 4-MP-1;   ● = 0.39 mol dm^{-3} 4-MP-1

Δ = 0.55 mol dm^{-3} 4-MP-1.

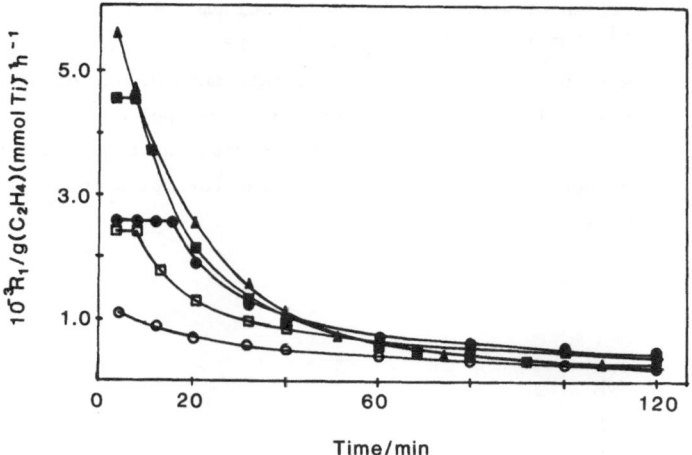

Fig.11. Typical rate-time profiles for the homopolymerization of ethylene and the copolymerization of ethylene and 4-MP-1 using a $MgCl_2/EB/TiCl_4-Al(i-Bu)_3$ catalyst system at 60 °C. Catalyst contained 1.35 wt% Ti. [Ti] = 0.030 mmol dm^{-3}; [Al] = 12.7 mmol dm^{-3}. Pressure = 1 atm.

○ = homopolymerization of ethylene; □ = 0.16 mol dm^{-3} 4-MP-1;
● = 0.24 mol dm^{-3} 4-MP-1; ■ = 0.39 mol dm^{-3} 4-MP-1
▲ = 0.47 mol dm^{-3} 4-MP-1

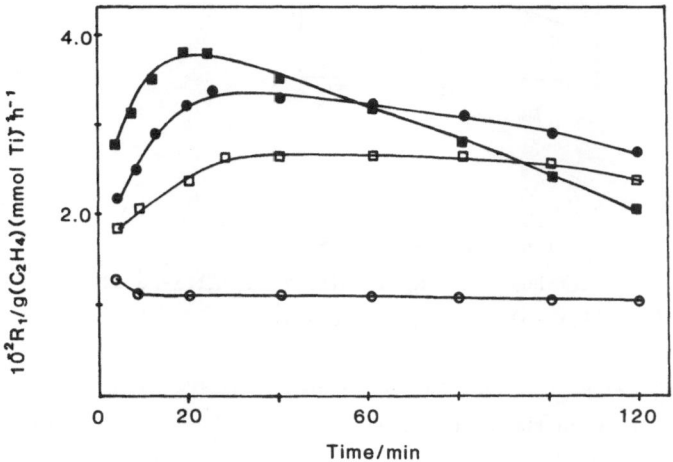

Fig.12. Typical rate-time profiles for the homopolymerization of ethylene and the copolymerization of ethylene and 4-MP-1 using a $MgCl_2/DIPB/TiCl_4-Al(i-Bu)_3$ catalyst system at 60 °C. Catalyst contained 4.0 wt% Ti. [Ti] = 0.13 mmol dm^{-3}; [Al = 12.5 mmol dm^{-3}. Pressure = 1 atm.

O = homopolymerization of ethylene; □ = 0.16 mol dm^{-3} 4-MP-1;
● = 0.332 mol dm^{-3} 4-MP-1; ■ = 0.47 mol dm^{-3} 4-MP-1

In order to investigate possible causes for the rate enhancement effects observed for the copolymerization of ethylene and 4-MP-1 polymerizations were treated with ^{14}CO at, or near, the time of maximum rate and the relevant active centre concentrations determined. Plots showing the variation in C_p^* with the concentration of 4-MP-1 are shown in Fig.13.

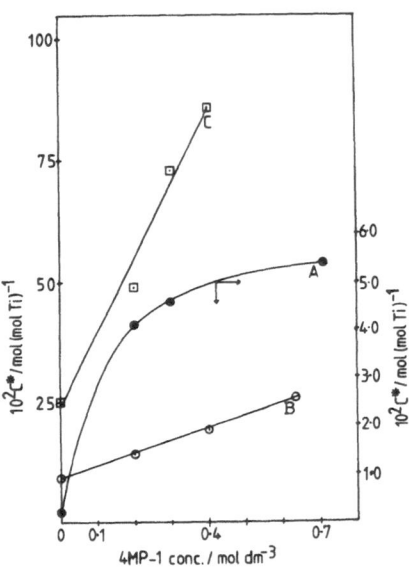

Fig.13. Variation of C_p^* with concentration of 4-MP-1.
A = δ-TiCl$_3$0.33AlCl$_3$-Al(i-Bu)$_3$; B = MgCl$_2$/DIBP/TiCl$_4$-Al(i-Bu)$_3$;
C = MgCl$_2$/EB/TiCl$_4$ - Al(i-Bu)$_3$. Contact time = 20 min.

An examination of the plots shown in Fig.13 shows the following:
(a) In homopolymerization the three catalyst systems have very different values for C_p^* :

$\quad$ δ-TiCl$_3$.033 AlCl$_3$ - Al(i-Bu)$_3$ $\quad$: $\quad$ C_p^* = 0.23% mol/mol Ti
$\quad$ MgCl$_2$/DIBP/TiCl$_4$ - Al(i-Bu)$_3$ $\quad$: C_p^* = 9% mol/mol Ti
$\quad$ MgCl$_2$/EB/TiCl$_4$ - Al(i-Bu)$_3$ $\quad\quad$: $\quad$ C_p^* = 25% mol/mol Ti.

(b) For all three catalyst systems C_p^* increases significantly with an increase in the concentration of 4-MP-1 in the range 0 to 0.4 mol dm^{-3}, the relative increase being much more pronounced in the case of the δ-TiCl$_3$0.33AlCl$_3$-Al(i-Bu)$_3$ catalyst system.(Table 4). The presence of 4-MP-1 in the case of the MgCl$_2$/EB/TiCl$_4$-Al(i-Bu)$_3$ catalyst system leads to a large percentage of the available titanium being used in polymerization centres. The enhanced rates of ethylene polymerization in the presence of 4-MP-1 clearly result from increased numbers of active centres.

Table 4. Rate Activation Factors in ethylene- 4-MP-1 copolymerization.

Catalyst system	4-MP-1 concentration mol dm^{-3}	R_1(max) Activation Factor	R_1(aver) Activation Factor
α-TiCl$_3$0.33AlCl$_3$ -Al(i-Bu)$_3$	0.39	11	16
MgCl$_2$/EB/TiCl$_4$ -Al(i-Bu)$_3$	0.47	4.7	2.1
MgCl$_2$/DIBP/TiCl$_4$ -Al(i-Bu)$_3$	0.47	2.8	2.7

Experimental details as in Fig 9-11

The presence of increased concentrations of active centres could arise from several causes.

(a) Fragmentation of catalyst matrix. Fragmentation of MgCl$_2$/phthalate/TiCl$_4$-AlEt$_3$ catalyst systems during prepolymerization with propylene followed by its removal and the introduction of ethylene has been demonstrated to take place (39). The use of a higher α-olefin is considered to produce the same effect and such prepolymerizations are used in commercial catalyst preparations. The importance and effects of catalyst disintegration have been described earlier in this paper.

(b) Active centre formation reactions. Although little is known about the actual initiation reactions, the monomer may be involved in the formation of active centres, or may be involved in the stabilization of such centres on formation. Thus different monomers could produce different concentrations of active centres.

(c) Displacement of adsorbed or complexed molecules. Displacement of molecules which otherwise would block sites may be effected by α-olefins.

$$S - D + M_1 \rightleftharpoons S - M_1 + D \qquad (6)$$
$$S - D + M_2 \rightleftharpoons S - M_2 + D \qquad (7)$$

α-olefins would be expected to be more effective in such displacement reactions than ethylene and so would generate greater concentrations of active centres.

(d) Diffusion phenomena. The occurence of diffusion effects (30) can also be expected also to lead to enhanced rates of copolymerization where the formation of copolymer segments break up the crystallinity of precipitated polymer and allows easier penetration of monomer to centres which would otherwise be dormant. A comparison of values for active centre concentrations in homopolymerization and in copolymerization systems requires, however, that these additional centres found in copolymerization systems do not function as centres at all in homopolymerizations and only become active in copolymerizations. The situation may however be dynamic with a smaller proportion of centres being used in homopolymerizations. Whilst the present indications are that (a) is likely to be a dominant factor, other factors can not be excluded, and further investigation is necessary before final conclusions can be reached.

FUTURE DEVELOPMENTS

The future promises to be as exciting as the present and the past. The demand for more accurate and meaningful methods for catalyst characterization can be expected to increase. There is a growing need for better models; models which can account for the following.

(a) Fragmentation of catalyst matrix.

(b) Active centre formation reactions.

(c) Active centre deactivation reactions.

(d) Centres of differing activity.

(e) Centres of differing stability.

(f) Diffusion phenomena.

ACKNOWLEDGEMENTS

This paper has incorporated results from a number of former and present research workers and I gratefully acknowledge the contribution of A A Akinbami, A L Burns, M O Jejelowo, J C Lea, W D McLellan and S Wang.

REFERENCES

1   Natta G (1959) J Polymer Sci 34: 21
2   Natta G, Pasquon I (1959) Adv Cat 11:1
3   Schnecko H, Kern W (1969) IUPAC, International Macromolecular Symposium,
     Budapest, p 365
4   Schnecko H, Kern W (1970) Chem Z 94: 229
5   Yermakov YuI, Zakharov VA (1975) in ; Chien JCW (ed) Coordination
     Polymerization: A Memorial to Karl Ziegler. Academic Press, New York, p 91
6   Tait PJT (1980) in: Lenz, RW, Ciardelli F (eds) Preparation and Properties
     of Stereoregular Polymers. Reidel Publishers Co, Dordrecht Holland, p 85
7   Tait PJT (1983) in: Quirk RP (ed) Transition Metal Catalyzed
     Polymerizations, Part A. Harwood Academic Publishers, New York, p115
8   Mejzlik J, Lesna M, Kratochila J (1987) Adv Polymer Sci 81: 84
9   Zakharov VA, Bukatov GD, Yermakov YuI, Demin EA (1972) Doklady Akademii
     Nauk 207:857
10   Burns AL, Caunt AD, Tait PJT (to be published)
11   Lea JC (1987) PhD Thesis, Manchester U K
12   Lea JC, Tait PJT (to be published)
13   Ajayi TT (1979) PhD Thesis, Manchester, U K
14   Ajayi TT, Caunt AD, Tait PJT (to be published)
15   Lesna M, Mejzlik J (1978) React Kinet Catal Lett Vol 9 No 1: 99
16   Mejzlik J, Lesna M (1977) Makromol Chem 178: 261
17   Bukatov GD, Zakharov VA Yermakov YuI (1978) Makromol Chem 179: 2097
18   Bukatov GD, Goucharov VS, Zakharov VA (1986) Makromol Chem 187: 1041
19   Tait PJT, Booth BL, Jejelowo MO (to be published) Makromol Chem, Rapid
     Commun.
20   Jejelowo MO, PhD Thesis, Manchester, UK
21   Burns AL (1976) PhD Thesis, Manchester, UK
22   Tait PJT (1980) Spec Per Rep, Macromol Chem, Roy Soc Chem 1 : 3
23   Tait PJT (1986) in: Keii T, Soga K (eds) Studies in Surface Science and
     Catalysis 25, Catalytic Polymerization of Olefins. Kodansha, Elsevier,
     Tokyo- Amsterdam- Oxford-New York, p 305
24   Tait PJT, Wang S (to be published), Brit Poly J
25   Wang S (1987) PhD Thesis, Manchester, UK
26   Tait PJT, McLellan DW (to be published)
27   Giannini U (1981) Makromol Chem Suppl 5: 216
28   Suzuki E, Tamura M, Doi Y, Keii T (1979) Makromol Chem 180: 2235
29   Kashiwa N, Yoshitake J (1982) Makromol Chem Rapid Commun 3: 211
30   Ray WH in (1983) : Quirk R P (Ed) Transition Metal Catalyzed
     Polymerizations Part A Harwood Academic Publishers New York, p191
31   Tait PJT, Abu Eid M, Enenmo EA 1985 ACS Meeting, Chicago, USA
32   Valvassoni A, Sartori G, Mazzanti G, Pazaro G (1963) Makromol Chem 61:46
33   Finogenova LT, Zakharov VA, Bunuyat-Zade AA, Bukatov GD, Plaksunov TK
     (1980) Polymer Sci USSR 22: 448
34   Kashiwa N, Yoshitaka J (1984) Makromol Chem 185: 1133
35   Lin S, Wang H, Zhang Q, Lu Z, Lu Y (1986) in: Keii T, Soga K (eds) Studies
     in Surface Science and Catalysis 25, Catalytic Polymerization of Olefins.
     Kodansha, Elsevier, Tokyo-Amsterdam-Oxford-New York, p91
36   Calabro DC, Lo Fy (1986) International Symposium on Transition Metal
     Catalysed Polymerization, Akron, Ohio, USA
37   Tait PJT, Downs GW, Akinbami A A (1986) International Symposium on
     Transition Metal Catalysed Polymerization, Akron, Ohio, USA
38   Akinbami AA (1985) PhD Thesis, Manchester, UK
39   Watkins ND (1987) (personal communication)

Steric Control of the Polymerization of α-Olefins in the Presence of Achiral Titanocenes-Methylalumoxane Homogeneous Catalysts

A. Zambelli and P. Ammendola

Dipartimento di Fisica, Università di Salerno, 84100 Salerno, Italy.

Ewen reported stereospecific polymerization of propene in the presence of homogeneous catalysts consisting of dicyclopentadienyldiphenyltitanium(IV) (CPT) and methylalumoxane (MAO) (1). Partially isotactic polypropylene is obtained in the presence of the above catalyst at low temperature (e.g. -40°C) and the stereochemical sequence of the configurations of the substituted carbons is in agreement with the Bernoullian statistical model of the stereospecific propagation proposed by Bovey (2). This fact is peculiar since the stereochemical sequence of the configurations of the substituted carbons of isotactic polypropylene obtained in the presence of either heterogeneous catalysts or homogeneous catalysts based on chiral stereorigid titanocenes or zirconocenes (1,3) is in agreement with the statistical model of the enantiomorphic sites proposed by Shelden (1,4,5). As suggested by Ewen (1) the Bernoullian model can be accounted for by assuming that the stereochemistry of the isotactic polyinsertion is controlled by the chiral chain end (1-3 like asymmetric induction (6)).

Syndiotactic polyinsertion of propene is also controlled by the chiral chain end (7) but the insertion of the monomer is secondary, whereas it is primary in the case of isotactic polymerization (1,7-10).

In order to check the proposed mechanism of the steric control we prepared polypropylene and poly-1-butene in the presence of CPT and MAO.

Enriched $Al(^{13}CH_3)_3$ (TMA) or $Al(^{13}CH_2CH_3)_3$ (TEA) were added to the catalyst hoping that the ^{13}C enriched aluminium alkyls appreciably exchange ligands with titanocene and that at least some macromolecules could initiate on the resulting titanium-enriched alkyl bonds (8). In fact in the spectra (Figs. 1 and 2) of the four polymer samples: a) polypropylene prepared in the presence of CPT-MAO-TMA; b) polypropylene prepared in the presence of CPT-MAO-TEA; c) poly-1-butene prepared in the presence of CPT-MAO-TMA and d) poly-1-butene prepared in the presence of CPT-MAO-TEA, one can observe the resonances expected for either the enriched methyls or the enriched methylenes of the end groups reported in the following scheme.

W. Kaminsky and H. Sinn (Eds.)
Transition Metals and Organometallics as
Catalysts for Olefin Polymerization
© Springer-Verlag Berlin Heidelberg 1988

Scheme   1

end group $\delta^{13}C$ (ppm from HMDS)

```
            C       C
            |       |
a1) ...C - C - C - C - ¹³C                    (20.5; 20.8; 21.4; 21.7)
```

```
            C       C
            |       |
b1) ...C - C - C - C - ¹³C - C                (27.5; 27.7; 28.4; 28.7)
```

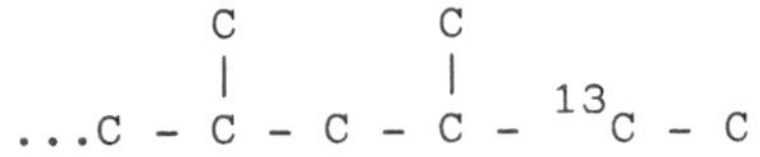

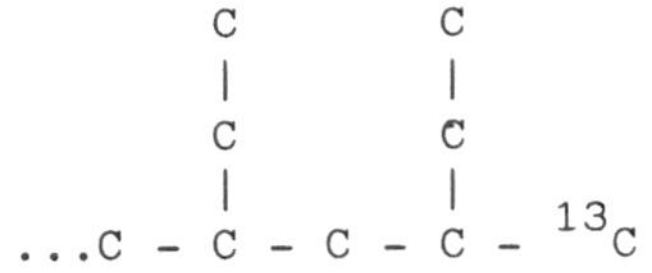

```
            C       C
            |       |
            C       C
            |       |
c1) ...C - C - C - C - ¹³C                    (17.7; 18.0)
```

```
            C       C
            |       |
            C       C
            |       |
d1) ...C - C - C - C - ¹³C - C                (24.2 and 24.5).
```

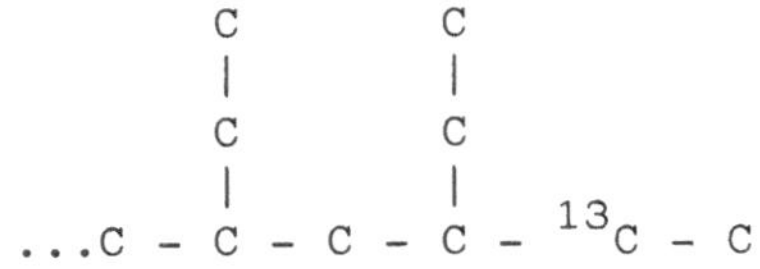

The resonances of the enriched carbons are splitted because of the effect on the chemical shift of ^{13}C of the mutual stereochemical arrangement of the substituent of the neighbouring monomer units (10), which (for the two first units) can be that shown in Fisher projection in Scheme 1 or that shown in Scheme 2.

Scheme   2

Syndiotactic end groups

```
            C                                       C
            |                                       |
a2) ...C - C - C - C - ¹³C        b2) ...C - C - C - C - ¹³C - C
                |                                       |
                C                                       C
```

```
            C                                       C
            |                                       |
            C                                       C
            |                                       |
c2) ...C - C - C - C - ¹³C        d2) ...C - C - C - C - ¹³C - C
                |                                       |
                C                                       C
                |                                       |
                C                                       C
```

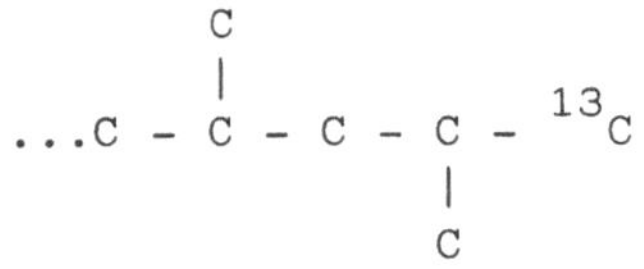

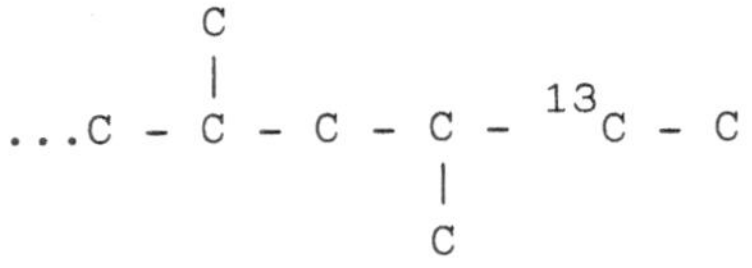

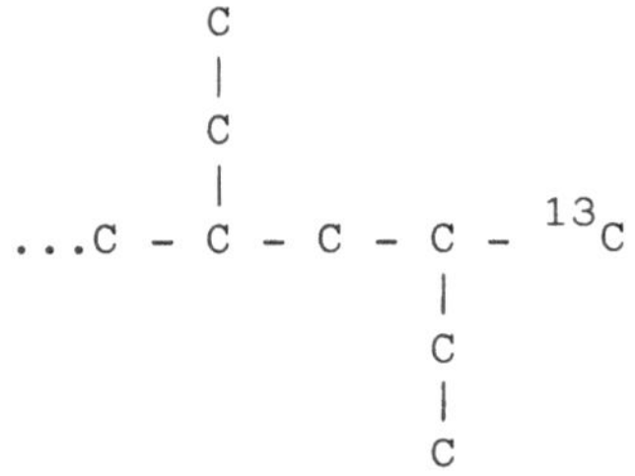

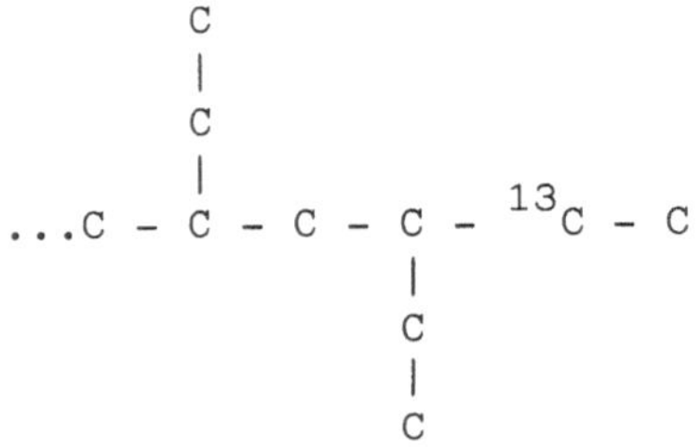

According to previous papers(7,8,10) the chemical shifts of the enriched carbons of the end groups having the stereochemical structure reported in Scheme 1, hereinafter called "isotactic", are centered at 20.5 and 20.8 ppm (a1), 27.5 and 27.7 ppm (b1), 17.7 ppm (c1), 24.2 ppm (d1).

The chemical shifts of the corresponding enriched carbons in the end groups of Scheme 2, hereinafter called "syndiotactic", are centered at 21.4 and 21.7 ppm (a2), 28.4 and 28.7 ppm (b2), 18.0 ppm (c2), 24.5 ppm (d2).

The intensities of the resonances of the enriched carbons of sample a (a1 and a2) are equal to each other as well as those of sample d (d1 and d2). Concerning sample b, the resonance of the enriched carbon of b1 (isotactic) is more intense than that of b2 (syndiotactic). In the spectrum of sample c, the resonance of the enriched carbon of c1 (isotactic) is less intense than that of c2 (syndiotactic).

These results are in agreement with the steric control of the growing chain end. In fact one can understand the experimental results by considering that the stereochemical structure of the considered end groups arises from insertion of two subsequent monomer units. The first insertion occurs either on $Mt-^{13}CH_3$ (Mt=metal) or on $Mt-^{13}CH_2-CH_3$ bonds. In both cases there are no chiral carbons on the alkyl bonded to Mt and therefore no steric control is expected. The insertion of the second monomer units occurs on :

$$
\begin{array}{ccc}
\quad\;\; C & & \quad\;\; C \\
\quad\;\; | & & \quad\;\; | \\
Mt - C - C - ^{13}C \quad (a) & or & Mt - C - C - ^{13}C - C \quad (b) \quad or \\
\end{array}
$$

$$
\begin{array}{ccc}
\quad\;\; C & & \quad\;\; C \\
\quad\;\; | & & \quad\;\; | \\
\quad\;\; C & & \quad\;\; C \\
\quad\;\; | & & \quad\;\; | \\
Mt - C - C - ^{13}C \quad (c) & or & Mt - C - C - ^{13}C - C \quad (d).
\end{array}
$$

In a and d there are no chiral carbons, and therefore the second insertion too is expected to be not stereospecific. On the contrary in b and c there is a chiral carbon and consequently it is expected that the second insertion will be more or less stereospecific. One can easily control that the like asymmetric induction actually would lead to the end groups b1 (isotactic) and c2 (syndiotactic).

A completely different pattern is observed when polymerization of propene and 1-butene is performed in the presence of stereorigid chiral titanocenes such as racemic diindenylethanedichlorotitanium(IV)/MAO and ^{13}C enriched $Al(CH_3)_3$ or $Al(C_2H_5)_3$. From the ^{13}C NMR spectra of the four samples (Figs. 3 and 4) one can observe that the end groups containing the enriched methylene (i.e. those of polypropylene and poly-1-butene prepared in the presence of enriched $Al(C_2H_5)_3$) are almost exclusively isotactic.

The end groups enriched on the methyl carbon (i.e. those of polypropylene and poly-1-butene prepared in the presence of enriched $Al(CH_3)_3$) are partially isotactic and partially syndiotactic. In this case it seems irrelevant whether the insertion of the first monomer unit leads to a chiral alkyl or not.

Therefore the placement of the enriched carbon is stereoregular when initiation occurs on metal-ethyl bonds while it is stereoirregular when it occurs on metal-methyl bonds. A similar trend was previously observed (7) for similar polymers prepared in the presence of heterogeneous isotactic specific catalysts and was explained by considering that in this case the isotactic steric control was due to the presence of a chiral stereorigid counter ion instead of being due to the chiral carbon of the last unit of the growing chain.

The different mechanism of the steric control for the two classes of catalysts is also confirmed observing (compare Figs. 1-2 and Figs. 3-4) that :

1) the polymers prepared in the presence of CPT are much less stereoregular than those prepared in the presence of diindenylethane-dichlorotitanium(IV).

2) Poly-1-butene prepared with CPT is much less stereoregular than polypropylene prepared with the same catalyst and poly-4-methyl-1-pentene (Fig. 5) is almost completely stereoirregular.

These results could be explained by considering that the extent of the 1-3 asymmetric induction from the last unit of the growing chain end to the incoming monomer, should depend on the relative hindrance of the substituents of the chiral carbon in the vicinity of the reactive metal carbon bond.

$$
\begin{array}{c}
R \\
| \\
Mt - CH_2 - C - P \\
| \\
H
\end{array}
$$

When turning from propene to 1-butene to 4-methyl-1-pentene the steric hindrance of the side substituent (R) increases more pronouncedly (methyl, ethyl, isobutyl) than that of the polymer chain (P) which, in first approximation, could be considered to be always equivalent to an isobutyl group. Accordingly, one could expect that asymmetric induction vanishes when R=i-butyl.

When the steric hindrance of R becomes larger than that of P one could also expect that the asymmetric induction might become unlike leading to more or less syndiotactic polymers.

We tested this idea by trying to polymerize vinylcyclopropane but the attempt was unsuccessfull since we did not achieve polymerization at all (11).

As reported previously a low yield of syndiotactic polystyrene was achieved in the presence of CPT-MAO (12). However it was observed that insertion of styrene into the active metal-carbon bonds of the syndiotactic specific catalyst tetrabenzyltitanium/MAO is secondary (13). As a consequence one can hardly consider that the mechanism of styrene polymerization is related to α-olefins polymerization at least in the presence of the catalysts here considered.

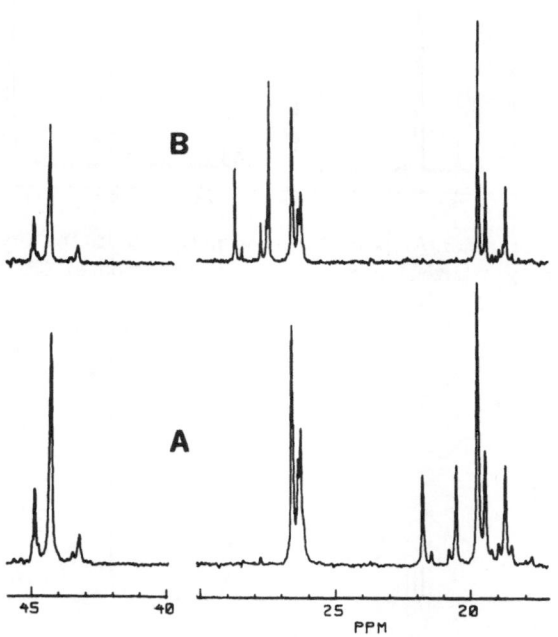

Fig. 1. ^{13}C NMR spectra of A) polypropylene prepared in the presence of CPT/MAO/TMA (sample a) and B) polypropylene prepared in the presence of CPT/MAO/TEA (sample b). HMDS scale.

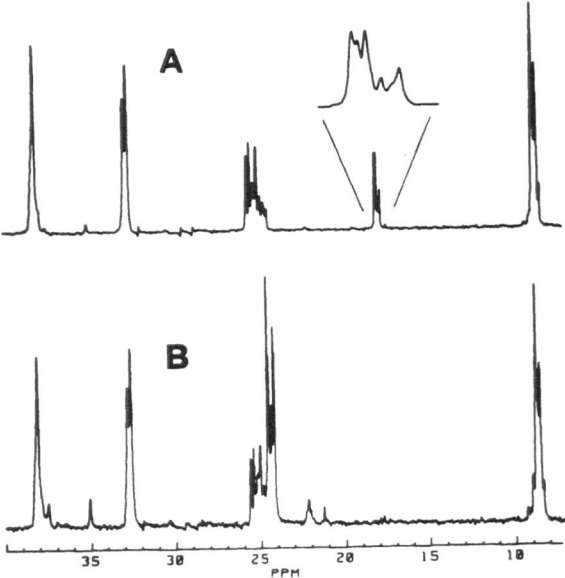

Fig. 2. ^{13}C NMR spectra of A) poly-1-butene prepared in the presence of CPT/MAO/TMA (sample c) and B) poly-1-butene prepared in the presence of CPT/MAO/TEA (sample d). HMDS scale.

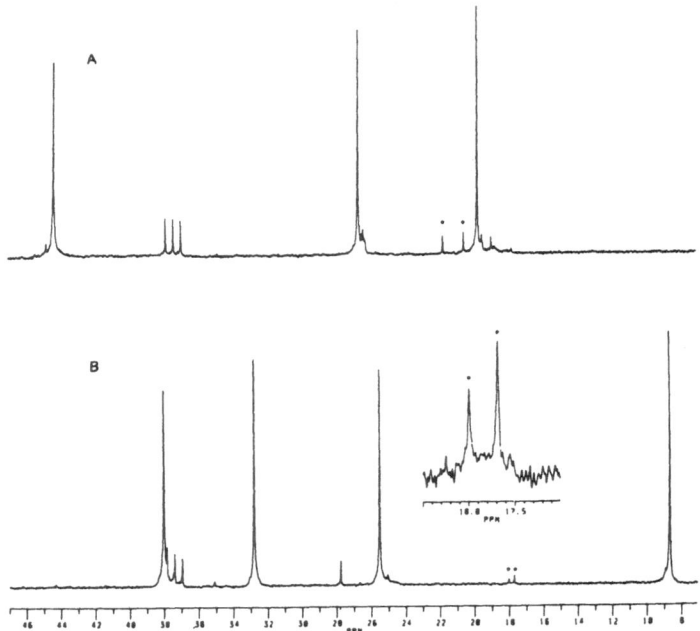

Fig. 3. ^{13}C NMR spectra of A) polypropylene prepared in the presence of diindenyl-ethanedichlorotitanium(IV)/MAO/TMA and B) poly-1-butene prepared in the presence of diindenylethanedichlorotitanium(IV)/MAO/TMA. HMDS scale.

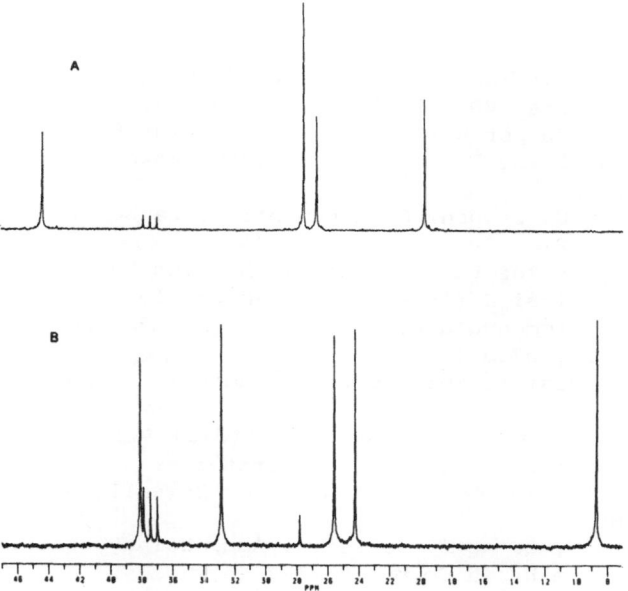

Fig. 4. ^{13}C NMR spectra of A) polypropylene prepared in the presence of diindenylethanedichlorotitanium(IV)/MAO/TEA and B) poly-1-butene prepared in the presence of diindenylethanedichlorotitanium(IV)/MAO/TEA. HMDS scale.

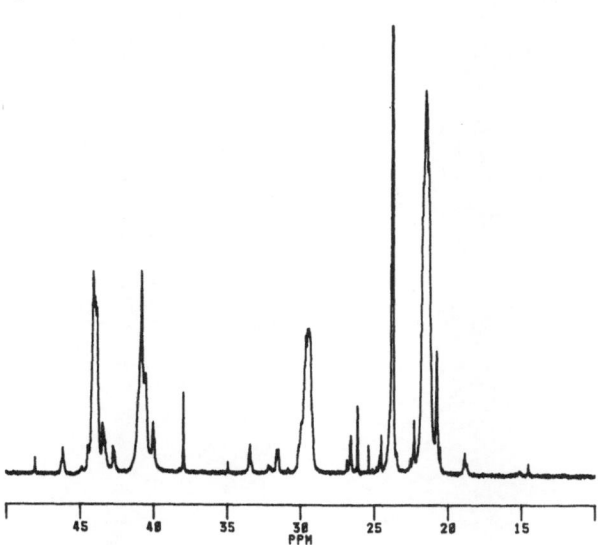

Fig. 5. ^{13}C NMR spectrum of poly-4-methyl-1-pentene prepared in the presence of CPT/MAO. HMDS scale.

REFERENCES

1)   Ewen JA (1984) J American Soc 106: 6355
2)   Bovey FA, Tiers GVD (1960) J Polym Sci 44: 173
3)   Kaminsky W, Kulper K (1985) Angew Chem Int Ed Engl 24: 506
4)   Shelden RA, Fueno T, Tsunetsugu T, Furokawa J (1965) J Polym Sci
     Part B 3: 23
5)   Wolfsgruber C, Zannoni G, Rigamonti E, Zambelli A (1975) Makromol
     Chem 176: 1121
6)   Seebach D, Prelog V (1982) Angew Chem Int Ed Engl 19: 857
7)   Zambelli A, Tosi C (1974) Adv Polym Sci 15: 38
8)   Zambelli A, Ammendola P, Grassi A, Longo P, Proto A (1986) Macro-
     molecules 19: 2703
9)   Longo P, Grassi A, Pellecchia C, Zambelli A (1987) Macromolecules
     20: 1015
10) Zambelli A, Bajo G, Rigamonti E (1978) Makromol Chem 179: 1249
11) Unpublished results from our laboratories
12) Ammendola P, Pellecchia C, Longo P, Zambelli A (1987) Gazz Chim
     It 117: 65
13) Pellecchia C, Longo P, Grassi A, Ammendola P, Zambelli A (1987)
     Makromol Chem Rapid Commun 8: 277

Possible Models for the Steric Control in the Heterogeneous High-Yield and Homogeneous Ziegler-Natta Polymerizations of 1-alkenes

Paolo Corradini, Vincenzo Busico and Gaetano Guerra

Dipartimento di Chimica dell' Università
Via Mezzocannone, 4
I-80134 Napoli (Italy)

INTRODUCTION

In our research group, we have been investigating the mechanisms presiding over the steric control in the Ziegler-Natta polymerizations of 1-alkenes to polymers of high or very high stereoregularity.

In this paper, we present recent results concerning the high-yield $MgCl_2$-supported catalysts as well as the homogeneous systems, based on titanium or zirconium compounds and alkyl-Al-oxanes, very recently disclosed.

For the former, in addition to possible models of active centers, experimental results (in particular, direct observations of initial polymer growth on $MgCl_2/TiCl_4$ single crystals) favouring such models are reported.

As for the homogeneous catalysts, we discuss the mechanisms of steric control for propene polymerization to isotactic polymer (in the presence of chiral systems) and preliminary mechanistic hypotheses on the polymerization of styrene to syndiotactic polymer.

POSSIBLE MODELS FOR THE STERIC CONTROL IN THE HETEROGENEOUS HIGH-YIELD ZIEGLER-NATTA POLYMERIZATIONS

We begin this section with a necessary brief mention of our past work on first generation heterogeneous Ziegler-Natta catalyst systems.

The active centers of violet-$TiCl_3$-based catalysts are generally thought to be located on the coordinatively unsaturated lateral faces, rather than on the saturated basal faces of the platelet-like $TiCl_3$ crystals (Fig. 1a) (1-5). This hypothesis is supported by microscopic observations of initial polymer growth on well-formed catalyst single crystals (6,7).

$TiCl_3$ layer

(a)

● Cl atoms up
○ Cl atoms down
● metal atoms

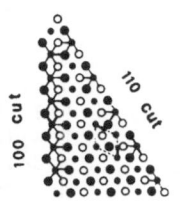

$MgCl_2$ layer

(b)

Fig. 1. Schematic drawings of the crystal lattice of violet $TiCl_3$ (a) and $MgCl_2$ (b). Possible lateral cuts of the structural layers are also indicated.

W. Kaminsky and H. Sinn (Eds.)
Transition Metals and Organometallics as
Catalysts for Olefin Polymerization
© Springer-Verlag Berlin Heidelberg 1988

The stereospecificity of these catalysts in the polymerization of 1-al
kenes is due, according to most authors (1-5), to interactions between
non-bonded atoms at chiral active centers.

In recent years, we have quantitatively evaluated possible models of
active centers by computing the non-bonded energies for all sets of in
ternal coordinates accessible to the atoms at the model centers (8-10).
Our calculations suggest that edges, steps and reliefs on lateral faces
of TiCl₃ crystals in a layered modification (α, γ or δ) behave similarly
in providing active centers able to exert a stereospecific control; po
lymerization should occur prevailingly at these more exposed centers
because of the much lower activation energies than those for centers on
plain surfaces.

A single general model emerges (Fig. 2): lower steric repulsions seem
to be implied for a coordination of the growing polymer chain in the
more hindered octahedral position of the two usually available at a sur
face Ti atom; this in turn imposes a chiral orientation of the first
C-C bond of the chain, which renders the center able to discriminate
between the two faces of a coordinated prochiral 1-alkene molecule.

As extensively discussed elsewhere (11-15), our model of catalytic cen
ters agrees well with a large number of experimental facts (e.g. type
of tacticity errors along a mainly isotactic chain; maintainance of iso
tacticity after an ethene insertion; stereospecificity of the initiation
reaction; relative reactivity of different 1-alkenes; degree of stereo
selectivity in the polymerization of chiral monomers), some of which
resulting from crucial experiments designed to test its validity (16,17).

In the last decade, more active systems based on TiCl₄ supported on MgCl₂
have largely replaced TiCl₃-based catalysts in the industrial production
of isotactic poly-1-alkenes (e.g. polypropene) (18-19).

The similar general behaviour of the two classes of catalyst systems
indicates that the nature of the active centers is the same (20-22).
On the other hand, in contrast with TiCl₃, MgCl₂-supported catalysts
require the use of suitable "conditioning" Lewis bases - as components
of the solid catalyst ("internal" base) as well as of the cocatalyst
("external" base) - in order to reach high stereospecificities (18,19,23,24)

Λ site

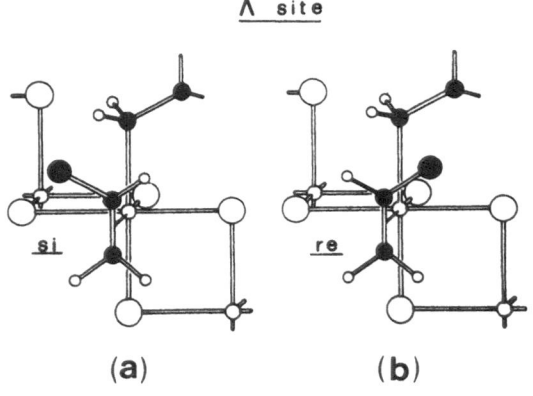

(a)   (b)

Fig. 2. Schematic drawing of
the local environment of a
model active center located
at a relief on the lateral
surface of a TiCl₃ crystal,
showing the two possible enan
tiomeric coordinations of a
propene molecule suitable for
its primary insertion (Λ site).
(a) si coordination (favoured for
insertion into the growing chain)
(b) re coordination

It is the aim of this section to show how (models of) active centers
resulting from the surface coordination of small titanium chloride units
to unsaturated faces of $MgCl_2$ crystals may have a structure which is
analogous to that of centers residing at unsaturated faces of $TiCl_3$, and
to discuss the role of the Lewis base modifiers in increasing catalyst
specificity.

Violet $TiCl_3$ and $MgCl_2$ have very similar crystal structures (Fig. 1)
(25-28), resulting from the stacking of structural layers according to
a close packing of the chlorine atoms. Each structural layer can be view
ed as a "sandwich" in which two planes of chlorine atoms embed a plane
of metal atoms occupying the octahedral interstices.

As already discussed for $TiCl_3$, the lateral faces of the $MgCl_2$ crystals
(and even of single, small $MgCl_2$ layers) must be coordinatively unsatu
rated for electroneutrality conditions. It seamed reasonable to suppose
that $TiCl_4$ coordination takes place more strongly to these faces, which would then
be the ones active in polymerization.

Up to now, on the other hand, no convincing experimental evidence of
the kind presented for $TiCl_3$ (6,7) had been reported for $MgCl_2$-supported
catalysts, likely for the difficulty in performing polymerization expe
riments on well-formed $MgCl_2$ single crystals, exceedingly hygroscopic.

We present here the first results obtained in our laboratory of the
gas phase polymerization of propene on single crystals of $MgCl_2$ on which
$TiCl_4$ had been deposited from the vapour phase (the experimental appa
ratus and procedures will be described elsewhere (29)). Polymer growth
was continuously followed using an optical microscope operating at high
magnification (1000X-2000X) and resolution (2 µm); the transmission image
was recorded on a videotape from a television camera mounted on the mi
croscope.

Figure 3 documents a typical process of initial polypropene formation
on a well-shaped $MgCl_2/TiCl_4$ single crystal (cocatalyst $Al(CH_3)_3$). The
crystal is oriented, as usual in these cases, with the large basal (001)
faces parallel to the window of the reactor (and to the plate of the
microscope). From the figure, it is seen that polymer growth is indeed
confined to the lateral faces, the basal faces remaining virtually
inactive.

Possible lateral cuts of $MgCl_2$ are shown in Fig. 1b, which correspond
to (100) and (110) faces. Both have been observed by optical and electron
microscopy on $MgCl_2$ crystals. These two kinds of lateral cuts imply very
different local situations, since electroneutrality conditions impose
an average coordination number of 5 for the Mg atoms on the (100) cuts,
of 4 for those on the (110) cuts.

Figure 4a shows three different models of epitactic placements of
$TiCl_4$ units, which turned out to be the most stable from electrostatic
energy calculations, on these two lateral faces; Fig. 4b illustrates
the same species after the reduction to $TiCl_3$ by the Al-alkyl cocatalyst.
(Experimental evidence for the formation of Ti_2Cl_6 dimers can be found
in the EPR data by Chien et al. (30) and by Sergeev et al. (31), poin
ting out that a considerable fraction of Ti^{3+} present on the $MgCl_2$ sur
face is EPR-silent). Alkylation of these species by the Al-alkyl would
turn them into active centers.

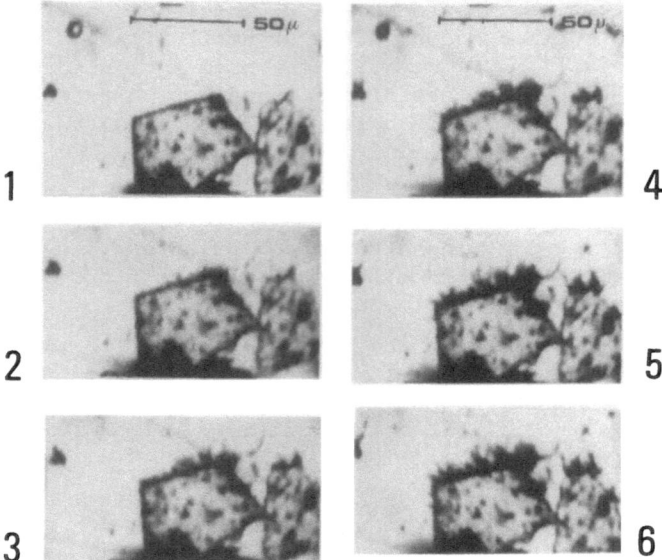

Fig. 3. Initial polypropene growth on a $MgCl_2/TiCl_4$ single crystal; polymerization takes place only on the lateral crystal faces, leaving the large basal faces practically unaffected. The pictures are taken every 30 seconds (from 0 s (1) to 150 s (6)).

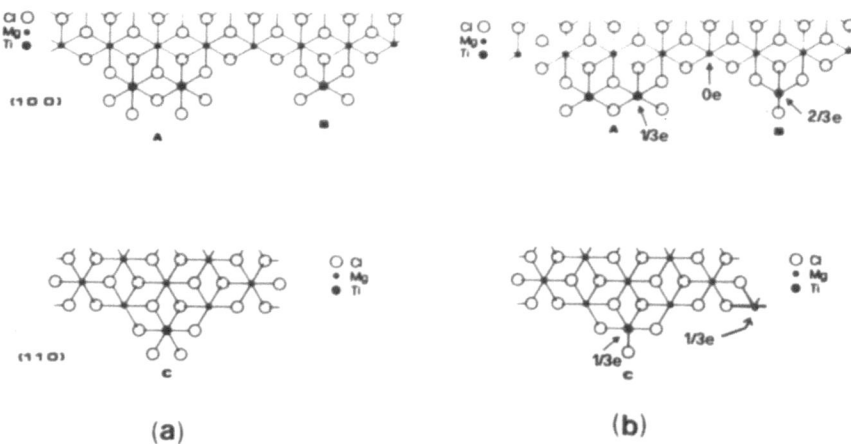

Fig. 4. (a) Models of epitactic placements of $TiCl_4$ units on the (100) faces (A,B) and (110) faces (C) of $MgCl_2$ crystals; (b) the same species after reduction to $TiCl_3$ by the Al-alkyl. The formal effective charges of surface Mg and Ti atoms are also indicated.

Considerations on the non-bonded interactions similar to those made for $TiCl_3$ (8-10) showed (32) that the mononuclear $TiCl_3$ surface coordination adducts (B,C in Fig. 4b) cannot provide stereospecific centers. (Here and in the following, we denote as "non-stereospecific centers" those active centers which are not highly stereospecific in propene polymerization).

On the contrary, active centers deriving from chiral Ti_2Cl_6 dimers on the (100) faces of $MgCl_2$ crystals (A in Fig. 4b) can have stereoregulating ability, the main factor determining their stereospecific behaviour being the fixed, chiral orientation into which the growing polymer chain appears to be forced (32). As is apparent from Fig. 5, such dimers are very similar to Ti_2Cl_6 reliefs proposed by us as precursors of stereospecific active centers on lateral (110) faces of violet $TiCl_3$ crystals (8-10); this would agree with the common behaviour of the stereospecific centers experimentally observed for the two classes of catalysts (20-22).

A possible explanation for the low stereospecificity of the $MgCl_2$-supported catalysts in the absence of Lewis bases is the existence of several types of active centers, only a fraction of which with stereoregulating ability. Indeed, according to our model, out of the three plausible precursors of active centers, two (B,C in Fig. 4b) are non-stereospecific.

In principle, (at least) two mechanisms can be hypothesized for the action of the Lewis bases in increasing catalyst specificity:
i) "Poisoning" of the non-stereospecific active centers, through coordination or some other chemical reaction (33,34);
ii) Conditioning of the $MgCl_2$ surface, in order to avoid the formation of non-stereospecific active centers.

What follows is a brief survey of experimental data (23,24) suggesting, in our opinion, that both mechanisms are active, the first one pertaining to the external base, the second one to the internal base. Esters of aromatic carboxylic acids were used both as internal and external bases; similar results, however, can be obtained for other classes of effective Lewis bases (e.g. hindered secondary amines, alkyl-alkoxy-silanes) (35).

A

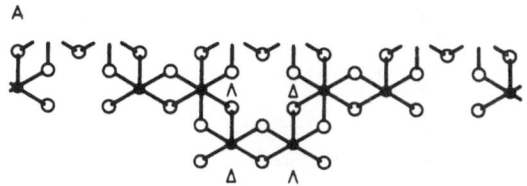

B

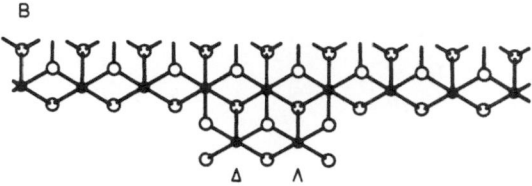

Fig. 5. Models of: (a) Ti_2Cl_6 reliefs on the (110) cut of $TiCl_3$; (b) Ti_2Cl_6 dimer epitactically placed on the (100) cut of $MgCl_2$. The chiral environments of the metal atoms are explicitly labelled.

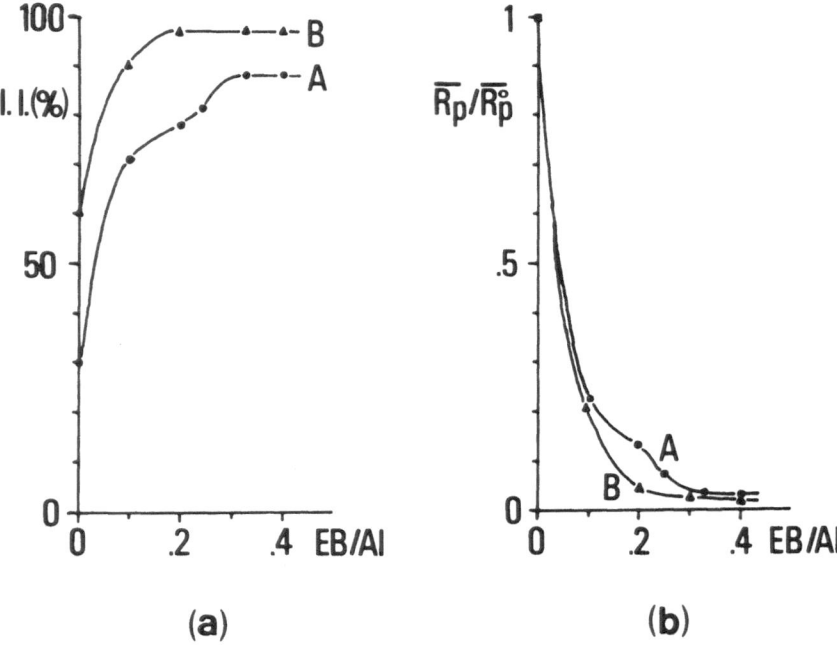

(a)   **(b)**

Fig. 6. Effects of the internal and external base on propene polymeriza tion in the presence of $MgCl_2$-supported catalysts: A - catalyst $MgCl_2/TiCl_4$; B - catalyst $MgCl_2$/ethyl benzoate/$TiCl_4$
(a) Isotactic index (I.I.) of polypropene vs. (ethyl benzoate/$Al(C_2H_5)_3$) mole ratio in the cocatalyst
(b) Normalized catalyst productivity of non-stereoregular polypropene fraction, $\bar{R}_p/\bar{R}_p^o$ vs. (ethyl benzoate/$Al(C_2H_5)_3$) mole ratio in the cocatalyst
For details on the experimental conditions, see ref. 24.

Table 1. Effects of the internal and external base on propene polymeri zation in the presence of $MgCl_2$-supported catalysts: $\bar{M}_w/\bar{M}_n$ values of selected non-stereoregular polypropene fractions. For details on the experimental conditions, see ref. 24.

Catalyst	(Ethyl benzoate/$Al(C_2H_5)_3$) mole ratio in the cocatalyst	$\bar{M}_w/\bar{M}_n$
$MgCl_2/TiCl_4$	0	5.0
	0.10	5.9
	0.33	6.1
$MgCl_2$/ethyl benzoate/$TiCl_4$	0	3.7
	0.10	2.7

Figure 6a reports the isotactic index (I.I., % by weight insoluble in boiling n-heptane) of the polypropene produced in the presence of two different catalysts (e.g. $MgCl_2/TiCl_4$ (A) and $MgCl_2$/ethyl benzoate/$TiCl_4$ (B)) as a function of the mole ratio (ethyl benzoate/$Al(C_2H_5)_3$) in the co catalyst. The corresponding normalized catalyst productivities referred to the non-stereoregular (n-heptane soluble) polymer fraction are shown in Fig. 6b. Table 1 lists typical values of $\overline{M}_w/\overline{M}_n$ for the non-stereore gular fractions of selected polymer samples.

The data suggest that, in the absence of internal base (catalyst A), mainly two types of non-stereospecific active centers are present, one of which of low reactivity with the external base; when internal base is used (catalyst B), mainly one type of non-stereospecific centers seems to be present, readily poisoned by the external base.

A tentative rationalization of the experimental data may be derived from simple electrostatic considerations on our model of catalytic sites. In this framework, the two types of non-stereospecific centers formed in the absence of internal base would derive from the surface complexes B,C in Fig. 4.

Using the procedure of Arlman and Cossee as applied to the crystal lattice of $TiCl_3$ (1,2) and extended by Goodall to $MgCl_2$ (33), a formal effective charge of Oe can be assigned to Mg atoms on the (100) faces, and of +(1/3)e to Mg atoms on the (110) faces of $MgCl_2$ crystals (Fig. 4b). This suggests that the (110) faces be more acid (in the Lewis sense) than the (100) faces - or, equivalently, that the (100) faces be more basic than the (110) ones. This qualitative statement is confirmed by energy calculations (in the electrostatic approximation) of $TiCl_4$ (a relatively strong Lewis acid) or Ti_2Cl_8 dimer binding to the two faces.

We believe that the role of the internal base, which is contacted with $MgCl_2$ prior to $TiCl_4$ (23,24) and is largely bound to Mg on the final catalyst (36,37), is to shield from $TiCl_4$ coordination the (110) faces (or similar sites) of the support, where only precursors of non-stereo specific active centers (such as C in Fig. 4) would be formed.

The reason why such centers are so detrimental to catalyst performance may originate again from electrostatic factors. Fig. 4b indicates the formal effective charges pertaining to the Ti atoms in the model precur sors of active centers: +(2/3)e for Ti atoms in species B, +(1/3)e for Ti atoms in species A,C, suggesting a higher Lewis acidity of the former. If so, the external base shall poison selectively the non-stereospecific centers derived from species such as B, but not those from species such as C, of Lewis acidity comparable with that of the stereospecific centers.

POSSIBLE MODELS FOR THE STERIC CONTROL IN THE HOMOGENEOUS ZIEGLER-NATTA POLYMERIZATIONS

As in the case of the heterogeneous catalysts, the evaluation of models of catalytic species for the homogeneous stereospecific Ziegler-Natta polymerizations was performed according to the following scheme:
i) Definition of stable (minimum energy) geometries for the possible stereo- and/or diastereoisomeric situations at the active center;
ii) Selection of those with conformations nearer to the transition state

(situations of proximity);
iii) Comparative evaluation of the conformational energy regions avai̲
lable for reactivity (transition situations and reaction paths).

In accordance with point ii), we start from conformational situations
in which the 1-alkene double bond is nearly parallel to the metal-poly̲
meryl bond; the reaction path goes further through the η^2 to η^1 defor̲
mation of the olefin and the nucleophilic attack to it of the polymeryl
chain.

This approach was first used for the polymerization of propene to syndio̲
tactic polymer in the presence of V-based catalysts; for details, the
interested reader is referred to ref. 38.

More recently, our attention has been drawn by the homogeneous isospeci̲
fic catalysts, based on titanocenes and zirconocenes in combination with
alkyl-Al-oxanes, newly discovered (39,40). Depending on the ligands
at the metal, the origin of the stereocontrol for such catalysts has
been attributed in some cases to the chirality of the catalytic complex
(39,41), in some other cases to that of the last inserted monomer unit
in the growing chain (39,42).

We present here a possible model for the chiral-site-controlled catalysis,
in the case of stereorigid chiral complexes with an ethene-bis-(1-inde̲
nyl) ligand (Fig. 7), as studied in our group (43).

We regard the coordination to the metal of three more ligands in addition
to the bis-indenyl (proposed by some authors (39,40)) as unlikely on the
basis of calculations of non-bonded interactions. In consequence, we
assume that the catalytic complex presents only two ligands (coordinated
monomer and growing polymer chain) besides the indenyl groups in the
stage preceeding monomer insertion (Fig. 7). This is possible, for in̲
stance, if the catalyst has an ion-pair nature, in analogy with the
cationic character of the catalytic species assumed for similar systems
(44-47):

$$[(\text{bis-indenyl ligand}) \, \text{Me} \, (\text{propene}) \, P_n]^+ \, [R(\text{AlCH}_3\text{O})_x]^- \longrightarrow$$

$$[(\text{bis-indenyl ligand}) \, \text{Me} \, P_{n+1})]^+ \, [R(\text{AlCH}_3\text{O})_x]^-$$

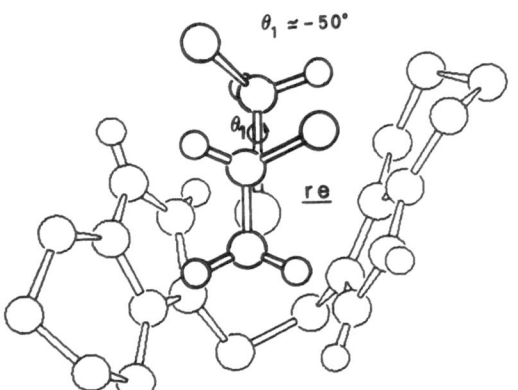

$\theta_1 \simeq -50°$

re

Fig. 7. Model catalytic site
for the chiral-site-controlled
homogeneous isospecific poly̲
merization, the ligands at the
metal being an ethene-bis-indenyl
group, a propene molecule and
an alkyl (polymeryl) group.
The minimum energy diastereoisomeric
complex is sketched with (R,R) chi̲
rality of coordination of the bis-
-indenyl ligand and re chirality
of propene coordination.

The non-bonded energies, minimized with respect to the orientation of
the olefin, for a (R,R) complex with an ethyl group and a propene mole
cule (for both re and si coordination of the latter) are shown in Fig. 8
as a function of the internal rotation angle, θ_1 around Me-C(ethyl)
bond (Me = Ti). The calculations indicate that the chiral environment
of the metal atom forces the growing chain to a chiral orientation (the
minimum at $\theta_1 = -50°$ being much deeper than those at $\theta_1 = +40°$) which
in turn favours the coordination of the prochiral monomer with one of
its two faces (the re coordination). The situation of energy minimum is
sketched in Fig. 7.

The similarity between this model and that proposed in the previous sec
tion for the heterogeneous isospecific catalysts is readily apparent
from a comparison between Fig. 7 and Fig. 2. In both cases, the chiral
environment of the metal atom forces the growing polymer chain to assume
a chiral orientation, which in turn discriminates between si and re
coordinated olefins.

The predicted behaviour of this model catalytic site seems to be in agree
ment with the available experimental data. In particular, the model is
non-stereospecific for monomer insertion into an initial metal-methyl
bond, as found by Zambelli and coworkers (41). Moreover, it is in accor
dance with the elegant results by Pino et al. (48) of analysis and opti
cal activity measurements on the saturated propene oligomers obtained,
under suitable conditions, with this kind of catalysts, proving that the
re insertion of the monomer is favoured in case of (R,R) chirality of
coordination of the bis-indenyl ligand.

Even more recently, the polymerization of styrene to syndiotactic polymer
with soluble Ziegler-Natta catalyst systems such as tetrabenzyltitanium
and methyl-Al-oxane has been reported (49,50). According to Zambelli et
al. (50), the type of monomer insertion is secondary, and polymer ste
reoregularity is very high, the fraction of rr triads being over 98%
(much higher than for syndiotactic polypropene obtained in the presence
of V-based catalysts).

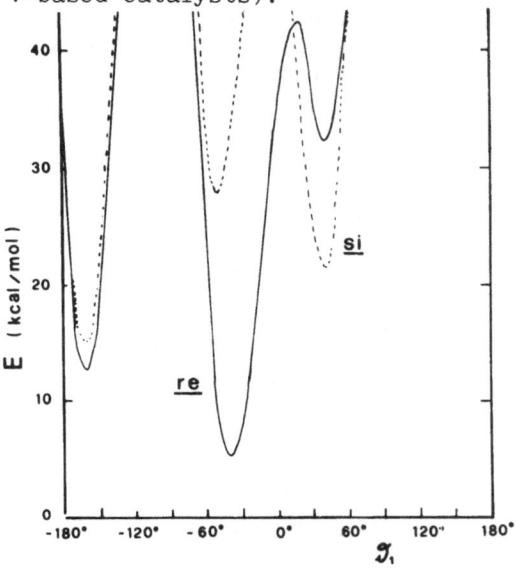

Fig. 8. Non-bonded energies,
minimized with respect to
olefin orientation, as a fun
ction of the dihedral angle
θ_1 (see Fig. 7), for the mo
del catalytic site of Fig. 7.
Solid line: re propene coordination
Dashed line: si propene coordina
tion.

The explanation of this high degree of stereospecificity must reside
in some kind of asymmetry in the catalytic complex, which gives rise
to a large difference in the activation energies for isotactic and
syndiotactic insertion.

According to Zambelli (50), the catalytic complex at which styrene poly
merization occurs is different from that acting for the polymerization
of ethene in the presence of the same catalyst system.

We may think of it as a cation in which the metal unsaturation is largely
compensated through the weak bonding at relatively large distances of
aromatic moieties, such as the benzene ring of the solvent, of the sty
rene monomer, of the growing polymer chain. It is well known, for in
stance, that the benzyl group may be bound to titanium not only through
the methylene group, but also with the aromatic group as if it were some
sort of a bidentate molecule (Scheme I) (51).

Scheme I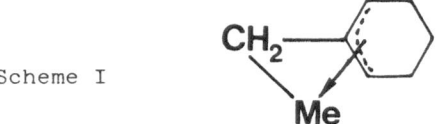

If the growing polymer chain interacts with the transition metal in the
catalyst through more than one carbon atom, a high degree of asymmetry
can be created thereby. This is shown in Fig. 9, which is intended as
an illustration of a possible mechanism to explain the observed stereo
specificity, rather than a really proved structure of the catalytic
complex.

The figure shows the hypothetical structure of a cation of the kind
[Ti L_2 (polymeryl) (styrene)]$^+$, in which the polymeryl group (behaving
as a near to bidentate benzyl ligand) and the aromatic ring of a monomer
molecule reside on the equatorial girdle of the complex. If the steric
situation is that shown in Fig. 9, it appears reasonable that, due to
restrictions provided by the bidentate binding of the chain, the two
enantiomeric faces of the double bond are differently apt to join the
growing chain. In particular, the further activation step could occur
more easily as shown under (a) rather than under (b), this giving rise
to a syndiotactic propagation.

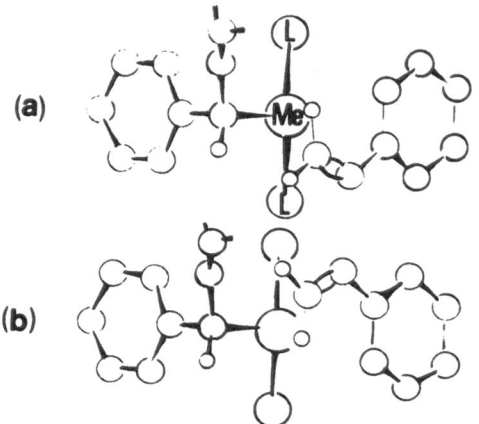

(a)

(b)

Fig. 9. Possible conformations
of a hypothetical complex, in
which a polystyryl chain and
a styrene molecule, both posi
tioned in the equatorial girdle,
interact with the metal through
the phenyl ring:
(a) Precursor to syndiotactic
    insertion
(b) Precursor to isotactic
    insertion (less favoured).

The provision of suitable asymmetric environments through multidentate coordination and the set up of steric restrictions to reaction may be operative, through similar mechanisms, for other polymerization systems as well as they are in some metal-aided highly stereospecific reactions occurring for low molecular mass compounds (52).

REFERENCES

(1) Arlman EJ, Cossee P (1964) J. Catalys. 3:99
(2) Cossee P (1967) in "The Stereochemistry of Macromolecules", Ketley AD Ed., vol. 1, chap. 3, M. Dekker (New York)
(3) Allegra G (1962) Nuovo Cimento 23:502
(4) Boor J Jr (1979) "Ziegler-Natta Catalysts and Polymerizations", Academic Press (New York)
(5) Kissin YV (1985) "Isospecific Polymerization of Olefins", Springer--Verlag (New York)
(6) Rodriguez LAM, Gabant JA (1963) J. Polym. Sci. Part C 4:125
(7) Rodriguez LAM, Gabant JA (1966) J. Polym. Sci. Part A1 4:1971
(8) Corradini P, Barone V, Fusco R, Guerra G (1979) Eur. Polym. J. 15:1133
(9) Corradini P, Guerra G, Fusco R, Barone V (1980) Eur. Polym. J. 16:835
(10) Corradini P, Barone V, Fusco R, Guerra G (1982) J. Catalys. 77:32
(11) Corradini P, Guerra G (1986) Proc. Int. Symp. on "Transition Metal Catalyzed Polymerizations", Akron (OH)
(12) Corradini P, Barone V, Guerra G (1982) Macromolecules 15:1242
(13) Ammendola P, Guerra G, Villani V (1984) Makromol. Chem. 185:2599
(14) Guerra G, Pucciariello R, Villani V, Corradini P (1987) Polymer Commun. 28:100
(15) Corradini P, Guerra G, Villani V (1985) Macromolecules 18:1401
(16) Zambelli A, Locatelli P, Sacchi MC, Tritto I (1982) Macromolecules 15:831; Zambelli A, Sacchi MC, Locatelli P, Zannoni G (1982) Macromolecules 15:211
(17) Sacchi MC, Locatelli P, Tritto I (1985) Makromol. Chem. Rapid Commun. 6:597
(18) Galli P, Luciani L, Cecchin G (1981) Angew. Makromol. Chem. 94:63
(19) Galli P, Barbè PC, Noristi L (1984) Angew. Makromol. Chem. 120:73
(20) Chien JCW (1979) in "Preparation and Properties of Stereoregular Polymers", Lenz RW, Ciardelli F Eds., Reidel (Dordrecht), p. 113
(21) Canova L, Mazzullo S, Giannini U (1980) Prepr. IUPAC Macro, Florence, 2:16
(22) Doi Y, Suzuki E, Keii T (1981) Makromol. Chem. Rapid Commun. 2:293
(23) Busico V, Corradini P, De Martino L, Proto A, Savino V, Albizzati E (1985) Makromol. Chem. 186:1279
(24) Busico V, Corradini P, De Martino L, Proto A, Albizzati E (1986) Makromol. Chem. 187:1115
(25) Natta G, Corradini P, Bassi IW, Porri L (1958) Atti Accad. Naz. Lincei (Ser. 8) 24:121
(26) Natta G, Corradini P, Allegra G (1961) J. Polym. Sci. 51:399
(27) Natta G, Corradini P, Allegra G (1959) Atti Accad. Naz. Lincei (Ser. 8) 26:155
(28) Ferrari A, Braibanti A, Bigliardi G (1963) Acta Cryst. 16:846

(29) Busico V, Corradini P, De Martino L, Di Rosa S, to be published
(30) Chien JCW, Wu JC (1982) J. Polym. Sci. Polym. Chem. Ed. 20:2461
(31) Sergeev SA, Poluboyarov VA, Zakharov VA, Anufrienko VF, Bukatov GD (1985) Makromol. Chem. 186:243
(32) Corradini P, Barone V, Fusco R, Guerra G (1983) Gazz. Chim. Ital. 113:601
(33) Goodall BL (1981) Proc. Int. Symp. on "Transition Metal Catalyzed Polymerizations: Unsolved Problems", Midland, A:355
(34) Zambelli A, Oliva L, Ammendola P (1986) Gazz. Chim. Ital. 116:259
(35) Busico V, Corradini P, De Martino L, Iadicicco A, to be published
(36) Sergeev SA, Bukatov GD, Zakharov VA, Moroz EM (1983) Makromol. Chem. 184:2421
(37) Terano M, Kataoka T, Keii T (1987) Makromol. Chem. 188:1477
(38) Corradini P, Guerra G, Pucciariello R (1985) Macromolecules 18:2030
(39) Ewen JA (1984) J. Am. Chem. Soc. 106:6355
(40) Kaminsky W, Kulper J, Brintzinger HH, Wild FRWP (1985) Angew. Chem. Int. Ed. Engl. 24:507
(41) Longo P, Grassi A, Pellecchia C, Zambelli A (1987) Macromolecules 20:1015
(42) Zambelli A, Ammendola P, Grassi A, Longo P, Proto A (1986) Macromolecules 19:2703
(43) Corradini P, Guerra G, Vacatello M, Villani V, Gazz. Chim. Ital., in press
(44) Dyachkowsky FS, Shilova AJ, Shilov AE (1967) J. Polym. Sci. 16:2333
(45) Lauher JW, Hofmann R (1976) J. Am. Chem. Soc. 98:1729
(46) Giannetti E, Nicoletti GM, Mazzocchi R (1985) J. Polym. Sci. Polym. Chem. Ed. 23:2117
(47) Eisch JJ, Piotrovsky AH, Brownstein SK, Gabe EJ, Lee FL (1985) J. Am. Chem. Soc. 107:7219
(48) Pino P, Cioni I, Wei J, Brintzinger HH, to be published
(49) Ishihara N, Seimija T, Kuramoto M, Uoi M (1986) Macromolecules 19:2465
(50) Pellecchia C, Longo P, Grassi A, Ammendola P, Zambelli A (1987) Makromol. Chem. Rapid Commun. 8:277
(51) See for example: Bassi IW, Allegra G, Scordamaglia R, Chioccola G (1971) J. Am. Chem. Soc. 93:3787; Latesky SL, McMullen AK, Niccolai GP, Rothwell IP, Hoffman JC (1985) Organometallics 4:902
(52) See for example: Sharpless KB, Woodard SS, Finn MG (1983) Pure Appl. Chem. 55:1823; Halpern J (1982) Science 217:401

Linear and Branched Polyethylenes by New Coordination Catalysts

K.A. Ostoja Starzewski*[1], J. Witte*, K.H. Reichert**, G. Vasiliou**

Bayer AG, Zentrale Forschung und Entwicklung
Wissenschaftliches Hauptlaboratorium, D-5090 Leverkusen*
Institut für technische Chemie, Technische Universität Berlin**

Novel bis(ylid)nickel catalysts exhibit remarkable properties in ethene polymerization (1-3). The special features and variability of the ylid bond system may be monitored using photoelectron spectroscopy (4-7).

$$Ni(0)/R_3\overset{+}{P}-\overset{|}{C}=\overset{|}{C}-\overset{-}{O}/R_3\overset{+}{P}\cdots\overset{-}{X} \qquad \text{polymerisation catalyst} \qquad [1]$$

This contribution is designed to at least touch on the topics: catalyst activation and selectivity, PE molecular weight control, linear und non-linear structures, short and long chain branching, simple copolymerizations and bifunctional catalysis.

The following text will focus primarily on systems in which one ylid ligand can coordinate structurally intact to the nickel, while the other has the potential to form a chelating PO ligand through P-to-Ni phenyl migration.

$$[2]$$

ylid <u>reori</u>

ylid <u>intact</u>

---

[1]Author to whom correspondence should be addressed

W. Kaminsky and H. Sinn (Eds.)
Transition Metals and Organometallics as
Catalysts for Olefin Polymerization
© Springer-Verlag Berlin Heidelberg 1988

The pure one-component (pre)catalysts can frequently be isolated from such bis(ylid)nickel reaction mixtures. In this case it is mainly multinuclear NMR methods which are used to optimize catalyst synthesis and to control purity. These methods also provide detailed information on the structure in solution and are sensitive probes for changes in the molecular and electronic structure, thus reflecting substituent effects.

The activity of ylid catalysts is superior to that of comparable phosphine derivatives, e.g.

$$NiPh(Ph_2PCHCMeO)(Ph_3PCH_2)$$
$$NiPh(Ph_2PCHCMeO)(Ph_3P)$$
[3]

Ylid-steered ethene polymers contain a lower proportion of oligomers and have higher intrinsic viscosities, DSC melt temperatures and molecular weights.

ESCA studies show that the $Ni(2P_{3/2})$ binding energy in the above ylid derivative is lowered by 1 eV compared with the phosphine analogue. The nickel center is reduced more strongly by the intact ylid coordination. The higher donor/acceptor ratio of ylid ligands in metal carbonyl model complexes (5,11) can be monitored by infrared spectroscopy since the ligand effect influences the carbonyl force constants (14). The differences between the two ligands triphenylphosphine-methylene and triphenylphoshine can be visualized by photoelectron spectroscopy since the C-localized ylid electron pair (IE_1 = 6.6 eV) is energetically far more easily available than the phosphorus lone pair (IE_1 = 7.8 eV). (6)

Changes in catalyst properties such as activity, activation temperature and selectivity are caused by the substituent effect in both ylid ligands.

An increase in activity, for example, is accompanied by a measurable effect on the P atom in the metallocycle which can be observed by P-31 NMR spectroscopy. This effect is a function of the substituents in the structurally intact ylid ligand in trans-position (2).

increasing catalyst activity ⟶

$\Delta\delta^{31}P_{coord}^{reori}$	+ 3.6 ppm	+ 4.3 ppm	+ 8.4 ppm
ylid $_{coord}^{intact}$	Me_3PCH_2	Ph_3PCH_2	Pr_3^iPCHPh
IE_1 (free ylid)	6.8 eV	6.6 eV	< 6.2 eV

[4]

The PE-spectroscopically correlated ordering of the free ligands with decreasing first ionization potential points to the link between activity and the electronic structure.

Activity is also influenced by suitable substituents in the
CO-stabilized ylid, i.e. at another position in the metallocycle. In
this way the turnover increases in the sequence

$$Ph_3PCHCHO < Ph_3PCHCMeO < Ph_3PCHCPhO \qquad [5]$$

to approximately $0.5 \cdot 10^5$ mole reacted ethene per mole nickel, while
maintaining the intact coordinated ylid ligand trimethylphosphine-
methylene (3).
An even more marked chemical change in the PO component dramatically
alters the selectivity for a certain PE molecular weight and leads to
a novel ligand-steered molecular weight control (Fig. 1). Chemical
fine tuning can be achieved by means of the other ylid ligand and
solvents of varying polarity. This means that we have access to practically all molecular weight ranges up to 10^6 g/mole and above;
ethene polymers from liquid α-olefins through soft and hard waxes up
to HDPE and even UHMW PE can be synthesized in this way (1).

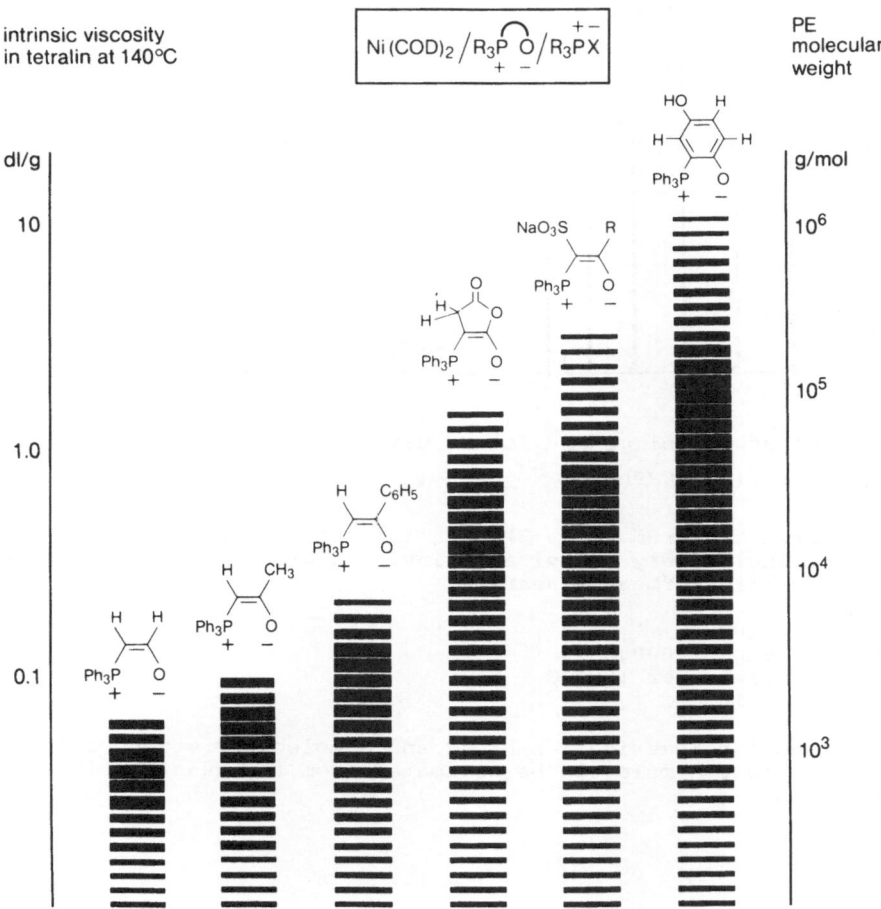

Fig. 1. Range of PE molecular weights and intrinsic viscosities
depending on the nature of the CO-stabilized ylid in ylid-nickel
catalysts

The "tailor-made macromolecule", however, also requires access to the branched structure. The selectivity of the bis(ylid)nickel catalysts frequently favours the formation of linear α-olefins. The FT infrared spectrum of this type of polymer and the GC of the oligomers can be used for characterization and as a reference for structural changes. Gas chromatography registers the homologous series of oligomeric α-olefins in the range C_4 to C_{40} (Fig. 2).

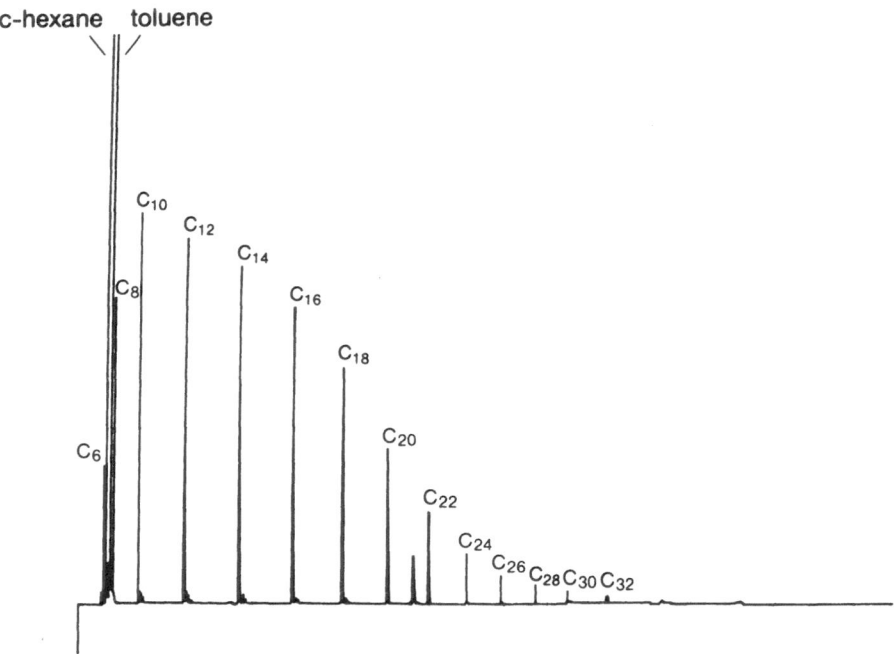

Fig. 2. GC of ethene oligomers formed using an ylid-nickel catalyst: $NiPh(Ph_2PCHCMeO)(Pr_3^iPCHPh)$

The FT infrared spectrum (Fig. 3) of a film of defined thickness shows almost exclusively methyl and vinyl end groups. Their presence in equal quantities proves linearity.

12   vinyl per 1000 C
0.2 vinylidene per 1000 C   [6]
0.4 trans-vinylene per 1000 C
12   methyl per 1000 C

A mean degree of polymerization of 42 and a molecular weight of approximately 1000 g/mole can be estimated from the quantitative end group determination.

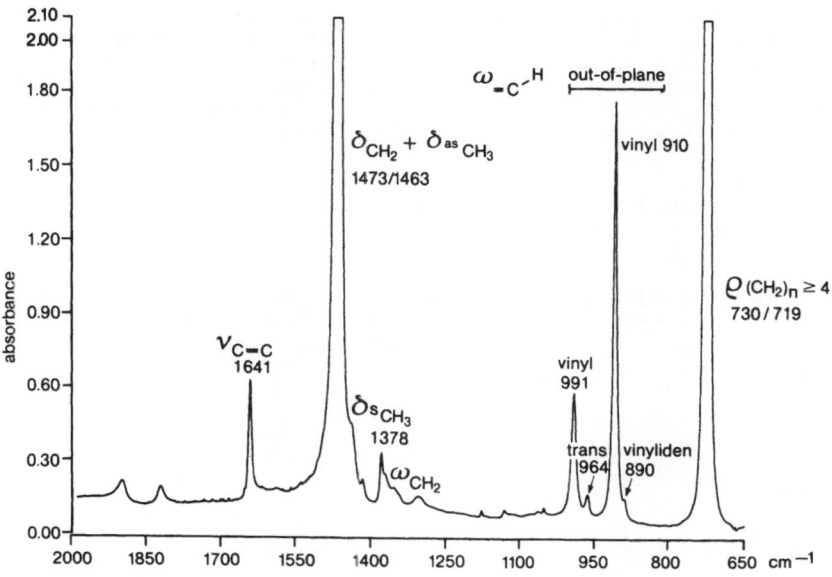

Fig. 3. FT IR spectrum of polyethene sample, prepared by using an ylid-nickel catalyst: NiPh(Ph$_2$PCHCMeO)(Pr$_3^i$PCHPh)

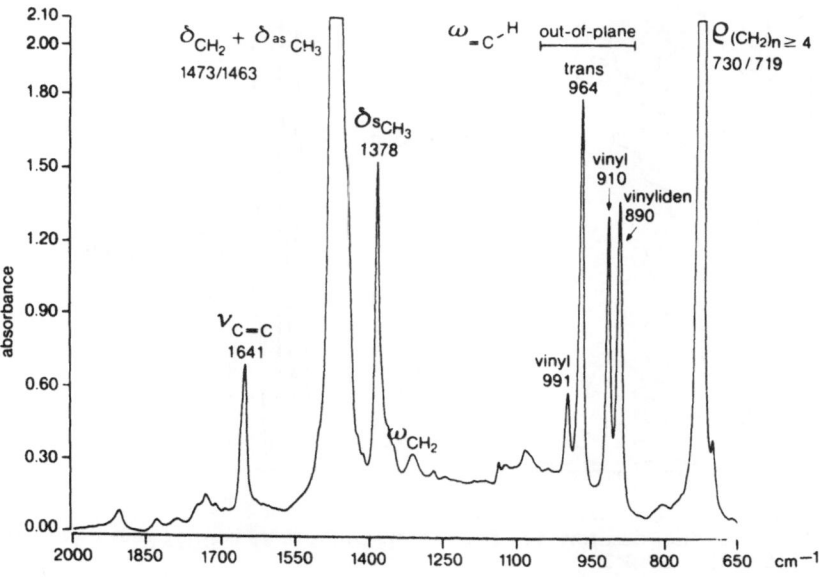

Fig. 4. FT IR spectrum of polyethene sample, prepared by using an ylid-nickel catalyst: Ni(O)/Ph$_3$PC(CMeO)$_2$/Ph$_3$PCH$_2$

Other selectivities can be adjusted by means of specific ligand combinations and/or high catalyst concentrations. In an extreme case, for example, almost equal quantities of vinylidene, trans-vinylene and vinyl are formed. The number of cis-vinylene double bonds cannot be determined because of overlapping bands, but they probably contribute to the high methyl content (Fig. 4).

$$
\begin{aligned}
&9 \text{ vinyl per } 1000 \text{ C} \\
&9 \text{ vinylidene per } 1000 \text{ C} \\
&9 \text{ trans-vinylene per } 1000 \text{ C} \\
&57 \text{ methyl per } 1000 \text{ C}
\end{aligned}
\qquad [7]
$$

The C-12, C-14 part of the complicated gas chromatogram (Fig. 5) is shown, and the tentative assignment of the prominent peaks by means of GC/MS coupling provides a plausible explanation for the unusual IR finding.

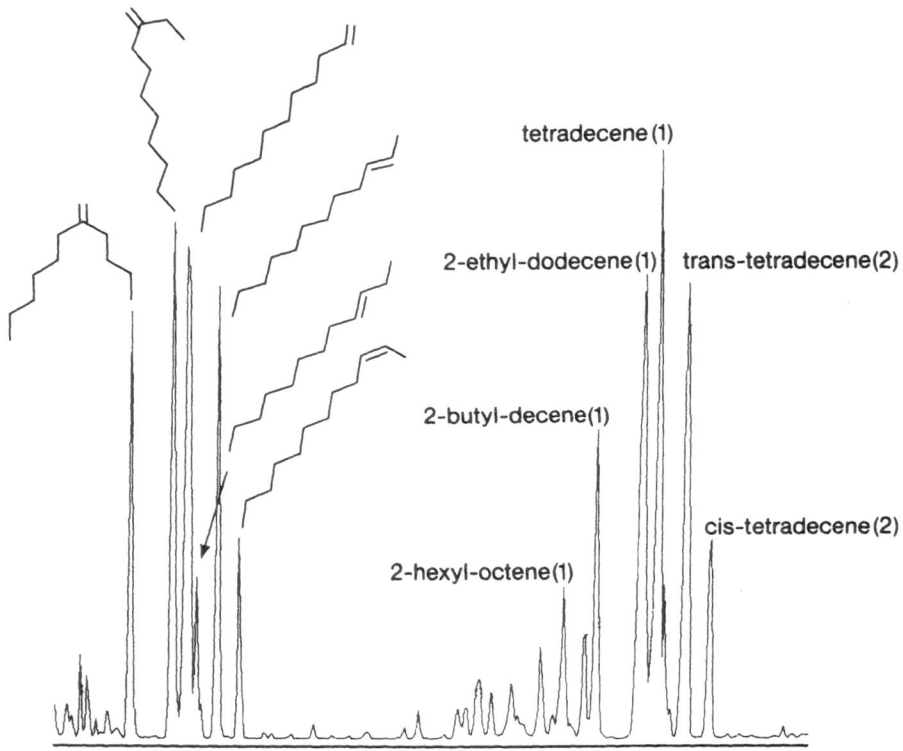

tetradecene(1)

2-ethyl-dodecene(1)   trans-tetradecene(2)

2-butyl-decene(1)

2-hexyl-octene(1)   cis-tetradecene(2)

Fig. 5. GC of ethene oligomers formed using an ylid-nickel catalyst $Ni(0)/Ph_3PC(CMeO)_2/Ph_3PCH_2$.
Tentative assignment of dominant C_{12} and C_{14} isomers by means of GC/MS coupling.

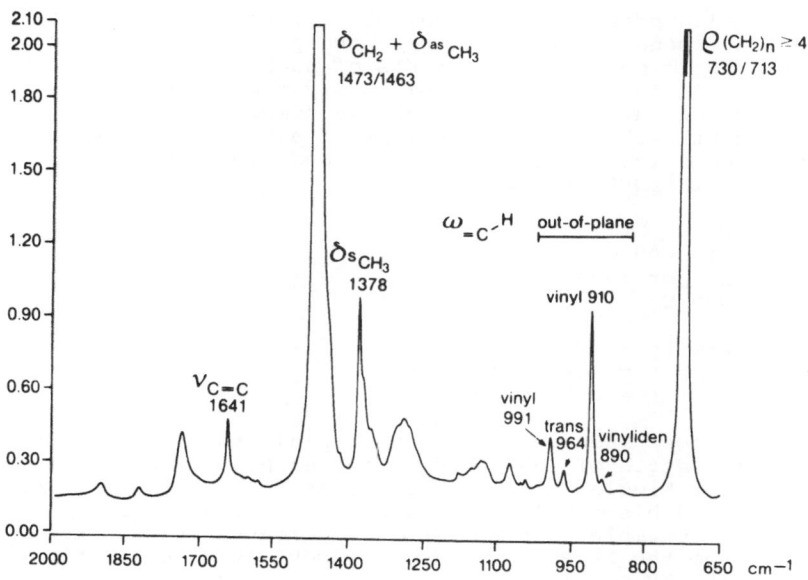

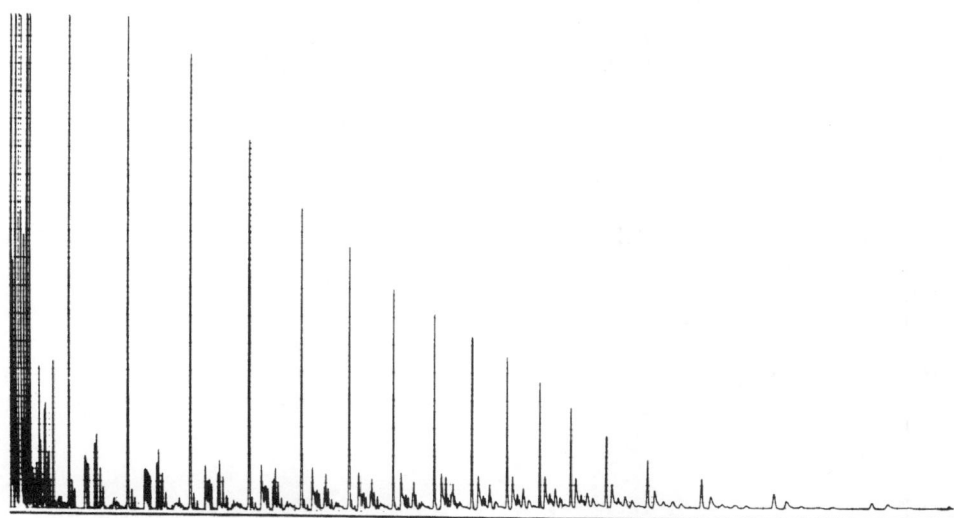

Fig. 6. top: FT IR spectrum of poly(ethene-co-propene) sample (6 mole % propene incorporated) using an ylid-nickel catalyst:

Ni(0)/Ph₃PC(SO₃Na)CPhO/Ph₃PCHCHCHPh

bottom: GC of ethene oligomers formed

A more conventional approach to the controlled formation of short-chain branches is ethene polymerization in the presence of a comonomer. GC and IR-spectroscopic finger prints of such an ethene/propene copolymerization using a bis(ylid)nickel catalyst are shown in Fig. 6. The increased number of methyl groups compared with vinyl groups is consistent with a propene content of approximately 6 mole %. Accordingly we find a lower melt temperature and lower density, i.e. behaviour typical of LLDPE structures.

6	vinyl per 1000 C
0.2	vinylidene per 1000 C
0.4	trans-vinylene per 1000 C
35	methyl per 1000 C

[8]

DSC melt temperature F_p = 121°C
intrinsic viscosity η = 0.4 dl/g
density ρ = 0.94 g/cm³

The amount of comonomer incorporated drops rapidly to 1 mole % and below as the chain length of the α-olefin increases. The copolymerization behaviour for Ziegler catalysts is occasionally somewhat more favourable. Ylid- and bismethylid-modified zirconocene/alumoxane catalysts (19), for example, can be used to increase the incorporation of propene from the LLDPE range (up to about 10 mole %) into the range of rubber-like EP copolymers (Fig. 7).

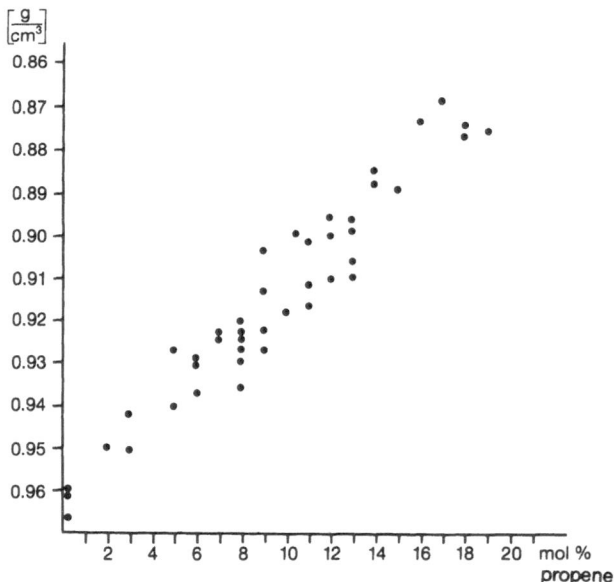

Fig. 7. R_3PCXY and R_2P(CH$_2$)Li/Zr/alumoxane catalyzed ethene-propene copolymerizations: polymer density vs amount of propene incorporated

But these systems are also largely unsuitable for the incorporation of long-chain α-olefins.

We therefore carried out experiments with surface chromium(II)/silica catalysts which are known to be highly active for the polymerization of long-chain α-olefins. (15-17)

The question was whether two compatible catalysts could be found, one of which oligomerizes ethene to long-chain α-olefins while the other copolymerizes them with ethene.

ethene $\xrightarrow{\text{catalyst I}}$ α-olefins

ethene + α-olefins $\xrightarrow{\text{catalyst II}}$ branched PE

[9]

Therefore homogeneous bis(ylid)nickel systems were combined with heterogeneous chromium catalysts which also work in the absence of aluminium alkyl cocatalysts. In this process the nickel complex is supported on the chromium contact, resulting in a new catalyst which is active in polymerization.

Oligomerisation:

Polymerisation:

[10]

The properties of the polymer formed in this way depend to a large extent on the Ni/Cr(II) ratio (Fig. 8): where there is a large excess of Ni only the known linear α-olefins are formed, the polymerization-active Cr centers are blocked. On reducing the amount of Ni, α-olefins and some high polymer are formed. And finally where there is an excess of Cr(II) relative to Ni, the transition metal ratio can be adjusted (18) so that almost exclusively high molecular weight PE with broad molecular weight distribution is produced.
Polymeranalytical, spectroscopic and rheological data (Fig. 9) compared with those from linear HDPE produced with a surface Cr(II)/silica catalyst strongly suggest that the bifunctionally produced polyethene (BiPE) described above has a long chain branched structure.

model

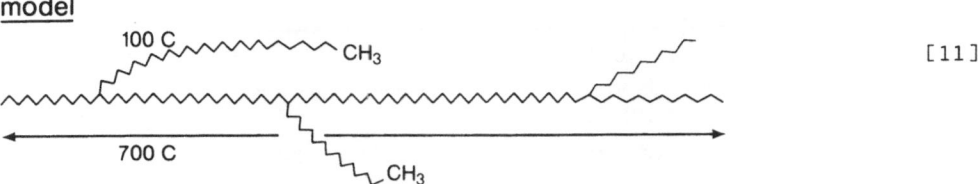

[11]

Variation of the Molar Transition Metal Ratio

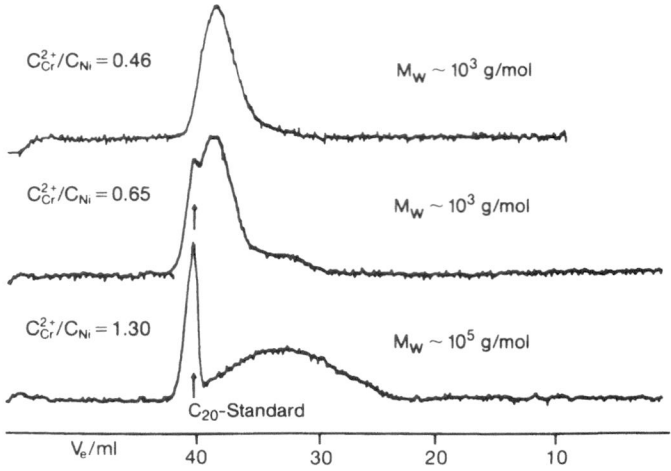

$C_{Cr}^{2+}/C_{Ni} = 0.46$ $M_W \sim 10^3$ g/mol

$C_{Cr}^{2+}/C_{Ni} = 0.65$ $M_W \sim 10^3$ g/mol

$C_{Cr}^{2+}/C_{Ni} = 1.30$ $M_W \sim 10^5$ g/mol

C_{20}-Standard

V_e/ml   40   30   20   10

Fig. 8. HPLC of polyethene samples, prepared by using bifunctional Ni/Cr catalysts:

NiPh(Ph$_2$PCHCPhO)(Me$_3$PCH$_2$) on surface Cr(II)/silica (800°C/350°C)
T = 100°C, P = 10 bar ethene, solvent: c-hexane/toluene

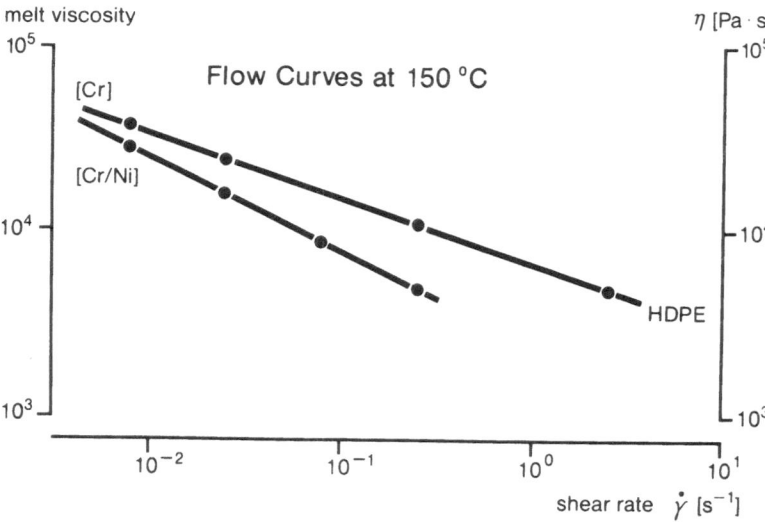

melt viscosity η [Pa · s]

10^5 10^5

[Cr]

Flow Curves at 150 °C

[Cr/Ni]

10^4 10^4

HDPE

10^3 10^3

10^{-2} 10^{-1} 10^0 10^1

shear rate $\dot{\gamma}$ [s^{-1}]

Fig. 9. Flow curves of [Ni/Cr]-BiPE and [Cr]-HDPE

	[Cr/Ni]-PE	[Cr]-HDPE
$[\eta]$:	1.30 dl/g	1.57 dl/g
ϱ :	0.96 g/cm^3	0.97 g/cm^3
Fp :	130 °C	136 °C
vinyl:	2.0 per 1000 C	1.6 per 1000 C
vinyliden:	0.10 per 1000 C	0.06 per 1000 C
trans:	0.04 per 1000 C	0.02 per 1000 C
methyl:	5.0 per 1000 C	1.5 per 1000 C

[12]

It is thus possible to synthesize branched PE from ethene alone, i.e.
without adding a comonomer, by means of bifunctional (Ni/Cr) catalysis (19). Variation of the nickel and chromium components open up
the possibility of producing a number of interesting polymers which
will contribute to the structural variety in transition metal-cata-
lyzed ethene polymerization with ylid-steering ligands.

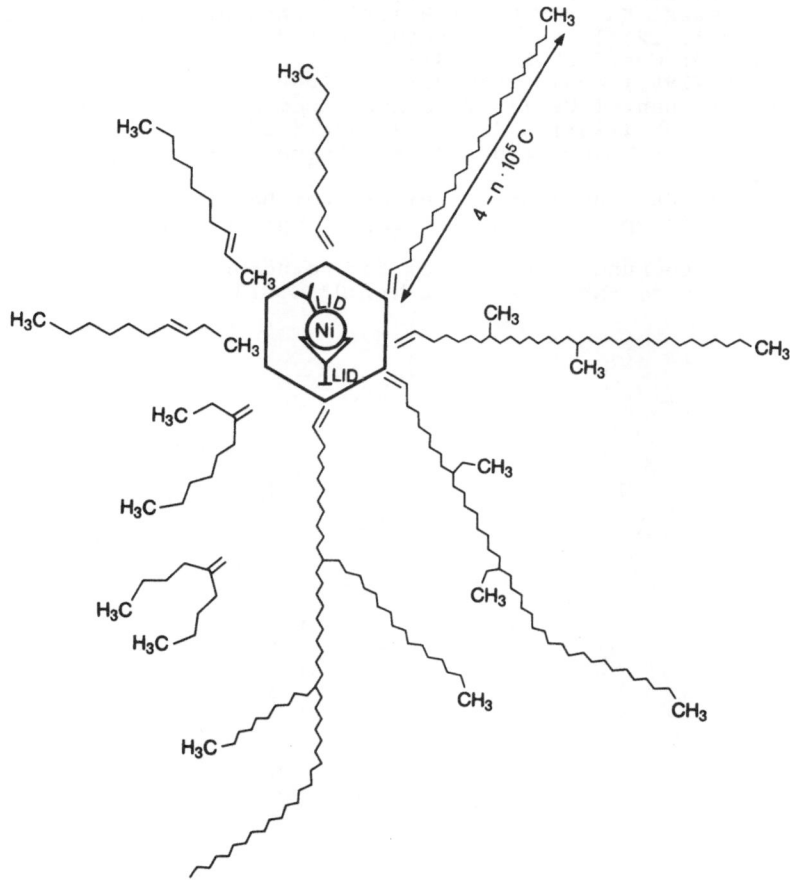

Fig. 10. Ylid nickel-steered ethene polymerizations

REFERENCES

1. Ostoja Starzewski KA, Witte J (1987) Angew Chem 99: 76-77
2. Ostoja Starzewski KA, Witte J (1986) Proceedings of the
    International Symposium on "Transition Metal Catalyzed Poly-
    merizations". Akron, Ohio, USA
3. Ostoja Starzewski KA, Witte J (1985) Angew Chem 97: 610-612
4. Ostoja Starzewski KA, Feigel M, Rieser J (1983) Phosphorus
    Sulfur 18: 448
5. Ostoja Starzewski KA, tom Dieck H (1979) Inorg Chem 18:
    3307-3316
6. Ostoja Starzewski KA, Bock H (1976) J Amer Chem Soc 98:
    8486-8494
7. Ostoja Starzewski KA, Richter W, Schmidbaur H (1976) Chem Ber
    109: 473-481
8. Keim W, Kowaldt FH, Goddard R, Krüger C (1978) Angew Chem 90:
    493; see also ref. 9, 10 and literature cited
9. Keim W, Behr A, Limbäcker B, Krüger C (1983) Angew Chem
    95: 505-506
10. Keim W, Behr A, Gruber B, Hoffmann B, Kowaldt FH, Kürschner U,
    Limbäcker B, Sistig FP (1986) Organometallics 5: 2356-2359
11. Ostoja Starzewski KA, tom Dieck H (1976) Phosphorus 6: 177-189;
    see also ref. 12, 13 and literature cited
12. Kaska WC (1983) Coord Chem Rev 48: 1-58
13. Schmidbaur H (1983) Angew Chem 95: 980-1000
14. Cotton FA, Kraihanzel CS (1962) J Amer Chem Soc 84: 4432-4438
15. Weiß K, Krauss HL (1984) J Catal 88: 424-430
16. Krauss HL, Weiß K, Langstein G (1986) German Patent Application
    DE-OS 3427319
17. Langstein G (1986) PhD Thesis, Universität Bayreuth
18. Vasiliou G (1987) part of PhD Thesis, Technische Universität
    Berlin
19. Supported by the Bundesministerium für Forschung
    und Technologie, BMFT - Projekt Nr. 03 C 213

Olefin Dimerization with a Homogeneous Titanium Catalyst and Polymerization Studies with Sterically Hindered Aluminium Co-Catalysts - Spin-Offs in the Development of Supported Catalysts for Propylene Polymerization

Brian L. Goodall

Koninklijke/Shell-Laboratorium, Amsterdam (Shell Research B.V.)
Badhuisweg 3, 1031 CM Amsterdam, The Netherlands

INTRODUCTION

Since the first commercial polypropylene plants of the 1950's enormous advances have been made in terms of engineering innovation, process development and, most importantly, catalyst technology, resulting in today's simple, energy- and monomer-efficient processes. The essential catalyst attributes which enabled the realization of these new processes are (a) high selectivity, thus obviating atactic-polymer removal, and (b) extremely high polymerization activity such that catalyst residues need not be removed from the polypropylene.

In an earlier publication (1) we described the development and performance of our Super High Activity Catalysts (SHAC) which are now used commercially in both liquid propylene ("bulk") and gas phase plants. The SHAC catalysts belong to the supported catalyst family comprising a titanium chloride "catalyst" on a magnesium chloride support. In parallel to the SHAC-related research and development work we also investigated two other possible routes to catalysts meeting the above-described criteria which involved studying homogeneous catalysts and attempting to substantially increase the activity of the more conventional titanium trichloride catalysts. It is these research programmes which form the basis of the present communication.

OLEFIN DIMERIZATION WITH HOMOGENEOUS TITANIUM CATALYSTS

Although it is now more than thirty years since Ziegler's discovery of transition metal polymerization catalysts, many of the mechanistic aspects concerning these catalysts are still debatable. Of the homogeneous catalysts especially the systems based on Cp_2TiCl_2 have received much attention since in 1957 Natta and co-workers (2) showed this titanium derivative to interact with aluminium alkyls to give a homogeneous ethylene polymerization catalyst. Recently the elegant low-temperature ^{13}C NMR studies of Fink (3) have elucidated many aspects of the ethylene polymerization reaction using this type of catalyst. We have studied the $Cp_2TiCl_2/AlEtCl_2$ system with the objective of establishing why the catalyst can effectively polymerize ethylene but not propylene and higher olefins.

In a typical ethylene polymerization experiment Cp_2TiCl_2 was dissolved in benzene to give a red solution, after saturating with ethylene (atmospheric pressure) ethylaluminium dichloride was added (Al/Ti ratio 12/1) giving an immediate colour change (to green) of the completely homogeneous system. A rapid exothermic reaction followed with oligomers or polyethylene being afforded, depending on the exact process conditions.

When propylene was employed under similar conditions the same colour change was observed, but the proton NMR study revealed the fate of the olefin to be twofold: the generation of a new olefinic product, and the Friedel-Crafts-type alkylation of the aromatic nucleus resulting in the formation of cumenes.

Although not anticipated, this alkylation reaction is perhaps not surprising considering the Lewis acidity of the catalyst system. Indeed, we note that Tazuma and

W. Kaminsky and H. Sinn (Eds.)
Transition Metals and Organometallics as
Catalysts for Olefin Polymerization
© Springer-Verlag Berlin Heidelberg 1988

362

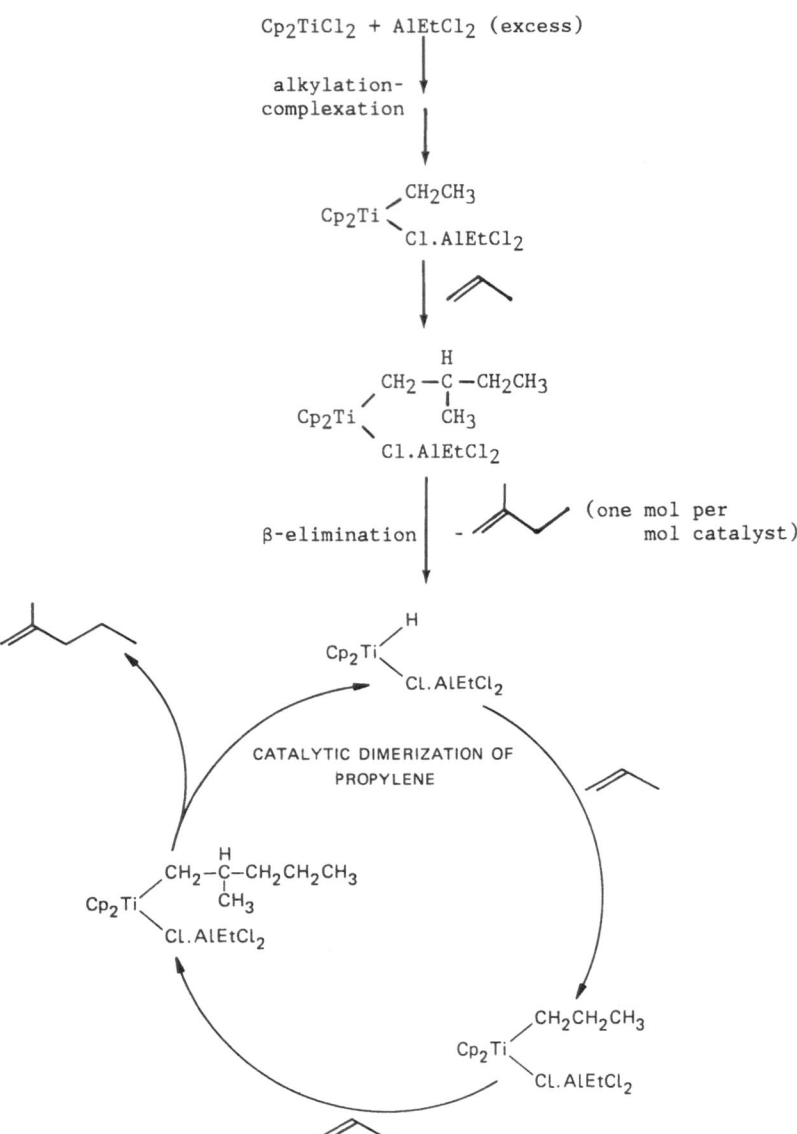

Fig. 1. Propylene dimerization

Kothari (4) reported the same side reaction in their olefin disproportionation work using a strongly Lewis acidic catalyst system (WCl_6/$EtAlCl_2$).

By introducing halogen substituents into the aromatic nucleus the solvent is deactivated towards the alkylation reaction, allowing the new olefinic product to be identified as 2-methylpentene-1, generated by the highly regioselective dimerization of propylene. The probable mechanism for this dimerization reaction is illustrated in Fig. 1; the insertion of propylene into a titanium-ethyl bond generates a branched alkyl group which, by virtue of its beta-tertiary carbon atom, readily undergoes beta-hydrogen elimination, liberating one mole of 2-methyl-1-butene and affording a titanium hydride, the active species in the propylene dimerization cycle. In the catalytic cycle itself every second mole of propylene inserted generates an iso-hexyl group which is branched in the beta-position and therefore undergoes the same rapid beta-elimination process generating the catalytic product, 2-methylpentene-1 (2-MP-1).

On scaling up the propylene dimerization reaction unexpected behaviour was observed. In a typical experiment the catalyst (Cp_2TiCl_2), 1 mmol) was dissolved in bromobenzene (350 ml) under propylene pressure (2 bar) at ambient temperature. Addition of the co-catalyst ($EtAlCl_2$, 15 mmol) initiated the dimerization reaction; after about 90 minutes the reaction was still proceeding smoothly, consuming about 300 mol propylene/(mol Ti.h). After this initial period of about 90 minutes there was a rapid increase in both temperature (from 25 to 35 °C) and propylene consumption (to about 4000 mol/(mol Ti.h)). After a total reaction time of 4 hours none of the expected 2-methylpentene-1 was observed; instead an oligomeric oil separated from the otherwise homogeneous reaction mixture.

We account for this unusual behaviour by considering that the dimerization product (2-MP-1) readily undergoes cationic polymerization under the influence of a Lewis acid (e.g. $EtAlCl_2$) in the presence of traces of moisture, to yield a co-oligomeric material from 2-MP-1 and propylene. Indeed, in control experiments it was found that the addition of small amounts of water to propylene dimerization reaction mixtures caused rapid and complete oligomerization or polymerization of the 2-MP-1.

We found that small amounts of an organic nitrogen base such as triethylamine or diazabicyclooctane (DABCO) completely supress the cationic polymerization process without influencing the rate of propylene dimerization (providing the molar ratio base/$EtAlCl_2$ is <1). Using this three-component catalyst system (Cp_2TiCl_2, tertiary amine, $EtAlCl_2$) it is possible to dimerize propylene to 2-MP-1 with a selectivity of >95 %, the remainder being traces of the corresponding trimer (no polymeric products are formed).

STERICALLY HINDERED CO-CATALYSTS

With the development of $TiCl_3$-type propylene polymerization catalysts of greatly improved activity it occurred to us that the target of a non-deashing polypropylene process could be achieved if only a chloride-free cocatalyst was found to replace the diethylaluminium chloride (DEAC). An obvious choice in this respect is the trialkylaluminium family; however, although these cocatalysts endow the resulting catalyst system with exceedingly high activity the stereoselectivity is invariably poor, this latter point being attributed to the "over-reduction" of the surface $TiCl_3$ sites to unselective Ti(II) centres. Over the years many research groups have investigated the effect of selected Lewis bases ("third components") such as ethers and esters in improving this stereoselectivity but with only limited success. For this reason we chose to investigate co-catalysts of the type R_2AlX, where R is an alkyl group and X is a monovalent group such as alkoxy, phenoxy, dialkylamino, etc.

Danusso (5) reported diethylaluminium phenoxide (Fig. 2.1) to be totally ineffective as a co-catalyst in propylene polymerization; the well-documented (6) self-association of this species into dimers and trimers must be considered responsible for this lack of activity. By introducing severe steric hindrance into this type of aluminium species it is possible to suppress the self-association tendencies, resulting in monomeric systems with very high reactivity. It is critical that the phenol is hindered in both the 2- and 6-positions and that the groups introduced in these ortho-positions are themselves very sterically demanding; generally, tertiary-butyl groups give excellent results, whereas isopropyl groups give marginal effects and ethyl groups are ineffective.

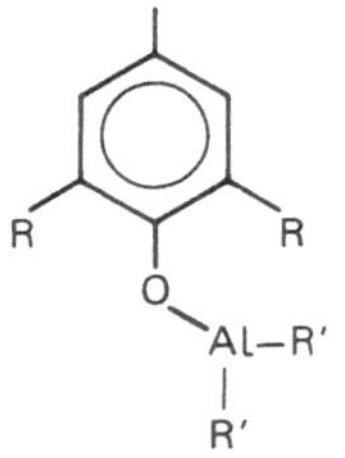

2.1 R = H, R' = C_2H_5

2.2 R = tert-C_4H_9
    R' = C_2H_5

2.3 R = tert-C_4H_9
    R' = iso-C_4H_9

Fig. 2. Dialkylphenoxyaluminium compounds

EXPERIMENTAL PART

For most of these studies we selected IONOL (4-methyl-2,6-di-tertiary-butylphenol) as the hindered phenol, which was simply mixed with a solution of the appropriate alkylaluminium compound to give the desired new co-catalyst derivative. Reaction was evidenced by gas evolution, the authenticity of the various compounds being confirmed by NMR methods.

The catalyst chosen for the propylene polymerization study was a highly active example of a so-called "Second Generation" $TiCl_3$ catalyst, such catalysts differing from the early $TiCl_3$ catalysts due to their small primary crystallite size, low $AlCl_3$ content and resulting higher (factor four or more) polymerization activity (7). The catalyst was prepared by the addition of a $TiCl_4$/DIAE (di-iso-amyl ether) solution in toluene to a triethylaluminium/DIAE solution in the same solvent at 20 °C, followed by heating to 110 °C and repeated washing with toluene.

The polymerization runs were carried out in the slurry mode, using iso-octane as diluent. All runs were carried out at 70 °C, 2.6 bar propylene pressure with a total reaction time of 4 hours. The productivity was determined by evaporating the resulting slurry to dryness and weighing (constant weight) the yield of dry, non-extracted polypropylene. This productivity value is used to calculate the catalyst activity, which is defined as the productivity divided by the number of run hours (four in all cases) and the propylene pressure (2.6 bar in all cases). The catalyst activity referred to in this paper has therefore the units g polypropylene/(g $TiCl_3$.h.bar), and is in fact the average polymerization activity over the whole reaction time (4 hours).

RESULTS AND DISCUSSION

We found the co-catalyst generated by reacting triethylaluminium with IONOL
(Fig. 2.2) to result in good polymerization activity but extremely poor stereoselectivity (>20 % xylene solubles; see Table 1). This disappointing result led us to
carry out a series of NMR studies in an attempt to determine the cause of the poor
selectivity. In order to better understand the interaction of catalyst and co-catalyst we used a homogeneous "catalyst model" for some of these investigations. The
tendency of titanium(IV) derivatives to undergo reduction, and the problems related
to the resulting paramagnetic species, led us to choose Cp_2ZrMe_2 for this purpose.
We observed that while co-catalyst 2.2 was stable indefinitely in solution, the
presence of the zirconium catalyst model caused a rapid redistribution reaction
(Fig. 3) This reaction (or "decomposition" of the co-catalyst) gives a plausible explanation as to why it is not possible to achieve high stereoselectivity, since the
free triethylaluminium generated in the reaction is known to result in low polymer
isotacticity.

Table 1. Propylene polymerization results obtained with IONOL-modified Co-catalysts

Entry	IONOL, mmol	Al^{i}Bu$_3$, mmol	AlEt$_3$, mmol	AlEt$_2$Cl, mmol	AlEtCl$_2$, mmol	Activity, g PP/(g TiCl$_3$.bar.h)	Xylene solubles, %w
1	-	-	-	9	-	141	7.8
2	4	-	4	-	-	158	>20
3	4	-	4	-	4	178	2.3
4	4	-	-	8	-	144	1.8
5	4	-	1	8	-	199	3.3
6	4	4	-	-	-	300	>20
7	4	4	-	-	4	206	1.3

Some results of this NMR study are listed in Table 2. It can be seen that compound
2.2 is stable in solution in the whole temperature range studied (-60 to +50 °C).
However, the presence of the zirconium "catalyst model" catalyzes the disproportionation of the aluminium species (Fig. 3) to such a degree that the reaction is very
rapid even at -60 °C. On warming the triethylaluminium released reacts to generate
new zirconium derivatives, which were not characterized.

Table 2. Proton NMR results (in perdeutero toluene)

Aluminium compound	Temp., °C	Free species tBu	In presence of Cp_2ZrMe_2 tBu	Cp	Me
$(RO)AlEt_2$	-60	1.50	1.60	5.60	-0.05
,,	+20	1.48	1.60	5.7 , 5.2	-0.1, -0.2
,,	+50	1.48	1.60	5.20	-0.1, -0.4
$(RO)_2AlEt$	+25	1.60			
$(RO)Al^iBu_2$	-60	1.52	1.52	5.66	-0.04
,,	-20	1.52	1.52	5.72	-0.1
,,	+20	1.50	1.50	5.76	-0.15
,,	+50	1.50	1.50	5.78	-0.18
,,	+50 (after 1 hr)	1.50	1.52, 1.60	5.78, 5.26	-0.17
$(RO)_2Al^iBu$	+25	1.60			
$AlEt_3$	+25	-	-	5.2	?
Cp_2ZrMe_2	+25	-	-	5.7	-0.15

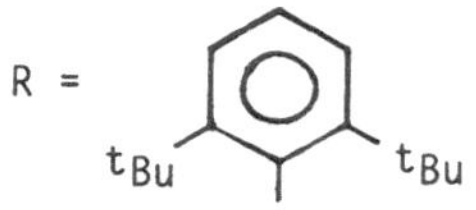

R =

Remarkably, we discovered that mixing equimolar proportions of 2.2 and ethylalumi-nium dichloride (catalyst poison!) gave rise to a new co-catalyst of significantly better performance than DEAC. As shown in Table 1 (entry 3) the activity is in-creased by 25 % while the stereoselectivity is greatly improved (2.3 % xylene solub-les). We attribute this excellent performance to the stabilizing effect of the chlo-ride bridges, which suppress the above-described redistribution reaction and conco-mitant elimination of triethylaluminium (Fig. 4).

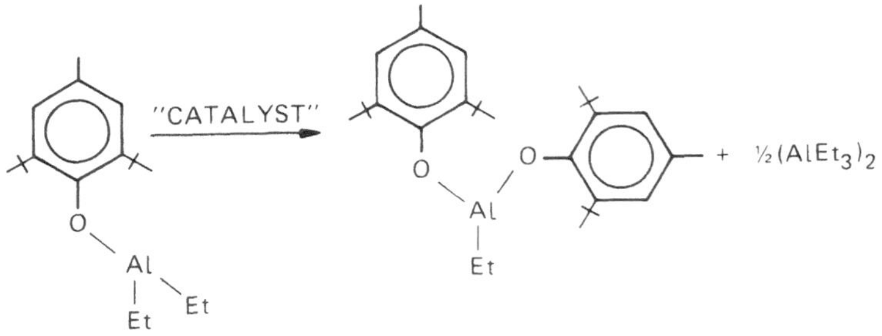

Fig. 3. Redistribution or disproportionation of sterically hindered aluminium species

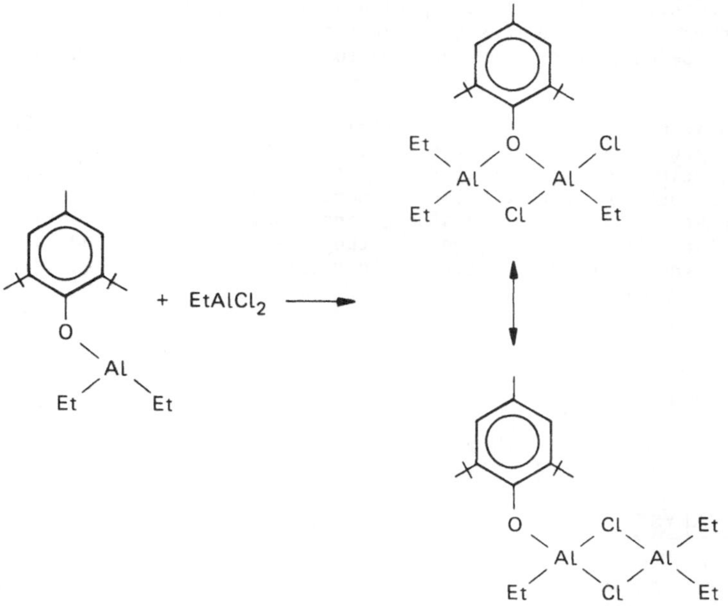

Fig. 4. The effect of chloride ions

Likewise a 2:1 mixture of DEAC and IONOL afforded a co-catalyst which combines good activity with excellent stereoselectivity (Table 1, entry 4). Furthermore the application of carefully controlled amounts of triethylaluminium to the DEAC/IONOL co-catalyst gave (compared to the standard DEAC run) an activity increase of >40 % with good selectivity (entry 5).

The improved activity of these sterically hindered phenoxy co-catalysts can be explained by considering that they selectively complex the sterically non-demanding catalyst poison (8), which is invariably present in these catalyst systems (Fig. 5).

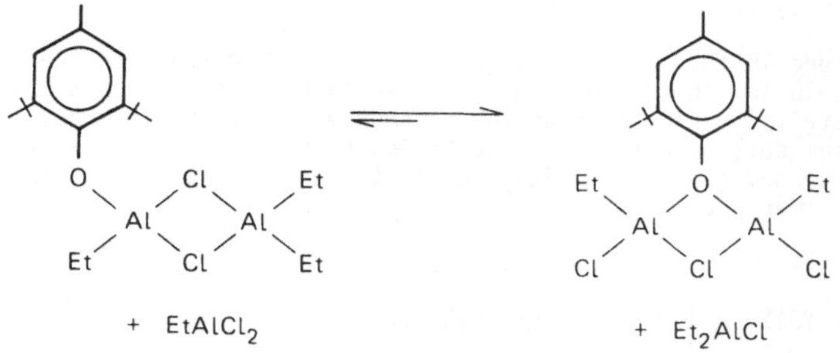

Fig. 5. Complexation of the catalyst poison, EtAlCl$_2$

The great improvement in stereoselectivity is rather surprising. On the basis of these results we assume that the ethylaluminium dichloride catalyst poison also has a strong negative effect on catalyst selectivity such that the complexation thereof gives the observed improvement. Another, in our opinion less likely, explanation is that the bulky phenoxy groups ligate and thus deactivate the atactic polymer generating active centres while being unable to do the same to the more sterically hindered isotactic sites.

We further investigated ways of suppressing the redistribution reaction (Fig. 3). It seems likely that the driving force for the reaction is the energy of dimerization of the tri-ethylaluminium released. In contrast to tri-ethylaluminium, tri-isobutyl-aluminium does not dimerize due to increased steric hindrance. Therefore, we prepared compound 2.3, which we reasoned would be stable since there appeared no driving force for redistribution. Unfortunately, an NMR study revealed that at >50 °C isobutylene is eliminated, and a similar redistribution reaction does indeed occur (Fig. 6). Some NMR data are listed in Table 2.

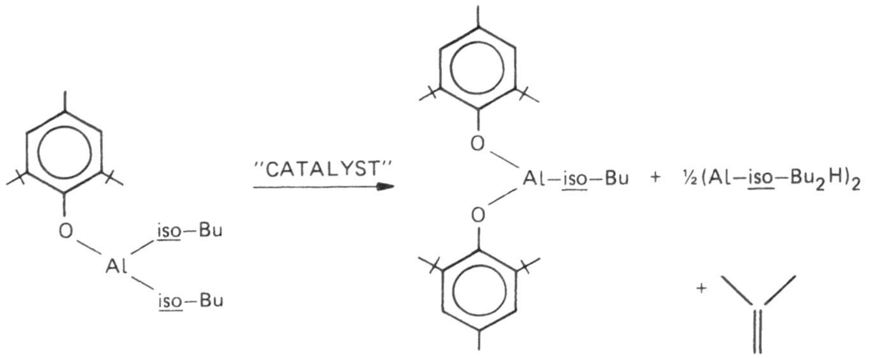

Fig. 6. "Decomposition" of sterically hindered co-catalyst

In a polymerization experiment the co-catalyst derived from tri-isobutylaluminium and IONOL (2.3) exhibited unprecedented activity (entry 6) but with very poor selectivity, the latter being attributed to the di-isobutylaluminium hydride released in the redistribution reaction described above. However, the equimolar mixture of the same derivative with ethylaluminium dichloride (entry 7) resulted in outstanding polymerization performance; an activity increase of roughly 50 % compared with DEAC and excellent stereoselectivity (1.3 % xylene solubles).

The catalyst activity recorded in this experiment (at 1.3 % xylene solubles) corresponds to a polypropylene yield in liquid propylene at 70 °C (4 hours residence time) of 24 kg polypropylene/g catalyst, which exceeds the polymer yield reached by many supported catalysts. The slow rate of catalyst deactivation ("decay") and excellent morphology control of such second generation $TiCl_3$ catalysts make this system particularly attractive for copolymer processes.

STERICALLY HINDERED CO-CATALYSTS IN ISOPRENE POLYMERIZATION

In the manufacture of "high-cis" isoprene rubber Ziegler catalysts are employed, usually beta-$TiCl_3$ (in contrast to the gamma- or delta-$TiCl_3$ used in polypropylene). We applied the same principle of sterically demanding phenol-modified co-catalysts in these systems in an attempt to improve catalyst performance, since it seems likely that alkylaluminium dichlorides also function as a poison in this catalyst system.

EXPERIMENTAL PART

The phenol-modified co-catalysts were prepared by reacting DEAC with the phenol (molar ratio 1:1) at ambient temperature for 2 hours. The resulting solution was added to the $TiCl_3$ catalyst (Al/Ti ratio 0.25) 10 minutes prior to initiation of the polymerization.

The beta-$TiCl_3$ catalyst used was prepared by premixing tri-isobutylaluminium with di-n-butyl ether and $TiCl_4$ for 1 hour at -20 °C and then allowing the mixture to react at ambient temperature for 1 hour (Al/Ti ratio 0.85, ether/Al ratio 0.25).

The polymerizations were run for one hour at 50 °C using 20 %w isoprene in isopentane to which DEAC had been added (0.5 mmol/l). The catalyst concentration used was 2 mmol $TiCl_3$/l. The additional co-catalyst components were dosed (10 minutes before initiation of the reaction), as indicated in Table 3.

RESULTS AND DISCUSSION

The results listed in Table 3 show that addition of the IONOL-modified co-catalyst to the beta-$TiCl_3$ catalyst gave an increase in polymer yield of roughly a third compared with a standard run. Less sterically demanding phenols, such as 2,6-di-isopropyl- and 2,6-di-methylphenol, had essentially no effect on the polymerization rate; neither had the addition of extra DEAC. Again we assume that the beneficial effect of the IONOL-modified co-catalyst must be attributed to its complexation of the alkylaluminium dichloride catalyst poisons present in the system.

Table 3. Isoprene polymerization results obtained with co-catalysts modified with phenols

Entry	Additional co-catalyst added	Isoprene conversion, %
1	NONE*	60
2	DEAC	56
3	DEAC + (2,6-di-tert-butyl-4-methylphenol) (1:1)	81
4	DEAC + (2,6-di-isopropyl-4-methylphenol) (1:1)	64
5	DEAC + (2,6-dimethylphenol) (1:1)	62

* Excess Al($\underline{\text{iso}}$-Bu)$_3$ used in catalyst preparation

CONCLUDING REMARKS

In the study with the homogeneous titanium system a new regioselective olefin-dimerization catalyst has been found, the catalyst being active under unusually mild conditions (ambient temperature and pressure). It can be concluded that for such Lewis acidic catalysts the rate of beta-hydrogen elimination is too high to permit effective polymerization of olefins higher than ethylene.

Tertiary amines must be used as "third components" in the catalyst system in order to suppress cationic polymerization of the olefinic dimerization products. This leads us to suggest that amines such as triethylamine, which are frequently described as being able to improve the isotactic index of polypropylene manufactured with TiCl3 catalysts, may fulfill a similar function, i.e. suppression of cationic (co)polymerization of olefinic oligomers formed during the polymerization, rather than the previously suggested preferential blocking of unselective active centres. Indeed we note that these amines are not effective in improving the isotactic index of polypropylene made using the less Lewis-acidic supported catalysts (where oxygen donors such as aromatic esters are preferred).

The tendency of sterically hindered aluminium alkyl compounds to undergo redistribution reactions in order to permit (partial) dimerization of the electron-deficient alkylaluminium species hampered our attempts at generating a halogen-free co-catalyst via this route. However, the introduction of sterically demanding phenoxy ligands into an alkylaluminium chloride results in substantial improvements in both catalyst activity and stereoselectivity. The performance of the resulting catalyst system matches or surpasses that of most supported propylene polymerization catalysts. The improved catalyst performance is ascribed to complexation of the (sterically non-demanding) RAlCl2 catalyst poison, and is also observed in the case of the beta-TiCl3 catalysed polymerization of isoprene.

ACKNOWLEDGEMENTS

I wish to acknowledge the experimental assistance of H. van der Heijden (olefin dimerization) and Ms. W. van der Linden and J.J.M. Snel (bulky-phenoxy-modified co-catalyst study). I also thank J.C. Chadwick, who carried out the isoprene rubber investigation.

REFERENCES

1. Goodall BL (1983) in: Quick RP (ed) Transition Metal Catalyzed Polymerizations: Alkenes and Dienes, Harwood, Part A, p 355
2. Natta G, Pino P, Mazzanti G, Giannini U, Mantica E and Perade M, J Polym Sci (1957) 26:120
3. Fink G (1983) in: Quick RP (ed) Transition Metal Catalyzed Polymerizations: Alkenes and Dienes, Harwood, Part B, p 495; Fink G, Fenzl W and Mynott R, Z Naturforsch B (1985) 40:158
4. Tazuma JJ and Kothari VM, Chem Eng New, Sept. (1970) 28:39; Ichikawa K et al., J. Catal. (1976) 44:416
5. Danusso F, J Polym Sci (1964) C4:1497
6. Mole T and Jeffrey EA (1972) Organoaluminium Compounds, Elsevier, Amsterdam, p 143-144
7. Goodall BL, J Chem Educ (1986) 63:191
8. Ingberman AK et al., J Polym Sci (1966) A4:2781

Nature of the Active Sites in Soluble Ziegler-Polymerization Catalysts

Generated from Titanocene Halides and Organoaluminum Lewis Acids*

J.J. Eisch, M.P. Boleslawski and A.M. Piotrowski

Department of Chemistry, State University of New York at Binghamton, Binghamton, New York 13901, U.S.A.

INTRODUCTION

That the active site in Ziegler-Natta polymerization catalysts is the transition metal-carbon bond receives strong support from the finding that pure transition metal alkyls, such as tetrabenzyltitanium [1], can themselves polymerize ethylene and propylene, even in the complete absence of any main-group alkyl. However, it is also evident that combinations of transition-metal salts or organometallics with main-group alkyls often display greatly enhanced catalytic activity and stereoregular polymerization in such processes [2]. The isotactic polymerization of propylene by $TiCl_3$ and Et_2AlCl [3] and the greatly accelerated polymerization of ethylene by Cp_2ZrMe_2 and $(MeAlO)_n$ [4] are cases in point.

In order to determine how the interaction of the transition-metal component with main-group metal alkyls leads to an active polymerization catalyst, we have avoided the heterogeneity inherent in most active Ziegler-Natta catalysts [5]. By choosing catalyst combinations that functioned in homogeneous media, we have been able not only to study the catalyst system by spectral and magnetic resonance techniques, but also to address the question of whether a solid surface is necessary for the activity or the stereoregulation by these catalysts. Our previous studies have examined the soluble, cyclopentadienyl titanium(IV)-based polymerization catalyst for ethylene, which is formed from titanocene dichloride and alkylaluminum halides [6]. Application of ^1H, ^{13}C and ^{27}Al NMR spectroscopy to a 1:1 mixture of titanocene dichloride ($\underline{1}$) and $MeAlCl_2$ ($\underline{2}$) permitted the formation of complex $\underline{3}$ to be monitored (eq. 1):

$$Cp_2TiCl_2 \ + \ MeAlCl_2 \ \rightleftharpoons \ Cp_2Ti\overset{..Cl..}{\underset{Cl}{\diagdown}}AlMeCl_2 \qquad\qquad [1]$$

$$\underline{1} \qquad\qquad \underline{2} \qquad\qquad\qquad\qquad \underline{3}$$

The isolation of $\underline{3}$ and the determination of its crystal structure corroborated the bonding suggested for this complex in solution. Then by allowing $\underline{3}$ to react with a surrogate for ethylene, namely trimethyl-(phenylethynyl)silane $\underline{4}$, the carbometallation product $\underline{6}$ was then isolated and its crystal structure also determined. The bonding in $\underline{6}$ makes evident that it arose by the attack of methyltitanocene(IV) tetrachloroaluminate ($\underline{5}$) on $\underline{4}$; $\underline{5}$, in turn, must have been generated from $\underline{3}$ (eqs. 2 and 3) [7]. Equations 1 and 2 can be viewed as two equilibria, lying in favor of $\underline{3}$ [8], since $\underline{5}$, cannot be detected by spectral means [9]. These steps generate the active catalyst $\underline{5}$, which

---

* Part 44 of the series, "Organometallic Compounds of Group III".

W. Kaminsky and H. Sinn (Eds.)
Transition Metals and Organometallics as
Catalysts for Olefin Polymerization
© Springer-Verlag Berlin Heidelberg 1988

possibly exists as a chloride-bridged, contact ion-pair [10]. Finally, equation 3 would be a model process for the initiation of ethylene poly-merization; by the use of 4, the sterically hindered 6 results and hence in this situation, further insertions of 4 or even ethylene cannot take place. Such an outcome permits the initiation step of the thwarted polymerization to be isolated and the bonding in 6 to be ascertained [7,10]. Once the pure insertion product 6 was identified by isolation, the original reaction of 3 with 4 in solution could be monitored by magnetic resonance spectroscopy and the conclusion reached that 6 is the first and only insertion product formed from this catalyst system and 4.

$$Cp_2TiCl_2 \cdot Cl_2AlMe \quad \rightleftharpoons \quad Cp_2\overset{+}{Ti}\text{-Me} \quad AlCl_4^- \qquad [2]$$

$$\underline{3} \qquad\qquad\qquad\qquad \underline{5}$$

$$\underline{5} \quad + \quad Ph\text{-}C\equiv C\text{-}SiMe_3 \quad \longrightarrow \quad \underset{Me}{\overset{Ph}{>}}C=C\underset{\overset{+}{Ti}Cp_2}{\overset{SiMe_3}{<}} \qquad [3]$$

$$\underline{4} \qquad\qquad\qquad\qquad\qquad \underline{6} \quad AlCl_4^-$$

In the present investigation we have applied this method of studying homogeneous polymerization catalysts to cyclopentadienyl titanium(III)-based systems in the hope of finding soluble catalysts capable of the stereoregular polymerization of alpha-olefins. Our detailed examination of the hydrocarbon-soluble η^3-allyl(bis-η^5-cyclopentadienyl)titanium(III) (7) has uncovered an interesting ethylene-polymerization catalyst, whose activity and mode of reaction are most sensitive to the cocatalytic action of Lewis acids and bases [11]. These findings shed considerable light on the multiple roles that the main-group alkyl component can play in Ziegler-Natta polymerizations. Moreover, the solvent effects observed for the reactions of 7 indicate that titanium(III)-carbon bonds can either lead to carbotitanation of unsaturated substrates or to electron-transfer reductions. This sensitivity to solvents means that titanium(III)-catalyzed polymerizations may occur by either a polar or a free-radical mechanism, depending upon the nature of the monomer and the reaction medium.

EXPERIMENTAL DESIGN

η^3-Allyl(bis-η^5-cyclopentadienyl)titanium(III) (7), prepared from titanocene dichloride and allylmagnesium bromide [12], forms violet-colored solutions in hexane or toluene. This compound is highly sensitive to moisture and oxygen, so all its reactions were conducted in carefully degassed solvents under an atmosphere of dry argon [13]. Reagent 7 was admixed individually with Me_nAlCl_{3-n}, dialuminoxanes or aluminoxanes and the resulting combinations evaluated as polymeriza-tion catalysts for ethylene under the same experimental conditions. The results of these studied are given in Tables 1 and 2.

The carbotitanating action of 7 by itself, or combined with a coreagent, Me_nCl_{3-n} or a donor solvent (THF), was evaluated toward two unsaturated substrates, trimethyl(phenylethynyl)silane (4) and methyl phenyl ketone (8). As in our previous studies, the behavior of a combination of 7 and Me_nAlCl_{3-n} towards 4 was viewed as a method of ascertaining the nature of the active site for the polymerizations of ethylene by $Cp_2Ti\cdot C_3H_5$ (7) and Me_2AlCl. Only this combination was catalytically active; both a combination of <u>7</u> with Me$_3$Al and of <u>7</u> with MeAlCl$_2$ did not polymerize ethylene (Table 1).

Table 1. Polymerization of ethylene by allyltitanocene(III) (<u>7</u>) and Me$_n$AlCl$_{3-n}$[a]

Me$_n$AlCl$_{3-n}$	Color of solution	Conc. of Ti[b]	Time h	Activity of catalyst (g PE/g Ti-h)
none	violet	23	24	0
MeAlCl$_2$	green	10	2	0
MeAlCl$_2$	green	50	24	0
Me$_2$AlCl	navy blue	10	2	8.4
Me$_3$Al	violet	10	2	0
Me$_3$Al	violet	30	24	0
Me$_3$Al	violet[c,d]	10	2	18

[a] Reaction conducted in 100 ml of hexane, under 3.0 atmospheres of ethylene at 70°C.
[b] Molar ratio of Al:Ti was 6:1, except in the last entry where Al:Ti was 1:1; mmol/l.
[c] An equimolar mixture of Me$_3$Al and <u>7</u> was most cautiously treated with one molar equivalent of H$_2$O.
[d] A solid separated after hydrolysis.

Samples of the polyethylene resulting from <u>7</u> and organoaluminum cocatalysts melted up to 137°C and thus were high-density polymer. Such samples were examined by mass spectrometry at 70eV (deep insertion probe) and their fragmentation pattern compared with polyethylene samples prepared from ethylene by a TiCl$_4$-Et$_2$AlCl catalyst system. Polyethylene from <u>7</u> displayed a much stronger m/e peak at 41, supporting the conclusion that the allyl group from <u>7</u> was now an end-group in the polymer.

RESULTS

Polymerization of Ethylene

Although η^3-allyl(bis-η^5-cyclopentadienyl)titanium(III) (<u>7</u>) does not lack for titanium-carbon bonds (albeit they are pi-bonds), it is completely inactive in polymerization. Also, admixing <u>7</u> with 6 molar equivalents of either Me$_3$Al or MeAlCl$_2$ does not yield a polymerization catalyst for ethylene. But admixing <u>7</u> and 6 equivalents of Me$_2$AlCl gives an effective polymerization catalyst (Table 1). It is noteworthy that the original violet color of <u>7</u> was unchanged by admixing with Me$_3$Al, became navy blue with Me$_2$AlCl and turned green with MeAlCl$_2$. Similarly, adding (MeAlCl)$_2$O, (Cl$_2$Al)$_2$O or (Me$_x$Cl$_{1-x}$AlO)$_n$ to <u>7</u> changed the violet color to navy blue or dark green and gave active polymerization catalysts (Table 2). Finally, the activity of the catalyst from <u>7</u> and Me$_2$AlCl was destroyed when 6 equivalents of THF were introduced.

Table 2. Polymerization of ethylene by allyltitanocene(III) ($\underline{7}$) and organoaluminoxanes[a]

Aluminoxane	Color of solution	Conc. of Ti[b]	Time h	Activity of catalyst (g PE/g Ti-h)
$(MeClAl)_2O$	navy blue	10^c	1.5	14.1
$(Cl_2Al)_2O$	navy blue	10^c	1.5	5.0
$(Me_xCl_{1-x}AlO)_n$	dark green	15	3.0	7.6
$(i\text{-}BuAlO)_n$	violet	25^d	36.0	0
$(EtAlO)_n$	violet	25^d	36.0	0

[a] Reaction conducted in 100 ml of hexane, under 3.0 atmospheres of ethylene at $70^\circ C$.
[b] Concentration in mmol/l.
[c] Molar ratio of Al:Ti was 6:1.
[d] Molar ratio of Al:Ti was 10:1.

Carbometallation of Trimethyl(phenylethynyl)silane ($\underline{4}$)

The pure titanium reagent $\underline{7}$ did not react with silane $\underline{4}$ during 12 h in refluxing hexane solution. Nor did a combination of $\underline{7}$ and 6 equivalents of Me_3Al react with $\underline{4}$. However, a 1:1 molar ratio of $\underline{7}$ and Me_2AlCl did react with $\underline{4}$ at $60^\circ C$ to yield (Z)-2-phenyl-1-trimethylsilyl-1,4-pentadiene ($\underline{9}$) (eq. 4):*

$$Ph\text{-}C\equiv C\text{-}SiMe_3 \quad \xrightarrow[\text{2. } H_2O]{\text{1. } \underline{7} + Me_2AlCl} \quad \underset{H_2C=CHCH_2}{\overset{Ph}{>}}C=C\underset{H}{\overset{SiMe_3}{<}} \qquad [4]$$

$$\underline{4} \qquad\qquad\qquad\qquad\qquad\qquad\qquad\qquad \underline{9}$$

Carbometallation of Methyl Phenyl Ketone ($\underline{8}$)

Although $\underline{7}$ alone did not carbometallate $\underline{4}$, it did react in toluene solution with methyl phenyl ketone. At $25^\circ C$ a mixture of cyclopentadienylation and allylation products was isolated after hydrolytic workup (eq. 5):

$$\underset{Ph}{\overset{Me}{>}}C=O \quad \xrightarrow[\text{2. } H_2O]{\text{1. } \underline{7}, \text{ PhMe}} \quad \underset{\overset{|}{Ph}}{\overset{\overset{Me}{|}}{Cp\text{-}C\text{-}OH}} \quad + \quad \underset{\overset{|}{Ph}}{\overset{\overset{Me}{|}}{H_2C=CH\text{-}CH_2\text{-}C\text{-}OH}} \qquad [5]$$

$$\underline{8} \qquad\qquad\qquad\qquad\qquad \underline{10} \qquad\qquad\qquad \underline{11}$$

---

* The stereochemistry of $\underline{9}$, corresponding to syn-carbometallation, was ascertained by the chemical shift of the vinyl proton alpha to the Me_3Si group [14].

Although 10 was the major product isolated at 25°C, the principal product obtained from the reaction of 8 and 7 in refluxing toluene was 11 (>70%); minor amounts of reduction products, 1-phenylethanol (15%) and 2,3-diphenyl-2,3-butanediol (12, 10%), were also formed (cf. infra). The change in the proportions of 10 and 11 with temperature indicates that the carbotitanation to form the precursor of 10 is readily reversible.

Reduction of Methyl Phenyl Ketone (8)

The principal reaction between 7 and methyl phenyl ketone (8) in refluxing tetrahydrofuran solution, especially for reactant ratios of 1:1 or >1:1 for 8:7, was no longer carbometallation but bimolecular reduction. Upon hydrolytic workup, a 60% yield of d,l-2,3-diphenyl-2,3-butanediol (12, mp 121-123°C) was isolated. It is noteworthy that little or none of carbinol 11 was detected (eq. 6):

$$
2 \quad \begin{array}{c} Me \\ \diagdown \\ Ph \diagup \end{array} C=O \quad \xrightarrow[\text{2. } H_2O]{\text{1. } 7, \text{ THF}} \quad
\begin{array}{c} Me \quad Ph \\ | \quad\quad | \\ Ph-C-C-Me \\ | \quad\quad | \\ OH \quad OH \end{array} \qquad [6]
$$

$$\underline{12}$$

Trapping of Allyl(chloro)methylaluminum from Allyltitanocene(III) (7) and Methylaluminum Dichloride

The reaction of 7 and $MeAlCl_2$ in hexane solution led to an allyl-chloride exchange with the formation of titanocene(III) chloride (13) and the allylaluminum compound 14. That this exchange took place (eq. 7) was verified by adding benzophenone to the reaction mixture and isolating the hydrolyzed allylation product 15, which would be expected to be formed from 14 and benzophenone (eq. 8):*

$$
Cp_2Ti \cdot C_3H_5 \quad + \quad MeAlCl_2 \quad \longrightarrow \quad Cp_2TiCl \quad + \quad
\begin{array}{c} H_2C=CH-CH_2 \\ \diagdown \\ \quad\quad Al-Cl \\ Me \diagup \end{array} \qquad [7]
$$

$$\underline{7} \qquad\qquad\qquad\qquad \underline{13} \qquad \underline{14}$$

$$
\begin{array}{c} Ph \\ \diagdown \\ Ph \diagup \end{array} C=O \quad \xrightarrow[\text{2. } H_2O]{\text{1. } 14} \quad
\begin{array}{c} H_2C=CH-CH_2 \quad OH \\ \diagdown \quad\quad \diagup \\ C \\ \diagup \quad\quad \diagdown \\ Ph \quad\quad Ph \end{array} \qquad [8]
$$

$$\underline{15}$$

---

* That 15 arose from the reaction of benzophenone with 14, and not from the action of 7 directly on benzophenone was proved in the following manner. Reagent 7 was allowed to react with Ph_2CO in hexane; the principal product obtained upon hydrolysis was not 15, but a dimer of the type $[Ph_2(C_3H_5)C]_2$, possibly with the 1,1-diphenyl-3-buten-1-yl groups partly coupled through a p-phenyl position [MS, ^1H NMR and IR data].

DISCUSSION

Mechanism of the Carbotitanation or Polymerization of Hydrocarbon Substrates

The interaction of allyltitanocene(III) (7) with various organoaluminum Lewis acids (Tables 1 and 2) proved to be most instructive about the role that main-group alkyls play in Ziegler-Natta polymerizations. Because of its pi-bonded ligands reagent 7 itself has almost a closed shell configuration about titanium (35e⁻). Hence, it lacks a sigma Ti-C bond and an available coordination site for a two-electron donor (ethylene or 4). For this situation, the main-group alkyl then can perform a further function than its usual alkylating role: by its Lewis acidity it can change the hapticity of the η^3-allyl group to η^1-allyl by complexing with it. In this complexation not only is a sigma Ti-C bond generated, but a coordination site is opened up on titanium (33e⁻). The greater Lewis acidity of Me$_2$AlCl over Me$_3$Al would therefore nicely explain why the former turns 7 into an ethylene polymerization catalyst, while the latter does not. Furthermore, in the same vein, it is evident why THF and other Lewis bases would destroy the catalytic activity of combinations of 7 with Me$_2$AlCl. The THF would coordinate with the Me$_2$AlCl and thereby destroy its requisite acidity.

Although MeAlCl$_2$ is an even stronger Lewis acid than Me$_2$AlCl$_2$, it is too strong and thereby destroys the Ti-C allylic bond completely (eq. 7) because of an allyl-chloride exchange. On the other hand, dialuminoxanes and some polyaluminoxanes appear to have sufficient Lewis acidity to change the hapticity of the Ti-C bond and to activate it for polymerization, without disrupting it in the manner of MeAlCl$_2$ (Table 2).

The capability of the active catalyst combination, 7 and Me$_2$AlCl, also to carbometallate the ethylene surrogate 4 suggests that the active agent for both ethylene polymerization and the carbotitanation of 4 resembles structure 16 (eq. 9). It should be recalled that mass spectral analysis of the polyethylene obtained gives support for the incorporation of the allyl group as a polymer chain end-group. The regiospecificity and syn-stereospecificity could be attributed to the monobridged isomer of 16, namely 17, initiating electrophilic attack on ethylene or 4 (eq. 10):

Hydrolysis of 18 would yield the observed product 9.

Mechanism of the Carbotitanation and Reduction of Ketone Substrates

In contrast with the inertness of pure reagent 7 towards ethylene or
the acetylenic silane 4, 7 readily carbometallates methyl phenyl ketone
(8) in toluene solution. Apparently, the greater Lewis basicity of 8
over 4 permits 8 to change the η^3-hapticity of 7 to form complex 19 and
the latter then leads to carbotitanation (20) through principally a polar
process (eq. 11):

$$\text{[11]}$$

A process competitive with the carbotitanation of ketone 8 is the
monomolecular or bimolecular reduction of the ketone. In fact, in THF
solution such reduction becomes the principal reaction between 7 and 8
(eq. 6). These reactions clearly involve electron-transfer processes,
as for example, equations 12 and 13:

$$\text{[12]}$$

$$\text{[13]}$$

The coupling of radicals like 21 with further 8 would be expected to
yield the racemic isomer of 22 for steric reasons [15].

The competition between a polar carbotitanation and an electron-transfer
reduction of ketones by reagent 7 could have a broad significance for
polymerizations initiated by titanium(III) catalysts. If the olefinic
monomer is relatively electron-rich (R = H,R'), then initiation of
polymerization may more readily occur by olefin coordination and a
polar carbotitanation (23).

If the olefinic monomer is relatively electron-poor (R = Ph, X), then coordination of the titanium may involve electron-transfer ($\underline{24} \longrightarrow \underline{25}$) and polymerization may then follow a free-radical pathway. In support of this suggestion are investigations of the polymerization of vinyl chloride by a variety of Ziegler-Natta catalysts. The preponderance of evidence points to a free-radical mechanism [16].

ACKNOWLEDGMENT

This research was carried out under the support of Grant CHE-8308251 from the National Science Foundation.

REFERENCES

1 Ballard DGH (1973) Adv Catal 23:263
2 Boor J Jr (1979) Ziegler-Natta Catalysts and Polymerizations. Academic Press, New York, pp. 670
3 Natta G (1955) Makromol Chem 16, No. 3:213
4 Sinn H, Kaminsky W, Vollmer HJ, Woldt, R (1980) Angew Chem Int Ed Engl 19:390
5 Natta G (1960) Makromol Chem 35:94
6 Long WP, Breslow DS (1960) J Am Chem Soc 82:1953
7 Eisch JJ, Piotrowski AM, Brownstein SK, Gabe EJ, Lee FL (1985) J Am Chem Soc 107:7219
8 Fink G, Zoller W (1981) Makromol Chem 182:3265
9 Long WP (1959) J Am Chem Soc 81:5312
10 Eisch JJ, Boleslawski MP, Piotrowski, AM (1986) Proceedings of the Symposium on Transition Metal Catalyzed Polymerization: The Ziegler-Natta and Metathesis Polymerizations, Cambridge University Press, New York, pp 29
11 Eisch JJ, Boleslawski MP (1987) J Organomet Chem, in press
12 Martin HA, Jellinek F (1967) J Organomet Chem 8:115
13 Eisch JJ (1981) Organometallic Syntheses, Vol 2, Academic Press, New York, pp 7-20
14 Eisch JJ, Manfre RJ, Komar DA (1978) J Organomet Chem 159:C13
15 Eisch JJ, Kaska DD, Peterson CJ (1966) J Org Chem 31:453
16 Boor J Jr (1979) Ziegler-Natta Catalysts and Polymerizations. Academic Press, New York, p 536

Living Polymerization of Olefins with Highly Active Vanadium Catalysts

Yoshiharu Doi, Naoko Tokuhiro, Masataka Nunomura, Hiroto Miyake, Sigeo Suzuki, and Kazuo Soga

Research Laboratory of Resources Utilization, Tokyo Institute of Technology, Nagatsuta, Midori-ku, Yokohama 227, Japan

INTRODUCTION

Living polymerization of olefins is of importance as a tool for the synthesis of tailor-made polyolefins. Several years ago, we demonstrated that a $V(acac)_3/Al(C_2H_5)_2Cl$ system produces the first well-defined living polypropylene (Doi et al. 1979). This living polypropylene has been used for the synthesis of block copolymers and terminally functionalized polypropylenes (Doi and Keii 1986a; Doi et al. 1987a). Recently, we found that tris(2-methyl-1,3-butanedionato) vanadium, $V(mmh)_3$, shows a very high activity (700g-PP/g-V h) for the living polymerizaiton of propylene at -40°C in a toluene solution of $Al(C_2H_5)_2Cl$ (Doi et al. 1986b). In the catalytic system, all of vanadium species functioned as active centers for the propagation of living polypropylene. In this paper we report that the highly active $V(mmh)_3/Al(C_2H_5)_2Cl$ catalyst is also useful for the synthesis of living copolymer of propylene with ethylene or 1,5-hexadiene as well as for the synthesis of living polypropylene. In addition, we report a new method for the preparation of a silica-anchored $V(mmh)_3$ and its catalytic properties for the polymerization of propylene.

RESULTS AND DISCUSSION

Living Polymerizaiton of Propylene

We have prepared various vanadium (III) compounds with β-diketones as ligands, and used as soluble catalysts in toluene for the polymerization of propylene. Among the vanadium compounds, $V(mmh)_3$ showed the highest activity for the living polymerization of propylene in the presence of $Al(C_2H_5)_2Cl$. As reported in a previous paper (Doi et al. 1986b), the polymerization was conducted in toluene, and a maximum activity of the catalyst was observed at -40°C. The living polymerization of propylene took place at temperatures below -40°C (see Table 1).

Here, we report solvent effects in the polymerization of propylene with the $V(mmh)_3/Al(C_2H_5)_2Cl$ catalyst. The polymerization was carried out in heptane at different temperatures. The results are given in Table 1. The $V(mmh)_3$ compound was insoluble in heptane, but it turned soluble on the addition of $Al(C_2H_5)_2Cl$. The hepane solution of $V(mmh)_3/Al(C_2H_5)_2Cl$ catalyst exhibited a red color, characteristic of V^{3+} species, at temperatures below -20°C. At temperatures above -20°C the color of the heptane solution changed from orange to yellow with increasing temperatures, which suggests the reduction of active V^{3+} to inactive V^{2+} species. In fact, a maximum yield of polypropylene was observed at around -20°C in heptane. As can be seen from Table 1, at

W. Kaminsky and H. Sinn (Eds.)
Transition Metals and Organometallics as
Catalysts for Olefin Polymerization
© Springer-Verlag Berlin Heidelberg 1988

temperatures below -40°C the polydispersities of polymers are relatively small (Mw/Mn = 1.3 - 1.4), but the moleculare weight distributions of polymers become broad (Mw/Mn > 1.5) at temperatures above -20°C. It can be concluded that a limiting temperature for the living polymerization of propylene with the soluble $V(mmh)_3/Al(C_2H_5)_2Cl$ catalyst is around -40°C in toluene or heptane.

In solution, the active V^{3+} species may react with each other to form inactive V^{2+} species at temperatures above -40°C, resulting in a decrease of the catalytic activity. Immobilization of $V(mmh)_3$ on a solid surface may prevent the unfavourable intermolecular reactions to give a thermally stable catalyst. Therefore, we have prepared a $V(mmh)_3$ anchored to functionalized silica, and used it as the catalyst for propylene polymerization.

Table 1. Solvent effect in propylene Polymerization with the soluble $V(mmh)_3/Al(C_2H_5)_2Cl$ catalyst at different temperatures[a]

Solvent	Temp. (°C)	[Al]/[V] (mol/mol)	Polmer yield (g/g of V)	$\bar{M}n \times 10^{-4}$	$\bar{M}w/\bar{M}n$	$[N]$[b] (mol/mol of V)
Toluene	-60	20	275	3.77	1.38	0.37
Toluene	-50	20	753	6.05	1.38	0.64
Toluene	-40	20	2070	10.5	1.39	1.01
Toluene	-20	20	1310	2.42	1.60	2.76
Toluene	0	20	680	0.97	1.68	3.61
Toluene	-60	100	616	2.98	1.22	1.05
Toluene	-50	100	2310	6.14	1.47	1.83
Toluene	-40	100	2370	4.18	1.75	2.89
Toluene	-20	100	1440	1.71	1.79	4.29
Toluene	0	100	361	0.86	1.71	2.14
Heptane	-40	10	834	4.78	1.38	0.89
Heptane	-20	10	872	2.72	1.72	1.63
Heptane	0	10	326	1.29	1.55	1.29
Heptane	-40	20	767	4.23	1.37	0.93
Heptane	-20	20	1188	2.51	1.75	2.41
Heptane	0	20	220	0.51	5.41	2.21
Heptane	-60	100	73	1.88	1.34	0.20
Heptane	-40	100	926	5.31	1.64	0.89
Heptane	-20	100	793	2.07	1.75	1.95
Heptane	0	100	393	0.83	2.10	2.42

[a]Polymerization conditions : $V(mmh)_3$ = 0.05 mmol, C_3H_6 = 830 mmol, toluene or heptane solution= 100 cm^3, and polymerization time= 3 h .

[b]Number of polymer chains produced per vanadium atom.

The procedures to prepare the silica-anchored V(mmh)$_3$complex $\underset{\sim}{3}$ are outlined in Fig. 1. A chlorinated silica $\underset{\sim}{1}$ was prepared by the condensation of chloromethyl-phenylethyltrichlorosilane (2 mmol) with the hydroxyl groups on silica (5g, Davison 952, 350 m^2/g, dried at 300°C in vacuo) in boiling heptane. The Cl and C contents in $\underset{\sim}{1}$ were 0.16 mmol/g and 1.9 mmol/g, respectively. The Cl content corresponds to 0.3 pendent functions per nm^2. The reaction of $\underset{\sim}{1}$ (5g) with 1,3-butanedionato sadium (4 mmol) was carried out for ~1 h at 40°C in acetone in the presence of NaI (0.8 mmol). The resulting precipitate was washed with aqueous HCl solution and dried in vacuo, giving 2-ethylbenzyl-1,3-butanedione-functionalized silica $\underset{\sim}{2}$ with C, 2.3 mmol/g and Cl, 0 mmol/g. To attach the vanadium complex on silica, the functionalized silica $\underset{\sim}{2}$ (5g) was treated with V(mmh)$_3$ (1.8 mmol) in toluene for 12 h at 25°C, followed by washing several times with toluene. The resulting silica 3 was brown in color and contained vanadium ions of 0.13 mmol/g.

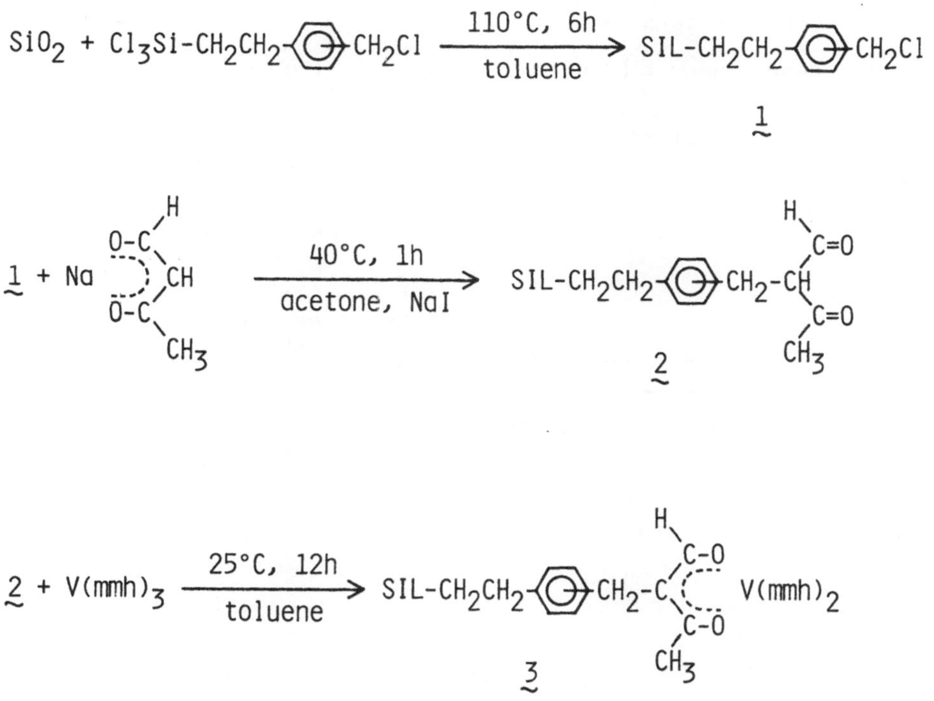

Fig.1 Preparation procedure of silica-anchored vanadium compound $\underset{\sim}{3}$. SIL= silica and mmh= 2-methyl-1,3-butanedionato.

The silica 3 was used as the catalyst for propylene polymerization in a toluene solution of Al(C₂H₅)₂Cl at temperatures of -70 to 0°C. The results are given in Table 2. When the anchored catalyst 3 was reacted with a toluene solution of Al(C₂H₅)₂Cl, the color of the solid catalyst changed from brown to red at temperatures of -70 to 0°C. The red color of solid catalyst remained unchanged even during the polymerization at 0°C. Any detectable vanadium species did not appear in toluene, indicating that vanadium species are bound on the surface of silica. The polymer yield at 3 h increases with raising temperature and attains a constant value at high temperatures above -40°C. At low temperatures below -60°C, the Mn of polypropylene increases with temperature and the polydispersities are relatively small (Mw/Mn = 1.3 - 1.5). This is the first example of solid catalyst which is capable of producing the polyporpylene with a narrow molecular weight distribution. At high temperatures above -40°C, the value of Mn decreases with temperature and the molecular weight distributions of polymers become broad (Mw/Mn = 1.9 - 2.2), indicating that chain terminating processes take place. The polypropylenes produced on the anchored catalyst 3 showed a lower syndiotactic regularity ([r] = 0.74) than that ([r] = 0.79) of polypropylene from the soluble V(mmh)₃ catalyst.

Comparing these results (Table 2) with those (Table 1) of the soluble catalyst, we have reached the following conclusions: (i) The anchored catalyst 3 shows a lower polymerization activity than that of the soluble V(mmh)₃ catalyst, but gives a relatively stable activity at high temperatures. (ii) A limiting temperature for the living polymerization of propylene over the anchored catalyst is still as low as -60°C. (iii) The living polypropylene bound on the surface of silica is very useful as a prepolymer for block copolymer synthesis.

Table 2. Results of polymerization of propylene with silica-anchored V(mmh)₃(3) in a toluene solution of Al(C₂H₅)₂Cl[a].

[Al]/[V] (mol/mol)	Temp. (°C)	Polymer yield (g/g of V)	Molecular weight	
			$\bar{Mn} \times 10^{-4}$	$\bar{Mw}/\bar{Mn}$
20	-60	162	5.31	1.36
20	-40	1040	4.13	2.24
20	-20	1030	1.98	1.94
100	-70	75	3.45	1.50
100	-65	101	4.62	1.51
100	-60	336	5.58	1.40
100	-40	583	3.25	2.00
100	-20	534	1.76	2.13
100	0	549	0.85	2.05

[a]Polymerization conditions : C₃H₆ = 830 mmol, V component = 0.03 mmol, toluene solution = 100 cm³, and polymerization time = 3 h.

Living Copolymerizaiton of Propylene with Ethylene

The copolymerizaiton of propylene with ethylene was performed at -60°C with the soluble $V(mmh)_3/Al(C_2H_5)_2Cl$ catalyst in toluene. The yield of copolymer increased proportionally with polymerization time. The activity was as large as 10 kg of EPR per g of V per h at -60°C, and almost fifty times higher than that of propylene homopolymerization. As given in Table 3, the copolymer obtained in 40 min shows a very high molecular weigh ($Mn = 10^6$) and a narrow molecular weight distribution ($Mw/Mn = 1.22$), indicative of the formation of a living copolymer of propylene with ethylene. As reported in a previous communicaiton (Doi et al, 1987b), the living copolymer of propylene with ethylene was a random terpolymer containing ethylene (1), primary inserted propylene (2), and secondary inserted propylene (3) units. It was found by analysis of the ^{13}C NMR spectrum that the mole fractions of three monomeric units in the living copolymer are: $N_1 = 0.60$, $N_2 = 0.35$, and $N_3 = 0.05$.

Table 3. Results of polymerization of olefins at -60°C with the soluble $V(mmh)_3/Al(C_2H_5)_2Cl$ catalyst [a]

No.	Monomer conc. (mol/dm^3)		Time	Polymer yield	$\bar{M}n \times 10^{-4}$	$\dfrac{\bar{M}w}{\bar{M}n}$	Propylene content
	$[C_2H_4]$	$[C_3H_6]$	(min)	(g/g of V)			(mol%)
1	0	8.3	180	0.62	2.98	1.22	100
2	0.3	8.3	40	6.5	102	1.22	40
3	0.3	0	60	1.5	—	—	0

a) Polymerization conditions : $V(mmh)_3=0.01mmol$, $Al(C_2H_5)_2Cl=5.0mmol$, and toluene solution$=100cm^3$.

Living Copolymerization of Propylene with 1,5-Hexadiene

The copolymerization of propylene with 1,5-hexadiene was performed with $V(acac)_3$ or $V(mmh)_3$ in a toluene solution of $Al(C_2H_5)_2Cl$. The results are listed in Table 4, together with the copolymerization conditions. The yield and Mn of copolymer increases almost proportionally with time in the copolymerization with the soluble $V(acac)_3/Al(C_2H_5)_2Cl$ catalyst at -78°C (see samples 2 and 3). The molecular weight distributions of copolymers were unimodal and their polydispersities were as small as 1.3, indicaitng that a living copolymerization of propylene with 1,5-hexadiene takes place.

Table 4. Results of copolymerization of propylene with 1,5-hexadiene in a toluene solution of vanadium component and $Al(C_2H_5)_2Cl$

sample no.	V-component	V (mmol)	Al (mmol)	C_3H_6 (mmol)	C_6H_{10} (mmol)	Temp. (°C)	Time (h)	Polymer yield (g/g of V)	$\bar{M}n \times 10^{-4}$	$\dfrac{\bar{M}w}{\bar{M}n}$	C_6H_{10} content[a] (mol%)
1	$V(acac)_3$	0.5	5	830	0	-78	3	51	1.50	1.2	0
2	$V(acac)_3$	0.5	5	830	25	-78	3	24	0.69	1.3	3.0
3	$V(acac)_3$	0.5	5	830	25	-78	6	55	1.57	1.3	4.3
4	$V(acac)_3$	1.0	20	83	25	-78	1	8	—	—	7.0
5	$V(acac)_3$	1.0	20	83	25	-78	4	24	—	—	7.4
6	$V(acac)_3$	1.0	20	42	50	-78	4	43	0.83	1.4	12.0
7	$V(mmh)_3$	0.05	5	830	0	-60	3	616	2.98	1.2	0
8	$V(mmh)_3$	0.05	5	830	25	-60	3	235	2.97	1.5	3.4

a) Determined from ^1H NMR spectra.

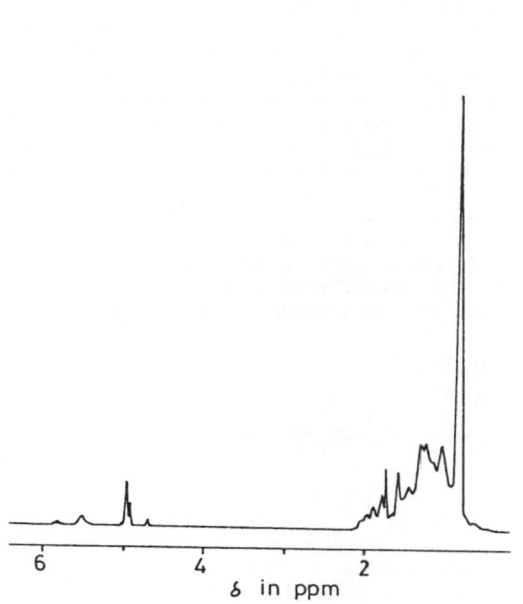

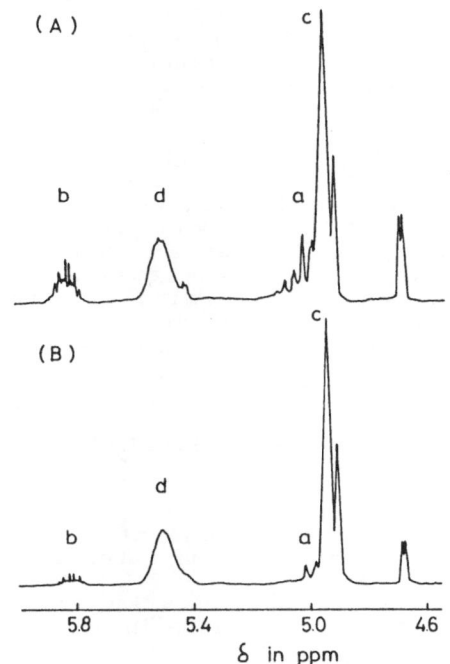

Fig. 2. 500 MHz ^{1}H NMR spectrum of copolymer sample 5 of propylene with 1.5-hexadiene.

Fig. 3. Expanded 500 MHz ^{1}H NMR spectra at δ=4.5-6.0 of copolymer samples 4(A) and 5(B) of propylene with 1.5-hexadiene

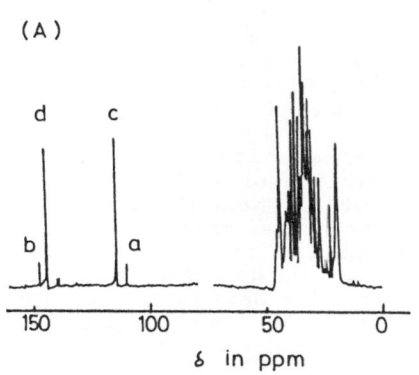

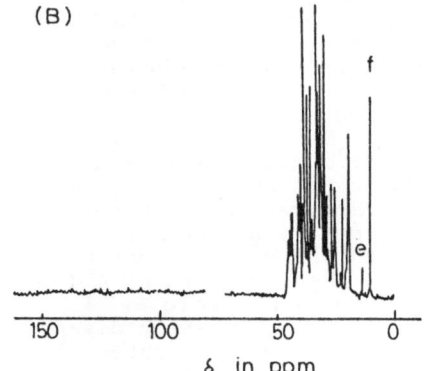

Fig. 4. 25 MHz ^{13}C NMR spectra of copolymer sample 6(A) and the hydrogenated sample(B).

Fig. 2 shows the 500 MHz ^{1}H NMR spectrum of copolymer samples 5. Except for the strong resonances at δ = 0.6 - 2.1, several weak peaks appear at lower magnetic field. Fig. 3 shows the expanded spectra of copolymer samples 4 and 5 at δ = 4.5 - 6.0. Four proton resonances a, b, c, and d at δ = 4.8 - 6.0 are assignable to the resonances of olefinic protons. The intensity ratios of peaks a to b and of c to d were 2.0, indicating the presence of two types of terminal olefins (CH_2 = CH-) in products. The relative peak areas of d to b were 2.6 for sample 4 obtained in 1 h, and 9.4 for sample 5 in 4 h. The relative intensity of peak b decreased as the copolymerization time increased, i.e., as the chain length of copolymer increased. From the above results, the signals a and b were assigned to the proton resonances of terminal olefins at the main-chain end, while the signals c and d were assigned to the proton resonances of terminal olefins at the side-chain end as:

$$\underset{a}{\underline{c} \ CH_2} \atop ...$$

$$CH_2 = CH-CH_2-CH_2-CH_2-CH_2 \overset{CH_3}{\underset{}{(}}CH-CH_2 \overset{\underline{c} \ CH_2}{\underset{\underline{d} \ CH}{)}}_m (CH-CH_2-CH_2-CH_2)_n \tag{1}$$

$$\underset{a}{} \qquad \underset{b}{}$$

The mole fractions of 1,5-hexadiene units in copolymers were determined from the intensity ratio of signals at δ = 4.5 - 6.0 to those at δ = 0.6 - 2.1. As given in Table 4, the 1,5-hexadiene content in copolymers increased with increasing the fraction of 1,5-hexadine to propylene in monomer feed.

In Eq (1) the structure of side-chain olefins of 1,5-hexadiene units in copolymer was proposed as an unusual vinyl group, since the proton resonance d (m, δ 5.46) appears at high magnetic field relative to the resonance b(m, δ 5.81). In general, the olefinic protons (CH_2 = CH-$\overset{\cdot}{C}$H-) bonding to methine carbon appear at higher magnetic field than those (CH_2 = CH-CH_2-) bonding to methylene carbon. The unusual side-chain structure of 1,5-hexadiene units is supported by the ^{13}C NMR spectrum of a hydrogenated copolymer.

Fig. 4 (A) shows the ^{13}C NMR spectrum of copolymer sample 6 (1,5-hexadiene units = 12 mol%). A complicated pattern of ^{13}C resonances appears at δ = 19 -50, which indicates a random copolymerization of propylene with 1,5-hexadiene. Four peaks of a (δ 109.4), c(δ 113.8), d(δ 143.7), and b(δ 146.6) at low magnetic field can respectively be assigned to carbons a , c, d, and b of terminal olefins in Eq (1).

The hydrogenation of copolymer sample 6 was carried out for 24 h at 100°C under 50 bar of H_2 in a toluene solution of $(PPh_3)_3RhCl$ catalyst to determine the side-chain structure of 1,5-hexadiene units in the copolymer. Recently, we reported that the soluble $(PPh_3)_3RhCl$ catalyst is very active for the hydrogenation of polybutadienes (Doi et al. 1986c). Fig. 4 (B) shows the ^{13}C NMR spectrum of the hydrogenated sample 6. In the spectrum four peaks a, b, c, and d at = 105 - 150 completely disappear, and new peaks e and f appear at 14.1 and 10.9, which indicates that all of the terminally olefin groups in sample 6 were hydrogenated as:

$$CH_3-CH_2-CH_2-CH_2-CH_2-CH_2 \overset{CH_3}{\underset{}{(}}CH-CH_2)_m \overset{\underset{\underline{f} \ CH_3}{CH_2}}{\underset{}{(}}CH-CH_2-CH_2-CH_2)_n \tag{2}$$

$$\underset{e}{}$$

The new peaks e and f of hydrogenated sample 6 can be assigned to the resonances of the methyl carbons e and f in Eq (2), since the intensity ratio of peaks f to e is almost consistent with that of peaks c to a. The ^{13}C chemical shift assignment of methyl carbons in the hydrogenated copolymer was made using the Lindeman and Adams relationship (1971). The calculated chemical shifts of methyl carbons e and f are respectively 13.86 and 11.36, which are in good agreement with the observed chemical shifts of 14.1 and 10.9. Here, let us calculate the chemical shifts of methyl carbons in different side-chain structures of hydrogenated 1,5-hexadiene units. The calculated ^{13}C chemical shifts of methyl carbons for propyl and butyl side-chains are respectively 14.35 and 13.86, which are far from the observed chemical shifts (10.9) of peak f. Thus, the ^1H and ^{13}C NMR analyses of copolymers revealed that the 1,5-hexadiene units in copolymers have an unusual vinyl structure as the side-chain as represented in Eq(1).

The structure of 1,5-hexadiene unit in the copolymer may be formed by the migration mechanism as:

$$
\begin{array}{c}
\overset{CH_3}{\underset{|}{*V-CH-CH_2-}}\!\!(P) \ + \ 1,5\text{-hexadiene} \longrightarrow \overset{\begin{array}{c}CH_2\\ \parallel\ 2\\ CH\\ |\\ CH_2\\ |\ 2\\ CH_2\\ |\ 2\end{array}}{\underset{|}{*V-CH-CH_2-}} \overset{CH_3}{\underset{|}{CH-CH_2-}}\!\!(P)
\end{array}
$$

$$\underset{\sim}{4} \hspace{7cm} \underset{\sim}{5}$$

$$
\longrightarrow \overset{\begin{array}{c}CH_2\\ \parallel\ 2\\ CH\end{array}}{\underset{|}{*V-CH-CH_2-CH_2-CH_2-}} \overset{CH_3}{\underset{|}{CH-CH_2-}}\!\!(P) \hspace{2cm} (3)
$$

$$\underset{\sim}{6}$$

In the above mechanism, 1,5-hexadiene monomer adds into a vanadium-polymer bond $\underset{\sim}{4}$ by a secondary insertion to give the structure $\underset{\sim}{5}$. The 1,5-hexadiene unit in complex $\underset{\sim}{5}$ is migrated along the side chain to form a complex $\underset{\sim}{6}$ which is active for the subsequent secondary insertion of propylene or 1,5-hexadiene.

EXPERIMENTAL SECTION

Tris(2-methyl-1,3-butanedionato) vanadium, $V(mmh)_3$, was synthesized by a reported method (Rohrscheid et al. 1967). The preparation of silica-anchored $V(mmh)_3$ was described in the text. The polymerization of olefins was carried out in a glass reactor (ca. 300 cm^3) or a steel autoclave (ca. 200 cm^3) with a stirrer. Propylene was condensed into toluene or heptane in the reactor kept at polymerization temperature. Then prescribed amounts of $Al(C_2H_5)_2Cl$ and vanadium compound were charged at the start of polymerization. For copolymerization ethylene or 1,5-hexadiene was admitted into the reactor. The polymerization was quenched at a given time by adding a methanol solution of hydrochloric acid kept at -78°C. The produced polymers were washed several times with methanol and dried under vacuum at room temperature.

Gel permeation chromatograms (GPC) of polymers were recorded on a Shodex 80 M column and o-dichlorobenzene as solvent at 140°C. The Mn and Mw were calculated from a molecular weight calibration curve of polypropylene. The 500 MHz ^1H NMR spectra of polymers were recorded at 27°C in $CDCl_3$ on a JEOL GX-500 spectrometer. The 25 MHz ^{13}C NMR spectra of polymers were recorded at 27°C in $CDCl_3$ on a JEOL FX-100 spectrometer under proton decoupling.

REFERENCES

Doi Y, Ueki S, Keii T (1979) Macromolecules 12: 814-819
Doi Y, Keii T (1986a) Adv.Polym.Sci 73/74: 201-248
Doi Y, Suzuki S, Soga K (1986b) Macromolecules 19: 2896-2900
Doi Y, Yano A, Soga K, Burfield DR (1986c) Macromolecules 19: 2409-2412
Doi Y, Hizal G, Soga K (1987a) Makromol.Chem. 188: 1273-1279
Doi Y, Tokuhiro N, Suzuki S, Soga K (1987b) Makromol.Chem., Rapid Commun. 8: 285-290
Lindeman LP, Adams JQ (1971) Anal.Chem. 43: 1245-1252
Rohrscheid F, Ernst RE, Holm RH (1967) Inorg.Chem. 6: 1315-1320

4. The Influence of the Reactor Design, Polyolefin Characterization

The Microreactor as a Model for the Description of the Ethylene Polymerization with Heterogeneous Catalysts

L.L. Böhm, R. Franke, G. Thum

Hoechst AG, P.O. Box 800320, 6230 Frankfurt/M. 80, FRG

ABSTRACT

In slurry as well as gas phase polymerization of ethylene with transition metal catalysts the catalyst particle is transformed in a polymer particle. The polymerizing system can be regarded as completely segregated with the catalyst and polymer particles as microreactors where the polymerization process takes place. Each particle represents a small individual reactor with its own material and energy balance. The actual knowledge concerning the relevant processes within these microreactors will be reviewed and compared with experimental investigations using modern high mileage catalyst/cocatalyst systems. It will be shown how polymer particle morphology can be influenced by catalyst design, and how catalyst design and polymerization conditions influence polymer properties.

INTRODUCTION

Polymerization processes are performed to transfer low molecular mass compounds into high molecular mass products. If this is done in a solution or bulk process viscosity usually increases by orders of magnitude which can lead to serious problems in technical plants (1,2). Such problems can be overcome by precipitation polymerization processes in which the polymer forms a seperate solid phase of small particles. In this case high polymer contents can be achieved under technical conditions (3). The slurry polymerization process of ethylene with transition metal catalyst/cocatalyst systems belong to this type of polymerization technology. A plant can be operated with polymer contents up to 40 v/v % which is a big advantage of this technology. This process can be considered as completely or nearly completely segregated (2,4) consisting of seperate phases as shown on Fig. 1.

Regarding the polymerizing system there are 3 phases: the liquid phase (diluent) usually an aliphatic hydrocarbon mixture, the gas phase (bubbles) formed by ethylene, hydrogen, comonomers, and some inert compounds, and the solid phase consisting of small catalyst or polymer particles. The catalyst particles are transformed into polymer particles in the course of polymerization, and these particles keep their identity over the whole polymerization period. They can be regarded as microreactors with their own material and energy balance as pointed out by Wicke an coworkers for heterogeneous catalytic processes (5-7). Advanced microreactor models for the ethylene and propene slurry, and also for the gas phase polymerization processes with heterogeneous

W. Kaminsky and H. Sinn (Eds.)
Transition Metals and Organometallics as
Catalysts for Olefin Polymerization
© Springer-Verlag Berlin Heidelberg 1988

catalysts were presented by Ray and coworkers (8-10). Also Reichert and coworkers (11) reported experimental results concerning the slurry polymerization of ethylene.

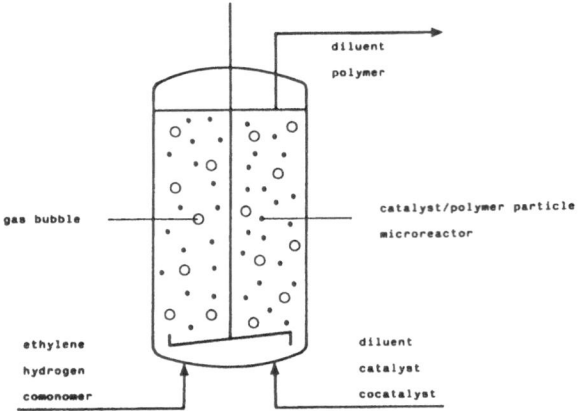

Fig. 1. Multiphase system within the slurry polymerization reactor (HDPE slurry process)

To get optimal results in respect to both polymer powder morphology (bulk density, average particle diameter, particle size distribution, particle shape), and polymer properties (catalyst content, average molecular mass, comonomer content, comonomer distribution, molecular mass distribution) the properties of the catalyst particle as the initial microreactor are of enormous importance and must be tailor-made, and the polymerization process must be performed under optimized conditions (3). The microreactor model teaches how the catalyst particle transforms into a polymer particle, and how this particle forming process can be influenced by catalyst design and process conditions. These results are discussed in relation to experimental data of the ethylene slurry polymerization using highly active Mg,Ti-catalysts (12,13) and triethyl- or isoprenylaluminum (14) as cocatalysts.

PARTICLE FORMING PROCESS AND MICROREACTOR MODELS

The particle forming process describes the transformation of a catalyst particle into a polymer grain. Polymer grain morphology depends on catalyst particle morphology and polymerization conditions which are well defined for each catalyst/cocatalyst system (3).

Usually one catalyst particle forms one polymer particle including that each particle keeps its identity over the whole polymerization time. Consequently, this unit can be regarded as a microreactor with its own material and energy balance. A catalyst particle produces a polymer grain increasing in diameter by a factor of 10 to 25 corresponding to a volume increase in the range of 10 to 10 . It is also known that this particle expands more rapidly at the beginning in relation to further volume increase (3). The particle size distribution of both catalyst and polymer usually remains unchanged as it is the case for particle shape (15-18). These facts are summarized on Table 1.

- One catalyst particle transforms into one polymer particle

- Average particle diameter increases by a factor of 10 to 25; Volume increases by a factor of 10^3 to 10^4

- Particle size distribution of catalyst and polymer is the same

- Particle shape of catalyst and polymer is the same

Table 1. Correlations between catalyst and polymer particle morphology

It is also known that catalyst particles are agglomerates of primary particles (small crystallites) with diameters down to 100 Å or even lower (17,19,20). Under polymerization such catalyst particles are des-integrated or fractured by the polymer generated at the surface of these primary particles. A realistic model for this process is the multigrain model presented and discussed in detail by Ray and coworkers (8-10).

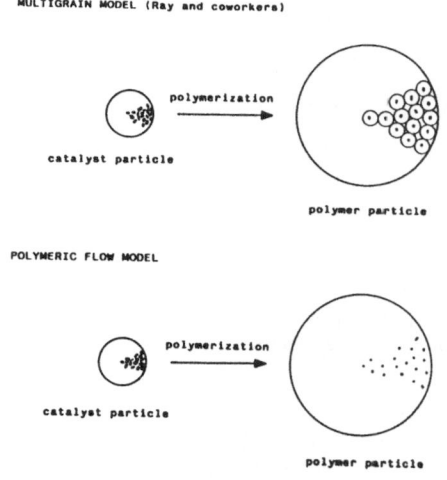

Fig. 2. Multigrain and polymeric flow model

As shown on Fig. 2 this model teaches that a polymer particle is com-posed of microparticles with a catalyst fragment (primary particle) at the center at which the polymer is formed. According to this model the physical and chemical processes within a polymer particle must be sepe-rated in both processes at the microparticles and within the macropar-ticle itself. On the other hand our investigation showed that this par-ticle forming process can be approximated by the polymeric flow model (21,3). This microreactor model is represented on Fig. 2 for compari-son. The polymer particle is assumed to be quasi homogeneous with the active sites distributed evenly over the whole particle. The physical values are averaged ones within the limits given by the diluent on one side and the semicrystalline polymer on the other. Although these two microreactor models look very different they lead to similar results in respect to the extent of monomer diffusion limitations and temperature gradients within the expanding microreactor.

Within the microreactor a fast and strongly exothermic polymerization reaction takes place (3,8-11) which may lead to temperature and concentration gradients within the particle and at the outer surface of the particle in respect to the surrounding diluent. From calculations starting with reasonable assumptions and data (3,8-11) for a technical process the critical period of this particle forming process can be detected. For a slurry process running at high polymer contents the start up period of polymerization is a critical situation because well pronounced concentration gradients within the particle and a sharp temperature rise of the particle in respect to the diluent may occur. This is schematically shown on Fig. 3.

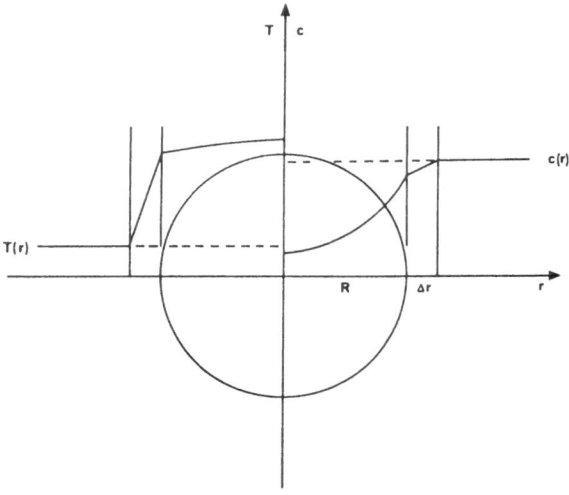

Fig. 3. Schematic representation of temperature and concentration gradients at the microreactor during the polymerization start up period

Concentration and temperature gradients occur if a rapid polymerization reaction takes place in a small microreactor volume at the beginning of the particle forming process. It can be estimated that temperature increase inside the particle is relatively small (3K). In contrast to this small temperature gradient there may occur a high overheating of the particle compared to the surrounding diluent in the range of more than 10 K. This can be regarded as hot spot formation caused by reduced heat transfer as a consequence of unsufficient particle flow within the slurry (11,22). It is quite favorate that overheating is reduced to some extent by diffusion limitation for the monomer inside the particle being at the maximum when polymerization starts.

As a concequence of the rapid volume increase of the microreactor both concentration and temperature gradients level out rapidly. It can be evaluated that the critical period usually has passed if the particle diameter has increased by a factor of 3 for catalyst particles with diameters below 50 μm (22). Although this happens within a short time in respect to average polymerization times as schematically indicated by the dotted area on Fig. 4, both temperature and concentration gradients must be avoided to generate polymer powder with acceptable morphology as an essential requirement to run a technical slurry process. This can be done by polymerization rate regulation as a function of time.

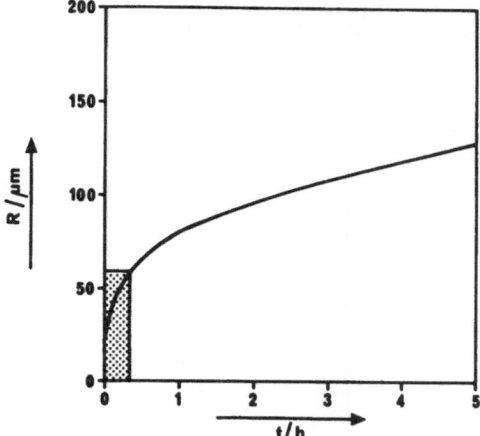

Fig. 4. Expansion of the microreactor as a function of time

POLYMERIZATION BEHAVIOUR AND ACTIVATION PROCEDURE

To get an active system for ethylene polymerization the catalyst must be contacted with the cocatalyst what can be done in different ways. This activation procedure influences the polymerization behaviour of the active system. The experimental investigations have been performed with a highly active catalyst and triethylaluminum as cocatalyst (12). Three different activation procedures are examined:

1. Catalyst and cocatalyst are contacted in the diluent at the polymerization temperature for at least 5 minutes before the polymerization starts by pressurizing the reactor with ethylene as reported elsewhere (19).

2. The catalyst is injected into the polymerization reactor which is held under polymerization conditions.

3. The catalyst is contacted with triethylaluminum to transform the total titanium into the trivalent state (97,5 mol% Ti^3) in a seperate step before used for polymerization. Excess triethylaluminum is removed by washing. This catalyst contains aluminum. The molar aluminum to titanium ratio is approximately 2 : 3. With this catalyst the polymerization is started as described under point 2.

To compare the polymerization rate versus time curves it must be proved that the temperature within the microreactor is the same for all experiments. As the avarage molecular mass is very sensitive to polymerization temperature it has been checked how average molecular mass depends on stirrer speed as described previously (19). All experiments were performed under conditions where average molecular mass becomes independent on stirrer speed which indicates the same temperature for the microreactor and the diluent. How the polymerization rate versus time curves are then influenced by the activation procedure is demonstrated on Fig. 5.

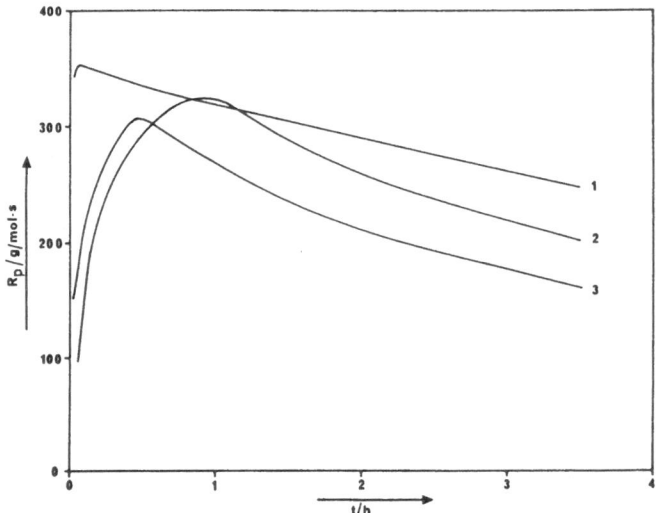

Fig. 5. Rate of polymerization (R_p) versus time (t) plots, polymerization conditions: Mg,Ti-catalyst (12), cocatalyst: triethylaluminum, 85 °C, ethylene pressure 1.5 bar, hydrogen pressure 1.2 bar, curve 1: activation procedure 1, Al/Ti-ratio 15, curve 2: activation procedure 2, Al/Ti ratio 4, curve 3: activation procedure 3, Al/Ti-ratio 3

Procedure 1 leads to an active system with maximum activity at time 0 decreasing slowly with increasing reaction time. If procedures 2 and 3 are applied the polymerization rate increases with time to pass a maximum with further decrease. The maxima are reached within one hour. This type of curve has been discussed by Ray and coworkers (22). They give the interpretation that such curves indicate diffusion limitation for the monomer at the beginning together with active site decay (diffusion limitated deactivating system). But this cannot hold for these curves because systems generated by activation procedure 1 doesn't show this behaviour. If these curves are compared with those calculated by Ray and coworkers (22) it can also be seen that this diffusion process has a longer time scale than monomer diffusion. As the activation process runs rapidly as shown for procedure 1 this behaviour can only be explained by a diffusion process of the cocatalyst through the polymer into the expanding microreactor to generate active sites in combination with a deactivation process. This is also supported by the fact that type 2 and 3 curves can be influenced and shifted by changing aluminumalkyl concentration. With increasing aluminumalkyl concentration the maximum is shifted towards shorter times until at very high concentrations time 0 is asypmtotically approached. By changing the molar ratio of the aluminum and transition metal compound (Al/Ti-ratio) the rate maximum can also be shifted up and downwards. This is shown on Fig. 6 comprising the Al/Ti-ratio realized in a technical process. If the Al/Ti-ratio is rised further the activity decreases as shown and discussed previously (19).

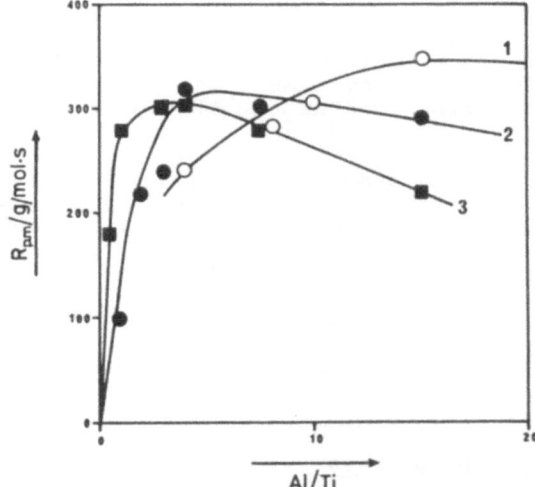

Fig. 6. Maximum rate of polymerization ($R_{p.m}$) versus Al/Ti-ratio, curve
1: activation procedure 1, curve 2: activation procedure 2, curve 3:
activation procedure 3

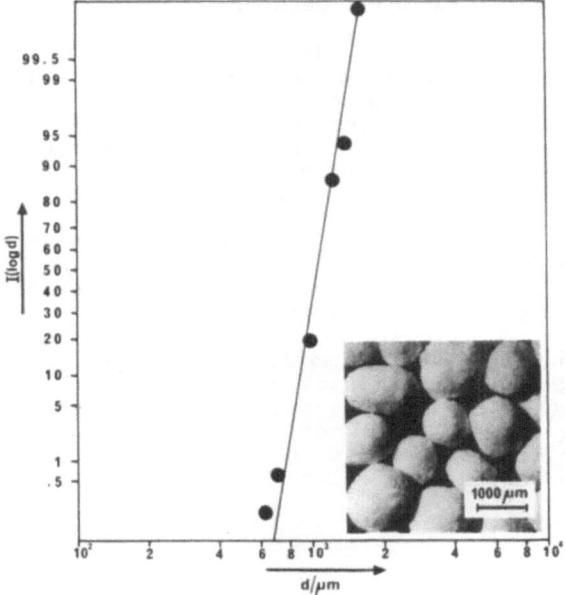

Fig. 7. Particle size distribution and particle shape of ball-like
polymer grains

In summary these investigations show that the polymerization behaviour within the microreactor can be regulated by the activation procedure in combination with the diffusion process of the cocatalyst through the polymer into the microreactor. This is very simple and effective to avoid concentration gradients and overheating at the beginning of the particle forming process.

If catalyst design, activation procedure and polymerization conditions are well optimized it is possible to generate ball-like polymer particles with approximately the same size for all polymer grains. An example is given on Fig. 7 (18).

The photo shows well shaped ball-like particles. The average particle diameter is 1110 µm and the particle size distribution is extremely narrow. The bulk density is 0.350 g/cm^3. This type of polymer powder is quite favorate to run a technical slurry process.

Catalyst particle design and polymer properties

In a technical process it is necessary to influence both polymer powder morphology and polymer properties. One of the most important property is molecular mass distribution. Polyethylenes prepared with Ziegler-catalyst/cocatalyst systems usually have very broad molecular mass distributions. There are some investigations (10,23-25) which show that such broad molecular mass distributions can only be caused by simultaneous polymerization of at least two types of active sites forming polymers with average molecular mass different by more than one order of magnitude under the same polymerization conditions. In previous papers (18,24) this type of catalyst has been described.

A highly active Mg,Ti-catalyst is coated with a further transition metal compound which covers the original catalyst like a shell. Again by changing the activation procedure and process conditions the polymerization behaviour of these catalytic systems can be influenced enormously as demonstrated on Fig. 8. If activation procedure 1 is used the polymerization behaviour is comparable to curve 1 given in Fig. 5. This is shown by the curves 1 and 2 which differ by the Al/Ti-ratio.

This behaviour changes if activation procedure 2 is used as represented by curves 3 and 4. These curves also depend on the Al/Ti-ratio. As shown by curve 4 it is possible to have a permanent increase of activity over 4 hours to reach the maximum of activity.

By changing polymerization behaviour average molecular mass as a function of time can be influenced. Experimental results are given on Fig. 9. It can be seen that these curves correspond nicely to the rate of polymerization versus time curves. Low activity at the beginning means strong average molecular mass dependence on time. It can also be realized that all curves approach the same value after long polymerization time.

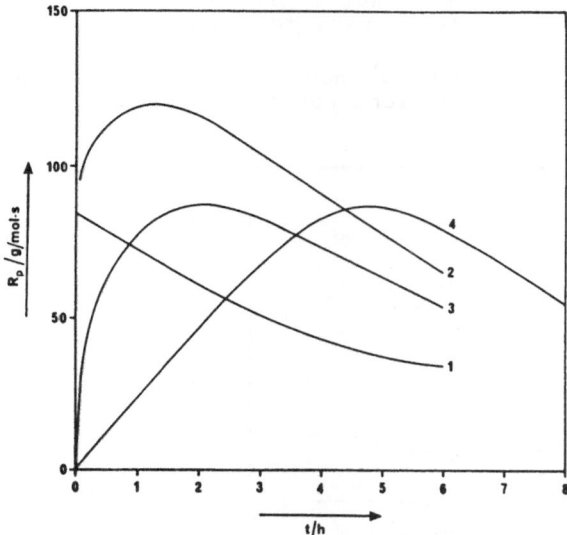

Fig. 8. Rate of polymerization (R_p) versus time (t) plots, polymerization conditions: Mg,Ti-catalyst (13), cocatalyst: isoprenylaluminum (14), 85 °C, ethylene pressure 3.0 bar, hydrogen pressure 5.3 bar, curves 1,2: activation procedure 1, Al/Ti-ratio = 110, 36, curves 3,4: activation procedure 2, Al/Ti-ratio = 36, 10

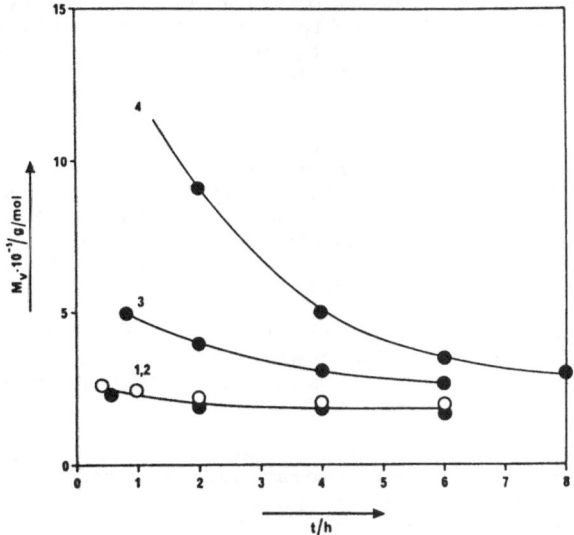

Fig. 9. Viscosity average molecular mass (M_v) versus time (t) plots, numbers see Fig. 8

If there is strong dependence of average molecular mass with time (curve 4 in Fig. 9) molecular mass distribution also changes simultaneously as represented on Fig. 10. At "short" time there is a pronounced bimodal distribution with high amounts of the high molecular mass fraction. With time the high molecular mass fraction is reduced in favour of the low molecular mass fraction. If average molecular mass is not dependent on time molecular mass distribution also is time independent. As represented on Fig. 10 time dependent molecular mass distribution approaches the time independent one after long reaction time.

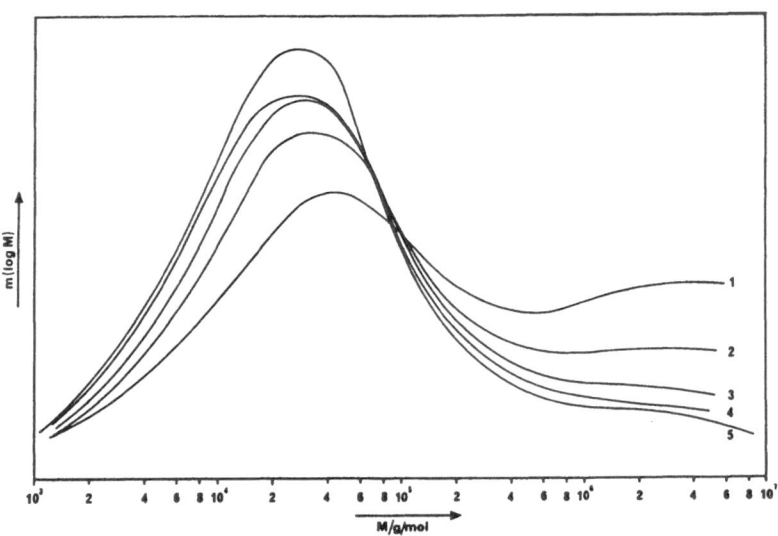

Fig. 10. Molecular mass distributions, polymerzation conditions see Fig. 8, curves 1-4 correspond to curve 4 in Fig. 8, polymerization time: curve 1, 2 h, curve 2, 4.5 h, curve 3, 7 h, curve 4, 14 h, curve 5 corresponds to curve 2 in Fig. 8, polymerization time: 6 h

In summary these results show the possibility to change molecular mass distribution via catalyst design, activation procedure, process conditions and time. Because there are two peaks with different average molecular mass remaining constant it can be concluded that two catalytic systems located within the same microreactor are polymerizing simultaneously. Fig. 11 represents two borderline cases of the particle forming process to explain these experimental results.

Fig. 11 teaches that each catalyst particle is composed of two types of primary particles, one type located at the center (filled points), covered by the other type (open points) like a shell. This is consistant with the catalyst preparation procedure.

If all centers are activated by filling the microreactor with the cocatalyst using activation procedure 1 before monomer addition polymerization starts at all active sites over the whole particle. Then activity is at the maximum just at beginning, (curve 1 in Fig. 8) average molecular mass is time independent (curve 1 in Fig. 9) as molecular mass distribution is (curve 5 in Fig. 10). In this case only a small fraction (~ 10 w/w %) of the high molecular mass compound is formed.

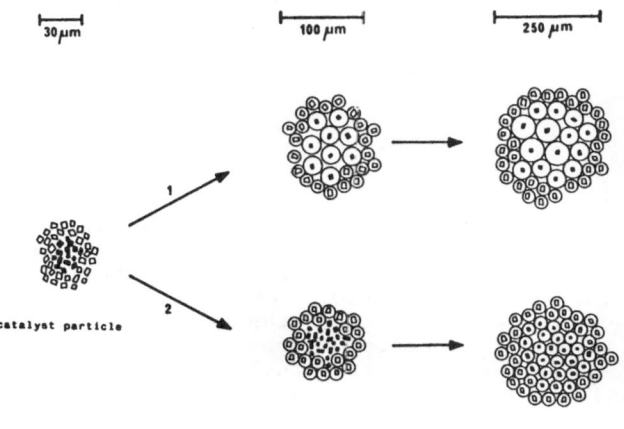

Fig. 11. Particle forming process of a two component catalyst: 1 : activation procedure 1, 2: activation procedure 2

If on the other hand activation procedure 2 is applied the cocatalyst diffuses into the microreactor. At the beginning active sites are formed only at the surface and start with polymerization immediately generating a polymer layer. This layer reduces diffusion of further aluminum organic compound into the microreactor. As a consequence rate of polymerization is low. These less active sites generate the high molecular mass fraction. Therefore the viscosity number or the average molecular mass is high at the beginning. As diffusion of the catalyst into the microreactor proceeds the number of active sites gradually increases but also sites with higher activity and better hydrogen response are generated in the center of the microreactor. Thus, the rate of polymerization becomes higher, the average molecular mass comes down and the low molecular mass fraction increases in relation to the high molecular mass fraction.

By changing catalyst design, type of cocatalyst, activation procedure, cocatalyst concentration, process conditions and polymerization time it is possible to change the polymerization behaviour and the molecular mass distribution within the borderline cases as discussed in detail. This is quite favorate for a technical process.

Discussion

The particle forming process as the most important process of the slurry or gas phase polymerization of ethylene with Ziegler-catalyst/ cocatalyst systems is well understood today (3,8-11). There are advanced models (multigrain and polymeric flow model) which describe qualitatively and quantitatively all relevant processes at the surface and within the particle including the critical situations to be avoided. The growing polymer grain can be regarded as a microreactor as first presented and discussed by Wicke and coworkers (5-7).

By changing catalyst design, type of cocatalyst, activation procedure, polymerization conditions and polymerization time it is possible to influence and regulate the important and relevant properties of the

polymer including both powder morphology and polymer data. On the basis
of these models considerable progress in catalyst and process development has been achieved.

EXPERIMENTAL

All polymerization runs were performed in a 200 dm³ stainless steel
batch reactor equipped with a stirrer and a flow disturber. The number
of revolutions could be chosen infinitely variable. The reactor was
equipped with a temperature control system. Temperature is monitored
continuously. The temperature deviation is in the range of 0.5 K. The
reactor could be purged with high purified nitrogen. The amounts of all
components (diluent, catalyst, cocatalyst, ethylene, hydrogen, comonomer) are measurable within narrow ranges. The ethylene flow as one of
the most important values is measured via a flowmeter, the total amount
of ethylene introduced into the reactor is controlled via a gasometer
and the amount of polymer. The gas phase within the reactor is monitored with a process GC. Hydrogen is further measured with a thermal flow
meter. Hydrogen and the comonomers are feed into the reactor to hold a
constant gas phase composition. There are different equipments to
introduce catalyst and cocatalyst into the reactor, to change activation procedure and start up conditions. During polymerization small
amounts of the slurry in relation to the total amount are removed for
testing. The test procedure are given elsewhere (19). The rate of polymerization versus time curves are reproducible within 10 %. The reactor is filled with 100 dm³ of a hydrocarbon diluent. The total amount
of polymer is in the range of 20 kg.

These investigations are made in the Research and Development Department Polyethylen of Hoechst AG, Frankfurt/Main.

REFERENCES

1 H. Gerrens, Chem. Ing. Techn. 52, 477 - 488 (1980)

2 H. Gerrens, Polymerisationstechnik in Ullmanns Encyklopädie der
  technischen Chemie, Verlag Chemie, Weinheim 1980, Band 19, p. 107 -
  165

3 L.L. Böhm, Chem. Ing. Techn. 56, 674 - 684 (1984)

4 E. Fitzer, W. Fritz, Technische Chemie, Springer Verlag, Berlin
  1975, p. 352 - 359

5 E. Wicke, G. Padberg, Chem. Ing. Techn. 40, 1033 - 1038 (1968)

6 P. Fieguth, E. Wicke, Chem. Ing. Techn. 43, 604 - 608 (1971)

7 E. Wicke, Chem. Ing. Techn. 46, 365 - 374 (1974)

8 S. Floyd, G.E. Mann, W.H. Ray, Heat and Mass Transfer Limitations and Catalyst Deactivation Effects in Olefin Polymerization for Gas Phase and Slurry Reactors in Studies in Surface Science and Catalysis 25, Catalytic Polymerizations of Olefins, Ed. T. Keii, K. Soga, Elsevier, Amsterdam, Kodanska, Tokyo 1986, p. 359 - 367

9 S. Floyd, K.Y. Choi, T.W. Taylor, W.H. Ray, J. Appl. Polym. Sci. 32, 2935 - 2960 (1986)

10 S. Floyd, T. Heiskanen, T.W. Taylor, G.E. Mann, W.H. Ray, J. Appl. Polym. Sci. 33, 1021 - 1065 (1987)

11 K.-H. Reichert, R. Michael, H. Meyer, Reaction Engineering Aspects of Ethylene Polymerization with Ziegler-Catalysts in Slurry Reactors in (8) p. 369 - 386

12 Belg. Pat. 737 778 (1968), Farbwerke Hoechst AG, Erf.: B. Diedrich, K.-D. Keil

13 EP 0 068 257 (1981), Hoechst AG, Erf. J. Berthold, B. Diedrich, R. Franke, J. Hartlapp, W. Schäfer, W. Strobel

14 H. Hoberg, H. Martin, R. Rienäcker, K. Zosel, K. Ziegler, Brennstoff-Chemie 50, 217 - 221 (1969)

15 P. Galli, L. Luciani, G. Cecchin, Angew. Makromol. Chem. 94, 63 - 89 (1981)

16 P. Galli, P.C. Barbé, L. Noristi, Angew. Makromol. Chem. 120, 73-90 (1984)

17 P.C. Barbé, G. Cecchin, L. Noristi, Adv. Polym. Sci. 81, 1 - 81 (1986)

18 R. Franke, L.L. Böhm, W. Strobel, G. Thum, U. Wolfmeier, High yield Catalysts for Tailor-made Polyethylene Production in the Slurry Process in Transition Metal Catalyzed Polymerization: Ziegler-Natta and Metathesis Polymerization, Ed. R.P. Quirk, Cambridge University Press, in press

19 L.L. Böhm, Polymer 19, 553 - 561 (1978)

20 R.P. Nielson, Active Center Generation in the Solvay Typ High Performance Titanium Trichloride Catalyst - An Interpretation from the Hermans-Henrioulle Patent in Transition Metal Catalyzed Polymerizations, Ed. R.P. Quirk, Harwood Academic Publ., Chur, London, New York, 1981, p. 47 - 82

21 T.W. Taylor, K.Y. Choi, H. Yuan, W.H. Ray, Physicochemical Kinetics of Liquid Phase Propylene Polymerization in 20, p. 191 - 224

22 S. Floyd, K.Y. Choi, T.W. Taylor, W.H. Ray, J. Appl. Polym. Sci. 31, 2231 - 3365 (1986)

23 L.L. Böhm, Angew. Makromol. Chem. <u>89</u>, 1 - 32 (1980)

24 L.L. Böhm, J. Berthold, R. Franke, W. Strobel, U. Wolfmeier, Ziegler Polymerization of Ethylene: Catalyst Design and Molecular Mass Distribution in (8), p. 29 - 42

25 Y.V. Kissin, Isospecific Polymerization of Olefins with Heterogeneous Ziegler-Natta Catalysts in Polymer/Properties and Applications 9, Springer Verlag, New York, Berlin, Heidelberg, Tokyo, 1985, Chapter IV.

LINEAR LOW DENSITY POLYETHYLENE PREPARED IN GAS PHASE WITH BISUPPORTED SiO$_2$-MgCl$_2$ ZIEGLER-NATTA CATALYSTS

R. SPITZ, V. PASQUET and A.GUYOT
CNRS - Laboratoire des matériaux Organiques
BP 24 - 69390 LYON-VERNAISON (France)

SUMMARY

Bisupported silica-MgCl$_2$-TiCl$_4$ catalysts suited for gas phase polymerization are used at 15 bars, 85-95° C for the synthesis of polyethylenes ranging from high density to linear low density. The main parameters of the reaction (temperature, pressure, monomer ratio) are studied in connection with the pecularities of the gas phase polymerization. The effect of the addition of a Lewis base to the solid catalyst on copolymerization reactivity and on the copolymer structure are studied.

INTRODUCTION

Ethylene- α-olefin copolymers synthesis was possible with the first titanium and chromium based catalysts (1, 2) but linear low density polyethylene (L.L.D.P.E.) appeared as a defined product in the middle of the 70's (3). The synthesis of products presenting better properties than the long branched polyethylene obtained at high pressure (4) requires a good control of the copolymerization reaction through the catalytic system and the copolymerization process. This was achieved in gas phase polymerization with the Unipol process for ethylene-butene copolymerization (5). Longer α-olefins are easier to polymerize in solution. L.L.D.P.E. can be also produced with a diluent or at high pressure (6, 7) but the gas phase polymerization permits the synthesis of copolymers covering all the range of densities from high to low and even very low (0.900) and all the range of melt-indices corresponding to all the possible uses of polyethylenes (8).

Catalyst

The gas phase polymerization, particularly in fluidized bed, requires an adaptation of the Ziegler catalysts : a controlled particle size distribution is necessary all along the polymerization process in order to avoid two opposit and both unwanted phenomena : the formation of fine particles and the plugging of the reactor. This can be avoided by a first prepolymerization step or by a convenient choice of the catalyst support. Silica supports are used for a long time as supports for chromium catalysts (2, 9) and have been adjusted to the requirements of the industrial gas phase polymerisation. But titanium chloride catalysts with titanium directly bound to an oxide support (silica or alumina) have limited activities (10, 11). Union Carbide bisupported catalyst keep the properties of MgCl$_2$ supported Ziegler

catalysts : silica is coimpregnated with a solution of MgCl$_2$ and

W. Kaminsky and H. Sinn (Eds.)
Transition Metals and Organometallics as
Catalysts for Olefin Polymerization
© Springer-Verlag Berlin Heidelberg 1988

TiCl$_4$ in tetrahydrofuran (12). Silica is then used as an inert
carrier for the bisupported catalyst.

We have used a different approach for this work. A support is prepared
by dehydration of silica coated with MgCl$_2$ in solution in water
(13). This allows variations in the preparation of the catalyst :
TiCl$_4$ impregnation leads to a Lewis base free catalyst but Lewis
base followed by TiCl$_4$ impregnation permits a controlled
introduction of a Lewis base. The composition of the solids used for
this study are summarized in Table 1.

Table 1. Caracterisation of supports and solids

	Surface area m^2/g	Pore radius[a] nm	% Mg weight	% Ti weight	% ethyl benzoate (E.B.)
Silica	300	10.5	0	0	0
Silica-MgCl$_2$ support	270	9.5	4.6	0	0
Silica-MgCl$_2$ TiCl$_4$ catalyst	210	-	4.6	2.5	0
Silica-MgCl$_2$ E.B. TiCl$_4$ catalyst	184	-	4.6	4.6	5

a) maximum of pore size distribution obtained by nitrogen desorption

Gas phase polymerization

Gas phase polymerization using a fluid bed are effective on Union
Carbide plants as soon as 1968 and different designs of fluidized or
stirred reactors have been developped since then (14). Fluid bed
polymerization reactors working closely to industrial conditions are
extremely difficult to control at laboratory scale. We have developed
a particular kind of stirred reactor approaching the conditions of
stirring of a fluid bed : all the reactor oscillates vertically at
high frequency (15). This kind of stirring avoids the grinding of the
growing polymer sometimes observed with a stirrer in the reactor. The
reactor allows the production of more than 100 g of polymer at
constant temperature and pressure with determination of the overall
kinetics and relative consumption of ethylene and butene. A convenient
adjustment of the gas mixture at the beginning of the polymerization
suppresses the compositional drift during the copolymerization.

Polymers ranging from high density (d = 0.960) to L.L.D.P.E (d < 0.920) are obtained varying the catalyst and the butene concentration in the reactor.

EXPERIMENTAL

Support preparation (SM)

10 g of silica GRACE 1952 are soaked at room temperature with 4.2 cm^3 of a saturated $MgCl_2$ water solution. The resulting solid is dried at 110° C under vacuum, mixed with 2 g NH_4Cl fluidized under argon stream and heated at increasing temperature (0.3 °C/min up to 200° C, 1.2° C min up to 550° C) then left 0.5 h at 550° C. After cooling under argon the support contains (by weight) 4.6 % Mg, 9 % Cl, 0 % N. The chlorine content is equivalent to 66.7 % Mg as $MgCl_2$.

Catalysts

SMT : the SM support is contacted with $TiCl_4$ (1 g/g SM) 2.5 hours at 65° C, washed with heptane and dried under high vacuum (10^{-8} bars) at room temperature. The color is pale grey. The titanium content of the catalyst is 2.5 % W/W.

SMCT : the SM support is first contacted at 40° C with ethylbenzoate (E.B. : 0.5 mole/mole Mg) diluted in heptane (0.1 mole/l). Surnatant heptane is filtered and $TiCl_4$ treatment occurs like for SMT.

SMDT : is prepared like SMT, dibutylphtalate (0.3 mole/mole Mg) replacing ethylbenzoate.

Polymerization : a stainless steel reactor (0.675 l) is stirred by vertical oscillations (6 cm, 7 H_z). Temperature (generally 95° C) is maintained constant (∓ 1° C). The reactor is continuously fed with a constant monomer mixture corresponding to the wanted polymer composition. The reagents are introduced at 20° C in the following order : isoprenylaluminium (IPRA : 0.5 mmol.) catalyst (10-20 mg) diluted with 2 g of an inert powder (KCl or polyethylene) in order to ensure stirring at the beginning of the polymerization. The reactor is then pressurized at 16 bars with an ethylene-butene mixture adjusted to avoid composition drift during the polymerization, H_2 is added and the ractor heated in order to reach the polymerization temperature (85° C for copolymerization) in 2.5 min. The composition of the gas phase is controlled by gas chromatography.

IR : the polymer are annealed under argon 24 h. at 10° C below their melting temperature. A Nicolet 20 SX FTIR is used for the determination of the ethyl branches and crystallinity.

Fractionation : the copolymers are extracted 4 hours with hexane at hexane boiling temperature (Kumagawa).

NMR : the ^{13}C NMR spectra of the hexane soluble polymer fractions are obtained on a Bruker 80 FT spectrometer at 20.1 MHz. The polymer is dissolved in a mixture 2 : 1 of tetrachlorethylene and hexadeutero benzene at 370 K. The methylene carbon of the isolated butene (EBE sequence) of the end of the butene sequences (BBE) and of the long butene sequences (BBB) are found respectively at 39.7, 37.25 and 34.95 ppm.

RESULTS AND DISCUSSION

Kinetics : the polymerization rate is constant at 95° C in the case of ethylene homopolymerization at 16 bars, IPRA being used as cocatalyst. In the case of butene copolymerization, the polymerization rate is faster at the beginning of the reaction but decreases continuously with time (Fig. 1). The stability of the catalyst is decreased when hydrogen pressure is increased or when aluminium alkyls with a reducing power greater than IPRA like trihexyl or trioctylaluminium are used. This suggests that the deactivation of the catalyst is due to a chemical effect. The fact that polymerization with constant rate for hours at rather high level (2400 g/g catalyst/h) is observed suggests that the polymerization is not diffusion controlled as it could be expected for a porous solid-gas reaction (16). Arguments against diffusion control can also be found in the fact that the molecular weights of the polymers are not time dependant. The increase of the polymerization rate from homo to copolymerization is observed with many catalytic systems based on different metals (17). In the case of gas phase polymerization, a rate increase associated to greater solubilities of the monomers in a copolymer presenting more amorphous phase could be a reasonable assumption. But similar activations are observed with the other polymerization processes and can be ascribed to the presence of more active centers in the case of copolymerization (18).

The reaction is first order versus ethylene pressure in a wide range (20-3 bars) of pressures . To obtain a result independant of the number of the active centers the catalyst is first contacted with monomer at the highest pressure (20 bars) and the rate is then measured at the wanted pressure. Without precaution an apparent order higher than one is measured. The same holds for the effect of temperature : to measure an activation energy independant of active center formation or deactivation, the reaction is started at the highest temperature (95° C) and the temperature decreased to the wanted value. The activity is extremely sensitive to temperature .

The activation energy measured in that way is 57 kJ/mole in the range of other reported values (19, 20). Such a high activation energy and the heat produced by ethylene polymerization : 95 kj/mole (21) contribute to make difficult the control of gas phase polymerization, especially near the polymer softening point.

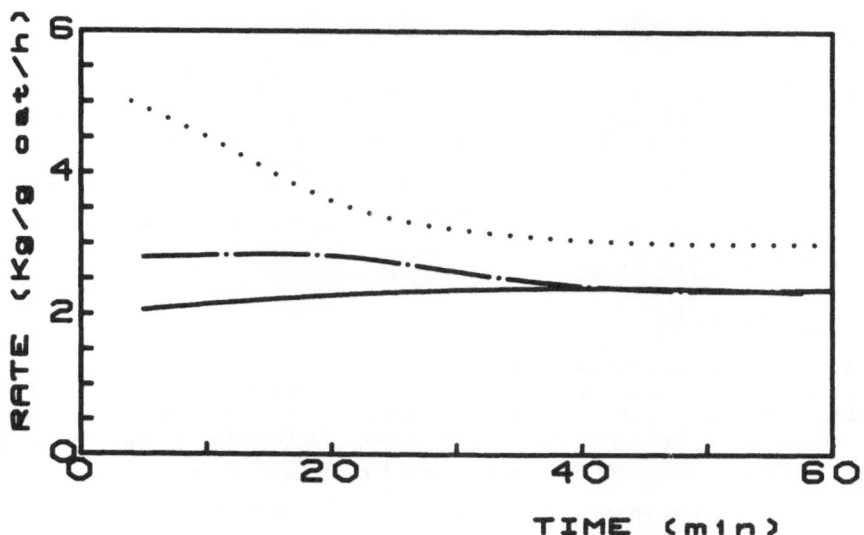

Figure 1 : Gas phase polymerization kinetics. Cocatalyst isoprenyl aluminium <u>homopolymerization</u> : 15 bars ethylene pressure 95° C catalyst SMT ——————
<u>copolymerization</u> : 15 bars monomer pressure 4.5 % molar butene in the monomer feed. 1 bar hydrogen pressure 85° C
catalyst SMT...   SMCT —— . ——

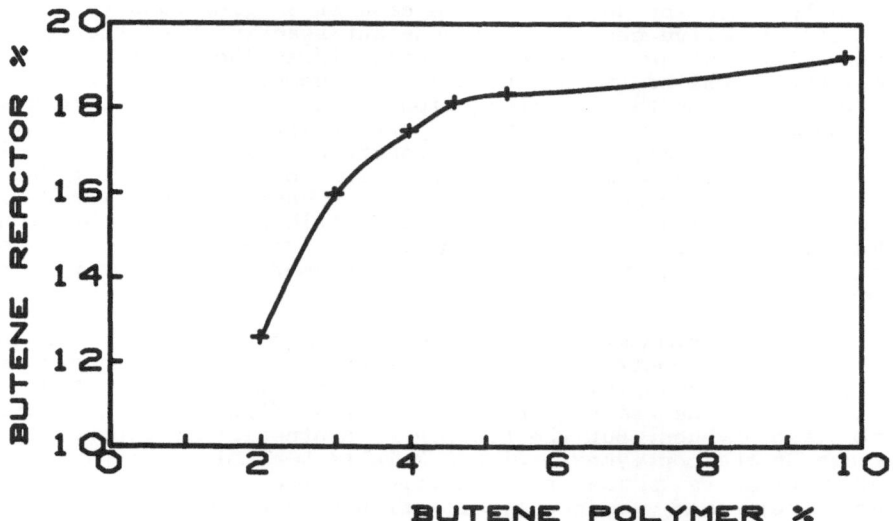

Figure 2 : Butene concentration (monomer mole %) in the reactor versus polymer composition (butene mole %)

Polymer productivity and catalytic residues

The stability of the activity compensates changes in the activity levels so typical polymers productions after 4 hours polymerization are of about 10,000 g/g catalyst for the SMT catalyst in ethylene polymerization, for the SMT and SMCT catalyst in the case of copolymerization. The corresponding catalytic residues are then Ti < 5 ppm, Mg < 5 ppm, Si ≃ 25 ppm, Al ≃100 ppm. The polymers have an apparent density of 0.4 g.cm^{-3} and a particle size distribution centered between 1 and 1.5 mm.

Copolymerization ratio

Copolymerization reactivities are generally described by methods developped for radical copolymerization. The corresponding model uses in the simplest case 4 constants corresponding to the addition of the two monomers to the two corresponding radicals (k_{AA}, k_{AB}, k_{BB}, k_{BA}). In fact, only two ratios ($r_A = k_{AA}/k_{AB}$, $r_B = k_{BB}/K_{BA}$) are then necessary to describe the copolymerization.

If the two constants are known, the composition and even the microstructure of the polymer can be computed for all copolymerization conditions (22). The two constants model is extensively applied in the case of Ziegler ethylene- α-olefin copolymerization (23-26) and terpolymerization (27). The different reactivity ratios published for ethylene and α-olefins doesn't desagree too much and indicate that ethylene is by far the more reactive monomer. But in fact, the application of a simple copolymerization equation, even if it appears to be useful in general cases is of restricted use in the case of L.L.D.P.E. Two conditions are generally fullfilled in the case of radical copolymerization and not in the case of Ziegler catalysis : 1) only one kind of active centers ; 2) constant reactivity of the active center in different polymerization conditions. The first condition leads to polymer chains having the same mean composition : copolymers synthetized with constant reactor composition cannot be fractionated regarding composition. In fact L.L.D.P.E. prepared with titanium catalysts are generally easy to fractionate by solvent extraction in two fractions with low and high comonomer content. More refined methods like temperature rising elution fractionation (T.R.E.F.) permits even to evidence a continuous distribution of compositions (28). The second condition means that the reactivities are not dependent on the composition of the copolymer. This point will be detailed below.

The fact that the polymer can be fractionated indicates, in the absence of diffusion control, that several kinds of active centers are present, probably a continuous distribution with a distribution of reactivities. The determination of reactivity ratios gives then only access to mean values. But the L.L.D.P.E. synthesis corresponds to a narrow range of the copolymerization domain (< 5 % mole butene) so the copolymerization is almost not sensitive to the reactivity constant r_B of the butene and the copolymerization with constant composition can be described with one parameter only.

The butene molar fraction in the reactor reaches a constant value even if the polymerization is started apart of equilibrium . In Fig. 2 is presented the variation of the polymer composition as a function of the reactor composition. The ratio r of the molar butene in the reactor to the molar butene in the polymer could then be a convenient parameter describing the formation of the copolymer. The smaller it is, the smaller will be the butene concentration required to produce a given copolymer composition. As it can be deduced form Fig. 2 this ratio decreases with increasing butene concentration in the reactor : the more there is butene in the copolymer, the easier does the butene copolymerize. The result seems in contradiction with the fact that the reported r_B values are by far under 1 in the case of propene or butene (25, 26, 29, 30). The determination of the reactivity ratios is generally not made in the same conditions and the polymerizations presented here are restricted to a narrow domain of compositions (low butene content) without compositions drift. The fact that the polymerization occurs in gas phase adds to the difficulty of the study : the composition of the gas phase is known, but the real mononomer composition near the active centers certainly depends on the monomers solubilities in the polymer (31) and also perhaps on the physical adsorption of butene in the pores of the growing polymer. An intermediate step of monomer adsorption on or near the active centers may also explain changes in the real reactivity ratios if there are changes in the oxidation state or in the coordination number of the active centers (32, 33).

The copolymerization reactivity is also sensitive to the catalyst composition. In Table 2 are presented the effect of variation of the catalyst composition on the reactivity ratio. Changes in Mg content or in the $MgCl_2$/MgO composition have only very little effect on butene reactivity. But addition of an aromatic ester, particularly ethylbenzoate favors butene incorporation in the polymer.

Table 2. Reactivity ratio (r = molar butene in reactor/molar butene in polymer) at 4.5 % monomer feed. Conditions : 85° C. 15 bars total monomer pressure - H_2 = 1 bar.

Catalyst Mg %	Lewis base	Mg as $MgCl_2$ %	r
4.6	no	30	4
4.6	no	66.7	4
2.1	no	55	4.3
10.8	no	79.4	3.9
4.6	Ethylbenzoate	66.7	3
4.6	Dibutylphtalate	66.7	3.5

Polymer heterogeneity :

Ethylene-butene copolymers with the same molar content of the comonomers may have different densities. At variance from some vanadium based catalysts (34, 35) the titanium based heterogeneous catalysts give rise to heterogeneous polymers as noticed above and this results in variations in the crystallization ability of the polymer and in poor properties in different domains : for instance copolymers presenting long butene sequences are sticky. Two kinds of heterogeneities are expected : polymer chains with different butene content ; different statistics of butene distribution inside the chains. Different approaches have been used to obtain a qualitative or even quantitative caracterization of the heterogeneities.

Melting point : the melting points are obtained by differential scanning calorimetry. They range between 126° C and 122° C when butene varies from 2 to 8 % molar. The melting points are associated to the fraction of the polymer with low butene content. After hexane extraction (see below) the melting point is the same for the insoluble residue but the peak is narrower. Homogeneous polymers with the same comonomer content have melting points near 100° C (35). The hexane soluble fraction presents some crystallinity with a broad melting peak near 90° C and accounts for the very wide DSC thermogram of the whole copolymer. The high temperature melting point can then be used as an indication of the presence of polymer chains with low butene content.

Table 3. Composition of the fraction and crystallinity of 3 typical copolymers

Catalyst	Fraction	Butene % molar	Crystallinity %
SMT	B	4.49	46.1
	I	3.08	49.8
	S	12.06	19.8
SMCT	B	5.74	41.8
	I	3.46	48.1
	S	12.44	18
SMDT	B	3.92	49
	I	2.90	51.1
	S	13.97	18.1

B : whole polymer ; I : hexane insoluble ; S : hexane soluble

Hexane extraction

The hexane extraction characterizes the amount of polymer with high butene content. The amount of hexane soluble polymer as a function of butene in the polymer is presented on Fig. 3. When the soluble fraction is computed as a function of two parameters : butene content and Melt Index, a good correlation is obtained for a mean square regression with :
Soluble polymer (%) = 1.711 + 0.683 [molar butene]^2 + 0.118 MI

The melt-index ranging here from 0 to 3, the effect of molecular weight appears to be rather negligible and the solubility varies as the square of the butene content. The ethyl branches contents of the different fractions are presented in Table 3. The soluble fractions have a molar butene content in a rather narrow range : 12-14 %. For a typical polymer with 0.920 g/cm^3 density the butene content is 4 % ; 12 % of the polymer is soluble in hot hexane with 14 % butene in the soluble fraction. The overall composition of the polymer is then :

<u>soluble fraction</u> : 12 % of the polymer, 14 % butene, accounting for 40 % of all the butene

<u>insoluble fraction</u> : 88 % of the polymer, 2.8 % butene, accounting for 60 % of all the butene.

Again, the polymers prepared with the SMCT catalyst are slighly less heterogeneous (Fig. 3 and Table 3) : the soluble fraction is reduced by 1 to 2 % with these catalysts.

<u>Crystallinity</u> : the crystallinities measured by the conventional method of Hendos and Schnell (36) using the IR spectra of annealed samples range from 16 to 24 % for the hexane soluble fraction, 48 to 53 % for the insoluble fraction and 36 to 50 % for the whole polymer with butene varying from 3 to 6 %. Typical examples are given on Table 3. The crystallinity of the homopolymer is found to be 68 % with the same method.

<u>Butene sequence distribution</u> : the butene sequence distribution can be obtained from ^{13}C NMR spectra. Quantitative study of the whole polymer is rather difficult at 4 % butene but the soluble fraction can be studied easily according to Ray and coll (36). For a typical hexane soluble fraction (SMT catalyst) with 14 % molar butene, the probabilities P_1 (butene insertion after an ethylene) and P_2 (butene after a butene) are respectively 0.129 and 0.114. Even in the butene enriched fraction, the sequence length of the butene units is short. Most of the butene units are isolated, less of 20 % of the butene are in sequences of two, the amount of butene in larger sequences is negligible (CH BBB triads at 34.95 ppm). In the insoluble fraction all the butene are isolated in the limit of the experimental error. So, referred to the whole polymer, not more than 8 % of the butene are in sequences larger than 1. The butene units have then no trend to range in long sequences. Little differences near the frontier of experimental error are found when the SMCT or SMDT are used :
P_1 = 0.14, P_2 = 0.10 to 0.11 corresponding to a slightly higher content of isolated butene units.

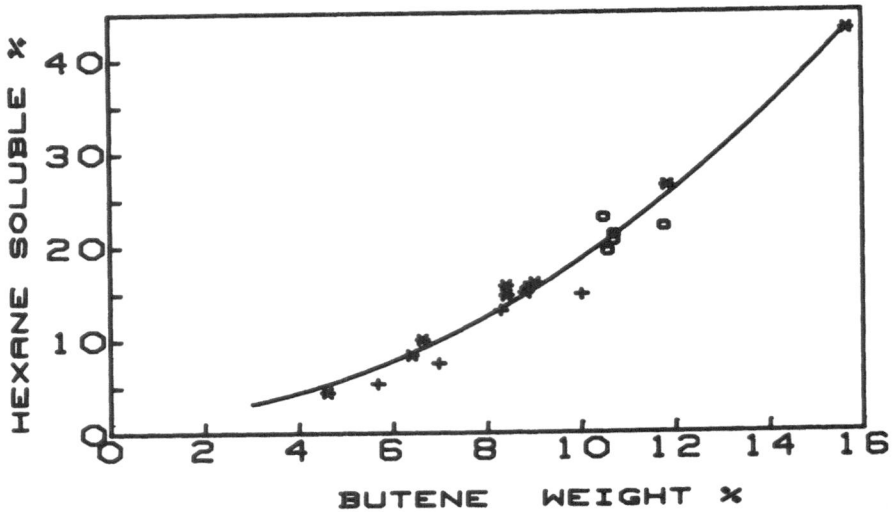

Figure 3 : Polymer fraction % soluble in boiling hexane versus butene percent in weight in the polymer   * catalyst SMT
+ catalyst SMCT
o catalyst SMDT

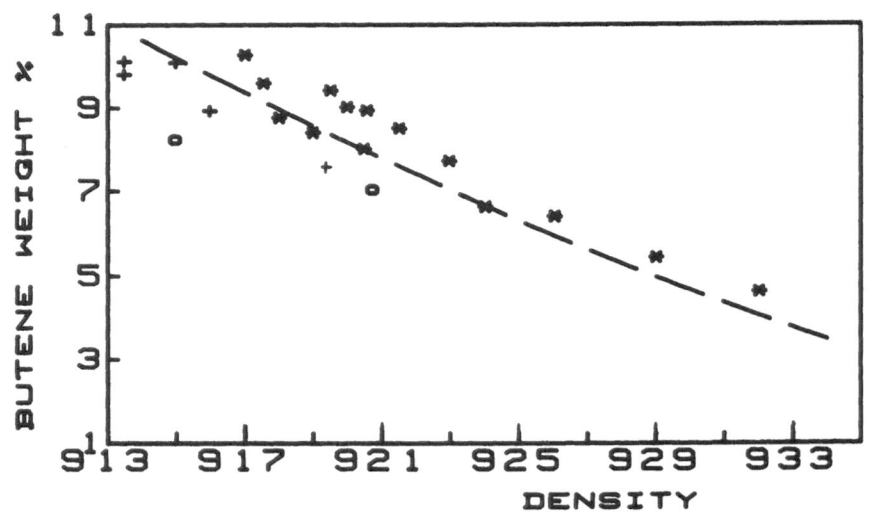

Figure 4 : Polymer butene percent in weight versus polymer density (kg/m^3) - catalyst * SMT   + SMCT   o SMDT (cocatalyst isoprenylaluminium)

Density : the density gives another approach of the overall heterogeneity of the copolymer. Considering, the variation of the density with butene content in the polymer, a frontier can be found between good and bad products as proposed by Elston (37). On Fig. 4 it can be seen that the copolymers of our study range above a curve except the polymers prepared with the SMCT and SMDT catalysts. The best products are clearly again obtained with the complexed catalyst.

REFERENCES

1. Belg. Patent 533,366 K. Ziegler (1955)
2. Belg. Patent 530,617 P. Hogan and R.L. Banks (to Phillips Petroleum Co.) (1955)
3. Belg. Patent 839,380 I.J. Levine and F.J. Karol (to Union Carbide Corp.) (1976)
4. D.E. James in "Encyclopedia of Polymer Science and Engineering - Second Edition" Vol. 6, p. 446-452 John Wiley and Sons Ed. (1986)
5. F.J. Kroland and F.I. Jacobson - Studies in Surface Science and Catalysis 25 - "Catalytic Polymerization of Olefins" T. Keii and K. Soga Ed., Kodansha Elsevier, p. 323 (1986)
6. a) ibid 4, p. 432
   b) A. Muzsay, G. Gyimesi and S.L. Felegyhazine - Int. Polym. Sci. Tech. 8 (10) 7 (1981)
7. L.I. Chiswell - Chem. Eng. Prog. $\underline{79}$ 84 (1983)
8. ibid ref. 4, p. 442 - 444
9. F.G. Karol, G.L. Karapinka, C. Wu, H.W. Dow, R.N. Johnson and W.L. Carrick - J. Polym. Sci. part Al $\underline{10}$ 2621 (1972)
10. K. Soga, T. Sano and R. Omishi - Polym. Bull. 4 157 (1981)
11. Yu.I. Yermakov, B.N. Kuznetsov and V.A. Zakharov in Studies in Surface Science and Catalysis 8 - "Catalysis by supported complexes" - Elsevier Ed. Amsterdam - Oxford New p. 199 (1981)
12. G.C. Goeke, B.E. Wagner, F.J. Karol - US Appl. 892,322 (1978) to Union Carbide Corp.
13. V. Pasquet Thesis - Lyon, 1985
    R. Spitz and V. Pasquet - FR. Pat. 85 09090 (1985)
14. N.F. Brockmeyer in "Encyclopedia of Polymer" ibid ref. 4 - vol. 7 p. 481 (1987)
15. G. Mabilon and R. Spitz - Eur. Polym. J. $\underline{21}$ 245 (1985)
16. S. Floyd, G.E. Mann and W.H. Ray - ibid ref. 5, p. 339
    W.H. Ray -paper presented at the International Symposium "Transition metal catalyzed polymerization" Akron, june 1986 (in press)
17. L.T. Finogenova, V.A. Zakharov, A.A. Buniyat-Zade, G.D. Bukatov and T.K. Plaksunov - Polym. Sci. USSR $\underline{22}$ 448 (1980)
18. R. Spitz, L. Duranel, P. Masson, M.F. Darricades-Llauro and A. Guyot in "Transition Metal Catalyzed Polymerization" R. Quirk Ed. - Akron Symp. June 1986 (in press)
19. J. Boor "Ziegler-Natta catalysts and polymerization" - chap. 18 Academic Press Inc. ed. (1979)
20. V.A. Zakharov, G.D. Bukatov and Y.I. Yermakov - Adv. in Polym. Sci. 51 63 (1983)
21. J.P. Hogan, D.R. Norwood and C.A. Ayres - J. Appl. Polym. Sci. Symp. $\underline{36}$ 49 (1981)
22. G. Odian "Principles of polymerization" Mc Graw Hill Book Co. Ed. p. 370-385 (1970)
23. Y.V. Kissin in "Transition Metal Catalyzed Polymerizations" Part B - R.P. Quirk Ed. Harwood p. 597 (1981)
24. Y.V. Kissin, D.L. Beach - J. Polym. Sci. Polym. Chem. Ed. 22 333 (1984)

25. L.L. Böhm - J. Appl. Polym. Sci. 29 279 (1984)
26. R. Quijada and A.M. Ramos-Wanderley - ibid ref. 5, p. 419
27. J.V. Seppala - Macromolecules 18 2409 (1985)
28. L. Wild, T.R. Ryle, D.C. Knobeloch and I.R. Reat - J. Polym. Sci.
    20 441 (1982)
29. Y. Doi, R. Ohnishi and K. Soga - Makromol. Chem. Rapid
    Commun. 4 169 (1983)
30. A.A. Baulin, S.S. Ivanchev, A.G. Rodionov, T.V. Kreitser
    and A.L. Goldenberg - Vysokomol. Soyed. A22 1486 (1980) - Polym.
    Sci. USSR 22 1630 (1980)
31. D.W. Van Krevelen "Properties of Polymers" - Elsevier Ed.
    Amsterdam - Oxford - New York p. 405-420 (1976)
32. K. Soga, T. Sano and R. Ohnishi - Polym Bull. 14 157 (1985)
33. K. Soga, T. Sano and Y. Doi - Polym Bull. 10 168 (1983)
34. Ibid. ref. 19, p. 566-577
35. B.K. Hunter, K.E. Russell, M.V. Scammell and S.L. Thomson -
    J. Polym. Sci. Chem. Ed. 22 1383 (1984)
36. H. Hendos and V. Schnell - Kunstoffe 51 69 (1961)
37. C.J. Ray, P.E. Johnson and J.R. Knox - Macromolecules 10 773
    (1977)

Morphological Characterization of Ziegler-Natta Catalysts and Nascent Polymers

A. Muñoz-Escalona, C. Alarcón, L. A. Albornoz, A. Fuentes and J. A. Sequera

Centro de Química, Instituto Venezolano de Investigaciones Científicas (I.V.I.C.), Apartado 21827, Caracas 1020-A, Venezuela

INTRODUCTION

The study of the structure and morphology of nascent polyolefins obtained with hetereogeneous and soluble Ziegler-Natta catalysts has attracted the attention of many researchers, being the subject of a considerable amount of publications appeared in the patent and scientific literature (1-2). From an industrial point of view, the interest arises from the necessity in controlling the growth of the polymer particles during polymerization due to the fact that their size, shape, density and texture influence the polymerization process and manufacture of the produced polymers.

According to the replication phenomena the control of the polymer morphology can be achieved by controlling the morphology of their parent catalysts particles. On the other hand, since early Natta's works (3), it was established that different morphologies of the catalysts, e.g. by grinding $TiCl_3$ catalyst, influence the shape of the kinetic curves and polymerization activities.

In this paper, the relationship between catalysts morphologies, polymerization kinetics and morphologies of the resulting polymers have been studied. Three groups of catalytic systems were used for polymerization of ethylene and co-polymerization with hexene-1 to obtain linear low density polyethylene (LLDPE). First of all, ethylene was polymerized with the soluble catalytic system formed by mixing Cp_2TiCl_2 with Et_2AlCl in the range from -29 to 70°C in order to find out whether the polymerization temperatures has any influence on the morphology of the nascent polyethylene crystals produced during polymerization. The morphologies of the polyethylene obtained with soluble catalysts could serve as a reference to the morphologies of the polymer produced with heterogeneous catalysts.

Among the heterogeneous Ziegler-Natta catalysts, the $TiCl_4$ supported on different silica and the high mileage catalysts synthesized by reduction of $TiCl_4$ with grignard compounds have been employed. These catalysts were used for the homo-polymerization of ethylene in heptane as reaction medium as well as for co-polymerization with hexene-1. Particular attention was paid to find out how the catalyst morphologies influenced the polymerization kinetic and morphologies of the resulting polymers.

EXPERIMENTAL

Catalytic Systems

For the preparation of the soluble catalytic system the Cp_2TiCl_2 supplied by Research Organic-Inorganic Chemical Co. (USA) and the co-catalyst Et_2AlCl kindly supplied by Ethyl Co. (USA) were used as received.

W. Kaminsky and H. Sinn (Eds.)
Transition Metals and Organometallics as
Catalysts for Olefin Polymerization
© Springer-Verlag Berlin Heidelberg 1988

For the synthesis of TiCl$_4$ supported catalysts different commercial silicas from Grace Davison (USA) and Crosfield (England) were used as carriers. Their physical characteristics are given in Table 1.

Table 1.

CARRIER CHARACTERISTIC USED FOR TiCl$_4$ SUPPORTED ZIEGLER-NATTA CATALYSTS

MANUFACTURER	TYPE	SURFACE AREA (m^2 x g.)	PORE VOLUME (ml x g)	AVERAGE PORE DIAMETER (Å)	AVERAGE PARTICLE SIZE (μ)
Davison	13-110SiO$_2$Al$_2$O$_3$	475	1.10	90	61
Davison	951SiO$_2$	600	1.0	65	58
Davison	56SiO$_2$	300	1.2	160	103
Davison	952SiO$_2$	300	1.65	200	70
Crossfield	EP-10SiO$_2$	331	1.92	235	--
Crossfield	SD-116SiO$_2$	350	1.97	297	--

They were choosen having a broad surface area, pore volumen and pore diameter in order to find out whether these morphological factors influence catalytic activities, abilities in incorporing high olefins as comonomers and control the nascet polymer morphology. Before impregnation with TiCl$_4$ they were dried at 150°C and 600°C under vacuum in order to eliminate the physically absorbed water and to generate two different types of hydroxyl groups, e.g. hydrogen bonded and isolated. Furthermore, silica were loaded with Ti by reaction with TiCl$_4$ supplied by Merck (W. Germany) after purified by vacuum distillation. The reaction was carried out in n-heptane under magnetic stirring, at room temperature during 3 h. using 0.7 molar solution of TiCl$_4$ and a ratio of 1 g. SiO$_2$/2 ml TiCl$_4$. After the impregnation was finished, the excess of solution was filtered off and the solid component washed with approximately 60 ml of n-heptane. Finally, the catalysts were dried at 60°C under vacuum. All procedures were carried out in a dry lab. in order to assure anaerobic anhydrous conditions. The total amount of supported Ti was determined as its peroxides by colorimetric methods using an UV Unicam SP 180 spectrometer.

To prepare the catalysts based on TiCl$_4$ and grignard compounds, the n-butil-MgCl was first synthesized in ether by well established methods. After the ether was eliminated by vacuum distillation, the grignard compound was placed in a 200 ml stirred vessel with n-heptane. Then an excess of solution of TiCl$_4$ in n-heptane was dropwised added. After the reaction was finished the solid was filtered, washed with n-heptane and dried under reduced pressure at 140°C. The solid catalyst was used without further treatment or, after been crushed or ball-milling during 5 h. and 27 h. in a porcelain mill.

Polymerization Procedure

Ethylene was polymerized with the soluble catalyst Cp$_2$TiCl$_2$-AlEt$_2$Cl using a schlenk type reactor. The polymerization temperature were controlled by placing the schlenk in a thermostatic bath. Low temperature, between -29°C to 0°C, were controlled by different eutectic mixture having constant temperature. As polymerization medium was used toluene to which, under magnetic agitation with teflon stirrer bar, Et$_2$AlCl was first added and then Cp$_2$TiCl$_2$. Polymerizations were conducted at normal ethylene pressure under stirring and also under quiescent conditions. In the later case, the polymerization take places on the gas-liquid interface. The polymer were recovered by filtration at the same polymerization temperature to prevent precipitation of lower molecular weight polyethylene which remain dissolved.

The polymerization with the heterogeneous catalytic systems were carried out in a batch 0.5 L stirred glass autoclave at 50°C under constant monomer pressure of 5 atm., 1200 rpm and n-heptane as reaction medium. The solid catalysts sealed in glass ampoules were introduced in the reactor containing 350 mL of n-heptane. Then, the reactor was pressurized at 5 atm. with ethylene followed by addition of the Et_3Al as co-catalyst. The $|Al|/|Ti|$ ratio was 30. The polymerization was timed after breaking the ampoule containing the catalyst and the polymerization rate was determined from the rate of monomer consumption. Details of the polymerization procedures have been described elsewhere (4). In case of co-polymerizations, 0.824 molar of n-hexene was firstly added to the n-heptane and the experiment carried out as previously described.

After polymerization were finished a solution of ethanol containing hydrochloric acid was injected into the reactor and the polymerization quenched. Then, the polymers were washed several times with water-diluted alcohol and dried at 50°C. The average specific polymerization rate (ASPR) was calculated over the whole polymerization time and expressed as g. poly./g. Ti x h.x atm.

Polymer Characterization

The melting behaviour of the polymers was determined by differential scanning calorimetry (DSC) using a DuPont DSC 900. The melting points were determined from the peaks of the DSC curves previous calibration of the abcissa with high purity standards. The amount of sample used range between 4 to 6 mg. and the scanning rate was 10°C/min. On the other hand, from the areas of the peaks the heat of fusion was calculated and the crystallinity of the samples determined using the expression:

$$\text{Crystallinity (\%)} = \frac{\Delta Hf}{\Delta H^*f} \times 100$$

where ΔHf is the heat of fusion of the polymer samples and ΔH^*f is the heat of fusion of a perfect crystalline polyethylene with folded-chain macroconformation and taken as 66 cal.g-1. Finally, two scans were recorded, corresponding to the nascent polymers and to the recrystallized samples from the melts.

The morphology of the nascent polymers were examined using electron microscope techniques. The particle size distribution were obtained directly from the SEM micrographs measuring a population of not less of 500 particles and with a serie of sieves. For catalyst observations a special device was developed allowing them to introduce into the SEM without deactivation. Finally, information of the texture inside the particles was obtained by sectioning them with a microtome after being embeded in epoxy resin.

RESULTS AND DISCUSSION

Morphology of polyethylene obtained with soluble catalyst

According to the Wunderlich's principle of crystallization during polymerization (5) under conditions where the polymerization rate is close to the crystallization rate a simultaneous polymerization and crystalization mechanism could take place, giving rise to fibrillar crystals with extended-chain macroconformations.

In Figure 1 the calculated crystallization rate of polyethylene as function of crystallization temperature is given.

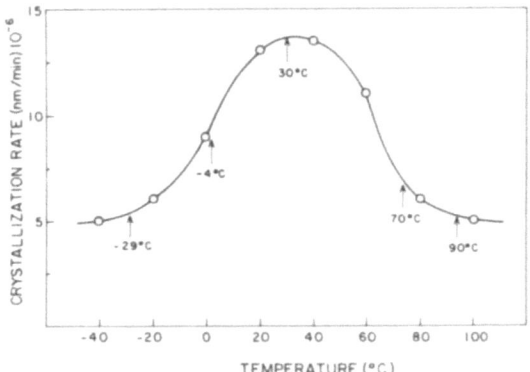

Fig. .1   CRYSTALLYZATION RATE VS. TEMPERATURE FOR HDPE (Tg = 85°C, Tm = 130°C)
ARROWS SHOWN POLYMERIZATION TEMPERATURE OF ETHYLENE USING
Cp$_2$TiCl$_2$ / Al Et$_2$Cl.

The choosen temperatures for polymerization of ethylene with the soluble catalytic system Cp$_2$TiCl$_2$-Et$_2$AlCl are also indicated. Although broad range of polymerization temperatures were used, only lamellare polyethylene crystals were obtained. These results indicate that polymerization and crystallization are two separate events. In this case the perfection of the resulting crystals during polymerization depend upon the length of the segment of polymer-chain before crystallize. When the temperature is low enough (-4°C) the polymerization rate is reduced so that the polymerization and crystallization can be brought as near as possible giving rise to well crystallized orthorhombic crystals (Fig. 2a), which are similar to the crystals obtained by crystallization of polyethylene from solution. When the polymerization is higher unperfect lamellare crystals having irregular form were obtained (Fig. 2b). Even more irregular polyethylene crystals were obtained by increasing the polymerization temperature.

a)   b)

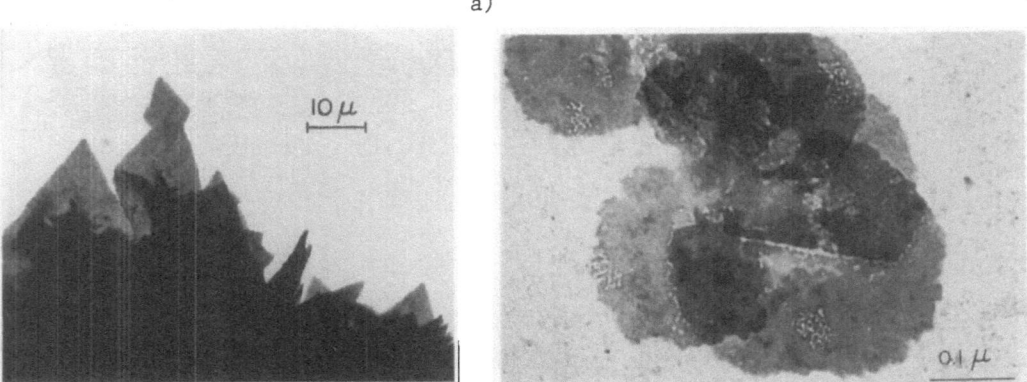

Fig. 2 Morphologies of PE produced with soluble catalyst at: a) -4°C,
b) 30°C

Further insight into the crystal structure of the sample was obtained by
DSC analysis. In Fig. 3, thermograms of nascent polymers (curve 1) together with curves of the same sample after been recrystallized from the
melt are given (curves 2). Only one melting peak was observed for all
samples, indicating that only one type of morphology is present.

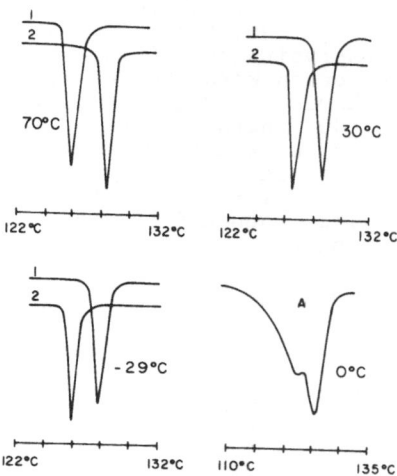

Fig. 3   DSC THERMOGRAMS OF AS—POLYMERIZED (1) AND RECRYS-
TALLIZED POLYETHYLENE SAMPLES OBTAINED AT DIFFERENT
POLYMERIZATION TEMPERATURES. SAMPLE A POLYMERIZED
AT 0°C BUT FILTRATED AT 27°C.

Giordiadis et al. (6), however, found two peaks for samples polymerized
at low temperature (0°C). This was probably due to the formation of an
additional morphology when the samples were filtered at temperatures
different to the polymerization temperature (see thermograme A in Fig.
3). Furthermore, it was observed that nascent polymers prepared at temperature below 30°C have better ordered structure than when crystallized
from the melts. This could indicate that at low temperatures the crystallization and polymerization rates lie very close to each other.
Therefore, as soon a certain segment of polymer chain has been produced,
it is incorporated into the nascent polymer crystal. At high temperature (70°C) the reverse situation is true giving rise to nascent crystals with less ordered structure than those obtained by recrystalliza-
tion from the melt.

Morphology of catalysts and polymers obtained with the catalytic systems TiCl₄ supported on silica

In case of supported catalyst the carriers used for the preparation of
the catalysts have a great influence on the morphology of catalyst and
on the catalytic activity. When using silica as supports, the catalysts
are synthesized by reacting their surface hydroxyl groups with TiCl₄.
As long as the shape, size and porosities of the silicas are not greatly changed by this reaction, the silica control the catalyst morphology.
Furthermore, due to the replication phenomena the catalyst act as a tem-
plate, controlling the final morphology of the nascent polymers.

Figure 4 shows the polymerization rates for homo-polymerization of ethylene with catalysts prepared by supporting TiCl₄ on different silicas

dried at 150°C. As a reference the polymerization rate with Stauffer TiCl$_3$-AA is also ploted. The importance of the porosity of the catalysts is indoubted. Only those silica having the highest pore volumen and diameter produce the most active catalysts (See Table 1), regardeless to their surface areas. In addition, all supported catalysts show acceleration type kinetic curves, with the exception of the catalyst based on silica Davison 952 which display a decay behaviour (7). The reason for this difference could be due to the morphological differences of this silica respect to the others. Thus, the silica 952 present hollows inside, which could serve as reservoirs for TiCl$_3$ formed during impregnation with TiCl$_4$ and further reduction with the co-catalyst. As it can be seen TiCl$_3$ display a decay behaviour. Higher activities were reached, when the silica were dried at 600°C before reaction with TiCl$_4$ (Compare Figs. 4 and 5). This result can be explained by assuming that single hydroxyl groups, produced at this temperature, are more active than the hydrogen bonded as published by Karol (8). On the other hand, the amount of TiCl$_4$ supported on the silica when dried at 600°C is lower and consequently the porosity of the original silica decrease less than dried at 150°C (See Table 2).

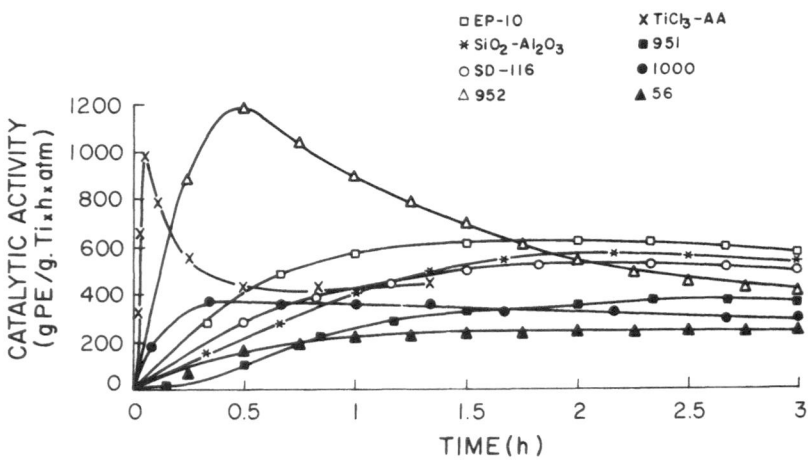

Fig. 4 Activities v.s. time for different silicas dried at 150°C. TiCl$_3$ is also included.

For co-polymerization of ethylene with 0.824 molar of hexene-1, only the catalysts prepared on porous silica and dried at 600°C were tested, due to the fact that they were the most active in homo-polymerizations. Fig. 6 shows the polymerization rate as function of time. It can be seen that all catalysts have lower catalytic activities in the co-polymerization (See also Table 2). It is also very interesting that catalysts based on silicas EP-10 and SD-116 change their kinetic behaviour from acceleration to slightly decay curves.

The main morphological feature of silicas Davison 951 and 952 have been previously reported (7). Both silicas are different. Thus, silica 951 is compact and very strong while silica 952 is brittle and exhibits hollows inside. The kinetic differences between both silicas has been, therefore, explained in base to their differences in the morphology (see above). On the other hand, both Crosfield silica EP-10 and SD-116 are very similar in shape and size. They are very irregular in shape and

rather soft. Fig. 7 shows a general view of the crossection of silica
EP-10 and details of the inner morphology of silica SD-116. Due to this
similarity, catalysts based on both silica exhibit quite the same behaviour in activity and ability in incorporing co-monomer.

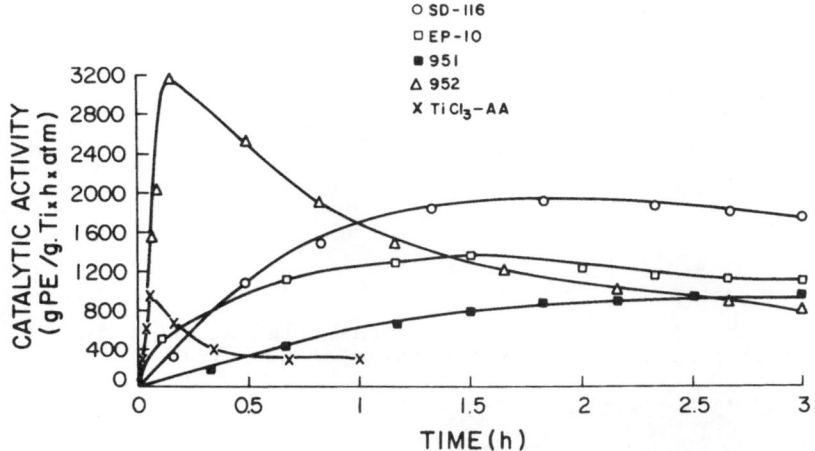

Fig. 5 Activities v.s. time for different silicas dried at 600°C. $TiCl_3$
is also included.

Table 2

CATALYTIC ACTIVITIES OF $TiCl_4$ SUPPORTED CATALYSTS ON DIFFERENT CARRIERS FOR HOMO- AND
CO-POLYMERIZATION OF ETHYLENE WITH HEXENE-1. POLYMERIZATION IN N-HEPTANE AT 50°C, 5 ATM.
DURING 3 H., Al/Ti = 30 CO-CATALYST $AlEt_3$

CATALYST	HEAT-TREATMENT UNDER VACUUM (°C)	SUPPORTED Ti (%)	AMOUNT OF PE (g)	ASPR [a] (gPE/gTi x h x atm)
		H O M O P O L Y M E R I Z A T I O N		
$TiCl_3$		24.11	133	201.0 [b]
11-110-SiO_2-Al_2O_3	150	5.62	120	479
951-SiO_2	150	6.93	77	305
56-SiO_2	150	4.60	51	249
952-SiO_2	150	4.85	86	661
EP-10-SiO_2	150	4.70	40	558
SD-116-SiO_2	150	4.79	71	499
951-SiO_2	600	3.04	83	753
952-SiO_2	600	2.53	65	1 483
EP-10-SiO_2	600	2.32	50	1 215
SD-116-SiO_2	600	2.26	62	1 450
		C O P O L Y M E R I Z A T I O N [c]		
951-SiO_2	600	3.04	37	447
952-SiO_2	600	2.53	14	412
EP-10-SiO_2	600	2.32	29	738
SD-116-SiO_2	600	2.26	35	822

a) Average specific polymerization rate. b) After 1 h. polymerization time.
c) Copolymerization with 0.824 molar of hexene-1.

Due to the replication phenomena, Davison silicas produce spherical polyethylene particles, when used for synthesis of supported catalysts
(Fig. 8a). In case of polyethylene obtained with Crosfield silicas,

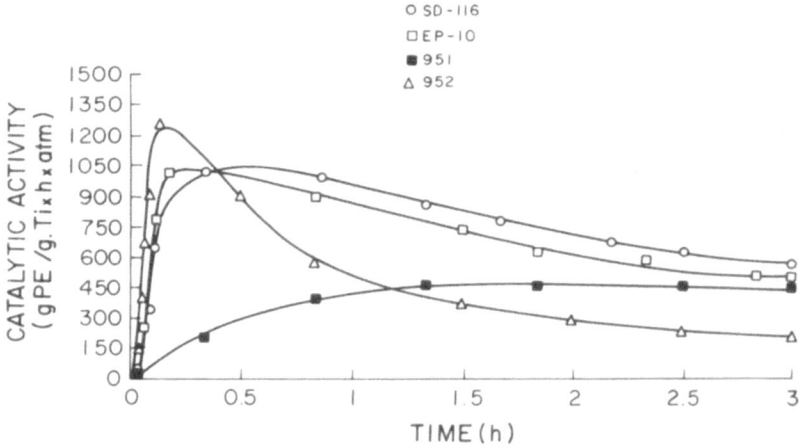

Fig. 6 Activities v.s. time of different silicas dried at 600°C in the co-polymerization of ethylyene with 0.824 molar of hexene-1.

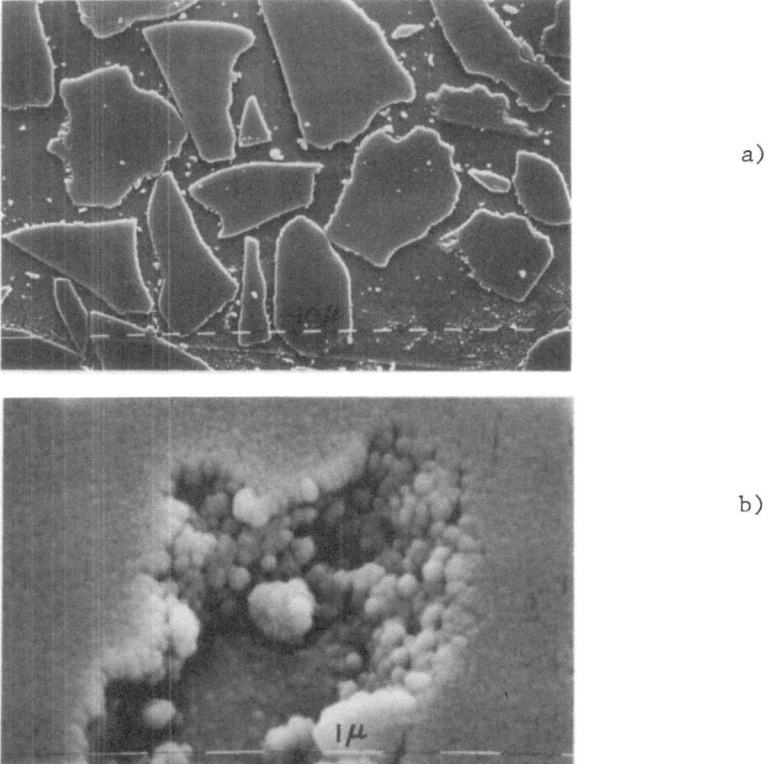

a)

b)

Fig. 7 SEM micrographs of a) crossection of silica EP-10 and b) details of silica SD-116

they exhibit rather irregular shape, like their parent catalyst synthesized with these type of supports (compare Fig. 7 and Fig. 8b). On the
other hand, due to the low mechanical strength of Davison silica 952,
it can undergo fracture during polymerization giving rise to fine polyethylene particles, changing therefore, the particle size distribution
when compared to the original silica. The grounding of the catalysts
during polymerization is more evident when the silica used for their
synthesis was dried at 600°C, due to the higher catalytic activities
(Fig. 9). The polyethylene particles produced with catalysts 952 are
bigger when synthesized using silica dried at 150°C, although their activities are lower.

a)

b)

Fig. 8 SEM micrographs of PE obtained with catalysts based on a) silica
Davison 952 and b) Crosfield EP-10

In case of Crosfield silicas, the particle size distribution of the
polyethylene reproduce very well the size distribution of catalysts
(Fig. 10). The replication factors achieved with supported catalysts
using silicas as carriers are between 6-10 depending upon the catalytic
activities and silica used.

Finally, the morphologies of the co-polymers produced with TiCl$_4$ supported Ziegler-Natta catalysts are very similar to the homo-polymers.
However, the particle of the co-polymers are smaller than the homo-

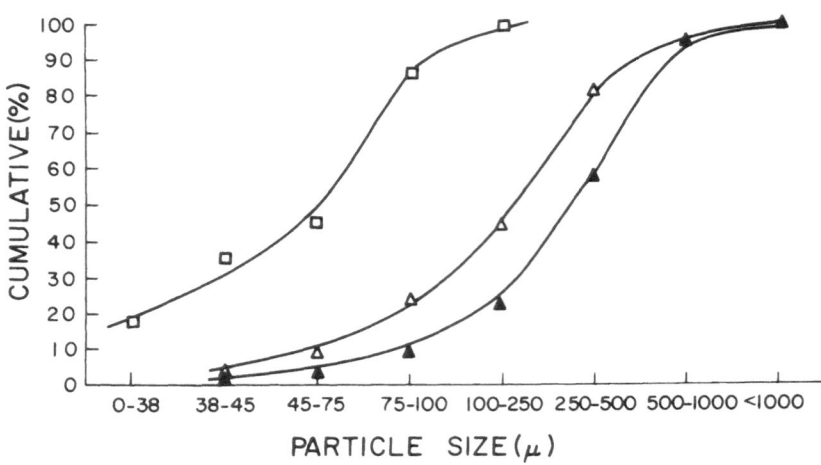

Fig. 9 Particle size distribution of ☐ silica 952, and produced PE with catalysts based on silica dried at: ▲ 150°C and △ 600°C.

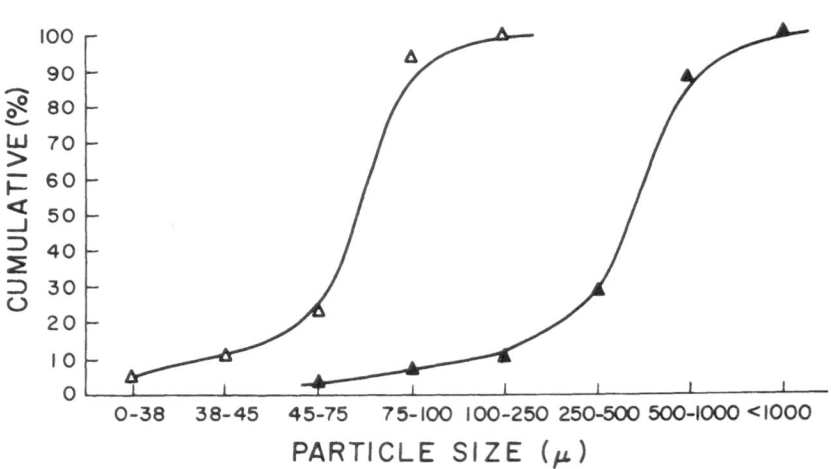

Fig. 10 Particle size distributio of (△) silica EP-10 and produced PE with catalyst based on silica dried at (▲) 600°C.

polymers due to the lower catalytic activities in the co-polymerization.

Morphology of catalysts and polymers produced with catalytic systems based on TiCl₄ and grignard compounds

Howard et al., (9) have published the synthesis of highly active catalysts for ethylene polymerization by reduction of TiCl₄ with grignard compounds. However, no information exists in the scientific literature concerning the morphology of this type of catalysts and their control on the morphology of the nascent polymers. In Fig. 11, the catalytic activity for these type of catalysts have been ploted. It can be seen, that these catalysts exhibit high activities for homo- and co-polymerization of ethylene with hexene-1. The effect of grinding can also be observed. The ground catalysts are more active, changing their profile of the kinetic curves from acceleration to decay. It seems to be, that unground catalysts increase their activities approaching to the stationary period of the milled one. This effect was first point out by Natta et al. (3) for the case of α-TiCl₃ catalyst. These results can be explained by admitting a mechanical action of polymer chains on the catalyst, according to Natta's theory of particle size adjustment.

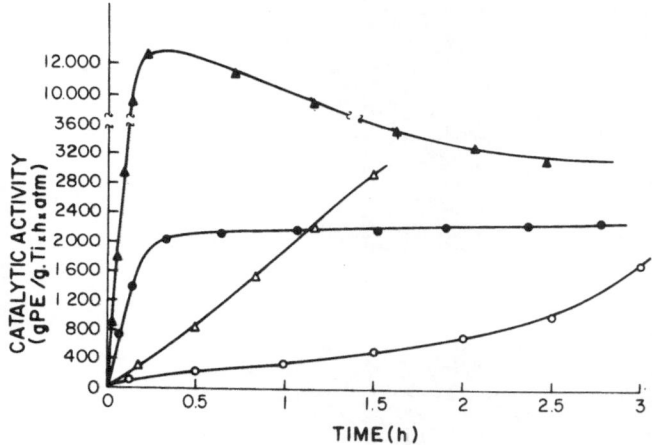

Fig. 11 Activities v.s. time for catalyst produced by reduction of TiCl₄ with grignard (O) homo-polymerization with unground catalyst; (Δ) co-polymerization with unground catalyst; (●) homo-polymerization with ground catalyst; (▲) co-polymerization with ground catalyst.

Furthermore, it is very surprising that these catalysts are more active for co-polymerization than for homo-polymerizations. The increase of catalytic activities in presence of the comonomer can be explained by proposing the formation of new active centers due to the coordination of the comonomer to the initial active centers and/or by decrease in the crystallinity of the forming polymer layer covering the catalyst particles. To prove the formation of new active centers in heterogeneous catalysts is a difficult task. However, it appears that this can be feasible, due to the fact that catalysts develop their activity just from the begining of the polymerization; even before a layer of polymer of considerable thickness has been formed on the surface of the catalysts particles. On the other hand, by DSC analysis it was able to prove that the crystallinity of the co-polymers is lower compared to the homo-polymers (10), so that difficulties for monomer diffusion can be in some extention overcome. Values of crystallinity of about 70-76% were determined for homo-polymers and 55-58% for co-polymers.

The mechanical activation by ball-milling of MgCl₂ in the synthesis of high active supported Ziegler-Natta catalyst has been investigated by using X-ray difraction techniques (11). Up to now, the effect of grinding on the morphology of the catalysts has been little investigated.

In Figure 12 general views of the catalyst prepared by reaction of grignard butil-MgCl with TiCl$_4$ before and after milling are shown. The particles of unground catalyst have sized up to 100 μ. By grinding they
were reduced in size, producing additionaly more uniform particles.

a)

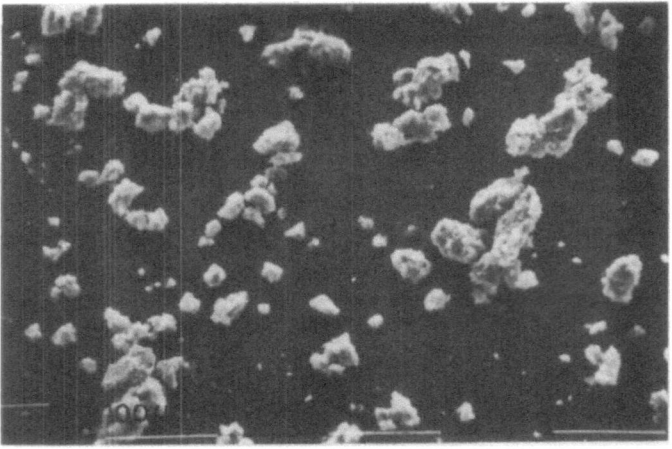

b)

Fig. 12 SEM micrographs of catalysts based on the reduction of TiCl$_4$
with grignard compound: a) unground; b) ground.

Details of the inner morphology could be observed at high magnification.
The catalyst are build-up by agglomeration of subparticles having flakes
shape (Fig. 13). The size of the flakes range between 0.2 - 1 μ. These flakes type particles are very dense and are probably formed in the
early stage of the catalyst synthesis. The porosity of the catalysts
are formed by the free spaces between the flakes. By milling up to 27
h. no significant change in the size of the flakes could be observed.

The effect of grinding was to separate the flakes each other producing smaller particles.

Fig. 13 Morphological details of catalyst based on the reduction of $TiCl_4$ with grignard compound.

By X-ray line scans of the three elements Cl, Ti and Mg their distribution along the line crossing the catalyst particles were obtained (see Fig. 14 a and b). It appears that the three elements are intimately dispersed along the catalyst. However, by using high resolution scanning trasmition electron microscopy it was able to detect phase separations between $MgCl_2$ and $TiCl_3$, although higher amount of $TiCl_4$ was employed for catalyst synthesis.

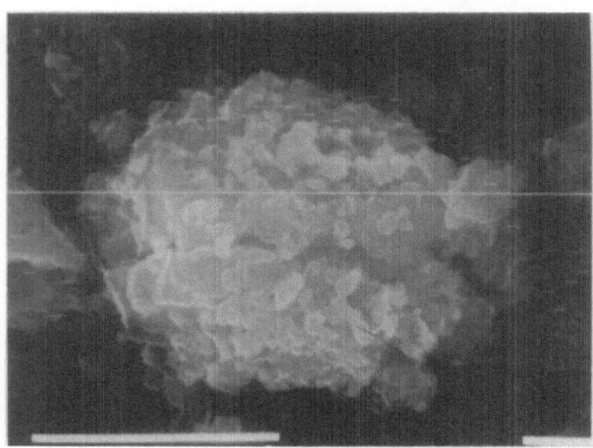

Fig. 14 a

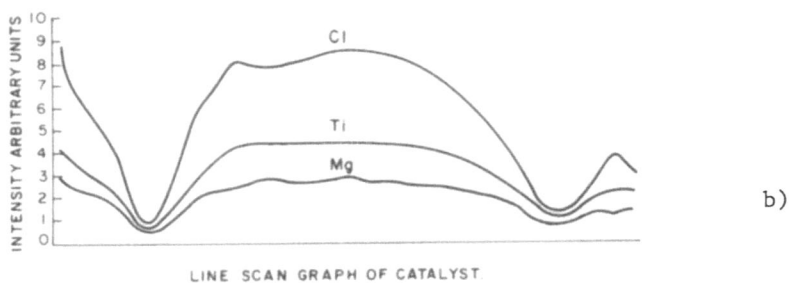

LINE SCAN GRAPH OF CATALYST

b)

Fig. 14 a) SEM micrographs of unground catalyst based on $TiCl_4$ reduced with grignard compound. Line indicate the place for X-ray line scan analysis, b) X-ray line scans for Cl, Ti and Mg.

Finally, the morphologies of the produced polymers can be seen in Fig. 15. The unground catalyst produce PE particles bigger irregular in shape and with a broad size distribution, while ground catalyst produce more uniform and smaller particles.

a)

b)

Fig. 15 SEM micrographs of PE obtained with catalysts based on $TiCl_4$ reduced with grignard compound: a) unground, b) ground

REFERENCES

1) Boor Z, (1979) Ziegler-Natta Catalyst and Polymerizations, Academic Press, New York.

2) Muñoz-Escalona A, Villamizar C and Frías P. (1981) Structure-Property Relationships of Polymeric solids. A. Hiltner Plenum Press, New York

3) Natta G and Pasquon I (1959) Advance Catalysis 11, 1

4) Muñoz-Escalona A, Hernández JG, Gallardo JA (1987) In Recent Advances in Polyolefins. T. Cheng (Eds) Plenum Press, New York

5) Wunderlich B. (1968) Advan. Polym. Sci. 5. 568

6) Giorgiadis T and Manley RJ, (1972) St. Kolloid-Z. u. Z. Polymers, 250 557

7) Muñoz-Escalona A, Gallardo JA, Hernández JG and Albornoz LA (to be published) In Transition Metal Catalyzed Polymerizations. R. P. Quirk (Eds) Plenum Press, New York

8) Karol FJ (to be published) In Transition Metal Catalyzed Polymerizations R. P. Quirk (Eds) Plenum Press, New York

9) Howard RN, Rooper AN and Fletcher KL, (1973) Polymer, 14, 365

10) Muñoz-Escalona A, García H and Albornoz LA (1987) J. of Appl. Polym. Sci. 34

11) Sivaram S and Srinivasan (1987) In Recent Advances in Polyolefins T. Cheng (Eds) Plenum Press, New York

Transmission Electron Microscopic Observation of Nascent Polypropylene Particles Using a New Staining Method

Masahiro Kakugo[*], Hajime Sadatoshi[*], Masakazu Yokoyama[**], and Keitaro Kojima[*]

* Sumitomo Chemical CO. LTD., Chiba Research Laboratory, 5-1 Anesaki Kaigan, Ichihara, Chiba, 299-01, Japan.

INTRODUCTION

Zieglar-Natta polymerization using heterogeneous catalyst is the only major process for the production of highly stereospecific poly-propylene, in which $TiCl_3$ and supported catalysts are mainly used as industrial catalyst. The typical catalysts are the 10-30 μm particles consisting of loosely bound agglomerates of fine catalyst crystallites, several tens to several handreds of angstrom in diameter[1-5]. The growth of the polymer particles on such catalysts[6-12] and their architecture[1,13-15] have aroused much interest widely, from the scientific and industrial point of view. It has been generally accepted that nascent polymer particles may replicate the original catalyst particle, including the internal structure[16].

However, the relationship between the primary polymer particle and the primary catalyst crystallite is still not completely elucidated due to the lack of a suitable observation method. In a previous paper we showed that a two-step staining method using 1,7-octadiene and OsO_4 is useful for visualization of saturated rubbers[17]. This paper discribes archtecture of the nascent polypropylene flakes observed in a transmission electron microscope by using this staining method.

EXPERIMENTAL

Catalyst

δ-$TiCl_3$(I) and (II) were prepared by reduction of $TiCl_4$ with $Al(C_2H_5)_2Cl$, and treatment of the reduced products with di-isoamyl ether and then with $TiCl_4$. $TiCl_3$ AA grade was supplied by Toho Titanium CO. LTD. A supported catalyst[19] was prepared by supporting $Ti(OR)_{0.5}Cl_{3.5}$ (OR; o-cresoxy) on $MgCl_2$. Crystallite sizes were determined by conventional X-ray diffraction measurement.

Polymerization

Polymerization was carried out in liquified propylene or n-heptane. The polymerization was terminated by purging propylene. The polymer flakes for scanning electron microscopic (SEM) observation were treated with methanol for 1 hr. at room temperature and filtrated,

** Sumitomo Chemical CO. LTD., Ehime Research Laboratory, 5-1 Sobiraki, Niihama, 792, Japan.

W. Kaminsky and H. Sinn (Eds.)
Transition Metals and Organometallics as
Catalysts for Olefin Polymerization
© Springer-Verlag Berlin Heidelberg 1988

followed by treatment with n-heptane for 5 min. at room temperature, filtration, and drying in a vacuum oven. The polymer flakes for transmission electron microscopic (TEM) observation were used without any treatment.

Electron microscopic observation

The polymer samples for SEM observation were coated with platinum by a conventional sputtering technique. The samples for TEM observation were immersed in purified 1,7-octadiene at room temperature for 2 hr., stained over 1% aqueous solution of OsO_4 for 3 hr. at 60 °C, and sectioned at -80 °C by means of ultramicrotome. The specimens were examined in a Hitachi H-500 electron microscope equipped with SEM device and energy dispersed X-ray microanalyzer (EXD).

RESULTS and DISCUSSION

Characterization of catalysts and preparation of samples

The catalysts subjected to the present experiment are listed in Table 1, where the dimensions of the catalyst crystallites are also shown. The polymers are shown in Table 2.

Table 1. Particle and crystallite size of catalyst.

Catalyst	Average particle size (μm)	Crystallite size (Å)		Polymerization activity (g-PP/g-cat)
		D_{300}	D_{003}	
δ-TiCl$_3$ (I)	19	108	185	1100[b]
TiCl$_3$ AA	30	120	100	1200[b]
δ-TiCl$_3$ (II)	18	75	75	4500[b]
		D_{110}	D_{003}	
Supported catalyst	12	40[a]	30[a]	19600[c]

a) not exact due to the very wide diffraction peaks.
b) polymerized at 65 °C for 2 hr. in liquid propylene.
c) polymerized at 65 °C for 1 hr. in liquid propylene by adding methyl-p-toluate as an electron donor.

Architecture of nascent polymer flakes

SEM photographs of the typical polymer flakes (A-4, B-3, and C-3) for the respective catalysts are shown in Fig. 1. Those polymer flakes show similar surface structure, irrespective of the catalysts, that is, aggromerates of the fine polymer globules about 1 μm in diamter. In addition, fibrillar structure having a width of about 1 μm is seen only in the polymer flake polymerized with the TiCl$_3$ AA catalyst system. Those morphological features are similar to those reported previously except that the size of the polymer globules is slightly different [1,13,14]. These polymer globules have been generally accepted to be the primary polymer perticles, each of which may grow surrouding the primary catalyst crystallite[13,18].

Table 2. Polymer samples.

Sample	Cat. system Ti-cat/Al-cat/donor (mmol/l)			Temp. (°C)	Press. (kg/cm²G)	Time (hr)	Medium	Yield (g-PP/g-cat)
	δ-TiCl$_3$(I)/DEAC[a]							
A-1	74	83		65	0	1	heptane[d]	12
A-2	5.5	17		60		0.5	propylene[e]	108
A-3	0.59	17		60		4	propylene[e]	880
A-4	0.15	13		65		5	propylene[f]	2030
	TiCl$_3$ AA /DEAC							
B-1	0.82	17		40		1	propylene[e]	84
B-2	0.82	17		60		4	propylene[e]	1280
B-3	0.13	13		65		5	propylene[f]	2140
	δ-TiCl$_3$(II)/DEAC							
C-1	0.18	12		60	10	1	heptane[f]	830
C-2	6.5	17		60		1	propylene[e]	1400
C-3	0.06	13		65		5	propylene[f]	8850
	supported catalyst /TEA[b]/MT[c]							
D-1	0.010	4.4	0.96	70	6	1.5	heptane[f]	8000

a) $Al(C_2H_5)_2Cl$, b) $Al(C_2H_5)_3$, c) methyl-p-toluate, d) polymerized in a 0.2 l glass flask, e) polymerized in a 0.1 l autoclave, f) polymerized in a 5 l autoclave.

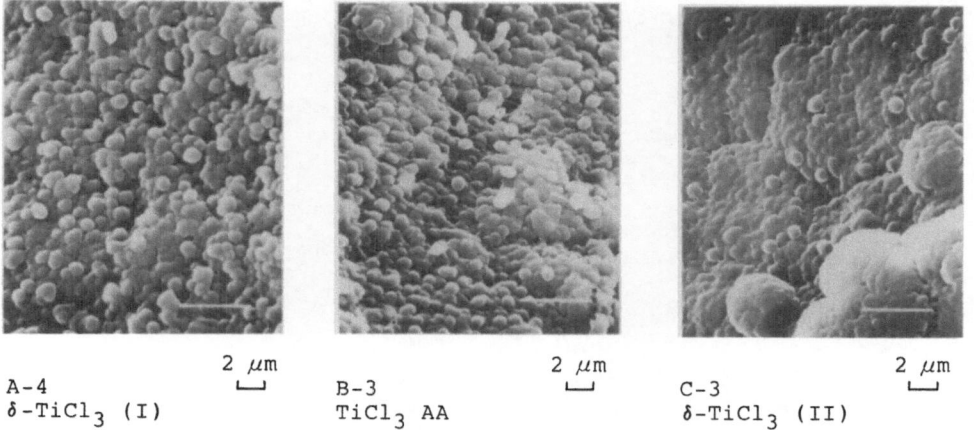

A-4
δ-TiCl$_3$ (I)
2 μm

B-3
TiCl$_3$ AA
2 μm

C-3
δ-TiCl$_3$ (II)
2 μm

Fig. 1. SEM photographs of typical polypropylene flakes

When this view is valid, an average diameter of the polymer globules (D) can be calculated from that of the catalyst crystallites (d) and the polymer yield (Y, g-PP/g-cat) by the fllowing equation:

$$D = d \times \sqrt[3]{(\rho_{cat} \times Y / \rho_{pp}) + 1}$$

where ρ_{cat} and ρ_{pp} are the densities of the catalyst and polypropylene. ρ_{cat} is taken as 2.7 g/cm^3 for TiCl$_3$ and ρ_{pp} as 0.9 g/cm^3. The average diameters thus calculated are shown in Table 3.

Table 3. Size of polymer globules.

Sample	size of polymer globule (μm)	
	observed	calculated[a]
A-4	1.0	0.27
B-3	0.9	0.20
C-3	1.1	0.22

a) see the text. The arithmetic mean of D_{300} and D_{003} was taken as the average size of the catalyst crystallites.

In the case of sample A-4, the observed average size of the polymer globules, about 1 μm in diameter, is approximately 4 fold larger than that calculated (0.27 μm). In the cases of samples B-3 and C-3, the obseved sizes of the polymer globules are also 4 to 5 fold larger than those expected. These results imply that the polymer globules seen on the surfaces of the polymer flakes may be not the primary polymer particles, but the secondary particles. In order to answer this question, we observed internal structure of the polymer particles by TEM. Fig. 2 shows TEM photographs of the portion near the surface of sample A-4 prepared with the δ-TiCl$_3$(I) catalyst system.

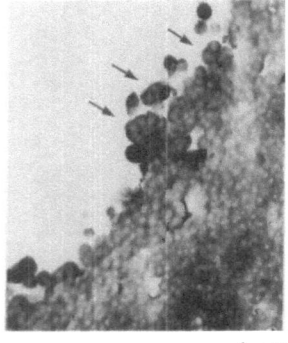 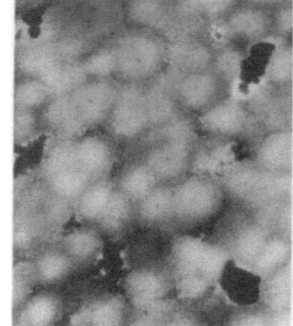

1 μm   0.1 μm

Fig. 2. TEM photographs of polypropylene sample A-4 prepared with the δ-TiCl$_3$ (I) catalyst system at different magnifications.

The agglomeration structure can be seen. To outward seeming, this picture resembles Fig. 1 very closely. It is, however, noteworthy that the diàmeters of the polymer particles in Fig. 2 are 0.1-0.4 μm, obviously smaller than those in Fig. 1. Furthermore, the large particles over 0.5 μm in diameter could be rarely found in other field, either. It is, therefore, most likely that the surface polymer globules are aggregates of the primary polymer particles, probably some tens of particles, judging from a difference in size. Aggregates indicated in Fig. 2 by arrows may be the surface polymer globules. In order to examine the polymer particles in more detail, we observed this specimen at a higher magnification. An electron micrograph is shown in Fig. 2, where one can see a number of the 0.2 to 0.35 μm polymer particles which contain one or sometimes two nuclei with a diameter of 80 to 170 Å at the core of each of them. Judging from their size, these nuclei are considered to be the catalyst crystallites. As shown in Fig. 3, similar pictures are observed in the polymer flakes prepared with the TiCl$_3$ AA, δ-TiCl$_3$ (II), or supported catalyst system.

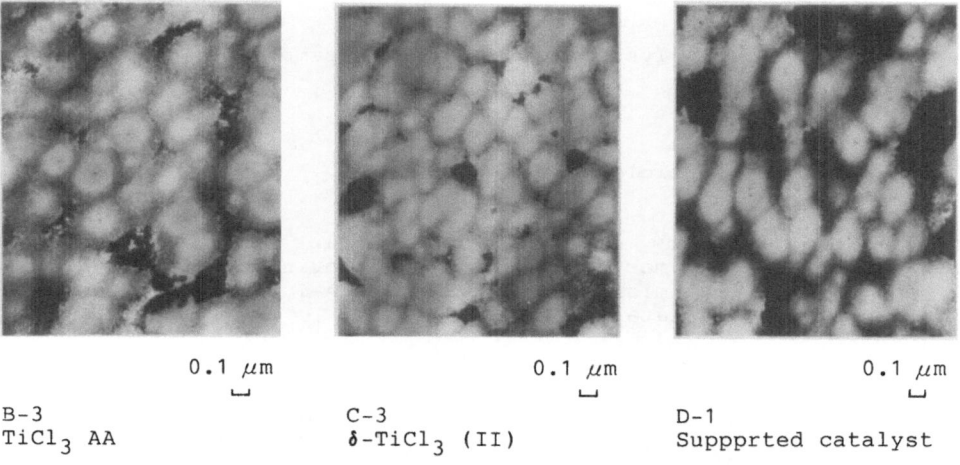

0.1 μm	0.1 μm	0.1 μm
B-3	C-3	D-1
TiCl$_3$ AA	δ-TiCl$_3$ (II)	Suppprted catalyst

Fig. 3. TEM photographs of polypropylene samples B-3, C-3, and D-1.

The nuclei observed in sample B-3 have diameters of 100 to 170 Å. The nuclei in sample C-3 are much smaller than those in sample B-3. In the case of sample D-1, polymer particles still contain some catalyst crystallites and the primary polymer particles could not be observed yet at this polymer yield. The diameters of these nuclei are very close to those of the catalyst crystallites. As a result, the nascent polymer flakes have basically a similar internal structure, irrespective of the sort of catalyst. Next, sample A-4 as the representative sample was examined by X-ray microanalysis. As shown in Fig. 4, titanium was detected from the nuclei A and B, which proves that they are the catalyst crystallites.

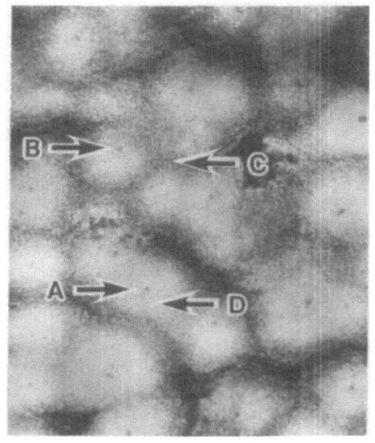
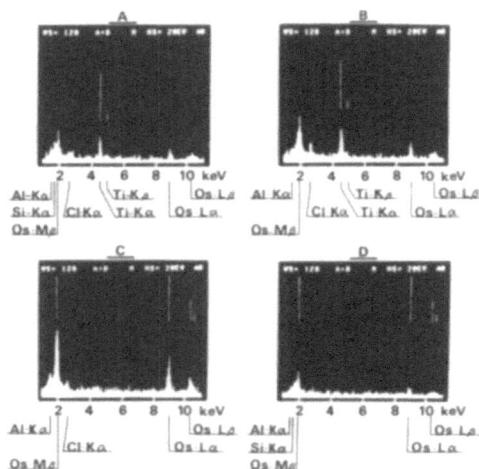

Fig. 4. X-ray microanalysis of polypropylene sample A-4 prepared with
δ-TiCl$_3$ (I).

Growth of polymer particles

In order to know how the polymer particles grow, we examined
structure of the polymer flakes widely differing in polymer yield.
Fig. 5 shows electron micrographs of samples A-1 through A-3 prepared
with the δ-TiCl$_3$ (I) catalyst system.

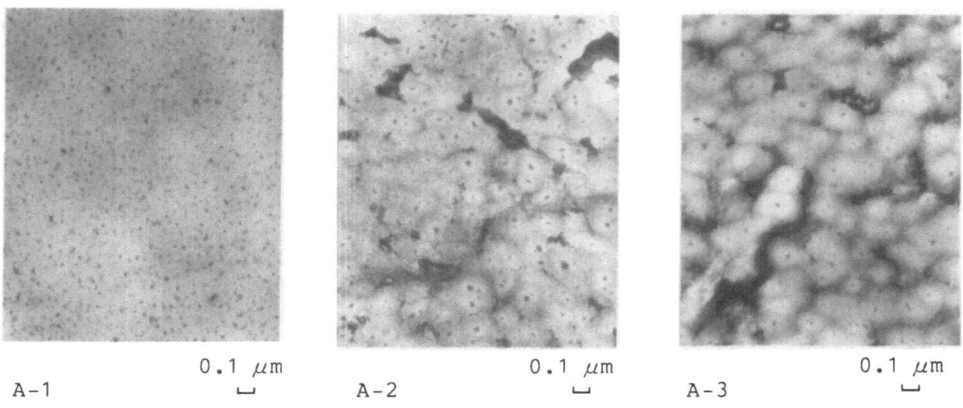

Fig. 5. TEM photographs of polypropylene samples A-1, A-2, and A-3.
Sample A-1 was not stained.

As seen from sample A-1 in Fig. 5, one can see that the catalyst crystallites are dispersing throughout polymer flake at the initial stage of polymerization. As the polymerization proceeds, the polymer particles 0.1 to 0.2 μm in diameter become visible. Such polymer particles contain some catalyst crystallites within each of them at intermediate polymer yield, i,e, at 108 (g-PP/g-cat). As the polymer yield increases further, the number of the catalyst crystallites contained in each of polymer particles decreases and eventually approaches one. At polymer yields over 880 (g-PP/g-cat), many primary polymer particles containig only one catalyst crystallite can be seen. Similar pictures, not shown in this paper, were obtained for the TiCl$_3$ AA and δ-TiCl$_3$(II) catalyst systems. The average sizes of primary polymer particles are plotted as a function of polymer yields on log-log graph paper in Fig. 6, though the primary polymer particles would be smaller in the case of the supported catalyst, as described above. These plots give straight lines with slopes close to 1/3, which shows that the primary polymer particles grow isotropically on the surface of each of the catalyst crystallites. This figure also shows that the size of primary polymer particles decreases with an increase in catalyst activity, compared at the same polymer yield, because the size of the catalyst crystallites decreases.

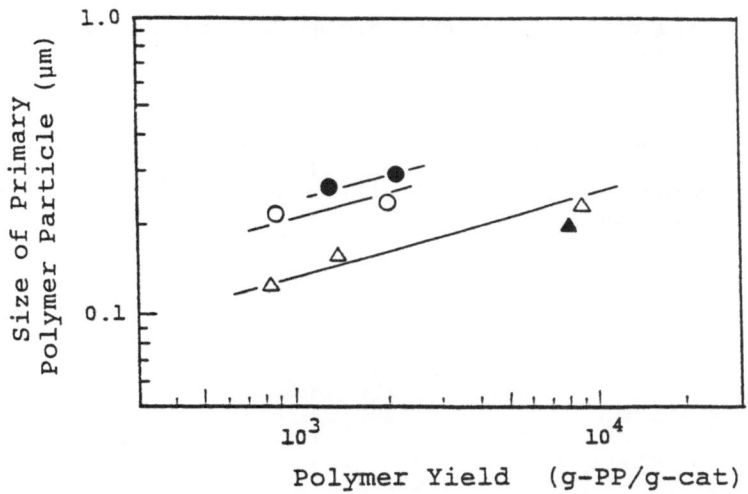

Fig. 6. Relationship between the size of primary particles of a polypropylene flake and the polymer yield;
○ δ-TiCl$_3$ (I) catalyst system, ● TiCl$_3$ AA catalyst system,
△ δ-TiCl$_3$ (II) catalyst system, ▲ Supported catalyst system.

We examined the surfaces of these samples by SEM. Fig. 7 shows electron micrographs of the polymer flakes prepared with the δ-TiCl$_3$ (I) catalyst system, varying in polymer yield. At lower polymer yield equal to 108 (g-PP/g-cat), the polymer globules are observed to have a broad size distribution ranging from 0.4 μm to 1.1 μm. At higher polymer yield the globules become round and uniform particles about 1 μm in diameter. In TiCl$_3$ AA and δ-TiCl$_3$ (II), not shown, similar polymer growth could be observed. As pointed out previously, these

fine polymer globules are the secondary particles much larger than
the primary polymer particles.

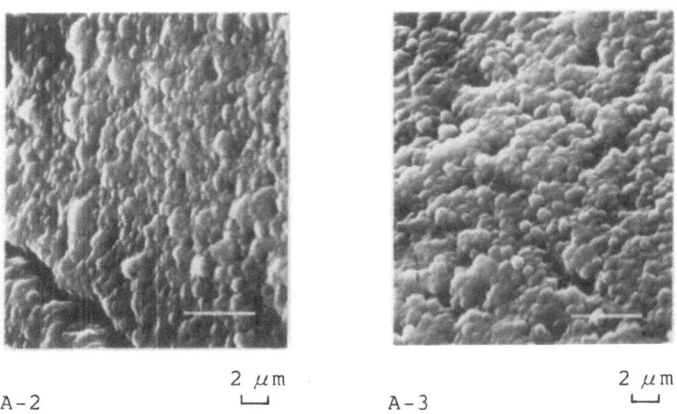

A-2   2 μm   A-3   2 μm

Fig. 7. SEM photographs of polypropylene samples A-2 and A-3 prepared
with δ-TiCl$_3$ (I) catalyst system.

In order to understand how such secondary polymer particles form, we
observed the replicas from the portions near the surface and core of
a polymer flake prepared with the δ-TiCl$_3$ (I) catalyst system. Fig. 8
shows transmission electron micrographs.

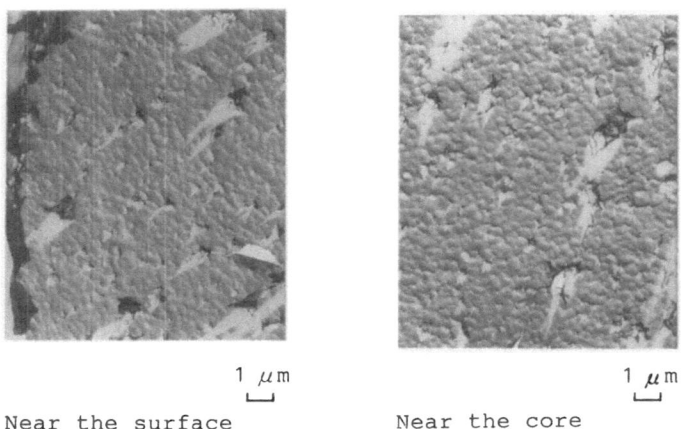

1 μm   1 μm

Near the surface   Near the core

Fig. 8. TEM photographs of polypropylene sample A-3 prepared with
δ-TiCl$_3$(I) catalyst system. A nascent polypropylene flake was sliced,
etched with n-heptane, and replicated with cellulose acetate film.

If a secondary polymer particle consists of the firmly bounded primary polymer particles, one might expect that the secondary particles would be visualized to a certain extent by this method. As seen from Fig. 8, both portions are almost identical, showing agglomerate structure of regular round particles. The size of the particles ca. 0.5 μm in diameter is obviously larger than that of the primary polymer particles but somewhat smaller than that of polymer globules observed on the surfaces of the flakes. This result indicates that the polymer flakes have a certain secondary structure both on the surface and in the interior. At the present time we can not, however, elucidate a difference in size between the secondary particles in Fig.7 and Fig.8.

The nascent polymer flake is a tertiary particle consisting of secondary polymer globules having diamieters of about 1 μm. And the polymer globles consist of some tens of much smaller primary polymer particles 0.2 to 0.35 μm in diameter, though these diameters may naturally vary, depending on the polymer yield and crystallite size of the catalyst.

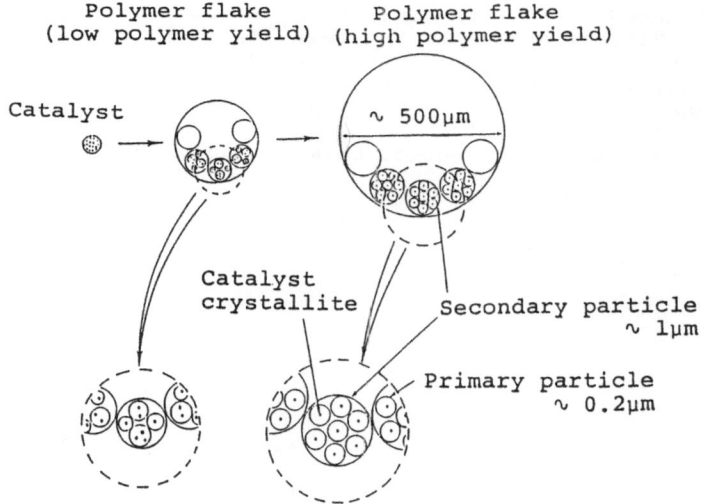

Fig. 9. Schematic model for polypropylene growth

From these observation results, we illustrate the growth of nascent polymer flakes in Fig. 9.

REFERENCES

1. C. W. Hock, J. Polym. Sci., A-1, 4, 3055-3064 (1966)
2. V. W. Buls and T. L. Higgins, J. Polym. Sci., A-1, 8, 1037-1053 (1970)
3. Z. W. Wilchinsky, R. W. Looney, and E. G. M. Tornqvist, J.

Catal., 28, 351-367 (1973)

4. R. T. Murray, R. Pearce, and D. Platt, J. Polym. Sci., Polym. Letter Ed., 16, 303-308 (1978)

5. P. C. Brabe, G. Cecchin, L. Noristi, Adv. Polym. Sci., 81, 1-81 (1987)

6. L. A. M. Rodriguez and J. A. Gabant, J. Polym. Sci., C, 4, 125, (1963); ibid. A-1, 4, 1971-1992 (1966)

7. P. Blais and R. ST. John Manley, J. Polym. Sci., A-1, 6, 291-334 (1968)

8. J. Y. Guttman and J. E. Guillet, Macromolecules, 1, 461-463 (1968)

9. A. Keller and F. M. Willmouth, Makromol. Chem., 121, 42-50 (1969)

10. H. D. Chanzy, R. H. Marchessault, Macromolecules, 2, 108-110 (1969)

11. V. W. Buls, and T. L. Higgins, J. Polym. Sci., A-1, 8 1037-1053 (1970)

12. R. T. K. Baker, P. S. Harris, R. J. Waite, and A. N. Roper, J. polym. Sci., Polym. Letter Ed., 11, 45-53 (1973)

13. R. J. L. Graff, G. Kortleve, and C. G. Vonk, J. Polym. Sci., Polym. Letter Ed., 8, 735-739 (1970)

14. J. Wristers, J. Polym. Sci., Polym. Phys. Ed., 11, 1601-1617 (1973)

15. R. Hoseman, M. Hentschel, E. Ferracini, A. Ferrero, S.Martelli, F. Riva, and M. Vittori Antisari, Polymer, 23, 979-984 (1982)

16. P. Mackie, M. N. Berger, B. M. Grieveson, and D. Lawson, J. Polym. Sci., Polym. Letter Ed., 5, 493-494 (1967)

17. M. Kakugo, H. Sadatoshi, and M. Yokoyama, J. Polym. Sci., Polym. Letter Ed., 24, 171-175 (1986)

18. G. Natta, I. Pasguon, and E.Giachetti, Chim. e Ind., 39, 1002-1012 (1957)

19. B. P. 2085016 (Sumitomo Chemical)